In the Light of Evolution

Volume V: Cooperation and Conflict

In the Light of Evolution

Volume V: Cooperation and Conflict

JOAN E. STRASSMANN, DAVID C. QUELLER, JOHN C. AVISE, and FRANCISCO J. AYALA, *Editors*

NATIONAL ACADEMY OF SCIENCES
THE NATIONAL ACADEMIES

THE NATIONAL ACADEMIES PRESS
Washington, D.C.
www.nap.edu

THE NATIONAL ACADEMIES PRESS **500 Fifth Street., N.W.** **Washington, DC 20001**

This volume is based on the Arthur M. Sackler Colloquium of the National Academy of Sciences, "Cooperation and Conflict," held January 7-8, 2011, at the Arnold and Mabel Beckman Center of the National Academies of Sciences and Engineering in Irvine, California.

The articles appearing in these pages were contributed by speakers at the colloquium and have been anonymously reviewed. Any opinions, findings, conclusions, or recommendations expressed in this volume are those of the authors and do not necessarily reflect the view of the National Academy of Sciences.

In the light of evolution / John C. Avise and Francisco J. Ayala, editors.
p. cm.
Vol. I based on a colloquium of the National Academy of Sciences, held December 1–2, 2006, in Irvine, California.
Includes bibliographical references.
ISBN-13: 978-0-309-21836-8
ISBN-10: 0-309-21836-5
1. Evolution (Biology)—Congresses. I. Avise, John C, 1948–. II . Ayala, Francisco José, 1934–
III . National Academy of Sciences (U.S.)
QH359.I55 2007
576.8—dc22
2007032455

Additional copies of this book are available from the National Academies Press, 500 Fifth St., N.W., Lockbox 285, Washington, DC 10055; (800) 624-6242 or (202) 334-3313 (in the Washington metropolitan area); Internet, http://www.nap.edu.

Cover image: Pictured is a Batik painting in woad, a plant-derived dye, on cotton. The two ants illustrate the phenomenon of cooperation, the focus of many of the chapters in this volume. This collection of articles explores recent developments in the study of the evolution of cooperation among all organisms from the level of genes to that to societies—from bacteria to humans. Image courtesy of Robin Paris, www.robinparis.co.uk.

Printed in the United States of America

THE NATIONAL ACADEMIES

Advisers to the Nation on Science, Engineering, and Medicine

The **National Academy of Sciences** is a private, nonprofit, self-perpetuating society of distinguished scholars engaged in scientific and engineering research, dedicated to the furtherance of science and technology and to their use for the general welfare. Upon the authority of the charter granted to it by the Congress in 1863, the Academy has a mandate that requires it to advise the federal government on scientific and technical matters. Dr. Ralph J. Cicerone is president of the National Academy of Sciences.

The **National Academy of Engineering** was established in 1964, under the charter of the National Academy of Sciences, as a parallel organization of outstanding engineers. It is autonomous in its administration and in the selection of its members, sharing with the National Academy of Sciences the responsibility for advising the federal government. The National Academy of Engineering also sponsors engineering programs aimed at meeting national needs, encourages education and research, and recognizes the superior achievements of engineers. Dr. Charles M. Vest is president of the National Academy of Engineering.

The **Institute of Medicine** was established in 1970 by the National Academy of Sciences to secure the services of eminent members of appropriate professions in the examination of policy matters pertaining to the health of the public. The Institute acts under the responsibility given to the National Academy of Sciences by its congressional charter to be an adviser to the federal government and, upon its own initiative, to identify issues of medical care, research, and education. Dr. Harvey V. Fineberg is president of the Institute of Medicine.

The **National Research Council** was organized by the National Academy of Sciences in 1916 to associate the broad community of science and technology with the Academy's purposes of furthering knowledge and advising the federal government. Functioning in accordance with general policies determined by the Academy, the Council has become the principal operating agency of both the National Academy of Sciences and the National Academy of Engineering in providing services to the government, the public, and the scientific and engineering communities. The Council is administered jointly by both Academies and the Institute of Medicine. Dr. Ralph J. Cicerone and Dr. Charles M. Vest are chair and vice chair, respectively, of the National Research Council.

www.national-academies.org

Arthur M. Sackler, M.D.
1913–1987

Born in Brooklyn, New York, Arthur M. Sackler was educated in the arts, sciences, and humanities at New York University. These interests remained the focus of his life, as he became widely known as a scientist, art collector, and philanthropist, endowing institutions of learning and culture throughout the world.

He felt that his fundamental role was as a doctor, a vocation he decided upon at the age of four. After completing his internship and service as house physician at Lincoln Hospital in New York City, he became a resident in psychiatry at Creedmoor State Hospital. There, in the 1940s, he started research that resulted in more than 150 papers in neuroendocrinology, psychiatry, and experimental medicine. He considered his scientific research in the metabolic basis of schizophrenia his most significant contribution to science and served as editor of the *Journal of Clinical and Experimental Psychobiology* from 1950 to 1962. In 1960 he started publication of *Medical Tribune*, a weekly medical newspaper that reached over one million readers in 20 countries. He established the Laboratories for Therapeutic Research in 1938, a facility in New York for basic research that he directed until 1983.

As a generous benefactor to the causes of medicine and basic science, Arthur Sackler built and contributed to a wide range of scientific institutions: the Sackler School of Medicine established in 1972 at Tel Aviv University, Tel Aviv, Israel; the Sackler Institute of Graduate Biomedical Science at New York University, founded in 1980; the Arthur M. Sackler Science Center dedicated in 1985 at Clark University, Worcester, Massachusetts; and the Sackler School of Graduate Biomedical Sciences, established in 1980, and the Arthur M. Sackler Center for Health Communications, established in 1986, both at Tufts University, Boston, Massachusetts.

His pre-eminence in the art world is already legendary. According to his wife Jillian, one of his favorite relaxations was to visit museums and art galleries and pick out great pieces others had overlooked. His interest in art is reflected in his philanthropy; he endowed galleries at the Metropolitan Museum of Art and Princeton University, a museum at Harvard

University, and the Arthur M. Sackler Gallery of Asian Art in Washington, D.C. True to his oft-stated determination to create bridges between peoples, he offered to build a teaching museum in China, which Jillian made possible after his death, and in 1993 opened the Arthur M. Sackler Museum of Art and Archaeology at Peking University in Beijing.

In a world that often sees science and art as two separate cultures, Arthur Sackler saw them as inextricably related. In a speech given at the State University of New York at Stony Brook, *Some reflections on the arts, sciences and humanities*, a year before his death, he observed: "Communication is, for me, the *primum movens* of all culture. In the arts . . . I find the emotional component most moving. In science, it is the intellectual content. Both are deeply interlinked in the humanities." The Arthur M. Sackler Colloquia at the National Academy of Sciences pay tribute to this faith in communication as the prime mover of knowledge and culture.

Contents

Preface to the *In the Light of Evolution* Series

Biodiversity—the genetic variety of life—is an exuberant product of the evolutionary past, a vast human-supportive resource (aesthetic, intellectual, and material) of the present, and a rich legacy to cherish and preserve for the future. Two urgent challenges, and opportunities, for 21st-century science are to gain deeper insights into the evolutionary processes that foster biotic diversity, and to translate that understanding into workable solutions for the regional and global crises that biodiversity currently faces. A grasp of evolutionary principles and processes is important in other societal arenas as well, such as education, medicine, sociology, and other applied fields including agriculture, pharmacology, and biotechnology. The ramifications of evolutionary thought also extend into learned realms traditionally reserved for philosophy and religion.

In 1973, Theodosius Dobzhansky penned a short commentary entitled "Nothing in biology makes sense except in the light of evolution." Most scientists agree that evolution provides the unifying framework for interpreting biological phenomena that otherwise can often seem unrelated and perhaps unintelligible. Given the central position of evolutionary thought in biology, it is sadly ironic that evolutionary perspectives outside the sciences have often been neglected, misunderstood, or purposefully misrepresented.

The central goal of the *In the Light of Evolution* (*ILE*) series is to promote the evolutionary sciences through state-of-the-art colloquia—in the series of Arthur M. Sackler colloquia sponsored by the National Academy of Sciences—and their published proceedings. Each installment explores evolutionary perspectives on a particular biological topic that is scientifically intriguing but also has special relevance to contemporary societal issues or challenges. Individually and collectively, the *ILE* series aims to interpret phenomena in various areas of biology through the lens of evolution, address some of the most intellectually engaging as well as pragmatically important societal issues of our times, and foster a greater appreciation of evolutionary biology as a consolidating foundation for the life sciences.

The organizers and founding editors of this effort (Avise and Ayala) are the academic grandson and son, respectively, of Theodosius Dobzhansky, to whose fond memory this *ILE* series is dedicated. May Dobzhansky's words and insights continue to inspire rational scientific inquiry into nature's marvelous operations.

John C. Avise and Francisco J. Ayala
Department of Ecology and Evolutionary Biology,
University of California, Irvine (January 2007)

Preface to *In the Light of Evolution, Volume V: Cooperation and Conflict*

This book is the outgrowth of the Arthur M. Sackler Colloquium "Cooperation and Conflict," which was sponsored by the National Academy of Sciences on January 7–8, 2011, at the Academy's Arnold and Mabel Beckman Center in Irvine, California. It is the fifth in a series of colloquia under the general title "*In the Light of Evolution.*" The first four books in this series were titled *Adaptation and Complex Design* (Avise and Ayala, 2007), *Biodiversity and Extinction* (Avise et al., 2008), *Two Centuries of Darwin* (Avise and Ayala, 2009a), and *The Human Condition* (Avise and Ayala, 2009b). The current volume explores recent developments in the study of cooperation and conflict, ranging from the level of the gene to societies and symbioses.

Any student of history knows that we humans can be a vicious lot, but paradoxically we are also among nature's great cooperators. Which of us, as an individual, can manufacture a cell phone or an airplane? Even our great conflicts—wars—are extremely cooperative endeavors on each side. Some of this cooperation is best understood culturally, but we are also products of evolution, with bodies, brains, and behaviors molded by natural selection. How cooperation evolves has been one of the big questions in evolutionary biology, and how it pays or does not pay is a great intellectual puzzle.

If nothing makes sense in biology except in the light of evolution (Dobzhansky, 1973), then for the first century after Darwin, cooperation and altruism did not make much sense. We could see that individual organisms sometimes helped others, even at a cost to their own fitness. It was clear that such behavior could benefit the group or the population, the species, or even other species and whole communities. However, it was not obvious how such effects would be heritable. All our mathematical models—the hard work of the modern synthesis—were about individuals with one allele out-reproducing those with an alternative. This process would favor individuals with higher reproduction but would not be expected to produce self-sacrifice. Yet, apparent cooperation was routinely attributed to the good of the group, species, or community. This situation changed in the first decade of Darwinism's second century. William D. Hamilton (1964a,b) argued that cooperation was important in nature, and that social evolution could be understood in terms of direct gains to the actor's own fitness or indirect benefits to the fitness of others who share the cooperation allele. There followed an intense period of exploring the indirect effects of cooperation and altruism, reinterpreting sexual selection and many other phenomena in terms of individual advantage, and understanding frequency-dependent effects via game theory, efforts that continue to the present.

The puzzle of cooperation was the dominant theme of research in the early years, whereas recent work has emphasized its importance and ubiquity. Far from being a rare trait shown by social insects and a few others, cooperation is both widespread taxonomically and essential to life. Major transitions in the hierarchy of life have often involved cooperation among lower-level units to the point where they evolve into higher-level organisms (Buss, 1987; Maynard Smith and Szathmary, 1995). Examples include the assembly of the eukaryotic cell with its symbiotic organelles, the evolution of multicellular organisms, and the organismal colonies of some social insects. Organisms are, at multiple levels, those units that have evolved to have, within their boundaries, extreme cooperation and minimal conflict (Queller and Strassmann, 2009; Strassmann and Queller, 2010). The depth of research on cooperation and conflict has increased greatly, most notably in the direction of small organisms. Microbes turn out to have highly developed cooperation (West et al., 2007a), and they, along with other model organisms, have proven instrumental in beginning to understand sociality at the genetic and molecular levels, the study of real selfish genes (Santorelli et al., 2008). The social evolution approach has given us new insights on diseases often caused by microbes (Foster, 2005). At the other end of the spectrum, we are getting a much better understanding of the cooperation and conflict that matter most to our species (Alexander, 1979). Cooperation has been central to humanity's spectacular success and will be central to our short-term and long-term fate.

Part I

THE FUNDAMENTALS OF EVOLUTIONARY COOPERATION

Although most of this book is about the new topics that are being treated as part of social evolution, such as genes, microbes, and medicine, the old fundamental subjects still matter and remain the object of vigorous research. The first four chapters revisit some of these standard arenas, including social insects, cooperatively breeding birds, mutualisms, and how to model social evolution.

There are many ways to think about and model social evolution. Inclusive fitness is one of the most venerable and most useful, and is the framework used by many authors in this book. In Chapter 1, David Queller revisits why inclusive fitness has been so useful and suggests ways to expand it to make it speak more directly to interactions besides kin selection. He delimits two other kinds of social selection that can be treated more explicitly in Hamilton's rule. "Kind selection," which involves synergisms between individuals expressing the same traits, groups together greenbeards (genes that in effect can identify the presence of copies of themselves in other individuals) and many cases of frequency-dependent games because these share the feature that individuals expressing the trait have different effects on other expressers compared to nonexpressers and they also share many differences from pure kin selection. "Kith selection" requires neither kin nor kind, but instead involves actors affecting partners in ways that feed back to the actor's fitness. Mutualism and manipulation are included in this category. The expanded version of Hamilton's rule with kin, kith, and kind could bring the advantages of Hamilton's methods to a broader range of social interactions.

Interactions between individuals of different species are a major type of kith selection, where individuals are selected to affect their partners in ways that ultimately benefit themselves (or their kin). Such interactions need not be cooperative, but when they are, they typically involve exchange of different services that one partner needs and the other can easily provide, so partners can be very different. Accordingly, in Chapter 2, Joel Sachs and colleagues explore associations or symbiosis among partners that are very different indeed, one being eukaryotic and the other prokaryotic. Such symbioses, by leading to mitochondria and chloroplasts, were responsible for the evolution of the eukaryotic cell itself. But additional symbioses are widespread and sometimes ancient. These authors use a combination of broad-scale phylogenetic analyses and case histories of particular systems to explore several transitions. They find, for example, that there is little phylogenetic signal to indicate that some bacterial groups are preadapted for eukaryotic symbiosis. Instead, the genes required appear to be quite widely available through horizontal transmission. Mutualistic interactions appear to arise from both parasitic and free-living ancestors. Once acquired, these mutualistic interactions seem to be quite stable, with few reversions to nonmutualistic forms. Given the tendency of vertically transmitted symbionts to degrade and the propensity of horizontally transmitted ones to cheat, this stability is somewhat surprising.

The social insects have long been viewed as the pinnacle of cooperation. This view is most tenable if one ignores the cooperation that goes on in transitions that are already complete, such as to multicellular animals or the eukaryotic cell. But some social insect colonies are so cooperative and integrated that they are viewed as superorganisms (organisms made up of other organisms). The motive force behind the evolution of these societies, which consist of close relatives, is kin selection (Hamilton, 1964a). In Chapter 3, Peter Nonacs points out that predictions from kin selection theory have been both successful and also disappointing. The difference, he suggests, is not due to chance. The successful predictions from sex-ratio theory and worker-policing theory work because the predicted behaviors can be achieved using simple environmental cues that correlate with kinship. It is easy to treat males differently from females, or workers from queens. The less successful kin selection predictions, such as parts of skew theory, may fail because they require genetic kin recognition mechanisms sufficient to detect closer from more distant relatives within colonies. This may not explain everything, because genetic kin recognition systems do exist, at least for distinguishing colony members from noncolony members. The interaction between environmental and genetic recognition systems has scarcely been explored, and Nonacs runs computer simulations showing how greenbeard loci can perturb the outcomes expected under pedigree relatedness alone.

After the social insects, cooperative birds and mammals have attracted the most attention. Many bird species have helpers at the nest, usually offspring from previous broods who have remained at their natal site (Cockburn, 2006). Kinship is important here too. Helping systems usually evolve from monogamous ones, and discrimination evolves in systems that show variation in relatedness (Cornwallis et al., 2010). But the story is more complicated, for two reasons. First, although, some helpers gain kin-selected benefits through helping close kin, others may gain direct benefits. Compared with the social insects, more research on birds has addressed the particular benefits of remaining at home and on the ecological constraints that may limit independent breeding. Variance in reproductive success has played a role in these discussions, but in Chapter 4, Dustin Rubenstein moves it to a more central position. He suggests that cooperative breeders may be bet hedgers, gaining advantage from a more uniform reproductive output in cooperative groups. Rubenstein draws on many years of his field data on starlings in Africa, where there is much variation in both time and space, and he finds support for several predictions of this hypothesis.

1

Expanded Social Fitness and Hamilton's Rule for Kin, Kith, and Kind

DAVID C. QUELLER

Inclusive fitness theory has a combination of simplicity, generality, and accuracy that has made it an extremely successful way of thinking about and modeling effects on kin. However, there are types of social interactions that, although covered, are not illuminated. Here, I expand the inclusive fitness approach and the corresponding neighbor-modulated approach to specify two other kinds of social selection. Kind selection, which includes greenbeards and many nonadditive games, is where selection depends on an actor's trait having different effects on others depending on whether they share the trait. Kith selection includes social effects that do not require either kin or kind, such as mutualism and manipulation. It involves social effects of a trait that affect a partner, with feedback to the actor's fitness. I derive expanded versions of Hamilton's rule for kith and kind selection, generalizing Hamilton's insight that we can model social selection through a sum of fitness effects, each multiplied by an appropriate association coefficient. Kinship is, thus, only one of the important types of association, but all can be incorporated within an expanded inclusive fitness.

Hamilton's rule and the associated concept of inclusive fitness (Hamilton, 1964a) have provided an extremely successful way of thinking about and modeling social evolution (West et al., 2007b).

Department of Biology, Washington University, St. Louis, Mo 63130. E-mail: queller@biology2.wustl.edu.

There are a number of reasons why this is true. It is simple, and therefore, users can apply its logic with ease; nevertheless, it is quite general. In some versions, it is exact, and even less exact versions are not necessarily a strong concern for field or comparative studies, where we can only measure crudely anyway. Crucially, it is often sufficiently independent of the genetic details, such as dominance and recessiveness, the number of genes, and their allele frequencies. This allows it to become an important tool of the phenotypic gambit (Grafen, 1984) and optimality approaches. It can be used for traits where we do not understand the underlying genetics, and, in fact, we never fully understand the genetics. It also conveniently separates selection into two kinds of summary terms: effects on fitness (costs and benefits) and population structure (relatedness). This separation makes the process easy to think about and the equations easy to apply. Inclusive fitness points to cause-effect relations, specifically to the various effects caused by the actor's behavior. This focus on what the actor can control allows us to tie into the long biological tradition of thinking of actors, or their genes, as agents. Additionally, it tells us that these agents should appear to be trying to maximize inclusive fitness.

Inclusive fitness is not perfect. It does not provide the most natural way to handle explicit dynamics. It usually takes population structure as a given, and when it does this, it may not yield insight into how population structure emerges. Although, in principle, it covers everything, its summary parameters may sometimes conceal interesting complexity. Even its treatment of social causation is incomplete. For example, although it would include any benefits from mutualism in with other effects on the actor's direct fitness, it does not usually separate out these effects or provide a causal treatment of them. Many or all of these deficits are fixable, although sometimes at the cost of making the models more complex and therefore, losing some of the advantages of the approach. In this paper, I will try to expand the types of social causation covered explicitly, while trying to maintain reasonable simplicity. For example, I will show how to specify mutualistic social effects in a category that I call kith selection, named after the largely archaic word for acquaintances, friends, and neighbors.

I will also argue that it is often worth distinguishing kin and kith selection from what I call kind selection, partly to properly capture social causality and partly because these forms of social selection act in very different ways. Inclusive fitness, developed by Hamilton (1964a), is closely associated with the process of kin selection, named by Maynard Smith (1964). However, they are not the same thing. Inclusive fitness is an accounting method and maximand. Kin selection is a process, and it can be described by other kinds of accounting. The obvious example is the neighbor-modulated approach that uses the same fitness partition as inclusive fitness but groups by effects received rather than effects given (Taylor and Frank, 1996). However, models with other fitness partitions, such as multilevel selection models, also often describe kin selection (Price, 1972; Hamilton, 1975; Wade, 1980; Queller, 1992c). Another reason is that inclusive fitness includes standard selection where there are no kin effects at all. Finally, kin selection, when interpreted as resulting from genomewide genealogical relatedness, does not cover all indirect effects. The most commonly cited examples are greenbeard genes (Dawkins, 1976b), which act based on their own identities rather than pedigree kinship. These are commonly grouped under kin selection, but I will argue that greenbeards are one example of the distinct phenomenon that I will call kind selection.

Specifically, I derive an expanded Hamilton's rule (1964a) or inclusive fitness effect (and neighbor-modulated fitness effect) as

$$-c + \sum b * r + \sum d * s + \sum m * f > 0. \tag{1}$$

The first two terms look like the standard Hamilton's rule but are not exactly the same, because some social effects have been split off into additional terms. Here, $-c$ is nonsocial direct fitness but does not include some social components of direct fitness. These fitness effects, m (for mutualism or manipulation), are multiplied by a feedback coefficient f to give the kith selection term. Also, kind effects d (deviation from additivity) multiplied by a kind coefficient s (synergism) are split off. These include greenbeard effects that are normally in indirect fitness and some frequency-dependent effects that are usually placed in direct fitness. This is an expanded form in two senses. First, it covers more kinds of social selection or at least, it covers more in a causal manner. Second, it expands out into the number of terms needed to describe this causation with two kinds of distinct terms: selection terms relating social actions to fitness components and association coefficients that essentially describe the relative heritability of those effects. I continue to call this a version of Hamilton's rule because of this key similarity.

In introducing kith and kind selection, I am not claiming to have discovered new forms of social selection. All of the social situations that I discuss have been explored in other ways. Nor should this treatment be viewed as invalidating the standard inclusive fitness approach; it can be viewed as a more detailed version of that approach. My goal here is to present a useful classification of social behaviors and derive a common theoretical framework that partakes of the many advantages of the inclusive fitness approach.

MODELING SOCIAL EFFECTS

In this section, I illustrate the method I use to partition different kinds of selection using the methods of Queller (1992a,b). The approach closely parallels the causal modeling approach pioneered by Lande and Arnold (1983), which is further developed for social traits in the indirect genetics effects approach (Moore et al., 1997; Bijma and Wade, 2008; McGlothlin et al., 2010). I begin with Price's (1970) equation 9 for the change in the average of some quantity—here, the average breeding value for a trait, $\overline{G}$, which can be for a single gene or multiple loci affecting a trait. Price's (1970) equation is an identity that always holds, but additional assumptions are often made. Here, I follow the common practice of ignoring its second term, which can incorporate effects like meiotic drive or change in environment, to focus on organismal selection and adaptation. Price's (Hamilton, 1964a) equation can then be written as

$$\overline{W}\Delta\overline{G} = \mathrm{Cov}(W,G), \tag{2}$$

showing that breeding value is expected to increase if it covaries positively with fitness. Now, consider a social trait where an individual's fitness is affected by both his own trait and the trait of a partner. For the moment, we will assume that we know each individual's genes for the trait, with a breeding value of G for the focal individual and G' for its partner. Fitness can be written in the form of a regression

$$W = \alpha + \beta_{WG'\cdot G}G + \beta_{WG'\cdot G}G' + \varepsilon. \tag{3}$$

The α is the intercept, and it can be conceived of as the base fitness before any social actions. The β symbols are partial regression coefficients for the effect of the focal individual's genes and the partner's genes on the focal individual's fitness, each holding the effect of the other individual constant. The ε is the residual or remainder, including the effects of any other causes and any truly random effects. The regression equation might make it seem that we are interested purely in estimation, but it is also gives us a model of fitness that, depending on the predictors, can be useful, useless, or even misleading.

Substituting Eq. (2) into expression (1) yields

$$\overline{W}\Delta\overline{G} = \mathrm{Cov}(\alpha,G) + \mathrm{Cov}(\beta_{WG\cdot G'}G,G) + \mathrm{Cov}(\beta_{WG'\cdot G}G',G) + \mathrm{Cov}(\varepsilon,G). \tag{4}$$

The first covariance drops out, because a constant has zero covariance. The last term drops out, because the residuals of a regression are uncorrelated with the predictor variables. If we are thinking in terms of a model, we assume that ε and G are uncorrelated. Next, we can pull the constant outside of the covariance terms to give

$$\overline{W}\Delta\overline{G} = \beta_{WG \cdot G'}\text{Cov}(G,G) + \beta_{WG' \cdot G}\text{Cov}(G,G'). \tag{5}$$

Average breeding value $\Delta\overline{G}$ will increase when $\beta_{WG \cdot G'}\text{Cov}(G,G) + \beta_{WG' \cdot G'}\text{Cov}(G',G) > 0$. Dividing through by the first covariance gives $\beta_{WG \cdot G'} + \beta_{WG' \cdot G}\text{Cov}(G',G)/\text{Cov}(G,G) > 0$ or

$$\beta_{WG \cdot G'} + \beta_{WG' \cdot G}\, \beta_{GG'} > 0. \tag{6}$$

This is Hamilton's rule, with the direct effect on fitness $\beta_{WG.G'}$, the indirect effect of a partner $\beta_{WG' \cdot G}$, and a regression coefficient of relatedness $\beta_{GG'}$. It is a neighbor-modulated fitness form of Hamilton's rule, which totes up effect on each individual, but it can be rearranged under quite general conditions to an inclusive fitness form that switches all of the primes and nonprimes in the second term and thus totes up the effects of each individual (Frank, 1998).

Because we assumed we knew the genes, this form is extremely general. It belies the claim that is occasionally made that inclusive fitness requires many assumptions (Nowak et al., 2010). Those claims are usually made about phenotypic versions that are used when we do not assume that we know the genetic basis of the traits, and the same limitations would generally apply to alternative models faced with that assumption. Therefore, proponents of inclusive fitness can rightly refute the claim of limited generality. However, one of the main appeals of inclusive fitness is that it can often be used without knowledge of the genes, and therefore, we will consider the phenotypic gambit shortly.

I have dwelled a bit on already published math (Queller, 1992a,c), because every subsequent derivation in this paper, for which I will not show the math, follows an exactly parallel procedure consisting of the following steps:

(*i*) Write a regression model for the actor's fitness.
(*ii*) Substitute that expression for fitness into the abbreviated Price's (1970) equation.
(*iii*) Divide the covariance into separate terms, one for each term of the regression.
(*iv*) Drop out the α (intercept) term.
(*v*) Drop out the ε (residual) term provided that $\text{Cov}(G,\varepsilon) = 0$.
(*vi*) Extract the regression coefficients from the covariances.
(*vii*) Ask when $\Delta\overline{G} > 0$.
(*viii*) Divide through by the covariance associated with actor's fitness.

We could stop at step *(vi)* to preserve a more general equation that predicts actual change in $\overline{G}$, but I will follow the customary step in inclusive fitness analysis of asking the more restricted question of when $\overline{G}$ increases. Either way, the crucial step turns out to be step *(v)*. This is the only step that invokes an assumption, which is $\mathrm{Cov}(G,\varepsilon) = 0$. This condition will, therefore, determine whether an exact Hamilton-type (1964a) result can legitimately be obtained. When it does drop out, we end up with an equation with the desired neat separation between fitness and structure terms, and therefore, I have called this the separation condition (Queller, 1992c).

CAUSALITY

There is nothing preordained about the predictors used in the derivation above. We could attempt to get a result from any equation predicting or describing fitness. Indeed, it was technically unnecessary to include the partner's breeding value. If we use only the focal individual's breeding value ($W = \alpha + \beta_{WG}G + \varepsilon$) and follow steps *(ii)*–*(viii)*, above, we show that $\overline{G} > 0$ when $\beta_{WG} > 0$. This does not take us far from Price's (1970) equation, but it has exactly the same level of validity and accuracy as the inclusive fitness result derived above. Why then do we prefer the inclusive fitness result? The first reason, to be treated shortly, is that leaving out the partner does not work when we try to play the phenotypic gambit. The other reason is that including the partner can provide some additional causal explanation. We are no longer just saying certain genes are associated with fitness; we are giving a breakdown of how that association is caused. It is this causal feature that I want to expand to include more than kin effects.

To illustrate the point about causality, consider another model of fitness based on the individual's breeding value *G* and the phase of the moon, represented by *M*. If we substitute $W = \alpha + \beta_{WG \cdot M}G + \beta_{WG \cdot G}M + \varepsilon$ into Price's (1970) equation, steps *(ii)*–*(viii)* lead us to the conclusion that $\Delta\overline{G} > 0$ when $\beta_{WG \cdot M} + \beta_{WM \cdot G}\,\beta_{GM} > 0$. The first term remains the effect of the actor's genes on its fitness, but the second term is now the effect of the moon phase and is multiplied by β_{GM}, a sort of moony relatedness linking breeding value and phase of the moon. This model is just as correct as the first two that we considered (the ε term must drop out, because *G* is one of the predictors); however, no one would consider it very useful, because moon phase is unlikely to have any causality. Even if the phase of the moon had some effect on fitness (in which case, we would need to take it into account for a full evolutionary explanation of the trait), the actor would still be a passive player. There is nothing the actor can do to alter the phase of the moon, and therefore, for optimality arguments, we can ignore it.

Any causes can be included (Queller, 1992a; Frank, 1998). In this respect, my approach is similar to that taken by the indirect genetic effects (IGE) school of social evolution, which can recover versions of Hamilton's rule (1964a) in very similar ways (Moore et al., 1997; Bijma and Wade, 2008; McGlothlin et al., 2010). IGE is an extension of quantitative genetics to social evolution, and quantitative genetics has always engaged in partitioning evolution into causal components. My interest here is not in all possible causes but in those that most clarify the role of selection on an actor's social behavior. Thus, in the same way that I exclude the moon phase from the model, I will not generally explore byproduct social effects. A lion that kills a zebra benefits local vultures, and this can influence their traits and fitness; therefore, the killing has a social aspect. However, the vultures do not influence the killing, and the evolution of that killing behavior (as opposed to the incidental effects on vulture traits) does not need to take vultures into account. An important exception is when there are byproducts with feedbacks on the actor's fitness.

PHENOTYPES AND SOCIAL CAUSES

Much of the value of inclusive fitness stems from its use in the phenotypic gambit (Grafen, 1984). If we know costs, benefits, and relatedness, we can usually make good predictions about what kinds of traits will be favored, even if we do not understand the underlying genetics. Such approaches are sometimes denigrated by theoreticians, who prefer precision and mathematical rigor over all else, but for understanding the real world, it is essential. To deny this is to deny that Darwin understood anything about adaptation, because all he had to work with was the fit of phenotypes to their environments and a knowledge that some form of heredity exists.

When kin are affected, the phenotypic gambit requires indirect effects. If we use only the actor's phenotype P to model its fitness ($W = \alpha + \beta_{WP}P + \varepsilon$) and follow steps (*ii*)–(*viii*), we predict that $\Delta\overline{G} > 0$ when $\beta_{WP} > 0$. This predicts that altruism cannot evolve, because a cost to self means a negative β_{WP}. However, we know that altruism can evolve. Mathematically, the reason that the phenotypic gambit fails here is step (*v*), the separation condition (Queller, 1992c). After the effect of the actor's behavior on its own fitness is removed, the residual ε is correlated with genotype G if the interaction involves relatives. The partner's fitness W is affected by the partner's behavior P', which is correlated with G', which, in turn, is correlated with G when the individuals are relatives.

The solution of inclusive fitness theory is to include the partner's phenotype in the fitness model: $W = \alpha + \beta_{WP.P'}P + \beta_{WP'.P}P' + \varepsilon$. Following steps (*ii*)–(*viii*) now yields

$$\beta_{WP\cdot P'} + \beta_{WP'\cdot P} \frac{\text{Cov}(G,P')}{\text{Cov}(G,P)} > 0. \tag{7}$$

Here, the two regression terms represent the cost and benefit like in (6), except that we now use phenotypes instead of breeding values. The relatedness coefficient is now a more complicated ratio of covariances (Michod and Hamilton, 1980). The ratio makes intuitive sense, however, particularly if we think of phenotype being one for performing the behavior and zero for not performing. Then, relatedness is essentially the ratio of the actor's breeding value when the partner performs the behavior to its breeding value when the actor performs the behavior. Switching this neighbor-modulated version to inclusive fitness gives

$$\beta_{WP\cdot P'} + \beta_{W'P\cdot P} \frac{\text{Cov}(G,P')}{\text{Cov}(G,P)} > 0. \tag{8}$$

Seger (1981) discusses the relationship among these regression coefficients.

KITH SELECTION

Hamilton's rule (Wade, 1980) is normally applied to kin selection, with the relatedness covariances arising from common descent (Hamilton, 1964a). However, there is nothing in the derivations that limits it to this case. The primary limitation, as I will show, is additivity of the two fitness components $\beta_{WP.P'}$ and $\beta_{WP'.P}$. Within that constraint, the gene-phenotype associations represented in the covariance ratio could have any cause. Queller (1985) pointed out that the phenotypic covariance ratio could also be used to describe reciprocity. Frank (1994, 1998) argued that mutualism or indeed, any correlated interaction could be described by a version of Hamilton's rule, and argued for a general informational view of relatedness coefficients. Fletcher and Doebeli (2009) further developed these themes and argued for abandoning genetic relatedness as the main key to cooperation in favor of correlated interactions. I will develop those themes here, grouping the mechanisms under the heading of kith selection—selection involving neighbors who need not be kin or similar in kind.

Fig. 1.1 illustrates the connections. Under kin selection, an arrow would connect G and G'. However, even if there is no kinship, P' and P can still be used to model or predict the focal individual's fitness W, resulting in expression (7) or (8). If the actor's phenotype predicts its partner's phenotype P' (heavy arrow), this generates a covariance between G and P' (or G' and P), making the relatedness coefficient in expressions (7) and (8) nonzero. However, we now allow P' to represent an entirely different behavior than P, coded for by different kinds of genes that possibly

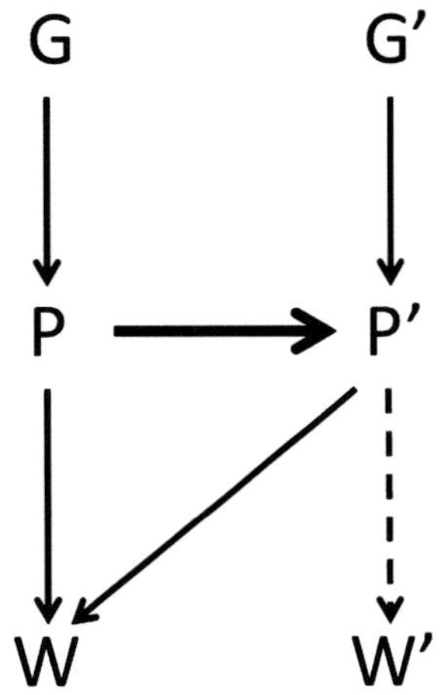

FIGURE 1.1 Kith selection. An actor's phenotype *P* can influence *P'*, its partner's phenotype (often for a different trait), by manipulation, partner choice, and partner fidelity feedback (heavy arrow). These components create an association between phenotypes *P* and *P'* and therefore, also between *P'* and *G* required in Eq. (7) [or *P* and *G'* in Eq. (8)].

belong to different species. *P* could be carbon production by an alga in a lichen, and *P'* could be nitrogen production by its fungal partner. The link between *P* and *P'* could come through several means, including the two kinds of mechanisms that can be involved in reciprocity and mutualism: partner choice and partner fidelity feedback (Sachs et al., 2004). If cooperators choose to associate with cooperators and reject noncooperators, this situation will generate a correlation between *P* and *P'*. The same will be true if individuals join at random, but those who give larger benefits induce their partners to return larger benefits. Finally, the actor could influence the partner's phenotype through pure manipulation.

Kin selection occurs through genetic identity, and can occur even if the partner does not express the behavior. Indeed, conditional helping of partners who do not help underlies some of the most important manifestations of kin selection, such as social insect workers helping queens. Kith selection, in contrast, requires phenotypic expression by the partner. The focal individual can affect its own fitness in kith interactions only through feedbacks. It affects the phenotype of the partner—whether by manipulation, partner fidelity, or partner choice—and the partner's phenotype feeds back on the actor's fitness. The essential role of phenotypes is brought out by modeling the partner's phenotype as $P' = \alpha + \beta_{P'P}P + \varepsilon$ and substituting it into the covariance ratio of expression (7):

$$\beta_{WP \cdot P'} + \beta_{WP' \cdot P} \frac{\mathrm{Cov}(G, \alpha + \beta_{P' \cdot P}P + \varepsilon)}{\mathrm{Cov}(G, P)} > 0. \qquad (9)$$

Splitting the covariance, dropping the α and ε terms, and extracting the β coefficient yields

$$\beta_{WP \cdot P'} + \beta_{WP' \cdot P}\, \beta_{P' \cdot P} > 0. \tag{10}$$

We now have Hamilton's rule with the usual effect on fitness of self ($\beta_{WP.P'}$) and partner ($\beta_{WP'.P}$), but instead of genetic relatedness, there is a structural feedback coefficient $\beta_{P'P}$ that tells how much the actor's behavior influences or is correlated with the relevant behavior of its partner. Remember that the phenotypes may be entirely different things (perhaps cooperative carbon production by an alga and cooperative nitrogen production by a fungus) but that a correlation can still exist between the two forms of cooperation. The actor's cooperation can pay, even if it pays a cost ($\beta_{WP.P'} < 0$), if its behavior causes or is associated with ($\beta_{P'P} > 0$) partner behaviors with positive benefits ($\beta_{WP'.P} > 0$).

If the partner is unrelated or in a different species, the standard Hamilton's rule (1964a) would simply require $\beta_{WP} > 0$, where β_{WP} includes any effects of the actor's behavior that operate by feedback through partners. That result is perfectly correct and does not need to be altered, but it does not capture the social causation. With expression (10) we can see that the actor increases its own fitness by a pathway that, like kin selection, involves social benefits and some kind of association.

Expression (10) is an expanded version of Hamilton's rule that captures kith selection, but it is a neighbor-modulated form, with effects on a focal actor rather than an inclusive fitness form that attributes all effects to a focal actor. Neighbor-modulated forms are often better for modeling (Taylor and Frank, 1996), whereas inclusive fitness forms are often better for intuition and insight. To obtain an inclusive fitness form that tells how actors value a partner's fitness, we need to include the partner's fitness.

I distinguish two cases. In the first case, the partner's fitness is incidental for the actor, affected only as a side effect of the actor's effect on the partner's phenotype (dashed arrow in Fig. 1.1). The effect of the actor's behavior on partner fitness is the product of its effect on the partner's phenotype $\beta_{P'.P}$ and the effect of the partner's phenotype on the partner's fitness $\beta_{W'.P} = \beta_{P'.P}\, \beta_{W'.P'}$. Therefore, $\beta_{P'.P} = \beta_{W'.P}/\beta_{W'.P}$, which can be substituted into expression (10) to give

$$\beta_{WP \cdot P'} + \beta_{WP' \cdot P} \frac{\beta_{W'P}}{\beta_{W'P'}} > 0$$

or, shifting the denominator,

$$\beta_{WP \cdot P'} + \beta_{W'P} \frac{\beta_{WP' \cdot P}}{\beta_{W'P'}} > 0. \tag{11}$$

Now, we have the actor's nonsocial effect on its own fitness and its kith effect on its partner's fitness $\beta_{W'.P}$ through its effect on the partner phenotype. The latter is multiplied by a regression ratio that tells how the actor values those fitness effects on the partner. This kith or feedback coefficient shows that the actor values effects on its partner's fitness (by P and P') only to the degree that they are associated with fitness returns to itself. This makes sense as a scaling factor for the actor, when it acts through affecting the partner's phenotype. The effect on the partner's fitness is incidental, but when the feedback coefficient is positive, the actor gets a positive feedback by aiding its partner. The feedback need not be positive. We could use the equation to describe manipulation that harms the partner but benefits the actor.

A second possibility is that the actor gains not so much by affecting some particular cooperative trait of the partner but by affecting its fitness in general. That is, effects on the partner's fitness are necessary for the feedback to the actor, not just an incidental effect. In a lichen, an alga that produces more carbon may make its fungal partner fitter, and fitter fungi may make more nitrogen that benefit the alga. Here, we write fitness as $W = \alpha + \beta_{WP.W'}P + \beta_{WW'.P}P' + \varepsilon$ and follow steps (*ii*)–(*viii*) to get a simpler result (12):

$$\beta_{WP \cdot W'} + \beta_{W'P}\,\beta_{WW' \cdot P} > 0. \tag{12}$$

Here, the actor affects its own fitness ($\beta_{WP \cdot W'}$) and the fitness of its partner ($\beta_{W'P}$), with the latter multiplied by a feedback coefficient of $\beta_{WW' \cdot P}$ that describes how much partner's fitness affects the actor's fitness, partialing out the nonsocial effects of the actor's behavior (which are included in the first term). This is a more intuitive result than expression (11), but it is really just a special case of it, where $P' = W'$. In both cases, the actor values its partner's fitness according to how it affects its own fitness, but in one case, it is mediated through some intermediate trait. The difference may be important for the evolution of complex mutualisms, which may be much easier to evolve when any benefits to partner's fitness feed back to the actor than if it occurs through only one or a few traits.

Expressions (10)–(12) provide Hamilton's rule forms to handle kith selection. As suggested previously, both reciprocity (Queller, 1985; Fletcher and Zwick, 2006) and mutualism (Frank, 1994, 1998; Fletcher and Zwick, 2006; Foster and Wenseleers, 2006) can be addressed using such results. The analysis here adds at least three features. First, manipulation can be added to the kinds of interactions treated. Second, the results can be expressed not just in terms of neighbor-modulated fitness but also in terms of inclusive fitness. Third, I have put these kinds of results into the common language of regressions and covariances used by quantitative geneticists. The regressions of phenotype on fitness are selection differentials. The coefficient that scales the second regression has to do with heritability; it is actually a ratio of the heritability of the nonsocial effect on self and the heritability of the social effect of, or on, one's partner. This has been shown previously for relatedness in the kin selection form, where the heritability of the indirect selection effect is lowered, because the partner is less likely to pass on the trait (Queller, 1992c). For kith selection, the heritability through social effects is lowered by the fact that the actor's phenotype does not perfectly predict the partner's phenotype.

KIND SELECTION

Another type of selection that is usefully considered separately from the other two is what I call kind selection (Strassmann et al., 2011a). The first example is the greenbeard gene, which has three effects: It produces a cue (like a greenbeard), perceives that cue in others, and directs some special action to those cue bearers (Hamilton, 1964a). Once viewed more as a thought experiment than as a real possibility (Dawkins, 1976b), real greenbeards are being identified with increasing frequency (Keller and Ross, 1998; Queller et al., 2003; Sinervo et al., 2006; Smukalla et al., 2008; Strassmann et al., 2011a). There are greenbeards that help others with the same trait, and there are greenbeards that harm others with different traits. There are both facultative greenbeards that take special actions to like or unlike interactants and obligate greenbeards that perform a general action that has different effects on like and unlike (Gardner and West, 2010).

Table 1.1 shows many differences between greenbeard or kind selection relative to kin selection (and also, for completeness, to kith selection). I will focus on greenbeards for the moment and come to other forms of kind selection later. The key difference is that, where kin selection works through genealogical kin of the actor, kind selection operates on those who specifically possess the same trait as the actor. Those two features can be correlated of course; kin tend to have similar genes that will tend to produce similar traits. However, in one case kinship is fundamental, and in the other case, phenotypic similarity is fundamental. Greenbeards can be favored even among nonkin. Conversely, kin selection can operate even in the absence of having actual traits in common; often, one kind of individual will express the trait, such as a worker bee's behavior, to benefit others who specifically do not express the trait (queens and males) but are nevertheless kin.

Where kin selection operates through cues that correlate with identity by descent, kind selection operates based on all identities (both by descent and by state). Indeed, identity by trait might be a better description; two separate loci producing the same greenbeard trait could work just as well as one. Because identity by descent is normally the same across the genome, kin-selected genes across the genome agree, and complex coop-

TABLE 1.1 Kin and Kind Discrimination

	Kin	Kind	Kith
Behavior	Action with partner	Interaction with partner	Feedback from partner
Key partner feature	Possession of same allele	Expression of same trait	Expression of other trait
Beneficiaries	Genealogical kin	Same trait or kind	Any
Role of genetic identity	By descent only	All identities (but really trait identity)	None
Kinship required	Yes	No	No
Genes	Often multigenic	Often one or linked complex	Often multigenic
Relatedness or genetic correlation	Same across genome	Higher at kind locus	None
Complex cooperation	Possible	Unlikely	Possible
Additive fitness effects	Yes	Usually no	Possible
Frequency dependence	Usually no	Usually yes	Possible

eration can easily be built. The situation with greenbeards is more complex. An altruistic greenbeard allele is related by $r = 1$ to its beneficiaries and therefore may give more aid compared with what would be favored at other loci not related to that degree. There has been some debate over whether greenbeards are outlaws with respect to the rest of the genome (Alexander and Borgia, 1978; Gardner and West, 2010). However, the important point here is that no other locus, unless very closely linked, would build on a greenbeard's identification of beneficiaries. More precisely, if it did build on this identification, it would only be to the extent that the greenbeard cue identified kin. As a result, we do not expect a lot of complexity from greenbeards—they are generally limited to simple traits.

The last two rows in Table 1.1 require a bit more explanation. Greenbeard traits depend on all identities, not just identity by descent, and therefore, they usually depend on the frequency of the trait in the population (facultative helping greenbeards can be an exception) (Gardner and West, 2010). Kin selection is typically frequency independent; the condition $-c + rb > 0$ includes no allele frequencies, because the fraction of alleles identical by descent is independent of gene frequency. However, kin selection models with costs and benefits that are nonadditive typically show frequency dependence. I will argue below that this is because these nonadditive models include a form of kind selection.

I will begin by comparing facultative and obligate greenbeards and then build an argument (Queller, 1984) that obligate greenbeards are insensibly different from more general forms of kind selection. In facultative greenbeards, the actor first classifies its partners and then performs the appropriate behavior. Fire ant workers identify queens lacking their greenbeard allele and then attack them (Keller and Ross, 1998). Obligate greenbeards, in contrast, perform a behavior to all interactants without prior identification, but the behavior has different effects on partners who are greenbeards versus those who are not. Bacteriocins provide many examples of obligate harming greenbeards (Riley and Wertz, 2002; Gardner and West, 2010). Many bacteria have several tightly linked genes that make a poison, which some cells release at times of stress, and also make an antidote to the poison, which they keep private (Riley and Wertz, 2002). Cells lacking the complex are killed by the poison, freeing up resources for those who have it. This greenbeard is obligate, because the cells produce the poison and antidote independently of who their partners are; however, the poison adversely affects only those that lack the gene complex (Gardner and West, 2010).

The key feature of a greenbeard is that it gives some fitness benefit to partners who share the trait that it does not give to partners who lack the trait. In a two-interactant payoff matrix, this can be represented as in Fig. 1.2. The simplest greenbeard effect does not require this full complexity. It could be represented with the d parameter alone; d is what a greenbeard cooperator gets when playing another greenbeard cooperator, and it is generally the sum of the cost of greenbeard cooperation and the benefit of being helped. These are not the c and b variables in the matrix, which instead represent any general costs and benefits, not specific greenbeard ones. Consider the *Ti* plasmid of *Agrobacterium tumefaciens,* an obligate helping greenbeard (Gardner and West, 2010). It harbors a number of genes that induce its plant host to produce a tumor and produce a food source in the form of opines (White and Winans, 2007; Platt and Bever, 2009). The costs of these behaviors are represented by c in Fig. 1.2—they apply whether there are nonbearers present or not. Any benefits that are public goods benefiting bearers and nonbearers alike—perhaps tumor production—are represented by b. However, the gains from opine pro-

	Cooperator	Defector
Cooperator	- c + b + d	-c
Defector	b	0

FIGURE 1.2 General payoff matrix for the two persons expressed in terms of general effects on self (*c*), general effects on partner (*b*), and an extra effect (*d*) that applies only when both partners perform the behavior.

duction are a greenbeard effect, because opine catabolism is also coded on the *Ti* plasmid; nonbearers do not benefit. This targeted benefit must be represented by *d*. Thus, greenbeard effects may be superimposed on nongreenbeard effects, and they can be viewed as nonadditive fitness parts. When you are both an actor and a recipient, the payoff is not the $-c + b$ that would come from adding the separate effects, but $-c + b + d$.

The payoff matrix in Fig. 1.2, required to represent greenbeard effects, is actually the general payoff matrix for two interactants (Queller, 1984). With three parameters plus zero, it covers exactly the same ground as any four-parameter matrix. Every such game can be expressed as a general effect on self *c*, a general effect on partners *b*, and a specific effect that applies only when both actor and partner perform the behavior *d*. That means that any nonadditive two-person game, one that requires the complexity of a *d* parameter, has the greenbeard-like character of giving some fitness gain (or loss) to those who share the trait but not to others (Queller, 1984). Additionally, most of the games that have occupied the interests of evolutionary theorists over the years are nonadditive. Long ago, I noted this similarity and toyed with the idea that all these games represent forms of greenbeard selection (Queller, 1984). That had the problem of subsuming the larger familiar category under the smaller—then nearly nonexistent—category of greenbeard. It might be more palatable to do the opposite (subsume greenbeard effects under the other type), but the problem here is that there really is no name for the process that underlies selection in these games. They are frequency dependent and nonadditive, but those terms do not capture the reason why the process works (the way kin selection does for affecting relatives). Kind selection does capture the feature, common with greenbeards, that an actor has different effects on its own kind than on different kinds.

Although I motivated this kind selection grouping using nonadditivity and frequency dependence, the similarities extend throughout Table 1.1. Most notably, the fitness increment (or decrement) represented by *d* depends on expression of the trait by one's partner. This involves all identities rather than just identity by descent, and kinship is not required. The similarity between partners receiving the *d* effect is generally higher than genealogical relatedness at the loci causing the behavior, but not at the rest of the genome. Cooperation that results from this single-trait similarity is expected to be relatively simple cooperation and not the highly complex cooperation that can lead to major transitions.

As an example, consider the Hawk–Dove game (Fig. 1.*3A)* (Maynard Smith and Price, 1973). There is some contested resource worth *V* fitness units at stake. Hawks fight, gaining all of the fitness units against doves, whereas two doves divide them peaceably. Two hawks will fight each other; a random one of them gets the resource, and the other gets injured, suffering fitness loss I. We can convert to the form of Fig. 1.3*B* by subtracting *V*/2 from all entries to get Fig. 1.3*B*. We can now see that being a hawk adds *V*/2 to your own fitness, subtracts *V*/2 from your partner's fitness, and subtracts an additional *I*/2 only when both partners are hawks. Thus, the *d* term here is negative, representing an antigreenbeard effect of harming one's own type. A negative *d* means negative frequency dependence, with strategies being more favored when rare, leading to the possibility of polymorphism.

An example of a positive *d* would be two ant foundresses cooperating in colony establishment. Groupers pay a cost of searching for other groupers (*c* term in Fig. 1.2) and may also impose a general cost on all potential partners as they negotiate or figure out who is a grouper and who is a loner (the *b* term, likely negative in this case). However, there are also synergisms that can apply to two groupers that associate. For example, if one dies before the first workers hatch out, the other inherits those workers, getting a double brood, an advantage that loners never get. Many such group effects, such as selfish-herd defense (Hamilton, 1971), can be viewed in this way.

Warning coloration in distasteful insects provides a more elaborate example (Queller, 1984; Guilford, 1985). A bright individual is more likely to be seen by a predator (*c* term). If eaten, it will teach the predator that insects like this taste bad. That might provide some general benefit *b* to both bright and dull forms, but warning coloration will not evolve for that reason. It is favored if an eaten bright bug teaches the predator specifically about bright bugs being bad. This is a positive *d*, a benefit that bright bugs confer only on other bright bugs.

I do not include all game theory under kind selection, only games between individuals with the same trait options, with nonadditive effects.

A

	Hawk	Dove
Hawk	(V-I)/2	V
Dove	0	V/2

B

	Hawk	Dove
Hawk	V/2-V/2-I/2	V/2
Dove	-V/2	0

FIGURE 1.3 Payoffs for the Hawk–Dove game in (*A*) conventional form and (*B*) the form of Fig. 1.2, emphasizing that there is a nonadditive effect of both partners performing the behavior, $d = I/2$.

Games between individuals with different roles, such as male and female, that express different traits are better considered as kith selection. Also excluded from kind selection are some frequency-dependent effects in multi-interactant games if the effects of an individual's behavior are the same on both like and unlike partners (Smith et al., 2010a).

How should we model kind effects? There are many ways, with game theory having been the most popular. Even within the inclusive fitness approach, there are multiple options. Frequency-dependent effects are often incorporated into direct fitness. Greenbeard effects, in contrast, are usually attributed to indirect fitness through the partner. This is odd given that these two kinds of effects are so similar, but it is because d effects are really joint effects of the pair, and the two different historical traditions that attributed them to one partner happened to choose differently. A third alternative is often better. If the effect comes from the joint behavior of both partners, the best causal representation would be joint one (Queller, 1984, 1985).

This can be accomplished, for the two-person game, by adding the joint phenotype $P \times P'$ as a part in the model (Queller, 1985, 1992b). This has a particularly clear interpretation when the trait is dichotomous and assigned values of 1 and 0. $P \times P'$ then equals zero unless both partners express the trait, and therefore, it becomes a variable indicating when that happens. Specifically, let $W = \alpha + \beta_{WP}P + \beta_{WP'}P' + \beta_{WPP'}PP' + \varepsilon$ (here, I omit the extra regression subscripts showing the partialed-out variables, but let it be understood that these are still partial regression coefficients). Now, follow steps (*ii*)–(*viii*) from earlier in the paper to find that ΔG increases when

$$\beta_{WP} + \beta_{WP'} \frac{\text{Cov}(G, P')}{\text{Cov}(G, P)} + \beta_{WPP'} \frac{\text{Cov}(G, PP')}{\text{Cov}(G, P)} > 0 \tag{13}$$

from the neighbor-modulated point of view or

$$\beta_{WP} + \beta_{W'P} \frac{\text{Cov}(G', P)}{\text{Cov}(G, P)} + \beta_{W'PP'} \frac{\text{Cov}(G', PP')}{\text{Cov}(G, P)} > 0 \tag{14}$$

for inclusive fitness with indirect effects on partners (Queller, 1992c). I have termed the second covariance ratio, which depends on when both partners perform the behavior, a synergism coefficient (Queller, 1984, 1985, 1992b).

There are two reasons for preferring these forms to simpler versions of Hamilton's rule (1964a) that bundle nonadditive effects into one of the other terms. The main reason is the same one that applies to the kith selection forms: It captures the social causality better. Instead of an undifferentiated average direct fitness that implicitly combines two kinds of direct fitness (some individuals get $-c$ and some get $-c + d$), the new forms split out those two effects and make the frequency dependence more explicit. It distinguishes true kin effects from effects that result from being similar in kind.

A secondary reason for preferring these forms is that they are sometimes more accurate than the simpler Hamilton's rule. As noted above, the strictly genetic version of Hamilton's rule (6) is always valid, but much of the value of Hamilton's rule lies in being able to apply the phenotypic gambit. The two phenotypic predictors in expressions (7) and (8) successfully capture the complexity of an additive game. Together, the two predictors define a plane as do the four fitness values in the two-person additive game. However, a plane cannot fit four nonadditive points. Adding $P \times P'$ as a predictor allows us to fit those points and explain more of the variance. However, more importantly, the simpler versions can sometimes be incorrect, biased in the same way that caused us to reject the simple direct fitness model in favor of inclusive fitness. Specifically, the crucial step (*v*) of our derivation procedure, dropping $\text{Cov}(G, \varepsilon)$, is not always possible for a model with only P and P'. Suppose, for example, that cooperation is multigenic. Cooperators all perform the behavior, but they can vary in their breeding values for the trait. Then, if partners are at least sometimes related, those actors with the highest breeding values will be more likely to have partners who also perform the behavior and therefore are more likely to get the d effect. Thus, the average G differs for actors who get $-c + b + d$ and those who get $-c + b$. One cannot simply average the two types any more than one could average eight fitness units given to a sibling and one unit given to a third cousin. In short, there are cases in phenotypic models where we cannot get away with two predictors. Synergism can be more complex than in the simple two-person game. When interactions occur in larger groups, additional terms may be needed to capture higher-order interactions (Smith et al., 2010a).

CONCLUSIONS

Although I have worked through kin, kith, and kind selection separately, the results can be combined in the expanded version of Hamilton's rule in expression (1). It covers more kinds of social selection in a causal manner. The inclusive fitness form would use terms from expression (8) for kin, expressions (11) or (12) for kith, and expression (14) for kind, whereas the neighbor-modulated form would use expressions (7), (10), and (13), respectively. These expanded social fitness results, like the traditional ones, separate out two kinds of distinct terms: selection terms relating social phenotype to fitness components and relative heritability terms that derive from associations of genes and phenotypes, or just phenotypes.

The model suggested here stakes out a middle position between standard inclusive fitness theory and more complex models (e.g., from population genetics). The goal has been to extend the advantages of inclusive fitness theory to a more explicitly causal analysis of social effects other than kin selection. I have chosen to still call the result Hamilton's rule because of the way both separate fitness terms from association or currency translation terms that measure relative heritability. This approach makes the phenotypic gambit a plausible strategy; we can ask how phenotypes affect fitness and then separately assess or measure the associations implied by relatedness, synergism, or feedback coefficients. Like standard inclusive fitness, these high-level summary variables can cut through much of the complexity of population genetic models, where often, a new model must be constructed and solved for a small change in assumptions. The result is, like standard inclusive fitness, an individual-centered analysis that allows us to use the intuition that comes from a simple model and viewing individuals as agents.

This model probably does not much change the view that standard inclusive fitness is maximized, although it does change it to some extent. Kith effects are simply cleaved off of the standard direct fitness term, and therefore, they do not alter total inclusive fitness. Kind effects are a bit more complex. The strictly genetic form of standard inclusive fitness (6) is always valid. However, the phenotypic form (8), which is often more useful in practice, is not always exactly valid under kind selection, and therefore, the expanded inclusive fitness with kind selection can differ somewhat from standard inclusive fitness. More work needs to be done on when these two forms differ and by how much, but I suspect that standard inclusive fitness will usually be a good approximation.

I have framed this paper largely in terms of the problem of cooperation, with positive costs to the actor and positive benefits to partners, that has intrigued biologists for the last several decades. However, of course, as with inclusive fitness, the equations here also apply when they have fitness terms of the opposite sign. If c is negative, we have selfish effects. If b is negative, harm falls on relatives. If d is negative, two actors together have more negative effects than one acting alone. If m is negative, the actor is harming its partner, which can be favored if it is coupled with a negative feedback coefficient—if negative effects on partners generate positive effects back to the actor. Predation is an extreme example.

One complication that I have not treated explicitly is kith selection with multiple partners. For example, mutualisms often involve a large partner of one species and many smaller (often microbial) partners in another species. Actions of one of the smaller partners may then feed back onto kin, so extra terms, with both feedback and relatedness, may be required (Frank, 1994, 1998; Fletcher and Zwick, 2006; Foster and Wenseleers, 2006). There could also be an interaction with kind selection if the fitness feedbacks affect actors and nonactors differently. For example, the *A. tumefaciens Ti* plasmid works this way, with the opine greenbeard effects operating through influence on the host plant.

No social model will perform all possible functions. There are tradeoffs in precision, realism, and generality (Levins, 1966) as well as in simplicity and elegance. Inclusive fitness does pretty well on most of these scores, but it does not tell us everything in either the standard or expanded forms. There is, for example, an increasing interest in how genetic relatedness patterns are generated in the course of selection, migration, and drift. Inclusive fitness typically (although not always) takes the relatedness pattern as given. This is true as well for the associations that underlie kith and kin selection. It is certainly useful to have more detailed models of how these associations are built up, and the paths may sometimes be too complex for such simple models to fully illuminate. However, the history of inclusive fitness suggests that it is also extremely useful to have summary models that cut through much of the complexity to illuminate crucial similarities and differences. Such models are especially useful to empiricists who do not usually know the genetics underlying their trait and prefer to work with a small number of parameters rather than many. These advantages should apply to the expanded social fitness model that includes and distinguishes kin, kith, and kind.

ACKNOWLEDGMENTS

I thank Francisco Ayala and John Avise for helping to organize the National Academy of Sciences Sackler symposium on cooperation and conflict. For comments on the manuscript, I thank two anonymous referees, Joan Strassmann, Michael Whitlock, Stuart West, Kevin Foster, Claire El Mouden, and the Oxford Social Evolution Group. Our research is supported by U.S. National Science Foundation Grants DEB 0816690 and DEB 0918931.

2

Evolutionary Transitions in Bacterial Symbiosis

JOEL L. SACHS,*† RYAN G. SKOPHAMMER,* AND JOHN U. REGUS*

Diverse bacterial lineages form beneficial infections with eukaryotic hosts. The origins, evolution, and breakdown of these mutualisms represent important evolutionary transitions. To examine these key events, we synthesize data from diverse interactions between bacteria and eukaryote hosts. Five evolutionary transitions are investigated, including the origins of bacterial associations with eukaryotes, the origins and subsequent stable maintenance of bacterial mutualism with hosts, the capture of beneficial symbionts via the evolution of strict vertical transmission within host lineages, and the evolutionary breakdown of bacterial mutualism. Each of these transitions has occurred many times in the history of bacterial–eukaryote symbiosis. We investigate these evolutionary events across the bacterial domain and also among a focal set of well-studied bacterial mutualist lineages. Subsequently, we generate a framework for examining evolutionary transitions in bacterial symbiosis and test hypotheses about the selective, ecological, and genomic forces that shape these events.

Ancestrally, bacteria and archaea persisted solely as free-living cells in terrestrial and aquatic habitats. Along with the evolution and diversification of animals and plants, the past 500 million years have also witnessed a massive radiation of bacteria. Bacterial

*Department of Biology, University of California, Riverside, CA 92521. †To whom correspondence should be addressed. E-mail: joels@ucr.edu.

lineages have evolved diverse mechanisms to gain entry and proliferate in the tissues and cells of multicellular eukaryotes (Merhej et al., 2009; Carvalho et al., 2010; Medina and Sachs, 2010; Toft and Andersson, 2010), and these symbionts vary in their effect on hosts from harmful to beneficial (Medina and Sachs, 2010; Toft and Andersson, 2010). Archaea have also evolved associations with hosts, but these interactions do not appear as diverse or ubiquitous. Bacterial symbioses (defined in the broad sense) include persistent, intimate associations between bacteria and other species and date back at least to the origins of eukaryotes (Sagan, 1967). Bacterial parasites range from infectious diseases that rapidly exploit hosts before infecting new individuals, to bacteria that are transmitted vertically from host parent to offspring and manipulate host reproduction to favor their own spread (Stouthamer et al., 1999). Parasitic bacteria have received intense focus from researchers over the last century because harmful infections represent a critical challenge to human health and economic interests. In contrast, except for a few early pioneers (Buchner, 1921), researchers have only recently focused on the biology of bacteria that enhance host fitness: bacterial mutualists (Sachs et al., 2011).

Bacterial mutualists are diverse (Williams et al., 2007, 2010; Merhej et al., 2009; Wu et al., 2009; Carvalho et al., 2010; Medina and Sachs, 2010; Philippot et al., 2010; Toft and Andersson, 2010) and exhibit a variety of lifestyles and coevolutionary relationships with eukaryote hosts (Sachs et al., 2011) (Table 2.1). First, beneficial bacteria vary in their degree of reliance on hosts for reproduction. Whereas some bacterial-derived organelles and endosymbionts cannot live independently of hosts, most bacterial mutualists retain extensive environmental phases and form infections that are facultative for the bacterium (Szathmáry and Smith, 1995; Nyholm and McFall-Ngai, 2004; Sachs et al., 2011). Second, bacterial mutualists inhabit diverse host tissues ranging from skin, mucosa, leaves, and roots to inter- and intracellular spaces. Some bacterial mutualists inhabit specialized structures in hosts (Becking, 1970; Savage, 1977; Sprent et al., 1987; Douglas, 1989; Bright and Sorgo, 2003; Currie et al., 2006; Nussbaumer et al., 2006; Visick and Ruby, 2006; Goettler et al., 2007; Vaishnava et al., 2008; Pflugfelder et al., 2009; Ran et al., 2010), whereas others range widely in host mucosa or other unstructured tissues (Hirose, 2000; Hirose et al., 2009; Kaltenpoth et al., 2009) (Table 2.1). Finally, bacterial mutualists provide a great variety of benefits to hosts, including nutrients (Becking, 1970; Sprent et al., 1987; Douglas, 1989; Hirose, 2000; Hooper et al., 2002; Ran et al., 2010), bioluminescence (Nyholm and McFall-Ngai, 2004), and antibiotic production (Currie et al., 1999; Kaltenpoth et al., 2005; Kost et al., 2007). Although bacterial mutualists by definition provide a net fitness benefit to hosts, they can also bear features that exploit hosts (Frank, 1996a,b; Sachs and Wilcox, 2006; Simms et al., 2006; Weeks et al., 2007; Oliver et al., 2008; Heath et al., 2010; Sachs et al., 2010a, 2011). As we detail later, each of these variables (degree of reliance on hosts, type of host habitat, and type of benefit provided to host) can modulate evolutionary transitions in bacterial symbiosis and can explain how and why transitions occur.

Here, we investigate evolutionary transitions that have occurred in the history of bacterial mutualism. We focus on (*i*) the origins of host association in bacteria (transitions in which environmental bacteria evolve to form intimate and persistent associations with hosts irrespective of effects on host fitness), (*ii*) the origins of bacterial mutualism from other types of bacterial lifestyles, (*iii*) shifts to the stable maintenance of bacterial mutualism, (*iv*) the capture of bacterial mutualists (via the evolution of strict vertical transmission within host lineages), and (*v*) the evolutionary breakdown of bacterial mutualism. Each of these events has occurred multiple times in the evolution of bacteria. Only symbiont capture possibly constitutes a "major evolutionary transition," defined as an integrating event in which partners lose the ability to replicate independently (Szathmáry and Smith, 1995). However, loss of independence often only occurs for the symbiont.

To study broad patterns and genetic drivers of transitions, we investigate phylogenomic data that span the bacterial domain (Williams et al., 2007, 2010; Merhej et al., 2009; Wu et al., 2009; Philippot et al., 2010; Toft and Andersson, 2010) (Fig. 2.1), and to study fine-scale patterns, we also analyze a focal set of bacterial mutualists (Table 2.1). Our domain-level data sources include a phylogeny with 350 bacterial taxa sampled from 20 phyla (Wu et al., 2009), coupled with phenotypic host-association data (Boussau et al., 2004; Merhej et al., 2009; Bright and Bulgheresi, 2010; Philippot et al., 2010; Toft and Andersson, 2010). The focal systems include beneficial symbionts chosen to represent host and bacterial diversity, breadth in symbiotic services, and variety in transmission modes. Our analysis of historical and selective scenarios that characterize transitions in bacterial symbiosis complements other work that has focused on genomic changes (Merhej et al., 2009; Carvalho et al., 2010; Medina and Sachs, 2010; Toft and Andersson, 2010). The phylogeny of Wu and colleagues (2009) and the review by Toft and Andersson (2010) are particularly germane to this study as they provide the domain-level dataset that we use to test hypotheses.

There are caveats to consider when inferring the evolutionary history of bacterial symbiosis at broad phylogenetic scales. First is the challenge of assigning host-association traits to bacterial species. Recent work suggests that fitness benefits provided by bacteria to hosts can be context dependent (Heath and Tiffin, 2007; Oliver et al., 2008; Heath et

TABLE 2.1. Fourteen Focal Bacterial-Host Mutualisms Analyzed

Symbiont, Host	Benefits Provided by Bacteria to Host	Host Localization	Transmission Among Hosts
Rhizobia [e.g., Sawada et al. (2003)], legumes	Nitrogen fixation (Sprent et al., 1987)	Nodules (Sprent et al., 1987)	Horizontal transmission (Sprent et al., 1987) with free-living stages (Sachs et al., 2009)
Frankia spp., actinorhizal plants	Nitrogen fixation (Becking, 1970)	Nodules (Becking, 1970)	Horizontal transmission with free-living stages (Huss-Danell and Frej, 1986)
Pseudonocardia spp. (fungus-growing ants)	Antibiotics (Currie et al., 1999; Kost et al., 2007)	Crypt structures on exoskeleton (Currie et al., 2006)	Vertical transmission to offspring ant colonies (Currie et al., 1999) and horizontal transmission with environmental pool (Mueller et al., 2008, 2010)
Endoriftia persephone, tubeworm	All nutrients (Nussbaumer et al., 2006)	Lobules in host trophosome (Bright and Sorgo, 2003; Nussbaumer et al., 2006; Pflugfelder et al., 2009)	Horizontal with free-living stages (Nussbaumer et al., 2006)
Burkholderia spp., stinkbugs	Unknown nutrients (Kikuchi et al., 2007)	Midgut crypts (Kikuchi et al., 2011)	Horizontal with free-living stages (Kikuchi et al., 2011)
Bacteroides thetaiotaomicron, humans	Nutrients (Hooper et al., 2002)	Crypt structures in gut (Savage, 1977; Vaishnava et al., 2008)	Horizontal transmission (Savage, 1977) with free-living stages (Carson et al., 2005)
Vibrio fischeri, bobtail squids	Bioluminescence (Nyholm and McFall-Ngai, 2004)	Deep crypts in light organ (Visick and Ruby, 2006)	Horizontal transmission with free-living stages (Visick and Ruby, 2006)

Host Association Origins	Mutualism Breakdown	Forces Stabilizing Bacterial Mutualism
Mutualist (Fig. 2.1)	Abandonment events (Sawada et al., 2003; Sachs et al., 2009, 2010a)	Partner choice (Kiers et al., 2003; Simms et al., 2006; Sachs et al., 2010b)
Mutualist (Fig. 2.1) (Normand et al., 1996)	No evidence	Unknown, host localization consistent with partner choice
Mutualist (Mueller et al., 2010)	Abandonment events (Mueller et al., 2010)	Byproducts (see discussion), no evidence of partner choice (Kost et al., 2007)
Ambiguous (Williams et al., 2010)	No evidence	Unknown, host localization consistent with partner choice
Parasite (Kikuchi et al., 2007)	Abandonment events (Kikuchi et al., 2011)	Unknown, host localization consistent with partner choice
Parasite (Fig. 2.1)	No evidence	Byproducts (*Discussion*), partner fidelity feedback (Wilkinson, 1999; Turnbaugh et al., 2009), partner choice (Vaishnava et al., 2008)
Parasite (Fig. 2.1)	Abandonment events (Nishiguchi and Nair, 2003)	Partner choice (Sachs et al., 2004, 2010b)

continued

TABLE 2.1. Continued

Symbiont, Host	Benefits Provided by Bacteria to Host	Host Localization	Transmission Among Hosts
Prochloron spp., didemnid ascidians	Photosynthates (Hirose, 2000)	Unstructured in cloacal cavity (Hirose, 2000; Ran et al., 2010)	Vertical transmission via physical transfer to larvae (Hirose, 2000). No known free-living state (Kojima and Hirose, 2010)
Coriobacterium glomerans, firebugs	Aid in digestion (Kaltenpoth et al., 2009)	Unstructured in guts (Kaltenpoth et al., 2009)	Vertical transmission via egg inoculation. Little potential for horizontal transmission or free-living stages (Kaltenpoth et al., 2009)
Streptomyces philanthi, beewolves	Antibiotics (Kaltenpoth et al., 2005)	Lobed antennomere reservoirs in antennae (Goettler et al., 2007)	Vertical transmission via brood provisioning of bacteria (Kaltenpoth et al., 2010). No known free-living state
"Mycetocyte" bacteria, diverse insects (Douglas, 1989)	Amino acids, vitamins (Douglas, 1989)	Unstructured in mycetocytes in diverse tissues (Douglas, 1989)	Vertical transmission via host transfer to oocytes, eggs, or larvae (Douglas, 1989)
Cyanobacterium spp., water fern	Nitrogen fixation (Ran et al., 2010)	Cavities in leaves (Ran et al., 2010)	Vertical transmission via bacterial motility, no free-living stage (Ran et al., 2010)
Plastids, plants Mitochondria, eukaryotes	Photosynthates Metabolism	Unstructured, intracellular	Transovarial, no free-living stage

Notes: Bacterial symbionts are indicated with genus and species when possible, and hosts are identified with common names. "Mutualist Benefits" specifies the types of resources or services that the bacterial symbionts provide to their hosts. "Host Localization" specifies the location that the bacteria inhabit during the majority of or key parts of their interactions with hosts and whether these locales are structured spatially. "Transmission Among Hosts" specifies transmission mode, and presence of free-living stages are identified. "Host-Association Origins" specifies the inferred ancestral condition at the origin of host association in the described lineage(s). "Mutualism

Host Association Origins	Mutualism Breakdown	Forces Stabilizing Bacterial Mutualism
Mutualist (Münchhoff et al., 2007)	No evidence of mutualism breakdown	Vertical transmission promotes partner fidelity feedback
Ambiguous (Kaltenpoth et al., 2009)	No evidence of mutualism breakdown	Vertical transmission promotes partner fidelity feedback
Mutualist (Kaltenpoth et al., 2006)	No evidence of mutualism breakdown	Vertical transmission promotes partner fidelity feedback
Parasite (Fig. 2.1)	No evidence of mutualism breakdown	Vertical transmission promotes partner fidelity feedback
Mutualist (Svenning et al., 2005)	No evidence of mutualism breakdown	Vertical transmission promotes partner fidelity feedback
Mutualist (Turner et al., 1999) Parasite (Williams et al., 2007)	No evidence of mutualism breakdown	Vertical transmission promotes partner fidelity feedback

Breakdown" specifies evidence of evolutionary transitions in bacterial lineages from mutualism to other lifestyles, with "abandonment" referring to transitions from mutualism to an environmental lifestyle. "Forces Stabilizing Bacterial Mutualism" specifies potential forces stabilizing cooperation in a bacterial mutualist lineage, divided into the three model classes [byproduct cooperation, partner choice, and partner fidelity feedback (Sachs et al., 2004)].

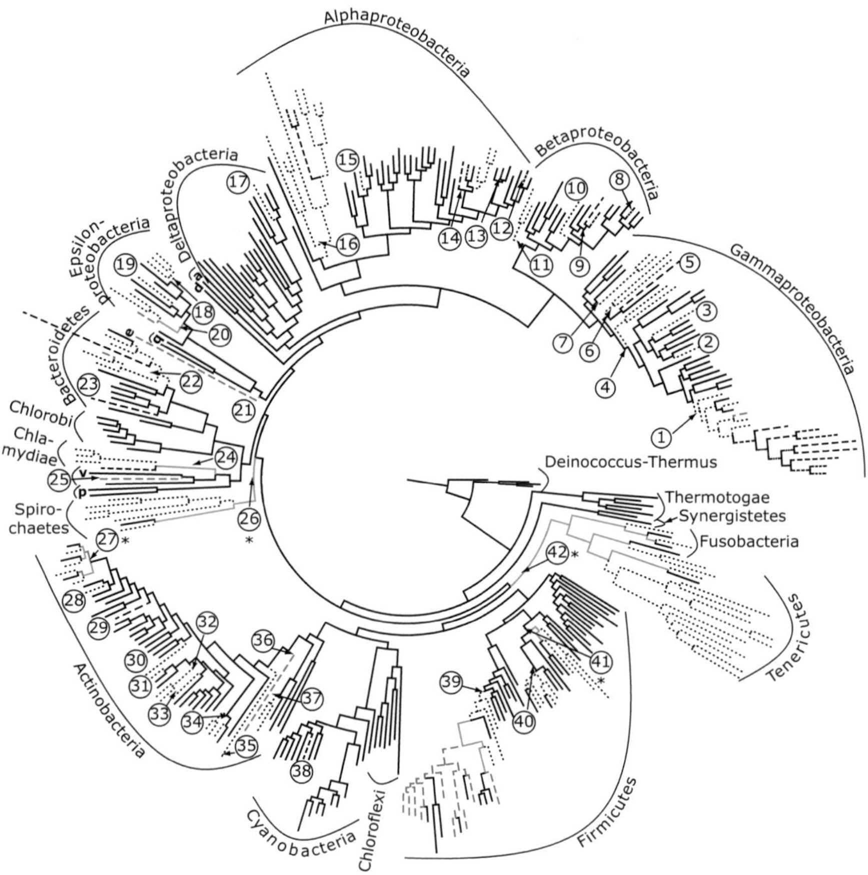

FIGURE 2.1 Inferred evolutionary history of bacterial host association. Ancestral states are inferred on a domain-level bacterial phylogeny modified from a previous study (Wu et al., 2009). The tree is a maximum likelihood reconstruction of a concatenated set of 31 single-copy genes from 350 bacterial species chosen to optimize phylogenetic sampling. Phyla and proteobacterial classes are labeled with their full names (e.g., Gammaproteobacteria; Firmicutes) or single-letter abbreviations (a, Acidobacteria; d, Defferribacteres; q, Aquificae; e, Elusimicrobia; v, Verrucomicrobia; p, Planctomycetes). Branch shades represent host-associated traits on the tips of the tree and inferred states on ancestral nodes (black, environmental; dashed gray, commensal; dashed black, mutualist; dotted, parasite). Host association traits were obtained from a prior review (Toft and Andersson, 2010). We inferred a minimum of 42 origins of host association (labeled 1–42). Origins at five nodes had equivocal parsimony reconstructions, noted with asterisks. Equivocal ancestral states are represented by gray branches. Additional origins are equally parsimonious at these nodes and provide an upper bound for global origins at 52.

al., 2010) and evolutionarily labile (Weeks et al., 2007; Sachs et al., 2010a, 2011), potentially blurring mutualist and parasite categories. Nonetheless, although striking exceptions exist (Weeks et al., 2007; Oliver et al., 2008), the majority of well-studied bacterial taxa can be unambiguously categorized into host-association categories (Moran and Wernegreen, 2000; Philippot et al., 2010; Toft and Andersson, 2010). Second is the challenge of accurately inferring past evolutionary events, which requires a robust and well-sampled phylogeny. The bacterial tree we use is well supported (Wu et al., 2009), but the sampling is sparse (relative to the domain of bacteria represented) and likely biased (only sequenced taxa are included). Finally, predictions about selective factors that drive transitions must be considered with caution, as phylogenetic comparisons often cannot distinguish evolution that predates the origins of host association from the consequences of these transitions. Our fine-scale analysis of the 14 focal symbioses serves as a complementary approach to help mitigate these challenges (Table 2.1).

ORIGINS OF HOST ASSOCIATION IN BACTERIAL LINEAGES

Origins of host association are transitions in which bacteria that live independently in the environment evolve to form intimate and persistent associations with hosts. To evolve host association, bacteria must be able to compete with other microbes on host surfaces, evade negative host responses, uptake novel resources on or inside the host, and ultimately gain transmission to new hosts. Considering these potential hurdles, one unanswered question is whether origins of host association are rare in bacterial lineages. Another question is whether certain bacteria taxa are more likely to evolve host association. In a phenotypic sense, this latter question addresses whether some bacteria bear preadaptations to host association.

Analyzing host association origins on a domain-level bacterial tree (Wu et al., 2009; Toft and Andersson, 2010) (Fig. 2.1 for taxon information), we inferred an environmental ancestral condition for the most recent common ancestor of bacteria and a minimum of 42 origins of host association across bacteria (*Methods*). An environmental ancestral condition is logical (as bacteria predate eukaryote hosts by at least 1 billion years) and is consistent with other analyses (Boussau et al., 2004). Origins of host association are diversely distributed across bacteria, emerging independently in at least 11 bacterial phyla. Proteobacteria, Actinobacteria, and Firmicutes each exhibit multiple origins of host association, whereas a few phyla such as Chlorobi, Chloroflexi, and Planctomycetes have never evolved host association (Madigan et al., 2009; Wu et al., 2009; Toft and Andersson, 2010).

Toft and Andersson (2010) predicted that bacterial preadaptations to host association might be ecological in nature, including access to mobile genes in soil and oceans and physical contact with diverse hosts, characteristics identified as common in Proteobacteria (Snel et al., 2002; Toft and Andersson, 2010). Although Proteobacteria exhibit 20 host-association origins, the evolutionary rate of host-association origins in this lineage (estimated as origins per adjusted branch length; *Methods*) is typical for eubacteria. Bacterial preadaptation to a host-associated lifestyle might also be genetically based, which is not mutually exclusive from ecological preadaptation. Several studies have begun to investigate genomic content changes correlated with transitions in host association, for instance by comparing phylogenetic relationships and genetic characteristics among bacterial mutualists, parasites, and related environmental species (Dale et al., 2001; Sawada et al., 2003; Horn et al., 2004; Frank et al., 2005; Ruby et al., 2005; Ma et al., 2006; Carvalho et al., 2010). The Rhizobiales represent an excellent case study, as these α-Proteobacteria include environmental bacteria, parasites, and mutualists (Sawada et al., 2003; Carvalho et al., 2010). Genomic comparisons of 19 species in this lineage uncovered a relatively small subset of loci unique to the host-associated species and revealed that these loci most often originated in host-associated lineages via horizontal transfer from other host-associated bacteria (Carvalho et al., 2010). Other lineages that encompass parasitic and mutualistic bacteria also show a similar pattern in which host-association loci exhibit evidence of horizontal gene transfer (Dale et al., 2001; Horn et al., 2004; Frank et al., 2005; Ruby et al., 2005; Ma et al., 2006). In summary, we found many origins of host association across bacteria and little evidence consistent with ecological or genomic predispositions to host association. The data suggest that transitions to host association might be constrained only by access to and compatibility with horizontally transferred loci that engender host-association traits (Toft and Andersson, 2010). Nonetheless, ecological constraints to host association cannot be ruled out; the bacterial taxa that have apparently never evolved host association might lack access to habitats with compatible hosts.

ORIGINS OF BACTERIAL MUTUALISM

Fundamental questions about the origins of bacterial mutualisms remain unresolved. Do bacterial mutualists evolve from parasitic ancestors or do they represent independent origins of host association (Ewald, 1987; Szathmáry and Smith, 1995; Corsaro et al., 1999; Moran and Wernegreen, 2000; Medina and Sachs, 2010)? If bacterial mutualists evolved from parasite ancestors, this predicts that transitions from parasitism to mutualism have occurred, whereas if mutualists originate separately from parasites, this predicts that mutualists have evolved directly from environmental taxa. Two scenarios have been suggested to resolve this issue. Ewald (1987) introduced a detailed hypothesis for the origin of bacterial mutualism in which (*i*) an ancestral parasite infects hosts via both horizontal and vertical transmission, (*ii*) a mutation knocks out the parasite's horizontal transmission pathway, and (*iii*) subsequent vertical transmission of the bacterium selects for reduced virulence and the enhancement of mutualistic traits [as vertical transmission can link reproductive interests of symbionts and hosts (Fine, 1975; Frank, 1996a,b; Sachs et al., 2004)]. This scenario is controversial because host-associated bacteria are thought to lack the genomic potential to easily switch from parasitism to mutualism (Moran and Wernegreen, 2000). The alternative hypothesis is that bacterial mutualists evolve directly from environmental bacteria, which is also problematic because it implies that free-living ancestors exhibited traits that could offer immediate benefits to hosts (Ewald, 1987).

We can empirically examine these alternative hypotheses by using the bacterial domain dataset (Wu et al., 2009; Toft and Andersson, 2010) and our focal systems (Table 2.1). At the domain level, many host-associated lineages are poorly sampled (Fig. 2.1), so this analysis must be considered preliminary. Bacteria on the domain-level tree include species classified as commensals, mutualists, and parasites (Toft and Andersson, 2010). Among the 42 host-association origins we reconstructed, 32 are inferred to have originated as parasites, 9 are inferred to have mutualist origins, and 1 origin is ambiguous (Fig. 2.1). Several mutualist taxa are nested in parasitic clades, consistent with three independent transitions from parasitism to mutualism (Fig. 2.1). It is unknown whether the evolution of vertical transmission drove these transitions because, in most lineages, the taxon sampling is poor and the order of events cannot be resolved. Among the nine mutualist lineages that evolved directly from environmental ancestors, six are nitrogen fixing. Consistent with Ewald's (1987) hypothesis, nitrogen fixation is an ancient bacterial trait (Raymond et al., 2004) that can potentially offer hosts immediate benefits. However, as we observed earlier for the origins of host association, nitrogen fixation loci are also prone to horizontal transfer as parts of genome islands. This creates a scenario in which bacterial mutualists can evolve de novo from environmental ancestors via the gain of a core set of symbiosis loci (Sullivan et al., 1995; Sachs et al., 2010a).

Among the 14 focal taxa, we can infer the host-association origins of 12 (Table 2.1). Three of the lineages that likely represent transitions from parasitism to mutualism are vertically transmitted (*Burkholderia* spp., "Mycetocyte" bacteria, mitochondria), consistent with the hypothesis that loss of horizontal transmission drove the origin of mutualism (Ewald, 1987). The history of the mitochondrion is somewhat ambiguous. Although some authors have suggested that mitochondria originated from a parasitic lineage of rickettsial bacteria (Moran and Wernegreen, 2000), no analysis of which we are aware has tested this hypothesis explicitly. In none of these cases can we resolve whether vertical transmission evolved before or after the transition from parasitism to mutualism. Seven of the symbioses are inferred to have originated as mutualists directly from environmental ancestors. As described earlier, these lineages carry traits that can offer immediate benefits to hosts, including antibiotic production, nitrogen fixation, and photosynthesis (Table 2.1). More detailed phylogenetic analysis is needed to resolve whether these cooperative traits predate the host association, as predicted by Ewald (1987). Finally, there are two symbioses that do not fit any of the aforementioned hypotheses. Both *Bacteroides thetaiotaomicron* and *Vibrio fischeri* are mutualists inferred to have evolved from parasites with no history of vertical transmission. For *B. thetaiotaomicron* (a dominant gut symbiont in humans), there is the possibility of pseudovertical transmission (Wilkinson, 1999; Turnbaugh et al., 2009). This is the hypothesis that hosts are more likely to transmit symbionts to kin, which approximates the effects of vertical transmission (Wilkinson, 1999). In summary, mutualist bacteria can evolve from environmental or parasitic ancestors. Bacterial phenotypes that offer immediate benefits to hosts are thought to promote origins of mutualism in environmental bacterial lineages, but well-studied cases implicate horizontal gene transfer (Sullivan et al., 1995; Sachs et al., 2010a) as an alternative. Vertical transmission is a predicted driver of transitions from parasitism to mutualism, but there is relatively little support for vertical transmission preceding the origin of mutualism (Weinert et al., 2009).

MAINTENANCE OF BACTERIAL MUTUALISM

In mutualist bacteria, it can be challenging to explain what prevents the spread of cheater mutants; symbionts that gain in fitness by exploiting hosts and giving little or nothing in return (Sachs et al., 2004). Three classes of models have been proposed for the maintenance of cooperation between species—byproduct cooperation, partner fidelity feedback, and partner choice (Axelrod and Hamilton, 1981; Bull and Rice, 1991; Sachs et al., 2004; Foster and Wenseleers, 2006)—and each of these models applies to bacterial mutualism. Byproduct cooperation occurs when the benefit provided by the symbiont to the host exists as an automatic consequence of a selfish trait, and thus byproduct cooperation carries no net cost for the symbiont (Brown, 1983; Connor, 1995). Partner fidelity feedback exists when fitness benefits delivered from a symbiont to its host feed back as returned benefits to the symbiont, such that beneficial symbionts are rewarded and harmful symbionts experience reduced fitness (Bull and Rice, 1991; Simms and Taylor, 2002; Sachs et al., 2004). Fitness feedbacks are only expected when symbionts and hosts interact repeatedly over time, such as occurs with vertical transmission. Partner choice occurs when hosts preferentially reward beneficial symbionts and or sanction cheaters, thus producing a selective advantage for symbiont cooperation (Bull and Rice, 1991; Denison, 2000; Sachs et al., 2004). To what degree is byproduct cooperation, partner fidelity feedback, or partner choice responsible for the maintenance of cooperative symbioses? These models can work independently or in concert with each other (Sachs et al., 2004; Foster and Wenseleers, 2006); however, little empirical research has compared their prevalence.

Among our 14 focal symbioses, byproduct cooperation can mostly be ruled out, such as in Rhizobia, in which nitrogen fixation is costly and occurs only during the symbiosis (Sachs and Simms, 2008). In contrast, we are not aware of examples in which byproduct cooperation has been demonstrated. Such scenarios are certainly possible. For instance, Actinomycete bacteria produce antibiotics on fungus-farming ants that keep the ants' fungal gardens pathogen-free (Table 2.1) (Currie et al., 1999). Antibiotic production is an anticompetitive function that benefits bacteria directly, whether on the surface of an ant or free in the soil, so it likely qualifies as a byproduct. Similarly, the symbiont *B. thetaiotaomicron* benefits humans by foraging and catabolizing compounds that the host cannot otherwise digest (Sonnenburg et al., 2005). The consumption of complex molecules and releasing of simpler compounds also must benefit *Bacteroides* directly. Byproduct cooperation is likely important for the origins of cooperative symbioses (Sachs et al., 2004), but when interactions have been established, hosts are expected to rapidly evolve traits to promote the infection and proliferation of beneficial symbionts (Connor, 1995; Foster and Wenseleers, 2006). For the *B. thetaiotaomicron*-human symbiosis, these host traits might include mechanisms to bias symbiont transmission to offspring [to maximize partner fidelity (Wilkinson, 1999; Turnbaugh et al., 2009)] or mechanisms to favor beneficial strains over more selfish ones [e.g., partner choice (Vaishnava et al., 2008)].

There is vigorous debate over the relative importance of partner fidelity feedback versus partner choice (Bull and Rice, 1991; Simms and Taylor, 2002; West et al., 2002a,b; Weyl et al., 2010; Archetti et al., 2011). Partner fidelity feedback is often equated with vertically transmitted symbioses, as vertical transmission tightly correlates symbiont and host reproductive interests (Sachs et al., 2004; Foster and Wenseleers,

2006). By this measure, partner fidelity is widespread across bacteria with multiple origins and diverse mechanisms of vertical transmission (Table 2.1). However, vertical transmission does not guarantee symbiont cooperation, as even rare opportunities for horizontal transfer or the potential to manipulate host reproduction can lead to parasitic bacterial phenotypes. For example, vertically transmitted parasites [such as some *Wolbachia* lineages (Weeks et al., 2007)] manipulate hosts to maximize their own transmission by biasing host sex ratio toward females (they are not transmitted to males) or by inducing cytoplasmic incompatibility (Stouthamer et al., 1999). On the contrary, most symbionts are horizontally transmitted (Nyholm and McFall-Ngai, 2004; Sachs et al., 2011). Under horizontal transmission, multiple symbiont genotypes often infect hosts, and, with rare exceptions (Sachs and Wilcox, 2006), partner fidelity is predicted to be weak (West et al., 2002a,b). Partner choice can efficiently select for symbiont cooperation under these conditions (Bull and Rice, 1991; Denison, 2000; West et al., 2002a,b; Foster and Wenseleers, 2006). Partner choice has been best demonstrated for legumes that form symbioses with nitrogen-fixing Rhizobia (Kiers et al., 2003; Simms et al., 2006; Sachs et al., 2010b) and squids that form symbioses with bioluminescent *V. fischeri* (Visick et al., 2000; Sachs et al., 2004). In both examples, hosts exhibit mechanisms to reward cooperative symbionts and punish cheaters. It can be difficult to experimentally distinguish partner-fidelity feedback from partner choice (Weyl et al., 2010). However, one approach is to assess if symbionts are spatially structured within the host. The degree to which hosts can spatially separate symbiont genotypes is a key prerequisite for partner choice mechanisms (Denison, 2000; West et al., 2002a,b; Sachs et al., 2004), but should have no bearing on partner fidelity feedback. Many hosts of horizontally transmitted bacteria have evolved specialized structures that can separate symbionts that vary in their fitness effects on the host and potentially aid in distinguishing beneficial strains from cheaters (Becking, 1970; Savage, 1977; Sprent et al., 1987; Douglas, 1989; Bright and Sorgo, 2003; Currie et al., 2006; Nussbaumer et al., 2006; Visick and Ruby, 2006; Goettler et al., 2007; Vaishnava et al., 2008; Pflugfelder et al., 2009; Ran et al., 2010) (Table 2.1 and Fig. 2.2). In most of these examples, there is no more than a correlation between symbiotic structure on hosts and the potential for partner choice. However, these data become powerful when coupled with phylogenetic and ecological information. Kikuchi and colleagues (2011) analyzed the presence and structure of midgut crypts among 124 species of stinkbugs that vary in diet as well as the presence of horizontally transmitted *Burkholderia* symbionts (Table 2.1 and Fig. 2.2). They found that (*i*) stinkbugs exhibit multiple *Burkholderia* genotype infections, a key prerequisite for partner choice;

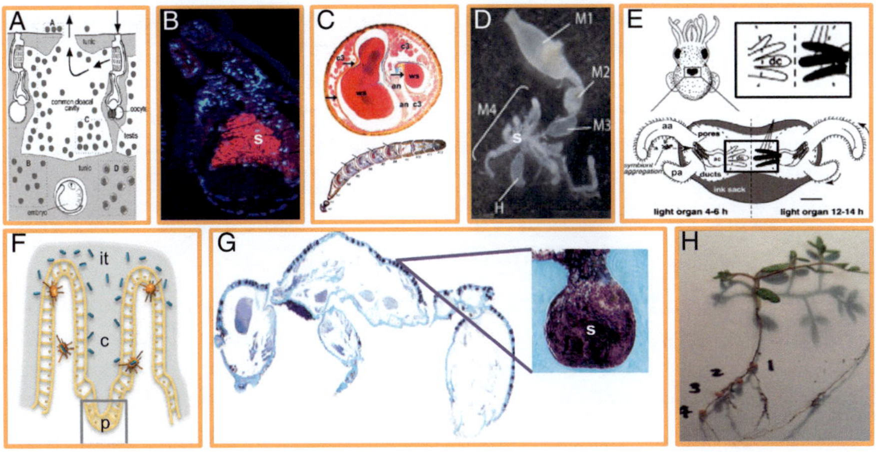

FIGURE 2.2 Symbiont housing structures and their potential to promote spatial structure. (*A*) Host Ascidian *Diplosoma* spp. and symbiont *Prochloron* spp. unstructured in host cloacal cavity [reprinted from Hirose et al. (2009)]. (*B*) Host hydrothermal tubeworm *Riftia pachyptila* with symbiont *Endoriftia persephone* (s) unstructured in host trophosome [reprinted from Nussbaumer et al. (2006)]. (*C*) Antenna of host beewolf *Philanthus triangulum* with symbiont *Streptomyces* (ws) housed in structured serial antennomere reservoirs (cross section above; longitudinal section below) [reprinted from Goettler et al. (2007)]. (*D*). Four-chambered midgut of host stinkbug *Dimorphopterus pallipes* with symbiont *Burkholderia* spp. (s) housed in structured crypts of fourth midgut section (m4) [reprinted from Kikuchi et al. (2011)]. (*E*) Juvenile squid host *Euprymna scolopes* during colonization by symbiont *Vibrio fischeri*, housed in structured deep crypts [dc; adapted from Visick and Ruby (2006)]. (*F*) Host mouse small intestine and symbiont *Bacteroides thetaiotaomicron* in structured crypts of Lieberkuhn (c) based with Paneth cells (p) [adapted from Vaishnava et al. (2008)]. (*G*) Dorsal cross section of host ant *Cyphomyrmex longiscapus* with *Actinomyces* symbionts (s) housed in structured crypts [reprinted from Currie et al. (2006)]. (*H*) Host legume *Lotus strigosus* with symbiont *Bradyrhizobium japonicum* structured in four numbered nodules (photo by J. L. Sachs).

(*ii*) the *Burkholderia* symbiosis has evolved in some, but not all, of the stinkbug species that exhibit midgut crypts; (*iii*) there is no evidence that the *Burkholderia* symbiosis has evolved in stinkbug species without such crypts; and (*iv*) crypts are not strictly correlated with different feeding habits of the bugs. These data suggest that crypts—which can potentially separate beneficial from harmful symbionts (Kikuchi et al., 2011)—are a key factor promoting stability in this bacterial mutualism. In summary, there is controversy over the relative importance of partner-fidelity feedback and partner choice as the key selective forces that maintain bacterial mutualisms (Bull and Rice, 1991; Simms and Taylor, 2002; West et al., 2002a,b; Weyl et al., 2010; Archetti et al., 2011). However, spatial separation among symbiont genotypes is a predicted indicator of partner choice (Denison, 2000; West et al., 2002a,b; Sachs et al., 2004), and such structure is common.

SYMBIONT CAPTURE

Symbiont capture occurs when bacteria that can replicate in the environment evolve to be strictly vertically transmitted within hosts and lose independent life stages. The most basal form of transmission is horizontal and likely occurs when bacteria are acquired from environmental pools (Huss-Danell and Frej, 1986; Nussbaumer et al., 2006; Mueller et al., 2008, 2010; Sachs et al., 2009; Barke et al., 2010). In other cases of horizontal transmission, the symbiont taxa can be found in the environment (Nishiguchi and Nair, 2003; Carson et al., 2005), but most transmission likely occurs among hosts (Savage, 1977; Wilkinson, 1999; Turnbaugh et al., 2009; Wollenberg and Ruby, 2009) with little contribution from environmental pools. Vertical transmission modes range from direct symbiont transfer within host germ lines to host behavioral mechanisms that supplement offspring with symbionts (Bright and Bulgheresi, 2010) (Table 2.1). Moreover, some bacteria cannot be easily categorized into horizontal or vertical transmission modes. For instance, some bacterial lineages are transmitted vertically, but in rare events, get horizontally transmitted to novel hosts, likely through vectors or predation (Russell et al., 2003; Dale and Moran, 2006). In most cases, captured lineages of bacteria are mutualists (our focus here), but obligate intracellular parasites such as *Wolbachia* and *Rickettsia* can also exhibit strict vertical transmission.

Symbionts with strict vertical transmission exhibit reduced effective population size and are subject to the accumulation of deleterious mutations and gene loss (Moran, 2003; Toh et al., 2006), transfer of DNA to host genomes (Martin and Herrmann, 1998), and obligate reliance on the host for basic nutrient synthesis (Shigenobu et al., 2000). Captured symbionts also experience reduced access to novel genetic material via horizontal gene transfer (Dale and Moran, 2006; Toft and Andersson, 2010), which limits the potential for novel functions to evolve and for recombination to restore function to degraded genomes. Such genome degradation tends to worsen over time (Moran et al., 2009) and ultimately cause loss of functions that are required for life outside of the host (Merhej et al., 2009). Hence, vertical transmission is often an irreversible evolutionary endpoint.

An unexplored question about symbiont capture is whether host, symbiont, or joint mechanisms are responsible for these evolutionary transitions. Although the evolution of vertical transmission can be costly to symbionts, hosts experience benefits including transmitting mutualists to offspring, minimizing symbiont diversity, and reducing mixing among symbiont genotypes, all of which promote symbiont cooperation (Frank, 1996a,b; Sachs et al., 2004). Thus, symbiont capture should be correlated with the evolution of host mechanisms to control transmission (Frank, 1996a). In some cases, hosts have specialized structures with no obvious function other than to transfer bacteria to offspring. Female stinkbugs bear organs on their ovipositors (Kikuchi et al., 2009) that transfer symbionts to their eggs. The ascidian *Diplosoma similis* (Hirose, 2000; Hirose et al., 2009; Kojima and Hirose, 2010) exhibits a specialized "plant rake," which it extends into its cloacal cavity during spawning and thus transfers bacterial symbionts to newly spawned larvae. In many cases, vertical transmission relies on specific host behaviors, such as when females smear symbionts onto eggs, egg cases, or cocoons of offspring (Douglas, 1989; Hirose, 2000; Kaltenpoth et al., 2005, 2010; Hirose et al., 2009; Kikuchi et al., 2009; Kojima and Hirose, 2010). However, bacterial mutualists can also promote their own vertical transmission. Among insect symbionts that inhabit mycetocyte structures within their hosts (Table 2.1), the bacteria sometimes migrate in the host from their mycetocyte structures to the host ovaries (Douglas, 1989). *Wolbachia* that infect *Drosophila* use the host microtubule cytoskeleton and transport system to maximize vertical transmission (Ferree et al., 2005). Moreover, the bacterial symbiont of the water fern *Azolla filiculoides* differentiates into a motile form and actively moves from adult plant leaves to infect the sporocarp of offspring plants (Ran et al., 2010). In all the examples in which the symbiont bears mechanisms to promote vertical transmission, there is no free-living existence and no potential for horizontal transfer (Table 2.1). Not surprisingly, when vertical transmission is the only mechanism to invade new hosts, symbiont traits are selected to enhance its efficiency. In summary, symbiont capture within host lineages involves a suite of deleterious effects that degrade symbiont genomes while providing benefits to hosts. As predicted by theory, the evolution of symbiont capture appears to be mostly driven by host mechanisms, but only a handful of bacterial–host interactions have been studied in detail (Bright and Bulgheresi, 2010).

BREAKDOWN OF SYMBIOSIS

There is debate about the evolutionary robustness of mutualisms, of which beneficial microbe–host interactions are a subset. Mutualist populations have been predicted to be prone to extinction (Vandermeer and Boucher, 1978), the spread of cheater mutants (Axelrod and Hamilton, 1981; Bull and Rice, 1991), and reversions to free-living existence (Vandermeer and Boucher, 1978; Keeler, 1985; Holland et al., 2004), but other research predicts that mutualisms are robust to these challenges (Doebeli and Knowlton, 1998; Ferrière et al., 2007; Douglas, 2008). Evolutionary transitions that result in the loss of mutualistic traits (Sachs and Simms, 2006) can be divided into transitions from mutualism to parasitism and transitions from mutualism to free-living status (i.e., abandonment of mutualism). Ancient bacterial mutualisms (Sagan, 1967; Moran, 2003; Keeling, 2010; Ran et al., 2010) serve as empirical examples of long-term robustness, but it is unknown whether such stability is common.

To what degree does mutualism breakdown occur in bacteria? We can investigate the evolutionary stability of bacterial mutualism by using the domainwide phylogeny (Wu et al., 2009; Toft and Andersson, 2010) (Fig. 2.1) and our focal symbioses (Table 2.1). The domainwide data can be considered only preliminary because of the paucity of dense taxon sampling. We could only infer two evolutionary transitions from mutualism to other lifestyles: one transition from mutualism to parasitism and one abandonment of mutualism. Nonetheless, this is a surprising paucity of transitions considering that we inferred 72 evolutionary transitions on the tree (Figs. 2.1 and 2.3).

Among the 14 focal systems, there is evidence of mutualism breakdown in four, all of which involve transitions from mutualism to free-living status in symbionts with extensive free-living stages (Table 2.1). Two particularly dynamic examples of mutualism breakdown have been uncovered in symbionts of ants (Mueller et al., 2010) and stinkbugs (Kikuchi et al., 2007, 2011). In the case of the ants, the symbionts are antibiotic-producing Actinobacteria that live in cuticular crypts supported by specialized exocrine glands (Currie et al., 2006). Lineages of these Actinobacteria have likely undergone multiple transitions between host-associated and environmental status based on the intermixing of symbiotic and environmental genotypes on a population-level phylogeny (Mueller et al., 2010). Similarly, a phylogeny of the *Burkholderia* bug symbionts encompasses many environmental isolates, consistent with multiple transitions from symbiotic to environmental status (Kikuchi et al., 2011). Evidence for abandonment of symbiosis has also been found among rhizobial lineages, some of which are related to plant and mammal parasites as well as environmental bacterial species (Sawada et al., 2003), suggesting the potential for multiple transitions among mutualism, parasitism, and environmental lifestyles (Sachs and Wilcox, 2006) likely driven by horizontal transfer events of symbiosis loci (Young and

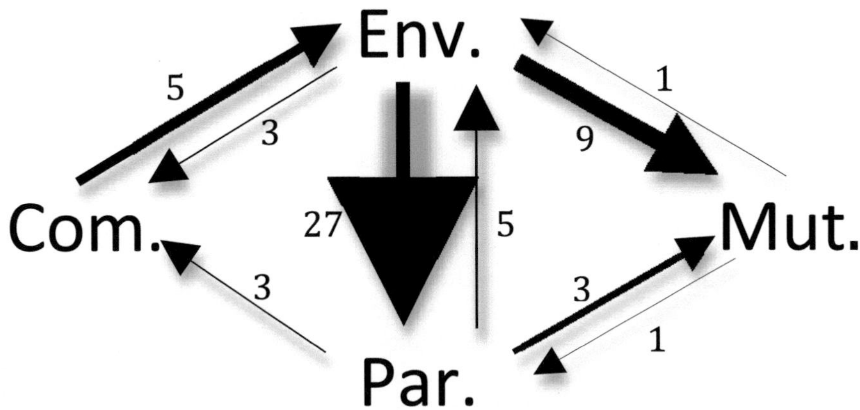

FIGURE 2.3 Path diagram of evolutionary transitions among bacterial host-association types. Transitions among four bacterial host-association types inferred in the tree by Wu and colleagues (2009) using lifestyle data from Toft and Andersson (2010). Thirteen transitions were undetermined on the tree as a result of ambiguity. There were zero transitions between mutualism and commensalism and zero transitions from commensalism to parasitism. Arrow sizes are scaled to the number of transitions between host-association types. *Note:* Com., commensal; Env., environmental; Mut., mutualist; Par., parasite.

Haukka, 1996). More focused analyses have inferred multiple events of evolutionary abandonment of mutualism within *Bradyrhizobium* populations (Sachs et al., 2009, 2010a), but found no evidence of transitions from mutualism to parasitism (Sachs et al., 2010a). In *Bradyrhizobium,* the abandonment of mutualism appears to be driven by degradation or wholesale loss of symbiosis loci encoded on a genome island (Sachs et al., 2010a). Finally, there is evidence of abandonment of mutualism within lineages of beneficial *V. fischeri,* with at least three evolutionary transitions from mutualism to environmental status (Nishiguchi and Nair, 2003) (Table 2.1). In summary, among different lifestyles that bacteria can exhibit, mutualism with hosts appears to be evolutionary stable with few transitions to other lifestyles. We found transitions from mutualism to free-living status, but virtually no evidence of transitions from mutualism to parasitism.

DISCUSSION

The evolutionary history of bacterial mutualism is rich and ancient. The origin of host association appears to be a readily surmountable step for bacteria. The commonness and near universality of this transition suggests that it is selectively advantageous and might be rarely affected by ecology. The evolution of bacterial mutualism is also common and phylogenetically diverse, and can occur via multiple routes. Bacterial mutualism most often appears to emerge from environmental ancestors. This can occur because the ancestral bacteria bear key traits (that can immediately benefit hosts) or by horizontal gene transfer of symbiosis loci (Sullivan et al., 1995; Sachs et al., 2010a), but neither mechanism is well understood. Bacterial mutualism can also arise from parasitic ancestors. It has been predicted that transitions from parasitism to mutualism are promoted by the evolution of vertical transmission (Ewald, 1987); however, more detailed work is needed to test this hypothesis. When bacterial mutualism has evolved, it can be stabilized via several selective mechanisms (Sachs et al., 2004). Partner choice, concomitant with the ability of hosts to spatially structure bacterial genotypes, is likely the dominant force maintaining bacterial mutualism.

Bacterial symbiosis first evolved with horizontal transmission, and several bacterial lineages have subsequently evolved strict vertical transmission. Some of the most ancient cases of bacterial mutualism exhibit vertical transmission, so this transition can promote the evolutionary stability of symbioses. We hypothesize that transitions from horizontal to obligate vertical transmission are host driven, as hosts (but not symbionts) most benefit from these transitions. Finally, evolutionary losses of bacterial mutualism are rare compared with other transitions in bacterial symbiosis. Evolutionary reversions from mutualism to environmental status occur in some bacterial lineages, potentially driven by the degradation or deletion of genes that encode symbiotic traits (Sachs et al., 2010a). In contrast, there is virtually no evidence in the phylogenetic record of transitions from mutualism to parasitism, thus refuting theory that predicts that mutualisms are vulnerable to fixation of cheater mutants (Axelrod and Hamilton, 1981; Bull and Rice, 1991; Sachs et al., 2004). The lack of transitions from mutualism to parasitism suggests that (*i*) bacterial mutualisms are evolutionarily robust or (*ii*) transitions from mutualism to parasitism are themselves unstable [and lead to extinctions or other stable states (Sachs and Simms, 2006)].

METHODS

We analyzed evolutionary transitions on a published 350-species bacterial phylogeny reconstructed by using a concatenated alignment of 31 proteins with maximum likelihood [PhyML (Guindon and Gascuel, 2003)] and an AMPHORA pipeline (Wu and Eisen, 2008; Wu et al., 2009) (Fig. 2.1, Table 2.1). Host-associated phenotypes were assigned based on a recent review (Toft and Andersson, 2010) that included host-association classifications of parasitic, mutualistic, commensal, or no interaction. We divided classifications into two characters: (*i*) association (host-associated or environmental) and (*ii*) type of host interaction (parasitic, mutualistic, commensal). Ancestral states were inferred by using parsimony [Mesquite 2.74 (Maddison and Maddison, 2010)]. When two equally parsimonious ancestral state reconstructions were found, we noted the ambiguity and listed a minimum estimate of transitions (Fig. 2.1).

To compare the relative frequencies of host-association origins among different bacterial lineages, we estimated the rate of origins over evolutionary time for each phylum and the complete tree. Rates were calculated by dividing the total number of origins of host association in a lineage by an adjusted sum of the taxon's branch length. The adjusted sum included only branches on which transitions from an environmental lifestyle to host association could occur (i.e., summed branch length of the taxon minus host-associated descendant branches of previously accounted origins and individual branches on which host association has been lost). The unit of branch length is the expected number of amino acid substitutions per site.

For focal symbiont taxa, we analyzed phylogenies containing the lineages of interest to assess whether host association originated from parasitic ancestors or free-living ancestors and to search for evidence of mutualism breakdown. Ancestral states for symbiotic lineage and evidence of mutualism breakdown were inferred by using parsimony on the available phylogenies (Normand et al., 1996; Turner et al., 1999; Nishiguchi and Nair, 2003; Sawada et al., 2003; Ruby et al., 2005; Svenning et al., 2005; Kaltenpoth et al., 2006, 2009; Kikuchi et al., 2007, 2011; Münchhoff et al., 2007; Williams et al., 2007, 2010; Sachs et al., 2009, 2010a; Mueller et al., 2010).

3

Kinship, Greenbeards, and Runaway Social Selection in the Evolution of Social Insect Cooperation

PETER NONACS

Social Hymenoptera have played a leading role in development and testing of kin selection theory. Inclusive fitness models, following from Hamilton's rule, successfully predict major life history characteristics, such as biased sex investment ratios and conflict over parentage of male offspring. However, kin selection models poorly predict patterns of caste-biasing nepotism and reproductive skew within groups unless kin recognition constraints or group-level selection is also invoked. These successes and failures mirror the underlying kin recognition mechanisms. With reliable environmental cues, such as the sex of offspring or the origin of male eggs, predictions are supported. When only genetic recognition cues are potentially available, predictions are not supported. Mathematical simulations demonstrate that these differing mechanisms for determining kinship produce very different patterns of behavior. Decisions based on environmental cues for relatedness result in a robust mixture of cooperation and noncooperation depending on whether or not Hamilton's rule is met. In contrast, cooperation evolves under a wider range of conditions and to higher frequencies with genetic kin recognition as shared greenbeard traits. This "excess of niceness" matches the existing patterns in caste bias and reproductive skew; individuals often help others at an apparent cost to their inclusive fitness. The results further imply a potential for greenbeard-type kin recognition to create arbitrary runaway social selection for shared genetic traits. Suggestive examples

Department of Ecology and Evolutionary Biology, University of California, Los Angeles, CA 90095. E-mail: pnonacs@biology.ucla.edu.

in social evolution may be alloparental care and unicoloniality in ants. Differences in kin recognition mechanisms also can have consequences for maintenance of advantageous genetic diversity within populations.

Seemingly overtly altruistic behavior, such as individuals accepting sterility, has puzzled evolutionary biologists since the time of Darwin. The first truly predictive framework for how reducing one's own reproduction could be adaptive came from the seminal work of W. D. Hamilton (1964a). His key insight was that fitness is "inclusive" of both an individual's direct reproduction and indirect gains arising through help provided to genetic relatives. Gains in indirect fitness, often labeled as "kin selection," recast altruism as an ultimately selfish act. Help relatives if the benefit provided (b), prorated by the genetic relatedness of the recipient (r), exceeds the cost to self (c). This is Hamilton's rule: Helping is adaptive if $br > c$.

More than any other taxonomic group, social Hymenoptera (ants, bees, and wasps) sit at an apparent peak of kin selection, with many species having morphologically sterile workers. Kin selection and applications of Hamilton's rule, however, extend far beyond the evolution of sterile castes to examine many aspects of cooperative (and noncooperative) behavior (Bourke and Franks, 1995). Thus, social insects have had a pivotal role in the development of kin selection theory and its elevation to being the dominant evolutionary paradigm for the study of cooperation and conflict. To date, there have been hundreds of tests of kin selection predictions in social insects (Abbot et al., 2011). However, despite this track record of remarkable utility, kin selection theory has recently become embroiled in controversy. The mathematics of inclusive fitness modeling have been directly challenged (Nowak et al., 2010). The evolution of cooperation is argued as better explained by group-level selection than by nepotism toward kin (Wilson and Wilson, 2007). Finally, the status of social insects as being a paramount example of kin selection has been questioned, with kin selection relegated to being a dissolutive force that primarily selects against cooperation and sociality (Wilson and Hölldobler, 2005). The response to these criticisms from defenders of inclusive fitness modeling and kin selection has been simultaneously vigorous and dismissive (Foster et al., 2006; Lehmann et al., 2007; West et al., 2007c, 2008; Abbot et al., 2011; Herre and Wcislo, 2011; Strassmann et al., 2011b).

The current conceptual maelstrom offers an opportunity for a critical appraisal of the effects of kin selection in the social Hymenoptera. Considering a model or hypothesis as either a failure or success is highly subjective. No single model can be expected to be 100% accurate for all taxonomic groups and in all situations. It is nevertheless fair to categorize a model as failing when a strong majority of studies reject its predictions. Conversely, models that are consistently supported by data are valued for accurately tracking evolutionary outcomes. By such standards, and contrary to its critics, kin selection theory has had major successes. However, contrary to its apologists, kin selection theory also has had some major failures. I will briefly review within the social Hymenoptera two successes, (*i*) sex investment ratios and (*ii*) conflict over reproduction by workers (i.e., worker policing), and two failures, (*i*) caste-biasing nepotism and (*ii*) reproductive skew theory. I will thereafter consider the mechanism for how kin are recognized as the driving force for the observed pattern of success and failure.

These four topics are chosen because I believe the underlying theory for the kin selective predictions is sound and that the possibility for kin nepotism to evolve is at least potentially present. This differs from two other cases, where kin selection predictions are suggested to have failed: the haplodiplody and monogamy hypotheses (Nowak et al., 2010). First, cooperative breeding has repeatedly evolved in the haplodiploid Hymenoptera. Haplodiplody creates a genetic asymmetry, such that a female is more related to her full sister ($r = 0.75$) than she is to her own offspring ($r = 0.5$). Therefore, if a singly mated mother produces a female-biased offspring sex ratio, it is genetically more advantageous for a daughter to help her mother raise more sisters. However, the balance of evidence from existing species where cooperative breeding is facultative finds that such species are not monogamous, do not predictably bias sex ratios toward females, or both (Bourke and Franks, 1995). Hence, the haplodiploidy hypothesis is not a robust test of kin selection because the required patterns of genetic relatedness likely did not exist in the putative ancestors of eusocial species (Nonacs, 2010). The second example is the "monogamy hypothesis," where cooperative breeding is predicted to be more likely to evolve in species where family groups are full siblings because of monogamy (Boomsma, 2009). However, a gene-based model for the evolution of cooperation found that helping actually often tended to spread more rapidly through populations with polygamy (Nonacs, 2011). This may be an instance where the underlying kin selection model actually produces erroneous predictions [as postulated by Nowak et al. (2010)].

SUCCESSES: BIASED SEX INVESTMENT RATIOS AND PARENTAGE OF MALES THROUGH WORKER POLICING

Another genetic consequence of the asymmetrical relatedness patterns attributable to haplodiploidy is that with one singly mated queen (i.e., monogynous with monandry), workers are more closely related to full sisters than to their brothers. This preference should drive female investment bias to the point where it is exactly offset by the relative mating advantage of the rare male sex (Trivers and Hare, 1976). Thus, if workers control sex investment ratios, a 3:1 female-to-male investment bias should result. Over 3 decades of evidence has solidly supported the core of the Trivers and Hare hypothesis (Nonacs, 1986; Bourke and Franks, 1995; Chapuis and Keller, 1999; Strassmann and Queller, 2007). Female-biased investment occurs frequently in species where the sister-brother relatedness asymmetry is present and far less often in species where it is absent. Within some species, there are both monogynous, monandrous colonies and others in which the relatedness difference between females and males is reduced or absent because of having multiple laying queens or one queen that uses sperm from multiple males. Extending Trivers' observation predicts that such populations should exhibit split sex ratios, with the monogynous, monandrous colonies favoring females and the others favoring males (Boomsma and Grafen, 1990). A recent review (Meunier et al., 2008) found that within-colony relatedness asymmetries do significantly affect bias in sex investment as predicted by kin selection.

Conflict over male parentage in Hymenoptera can also be present because workers retain the ability to produce haploid eggs in many species, and therefore can produce sons (Bourke and Franks, 1995). Although a worker's sons and nephews ($r = 0.375$) are more closely related than brothers ($r = 0.25$), males produced by half sisters are less related ($r = 0.125$). Thus, when more than 50% of the workers are half sisters, a nonlaying worker is more related to a queen-produced male than to the average worker-produced male. Maximization of inclusive fitness would therefore predict that workers replace (i.e., police) worker-produced eggs with queen-produced eggs (Ratnieks et al., 2006). Policing occurs in all examined species where workers are more related to the queen's male offspring than to the average worker-produced male (Wenseleers and Ratnieks, 2006a). Contrary to expectations generated from the relative genetic value of brothers vs. nephews, however, worker policing also occurs in species where workers are more related to worker-produced males than to queen-produced males (Hammond and Keller, 2004). Thus, worker policing may also often reflect group-level advantages, such as the replacement of less viable worker eggs with more viable queen-laid eggs (Nonacs, 2006a). Despite these notable exceptions, extensive phylogenetic analyses support broad predictions of kin selection theory (Ratnieks et al., 2006; Wenseleers and Ratnieks, 2006b): (*i*) Worker policing occurs more frequently in species where queen-produced males have higher mean relatedness to workers; (*ii*) frequencies of worker-produced males correlate to increasing relatedness between worker-produced males and workers; (*iii*) frequency of worker laying negatively correlates to effectiveness of policing; and (*iv*) worker policing is less prevalent after queen death in colonies with lower between-worker relatedness [i.e., a reversal of the pattern found in (*ii*) in the presence of viable queens].

FAILURES: PATRILINE NEPOTISM IN CASTE BIAS AND REPRODUCTIVE SKEW

In the eusocial Hymenoptera, sterile female workers primarily raise the offspring (Bourke and Franks, 1995). These offspring can have a wide range of relatedness to the tending females as a result of queens mating with multiple males or multiple fertile queens in colonies. The former creates patrilines of full sisters within the colony, and the latter can create worker cohorts ranging in relatedness from nieces (if queens are sisters) to totally unrelated. Therefore, any worker cohort that manages to have its full sisters preferentially raised as the future reproductives should significantly increase its fitness. Careful observation, however, has yet to find any significant conflict or favoritism over care giving under a wide range of scenarios. Honey bees (*Apis mellifera*), whose queens mate with many males, have been studied most extensively for evidence of queen-rearing nepotism, but none has been conclusively demonstrated (Tarpy et al., 2004; Ratnieks et al., 2006). Similar studies in other social Hymenoptera have also failed to find nepotism in colonies with multiply mated females or multiple queens (Keller, 1997; Gamboa, 2004).

If facultatively cooperative groups are more productive than the expected cumulative output of all their individual members, it is possible for reproduction to be shared so that all individuals have higher fitness through cooperation (Nonacs and Hager, 2011). A fitness-maximizing division of the reproduction (i.e., the reproductive skew within the group) can be predicted through an inclusive approach. Groups can be stable if all individuals accrue fitness that is equal to or greater than their expected fitness from reproducing on their own (Nonacs, 2006b). Although there are numerous variants of skew models, all share the general features that predicted skew within groups ought to be affected by the genetic relatedness of the group members, their relative ability to compete for reproductive shares, and the relative productivity of groups vs. solitary individuals (Nonacs and Hager, 2011). In an extensive review of experimental tests of skew theory, none of these predicted relationships were consistently found (Nonacs and Hager, 2011). In 21 of 27 studies, there was no significant correlation between genetic relatedness and skew across groups within populations. In 13 of 18 cases, correlates with competitive ability had no significant effect on skew, and in only 3 of 18 cases did skew significantly correlate to factors likely to cause ecological constraints. Finally, 3 of 4 studies estimating inclusive fitness of group members and solitary individuals found it unlikely that cooperation was favorable for all group members. Individuals of low relatedness would probably gain higher fitness by reproducing solitarily rather than being subordinate group members. Therefore, the overall evidence suggests that individuals rarely modulate reproductive shares in response to the intrinsic characteristics of other group members, such as relatedness (Nonacs, 2006b; Port and Kappeler, 2010; Nonacs and Hager, 2011).

The failure of reproductive skew models to predict the behavioral dynamics between group members within populations contrasts with more accurate predictions at the population or species level. In these comparisons, differences in mean within-group relatedness or differences in environmental constraints for being solitary do accurately predict which population or species should exhibit the greater reproductive skew (Reeve and Keller, 1995; Nonacs and Hager, 2011). For example, if two populations significantly differ in mean within-group relatedness, subordinate individuals in low-relatedness populations stand to gain more fitness by reducing reproductive skew. Thus, there would be greater selective pressure with overall low relatedness to share reproduction more equally, independent of whether individual-level relatedness can be recognized (Reeve and Keller, 1995).

MECHANISMS OF KIN SELECTION

"Failure" is a semantically loaded word and should not imply that there are no evolutionary explanations for the observed outcomes in caste rearing and reproductive skew. However, these explanations invoke elements that are added to kin selection theory and do not follow from it. For example, caste-rearing nepotism would be absent if workers are constrained by an inability to distinguish their own patrilines from others. Without some means of kin recognition, kin selection could not "fail" as a hypothesis because it simply would not be relevant to the phenomenon at hand. Alternately, caste-rearing nepotism could create such within-colony conflict that overall colony productivity or survival is seriously reduced. Hence, conflict would not be selectively advantageous. This alternative explanation, again, does not follow from kin selection theory. No matter the overall cost, successful nepotists would always have a selective advantage relative to nonnepotists within colonies. It is only selection on the across-group level that could favor this ergonomic efficiency explanation for the absence of nepotism. The same two possibilities, constraint or ergonomics, could also explain why reproductive skew models do not adequately predict patterns of cooperative breeding (Nonacs and Hager, 2011). Nevertheless, recent work casts doubt on the constraint hypothesis as a broadly viable explanation. For example, individual ants produce genetically heritable hydrocarbon profiles that could, theoretically, be used to identify kin (van Zweden et al., 2010). However, such markers are readily transferred to create effective nestmate recognition rather than within-nest nepotism. This leaves only the group selection hypothesis as a plausible, if unsatisfying, explanation. It is unsatisfying because it is untestable in many cases; that is, if species A never exhibits caste-rearing nepotism, how can it be shown that it is because such behavior reduces overall colony productivity? Therefore, instead of relegating the solution as only explainable by difficult-to-test group selection, it is useful to reexamine kin selection predictions relative to the mechanistic aspects of exactly how individuals recognize or define other group mates as genetic kin. In essence, kin selection theory may accurately predict the outcomes for caste rearing and reproductive skew, but the predictions themselves may differ from earlier expectations.

There are only two ways that kinship can be assigned, either through recognizing genetic similarity or by using environmental cues that accurately predict genetic similarity. The latter can lead to simple and relatively effective rules-of-thumb for cooperation. For example, in a population where most females mate with only one male, two female wasps maturing at the same time on a nest are likely to be full sisters and can behave according to predictions from Hamilton's rule. However, such rules-of-thumb are not absolutely reliable: The females could also be half sisters, cousins, or unrelated because of multiple mating, multiple foundresses, or nest usurpation. Errors in estimation of relatedness could be common (Nonacs, 2006b).

In contrast to using environmental cues, individuals could recognize kin through a shared phenotypic trait that uniquely represents the presence of identical alleles. To function as an effective kin selective mechanism, genetic recognition must have three elements: expressed phenotypic trait(s), an ability to recognize traits in other individuals, and the proclivity to direct aid toward those others if a suitable opportunity arises (Gardner and West, 2010). These three elements can either be genetically linked as a single "greenbeard" system or be a set of genetic markers monitored by a single perception locus that triggers cooperative behavior (Lacy and Sherman, 1983). In the second scenario, recognition and action need not be linked to any cue alleles. Although the basic premise of greenbeard kin selection is sound, functioning greenbeards have been thought likely to be rare for several reasons (Gardner and West, 2010). A greenbeard system must always co-occur within individuals. For example, if the willingness to aid becomes decoupled from the phenotypic trait, a "falsebeard" individual would result. Such individuals would receive benefits from other greenbeards but would never provide benefits (and thus never incur any costs for being helpful). Under a wide range of conditions, cheating falsebeards could be at a selective advantage over true greenbeards and prevent widescale cooperation (Gardner and West, 2010). A second problem is that successful greenbeard genetic systems would encounter Crozier's paradox (Crozier, 1986). As the greenbeard system increases in frequency, it becomes less useful for identifying kinship (Rousset and Roze, 2007). As it approaches fixation, it can no longer function for kin nepotism because almost all individuals in the population would be viewed as identically related. A final problem for a greenbeard system, but not for phenotype matching, is that individuals are either highly related at a locus (both have identical alleles) or totally unrelated. Thus, if the rest of the genome is Hamiltonian in the sense of estimating relatedness relative to environmental cues, this could lead to different estimates of r as well as intragenomic and interlocus conflict. Assuming that disproportionally more loci are Hamiltonian in driving behavior, this is predicted to lead to suppression of any greenbeard favoritism (Helanterä and Bargum, 2007); however, an alternative argument is supported by Gardner and West (2010).

Despite the theoretical objections, more examples of greenbeard behavior have been found recently in hydrozoans (Grosberg et al., 1985), slime molds (Queller et al., 2003), yeast (Smukalla et al., 2008), and ants (Keller, 2007). The most compelling example is in side-blotch lizards (*Uta stansburiana*), where blue-throated males preferentially establish territories next to each other (Sinervo et al., 2006). These aggregations form with respect to the blue-throat trait and are not predicted by whole-genome relatedness. Neighboring blue-throats are more successful at mate guarding against larger aggressive orange males. However, these benefits accrue only to blue males not next to an orange male; blue males adjacent to orange males suffer a cost. Possibly in reaction to a perception that such a greenbeard system could not be evolutionarily stable, Sinervo et al. (2006) insightfully comment, "A proximate explanation for kin altruism is not that kin share a fractional number of genes; rather, kin altruists share key genes for signal, self-recognition, and donation behavior" (p. 7376). In this context, it is seen that kin nepotism through genetic recognition differs from existing models of greenbeard nepotism not in process but simply in the number of genes involved. This view argues that kin selection can result from a collaboration of multiple greenbeards sharing a common interest.

SIMULATING DIFFERENT PATHWAYS FOR THE EVOLUTION OF COOPERATION

Although the evolutionary dynamics of single greenbeard systems have been extensively examined (Gardner and West, 2010), multiple greenbeards evolving synchronously have not received similar attention. I consider this latter situation by simulating the evolution of helping behavior under three scenarios:

(*i*) Help is allocated so as to maximize inclusive fitness as predicted by Hamilton's rule.

(*ii*) Help is allocated relative to recognized shared alleles, with the helper and helped having the same greenbeard allele(s). In this scenario, falsebeard mutants can also arise that induce cooperation from greenbeard alleles but never extend help.

(*iii*) A mixed system exists where some loci favor cooperation as predicted by Hamilton's rule and others behave as greenbeards or falsebeards. This tests the degree to which interlocus conflict can suppress or mask greenbeard effects.

Phenotype matching (Lacy and Sherman, 1983) is a hybrid version of greenbeard and Hamiltonian kin recognition, where a number of cues act as greenbeards but there is only one recognition locus. This locus collates matches from all the cues to generate an estimate of r that is used in Hamilton's rule to determine if cooperation occurs. I do not specifically evaluate phenotype matching, although scenario (*i*) can be viewed as a version of the model that is completely accurate for estimating relatedness. The model is written in TrueBasic.

All simulations assumed a diploid population and started with 500 mothers, each mating with two males. All mothers produced two offspring, a potential helper and helped offspring that could be full siblings or half siblings. All offspring genotypes were randomly determined with respect to father and which allele was contributed by each parent. Loci were not linked and segregated independently. Only offspring and no mothers, helpers, or helped survived to the next generation. For the first two scenarios, all offspring had 1–15 loci (depending on the simulation) at which helping behavior could independently evolve. All alleles at these loci were initially "null" and neutral with respect to whether help was offered. Helping alleles were randomly mutated into the population at the rate of 0.01 mutations per offspring. If no cooperation occurred, both offspring mated twice and produced two offspring of their own, which were added to the pool from which the next generation's mothers and fathers were chosen. Thus, the inclusive fitness of noncooperation (IF_{NC}) equaled $2(0.5) + 2r$, where r is the relatedness of the sibling's future offspring calculated at the level of the entire genome. If cooperation occurred, the helper did not reproduce and its sibling produced 2–12 offspring (b) depending on the conditions of the simulations (Table 3.1). This results in $IF_C = rb$. A Hamiltonian allele would vote "yes" if $IF_C > IF_{NC}$ and "no" otherwise. (Note that to be completely consonant with Hamilton's rule,

TABLE 3.1 Payoffs for Helping Behavior

Mean Benefit (b) Levels	Hamiltonian		Greenbeard			
	Half Sib	Full Sib	Helper/Helped			
	r = 0.125	r = 0.25	1/1	1/2	2/1	2/2
Good						
Low: b = 4	−0.75	−0.5	0	1	−1	0
Medium: b = 8	−0.25	**0.5**	**2**	**5**	**1**	**4**
High: b = 12	**0.25**	**0.5**	**4**	**9**	**3**	**8**
Moderate						
Low: b = 3	−0.875	−0.75	−0.5	0	−0.5	−1
Medium: b = 6	−0.5	0	1	**3**	0	**2**
High: b = 9	−0.125	**0.75**	**2.5**	**6**	**1.5**	**5**
Poor						
Low: b = 2	−1	−1	−1	−1	−2	−2
Medium: b = 4	−0.75	−0.5	0	**1**	−1	0
High: b = 6	−0.25	0	**1**	**3**	0	**2**

Notes: The mean number of offspring produced by helping (b) varies from good to poor, with a lower, medium, or higher payoff equally likely under each overall condition. For Hamiltonian loci, relatedness (r) to the offspring of potential helped individuals is calculated at the level of the entire genome. The inclusive fitness of noncooperation (IF_{NC}) always equals two own offspring plus two nieces or nephews, and the inclusive fitness of cooperation (IF_C) always equals rb. Hamiltonian loci vote to provide help (shown in bold) whenever $IF_C - IF_{NC} > 0$. For greenbeard loci, potential helpers can have either one or two greenbeard alleles (first number in pair). They may help only individuals that also have one or two identical greenbeards or the matching falsebeard alleles (second number). Greenbeard loci vote to help (in bold) only when the expected number of alleles transmitted by a helped individual, minus the sum of helper and helped reproducing noncooperatively, exceeds zero. Therefore, a "yes" or "no" vote can depend on the zygosity of both potential helper and helped individuals.

the number of offspring raised by the helped offspring without a helper ought to be subtracted from both IF values. For mathematical simplicity, I allow this value to cancel out during calculations.) I assume that Hamiltonian alleles are completely accurate in discriminating full sibs and half sibs. This could occur as a modified greenbeard system where a single recognition/action system simultaneously monitors multiple phenotypic or environmental cues (unspecified in this model) to estimate genetic relatedness reliably. Unlike a single greenbeard, the components would not have to be linked to each other. Similarly, I assume that although b varies stochastically, the level is always recognized. Thus, Hamiltonian alleles vote identically across all loci, giving either a 100% yes or no vote for helping among themselves. Therefore, depending on whether overall environments were good, moderate, or poor for producing benefits from cooperation, Hamiltonian alleles would favor helping in ~50%, 17%, or 0% of the possible cases (Table 3.1).

In contrast, at each greenbeard locus, there are four different possible greenbeard alleles (each producing a hypothetical unique phenotypic cue) that can mutate into the population. For each greenbeard allele, a matching falsebeard allele can also mutate into the population. A greenbeard allele in a helper could vote yes if the helped sib had identical greenbeard or matching falsebeard allele(s) at the same locus. It would vote yes, however, only if a helped individual would be expected to transmit more greenbeard alleles to the offspring generation than the combined reproduction of noncooperating helpers and helped (Table 3.1). It would vote no if noncooperation was expected to transmit more alleles. Thus, the vote of a greenbeard locus depended on b and the heterozygosity or homozygosity of the helper and helped with respect to greenbeard and matching falsebeard alleles. Different greenbeard alleles at the same locus do not help each other and would vote no. Falsebeard alleles in a helper always vote no. Thus, across greenbeard loci, there could be a mixture of yes and no votes and probability of helping occurs relative to the proportion of yes votes (e.g., 12 yes votes and 3 no votes would result in an 80% probability of helping).

In the third scenario, five loci evolved cooperation as Hamiltonian and one to five loci evolved as greenbeards, both as described above (resulting in a range of a 5:1 advantage for Hamiltonian loci to 5:5 equal weighting). Helping occurs with a probability that is the proportion of yes votes across all votes by Hamiltonian and greenbeard loci. All scenarios were simulated for 2,500 generations with 20 replicates. Random mutation was present for the first 2,000 generations to allow all possible alleles to enter the population. For the last 500 generations, greenbeard alleles were prevented from mutating into the population. Therefore, at the end of the simulations, the frequency of greenbeard alleles was more reflective of their relative selective advantage (i.e., they could be selectively eliminated from populations).

RESULTS

The invasion of helping is charted as the proportion of mothers (out of 500) that produced pairs of offspring that decided to cooperate. Hamiltonian alleles rapidly invade a noncooperative population when cooperation is at least sometimes advantageous (Fig. 3.1). The resulting level of sociality is commensurate with how often $IF_C > IF_{NC}$ (e.g., 50%, 17%, or 0%). The number of loci that could be involved has no effect on the rate of spread of helping behavior. Overall, the system consistently evolves rapidly to maximize inclusive fitness.

Sociality also often evolves with only greenbeard loci but differs from Hamiltonian patterns. Unlike with Hamiltonian alleles, a signifi-

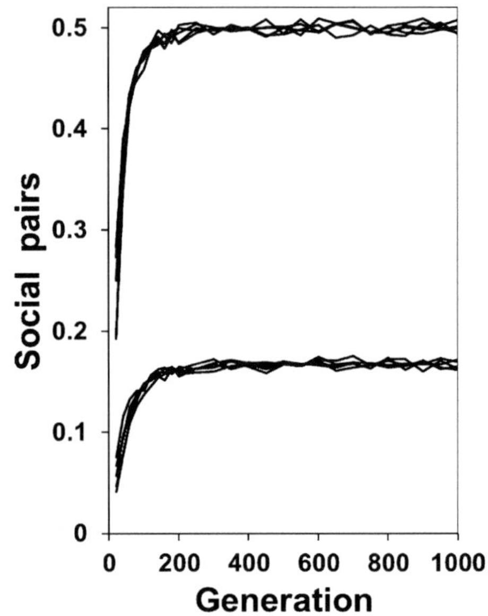

FIGURE 3.1 Increase in sociality with Hamiltonian alleles that help if $IF_C > IF_{NC}$ for full or half sibs, as measured by the proportion of offspring pairs that help each other. The upper set of lines is for simulations where helping often produces large benefits (*b* = 4–12), and the lower set is for simulations where helping has moderate benefits (*b* = 3–9). No sociality results under conditions where helping provides poor benefits (*b* = 2–6). Lines in each set represent 1, 3, 7, 11, or 15 loci and are shown for only the first 1,000 generations of the simulation.

cant fraction of the population can exhibit cooperation under conditions where helping never produced higher inclusive fitness at a genome level (Fig. 3.2*A*). Although this level of cooperation was often enhanced through continual input of greenbeard alleles by mutation, greenbeards continued to persist for many generations in the population when their entry by mutation was turned off (Fig. 3.2). Moreover, higher levels of cooperation evolved and greenbeard alleles persisted longer in the absence of mutation as the number of greenbeard loci increased across all levels of *b*.

At most loci, a single greenbeard allele was strongly numerically dominant in frequency no matter the level of *b* or the number of loci involved. Thus, Crozier's paradox (Crozier, 1986) often occurs at individual loci, where one greenbeard allele rises to high frequency at the expense of all other possible alleles. Although falsebeard alleles occasionally reached high frequency, they did not predominate at the majority of loci for any combination of *b* and loci number. This result is somewhat surprising because falsebeards contribute to producing nonhelping phenotypes by always voting against cooperation. Thus, conditions where cooperation was never beneficial at a whole-genome level, $IF_{NC} \geq IF_C$ (Table 3.1), might have been expected to tip the balance of selection toward a falsebeard allele, but this happened only at a minority of loci. As more loci are interacting or *b* becomes moderate or good, allele populations are almost uniformly dominated by a single greenbeard allele.

Across most individual simulation trials, there appeared to be little interaction between greenbeard and falsebeard alleles at a given locus. Occasionally, an increase in the frequency of a matching falsebeard could tip the balance of selection from one greenbeard allele to another and lead to a replacement as the most frequent allele in the population.

The high frequency of greenbeard alleles under many conditions results in more cooperative behavior between siblings than predicted by Hamilton's rule for any given *b* value. For example, even with a large

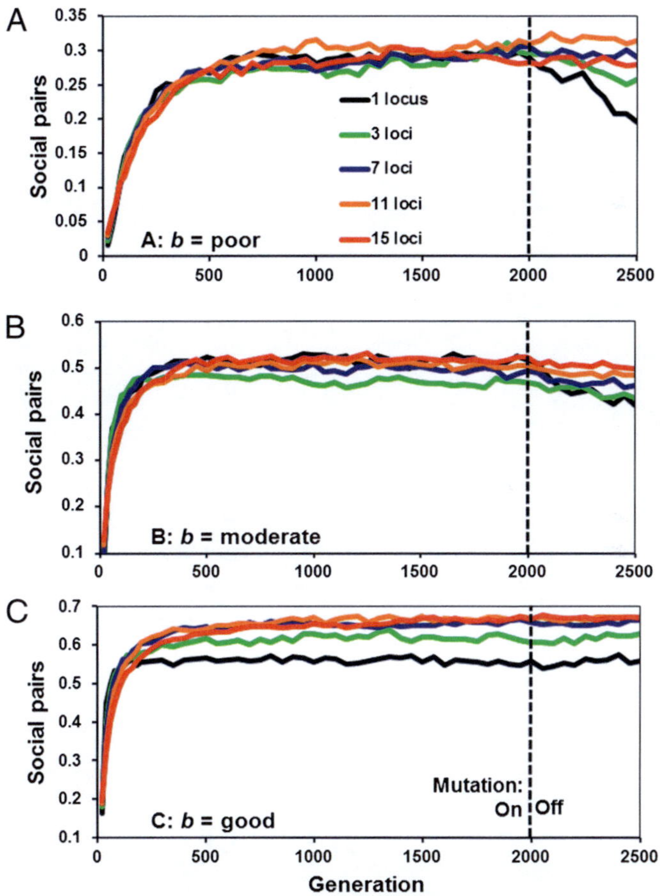

FIGURE 3.2 Proportion of social pairs with loci that favor helping if individuals share the same greenbeard alleles or matching falsebeards. Simulations are where helping provides poor benefits (*A*, *b* = 2–6), moderate benefits (*B*, *b* = 3–9), or large benefits (*C*, *b* = 4–12). For the last 500 generations, no mutations that produce greenbeard alleles were allowed.

mean payoff to cooperation, applying Hamilton's rule predicts that only ~50% of the helpers' decisions ought to be to cooperate (Fig. 3.1). Compared with this prediction, cooperation through greenbeards results in many "mistakes" (Fig. 3.3). Most of the errors are of individuals being helped when $IF_C < IF_{NC}$ predicts no helping.

The results also show that there can be interlocus conflict across Hamiltonian and greenbeard loci. These two regions of the simulated genomes can disagree as to whether cooperation is advantageous. The

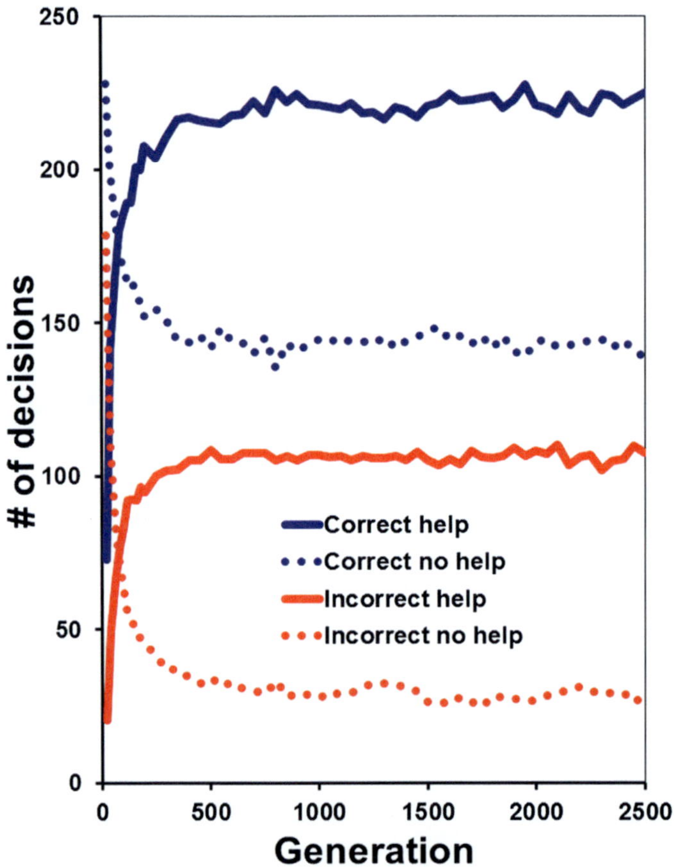

FIGURE 3.3 Outcome of decisions made by individuals with greenbeard loci. "Correct" and "incorrect" are determined for 500 pairings relative to whether $IF_C > IF_{NC}$ at the whole-genome level holds with half ($r = 0.125$) or full ($r = 0.25$) sibs. For the last 500 generations, no mutations that produce greenbeard alleles were allowed. Only the situation with seven greenbeard loci and a large level of benefit ($b = 4–12$) is graphed.

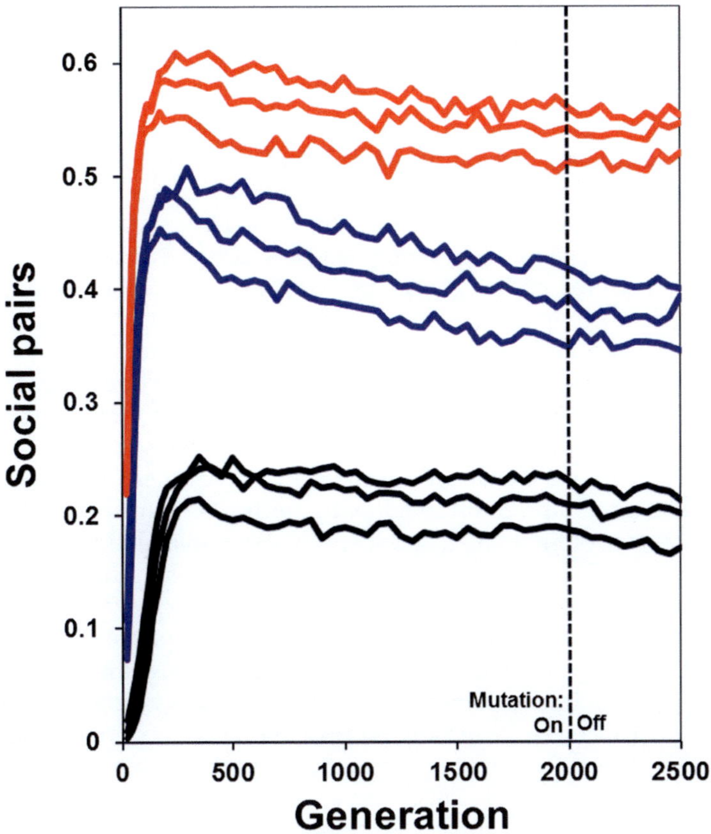

FIGURE 3.4 Proportion of social pairs when individuals have both Hamiltonian and greenbeard loci. The three sets of lines represent conditions with good (top series, b = 4–12), moderate (middle series, b = 3–9), or poor (bottom series, b = 2–6) benefits for helping. In all simulations, there are five Hamiltonian loci. From the lowest to highest line in each set, the numbers of greenbeard loci are one, three, or five. For the last 500 generations, no mutations that produce greenbeard alleles were allowed. In the absence of greenbeard loci, the three levels of benefit predict that a pure Hamiltonian population would have approximated 50%, 17%, and 0% social pairs.

resulting levels of sociality appear to be an approximate balance between the relative number of loci that select for Hamiltonian or greenbeard levels of cooperation (Fig. 3.4). Thus, cooperation sometimes occurs under conditions where combinations of sibling relatedness and benefit predict it should not; however, cooperation also sometimes does not happen in the presence of shared greenbeard loci across offspring.

DISCUSSION

Kin Recognition Mechanism and the Expression of Cooperation

In predicting patterns of life history traits, robust theory must combine with a realistic mechanism for action. The distribution of success and apparent failure in past kin selection predictions for social insects illustrates the importance of this connection. Thus, genetic mechanisms for identifying kin (e.g., a hypothetical series of greenbeard loci) are less likely to maximize an actor's inclusive fitness than environmental cues that provide only a probability level of sharing genotypes. That kin selection theory is accurate only with probabilistic environmental cues helps to explain patterns of behavior within social insects. For example, the sex of a given offspring may be hidden for some time through development, but it must eventually become recognizable (Nonacs and Carlin, 1990). This gives workers a clear cue for biasing investment toward females and away from males. Similarly, having multiple matrilines or patrilines within a single colony may give out an unmistakable diversity signal (van Zweden et al., 2010). Therefore, kin selection models for optimal sex ratio investment have usable proximate cues and function well in predicting ultimate allocation patterns at both the colony and population levels.

Conflict over and suppression of worker laying similarly have an available proximate mechanism. Although both workers and queens can produce male eggs, an increasing body of evidence shows that queen eggs are recognizably different from worker eggs in morphology and chemical signatures (Ratnieks et al., 2006; van Zweden et al., 2009; Meunier et al., 2010). As a result of nutritional differences, queens may always be able to imbue their eggs with specific signals that cannot be faked by workers. Thus, workers can discriminate between eggs and maximize their inclusive fitness through manipulating which individuals will produce the males.

In comparison, there is almost no supportive evidence for within-colony nepotism in the production of new queens (Keller, 1997; Gamboa, 2004; Tarpy et al., 2004; Ratnieks et al., 2006). This is despite a potentially huge boost in inclusive fitness for eusocial Hymenoptera (e.g., the replacement of an unrelated female, $r = 0$, or half sister, $r = 0.25$, with a full sister, $r = 0.75$). From a mechanistic perspective, however, the only cues that workers could use to behave nepotistically would be through recognizing one or more shared alleles. Similarly, reproductive skew models require that group members estimate kinship in apportioning reproductive shares. Although unstated in the models, this assumes a type of greenbeard kin recognition and may be why they fail to predict skew across groups within populations. There is no obvious nongenetic cue that would, for instance, be able to differentiate emerging adults on the same nest as full sisters rather than cousins. However, evolved patterns of reproductive sharing at population or species levels ought to reflect the expected mean levels of relatedness or ecological constraints, regardless of any ability for individual-level recognition of genetic relatedness. Hence, skew models are relatively more successful at predicting population-level differences in reproductive sharing (Nonacs and Hager, 2011).

An interesting result from the simulation models is that greenbeard nepotism results in seemingly overly cooperative populations (Fig. 3.3), which is exactly how caste nepotism and reproductive skew appear to fail. In both cases, individuals are too nice. In caste determination, matrilines do not discriminate across females in how they are treated. Close kin, distant kin, and nonkin are equally helped. Reproductive skew on wasp nests tends to be very high, indicating that subordinates willingly cooperate to their apparent fitness detriment (Nonacs et al., 2006). Most paradoxically, fertile unrelated wasps join (and are allowed to join) groups with no differential treatment (Queller et al., 2000; Leadbeater et al., 2010). Relatedness has no predictive role in establishing aggression patterns or dominance hierarchies on nests (Nonacs et al., 2006; Leadbeater et al., 2010).

Runaway Social Selection Through Greenbeard Nepotism

The significance of greenbeard kin nepotism in evolutionary biology is controversial (Gardner and West, 2010; Leigh, 2010). However, in the models presented here, two of the three major objections to greenbeards often fail to prevent the evolution of helping. A falsebeard cheating genotype that accepts cooperation but does not reciprocate rarely destabilizes a greenbeard kin recognition. The inability of a falsebeard to predominate may result from several simultaneous processes. First, positive kin assortment into sibling pairs often imposes a cost onto cheating. Although there is the probability of drawing benefits from unrelated greenbeards, there can be the relatively greater chance of not helping siblings that share identical falsebeard alleles when such help would increase allele frequencies. A second process occurs as a greenbeard allele reaches higher frequencies. At this point, more of the helper/helped pairings involve homozygous individuals, which increases the net gain for helping (Table 3.1). This may generate positive feedback in more helping and continued increases in allele frequency. Finally, when greenbeard alleles predominate at multiple loci, they will increase commonality of interest. This could result in greenbeard alleles forming a "voting block" as regards cooperation and depress the selective advantage of a falsebeard at any individual locus. Altogether, the results are strongly suggestive that models of greenbeard recognition systems based on single-locus dynamics (Gardner and West, 2010) may predict very different outcomes from multilocus situations. These ramifications need to be explored more extensively in future work. Also, the models here assume idealized versions of Hamiltonian and greenbeard alleles in that all alleles "know" the payoff for helping, know the cost for abandoning direct reproduction, and properly identify shared alleles or correctly measure *r*. The effects of mistakes in these estimates could be valuable to consider (Nonacs, 2006b).

The second objection is that greenbeards will sometimes function to the detriment of the much larger nongreenbeard genome, and therefore are expected to be suppressed (Helanterä and Bargum, 2007). However, suppression is only partial in this model. Shared greenbeard alleles increase in populations and raise levels of cooperation proportional to their abundance relative to Hamiltonian loci (Fig. 3.4). This supports the view that greenbeard alleles are not intragenomic "outlaws" but are under similar selection pressures as the rest of the genome (Gardner and West, 2010).

An effect of greenbeard nepotism is that it quickly increases the frequency of arbitrary phenotypic traits in social settings in what is very much a "runaway" process (i.e., in the absence of social interactions, the traits convey no benefit to their bearer). This outcome is relevant to West-Eberhard's (1983) proposal that parents might allocate resources biased toward offspring with particular traits that have no effect other than being "attractive." If such attractive traits are greenbeards, their bearers would benefit from the genetic correlation between trait and preference as in runaway sexual selection. The process could be further enhanced by the feedback between the commonness of an allele and increased payoffs for cooperation. Such runaway social selection is proposed for bright coloration and plumage in young birds, which appears to function in attracting parental care (Lyon et al., 1994; Ligon and Hill, 2010) but, interestingly, not in identifying kin (Shizuka and Lyon, 2010). Moreover, rapidly reaching Crozier's paradox (i.e., fixation at the loci involved) may leave no trace in present-day behavior of past genetically based favoritism across offspring. This suggests that some morphological and behavioral traits in social species without clear adaptive value may have evolved through greenbeard nepotism.

In this suite of traits arising from greenbeard nepotism could be the initial evolution of cooperative breeding itself. Specifically, greenbeard traits can produce low levels of social behavior under conditions where Hamilton's rule would predict no cooperation (Fig. 3.2). If, for example, offspring dispersal is limited, positive kin assortment would result among neighboring individuals. This would create preconditions where greenbeard alleles producing alloparental behavior could be selectively favored. If alloparental care is as simple as feeding a nearby hungry mouth (Jamieson, 1989), a greenbeard trait for feeding a "mouth like mine" could evolve even if it did not appear to increase kin selective inclusive fitness (Emlen et al., 1991; Komdeur, 1996). Once social groups are established, more elaborated forms of cooperation could evolve with enhanced benefits provided to helped individuals that are potentially in line with Hamilton's rule.

Another trait that could result through runaway social selection is the phenomenon within ants of unicoloniality. Unicoloniality occurs when adjacent nests show atypically low or no aggression toward each other and is commonly associated with invasive ant species, where supercolonies can arise that extend over thousands of kilometers (Helanterä et al., 2009). Unicolonial associations are genetically homogeneous, with the majority of ants having low relatedness to each other. The reduction in intraspecific aggression appears to be attributable to a loss of genetic diversity at recognition loci (Suarez et al., 2008). Although unicolonial behavior may be enhanced by genetic bottlenecks, it is also present in situ in native habitats and apparently can evolve within large populations (Pedersen et al., 2006; Wang et al., 2010). In other words, unicoloniality can be selectively favored and is not simply always a byproduct of genetic bottlenecks from introductions. Interestingly, unicoloniality could be categorized as workers showing excessive cooperation by accepting and raising unrelated individuals, exactly as predicted by the model of greenbeard cooperation. The suggestion would be that if greenbeard alleles arise in a species at their recognition loci, it would be possible for such alleles to sweep through a population, carrying unicoloniality in their wake. This would lead to high genetic similarity at recognition loci in populations where within- and across-nest relatedness is almost identical [which is indeed observed in unicolonial ants (Brandt et al., 2009)]. Finally, it is a suggestive coincidence that one of the known greenbeard allele systems (Gp-9 in fire ants) appears to have arisen in an exotic species as that species was simultaneously evolving a unicolonial population social structure (Keller, 2007).

Genetic Diversity, Relatedness, and Social Heterosis

Controversies about kin selection often concern the best methods to model the effects of genetic relatedness (Wilson and Hölldobler, 2005; Foster et al., 2006; Lehmann et al., 2007; West et al., 2007c; Wilson and Wilson, 2007; Nowak et al., 2010). This tends to obscure the more salient point of how important the level of relatedness is for favoring cooperation (Nonacs, 2011). The dynamics of social groups can be dominated by either their genetic relatedness or their genetic diversity, because both bring evolutionarily selective advantages. High relatedness means that group benefits will tend to be exclusively shared by alleles identical by descent. It becomes relatively less important which individuals are the reproducers and which individuals are the helpers (Nonacs and Hager, 2011). In contrast, higher genetic diversity can create social heterosis, or the ability to exploit a wider range of resources more effectively, and thus increase total group benefits (Nonacs and Kapheim, 2007). Although more restrictive on the possible skew between reproducers and helpers, within-group outcomes become less important relative to across-group competition. The evolutionarily inescapable point is, however, that nepotism, as predicated on Hamilton's rule, must come at the expense of genetic diversity. Similarly, selection for genetic diversity may require behavioral biases against closer genetic relatives. Relatedness and diversity have an unavoidable tradeoff between them.

The model presented here gives no advantage to genetic diversity; therefore, cooperation evolves to maximize population-level genetic similarity. Alleles go to high frequency or fixation acting either as greenbeards or to maximize genome-level inclusive fitness. Nevertheless, one might expect that the diversity/ relatedness tradeoff could be different for greenbeard or Hamiltonian alleles. Any natural system where the benefits of cooperation are primarily directed to close relatives would strongly select against genetic diversity. In contrast, cooperation could evolve with considerably less of a tradeoff with a greenbeard kin recognition mechanism. The benefits of cooperation would not necessarily only flow to the closest relatives. As argued above in the case of ant unicoloniality, selection for nepotism based on greenbeard similarity could potentially affect only a limited part of the genome. Social heterosis could simultaneously select for genetic diversity at the remainder of the genome, with the result being a patchwork genome of regions of low and high genetic diversity (Nonacs and Kapheim, 2007). It is difficult to imagine how such opposing selective processes could simultaneously operate when one or more traits are being selected relative to their probabilities of being shared. Kin nepotism following Hamilton's rule will always work to reduce genetic diversity. It is the dynamic evolutionary consequences of selection for kin vs. selection for genetic diversity that should draw the future attention of both theoreticians and experimentalists.

ACKNOWLEDGMENTS

I thank J. Field, D. C. Queller, T. Wang, and an anonymous reviewer for helpful comments on the manuscript.

4

Spatiotemporal Environmental Variation, Risk Aversion, and the Evolution of Cooperative Breeding as a Bet-Hedging Strategy

DUSTIN R. RUBENSTEIN

In cooperatively breeding systems in which some individuals delay reproduction to help raise others' offspring, environmental variation in space and time influences individual reproductive strategies as well as interspecific patterns of sociality. Although most environmental explanations for cooperative breeding emphasize the mean fitness gains of living socially, the fittest individuals are not always those that produce on average the highest number of offspring. At times, variance in fecundity can influence fitness as much as mean fecundity, particularly in small populations like those of cooperative breeders. Cooperative breeding behavior could therefore be a risk-averse strategy to maximize fitness by reducing environmentally induced fecundity variance. Such a within-generation bet-hedging hypothesis for social evolution predicts that (*i*) variance in reproductive success should be related to environmental variation, (*ii*) variance in reproductive success should be related to the potential for cooperation in a group, and (*iii*) the potential for cooperation should be related to environmental variation. Using data from a 10-year study of cooperatively breeding superb starlings (*Lamprotornis superbus*) living in a temporally and spatially variable savanna ecosystem, I found that variance in reproductive success declined with increasing environmental quality (temporal variation), increasing territory quality (spatial variation), and increasing group size (potential for cooperation), which is itself

Department of Ecology, Evolution and Environmental Biology, Columbia University, New York, NY 10027. E-mail: dr2497@columbia.edu.

related to environmental variation. To understand the adaptive value of cooperative breeding behavior in variable environments, researchers must consider both mean and environmentally induced variance in fecundity. Determining how spatiotemporal environmental variation drives risk-averse strategies may provide insights into the evolution of complex social behavior.

Kin selection, or reproductive strategies that favor an organisms' relatives, is often invoked to explain the evolution of cooperation and the formation of complex animal societies (Hamilton, 1964a; West-Eberhard, 1975). In cooperatively breeding systems in which some individuals delay independent breeding to help raise the offspring of others, the inclusive fitness benefits of helping genetic relatives may outweigh the potential costs of trying to breed independently (Brown, 1987). Recent theoretical (Boomsma, 2007, 2009) and comparative work in both invertebrates (Hughes et al., 2008) and vertebrates (Cornwallis et al., 2010) suggests that high relatedness among group members may be critical to the evolution of complex animal societies. However, despite renewed interest in determining how genetic relatedness among group members can influence social interactions and the evolution of family groups (Boomsma, 2007, 2009; Hughes et al., 2008; Hatchwell, 2009; Nam et al., 2010; Sharp et al., 2011), relatedness alone cannot explain why some individuals in a group breed whereas others do not, or why some species breed cooperatively whereas other closely related ones do not. In other words, relatedness may set the stage for cooperation in animal societies, but it is not sufficient to explain many individual differences in reproductive strategies or interspecific patterns of social diversity (Rubenstein and Lovette, 2007; Jetz and Rubenstein, 2011).

Environmental factors are known to influence complex vertebrate social behavior (Alexander, 1974; Jarman, 1974), as well as explain many of the individual differences in reproductive strategies (Emlen, 1982a; Komdeur, 1992; Covas et al., 2004; Rubenstein, 2007a) and interspecific patterns of sociality (Rubenstein and Lovette, 2007; Jetz and Rubenstein, 2011). The role of environmental factors in shaping animal societies is central to the ecological constraints hypothesis (Emlen, 1982a), which argues that when barriers to dispersal are high, offspring will be selected to delay dispersal and remain at home as part of a group because the probability of reproducing successfully outside the group is low. The ecological constraints hypothesis (Emlen, 1982a) and its other derivations (Koenig and Pitelka, 1981; Koenig et al., 1992) predict the environmental conditions under which delayed dispersal is likely to occur (Hatchwell and Komdeur, 2000). These conditions include a shortage of vacant breeding territories (i.e., habitat saturation), the costs of dispersal, difficulties in finding a mate, and a low chance of successful reproduction once a territory is established [reviewed in Hatchwell and Komdeur (2000)]. Although each of these conditions represents an external constraint or cost associated with dispersal and independent breeding, such conditions represent only one side of the cost–benefit equation of social living. An alternative theory, the benefits of philopatry hypothesis, instead argues that delayed dispersal is the result of intrinsic benefits gained by remaining on the natal territory (Stacey and Ligon, 1987, 1991). These benefits include enhanced survival, indirect fitness gains from helping relatives, opportunities for obtaining a nearby breeding vacancy in the future, and the chance to inherit the natal breeding territory itself [reviewed in Cockburn (1998)]. Despite much initial debate over the relative importance of these two hypotheses in shaping cooperative groups (Emlen, 1994), it is now widely accepted that they are more similar than they are dissimilar, because they place different emphasis on the costs of dispersing vs. the benefits of not dispersing as a result of environmental constraints (Emlen, 1994, 1997a; Hatchwell, 2009).

Nearly all of the early environmental hypotheses for cooperative breeding behavior have focused primarily on the costs and benefits associated with breeding on territories of varying quality, or the fitness consequences of living in a spatially heterogeneous landscape where suitable territories are limiting (Emlen, 1982a; Stacey and Ligon, 1991; Koenig et al., 1992). However, spatial constraints on dispersal (i.e., habitat heterogeneity) are not the only form of environmental variation that can influence cooperative breeding behavior. Environmental variation in time can also influence social behavior, including dispersal decisions and the adoption of different breeding roles. Although the ecological constraints hypothesis is generally used to emphasize the role of habitat heterogeneity in influencing dispersal decisions, its original description also recognized the importance of environmental unpredictability in driving cooperative breeding (Emlen, 1982a), although this idea went largely untested for decades [but see Curry (1989) and Curry and Grant (1990)]. Recent work in cooperatively breeding birds living in unpredictable environments suggests that erratic and variable climatic patterns can also influence social complexity. Temporal environmental variation resulting from unpredictable patterns of rainfall (i.e., climatic uncertainty) has been shown to influence not only individual behavioral decisions and reproductive roles (MacColl and Hatchwell, 2002; Canario et al., 2004; Rubenstein, 2007a; Covas et al., 2008) but also interspecific patterns of sociality on continental and global scales (Rubenstein and Lovette, 2007; Jetz and Rubenstein, 2011). Together, spatial and temporal patterns of environmental variation explain many of the individual-level cooperatively breeding behaviors, as well as the broadscale interspecific patterns of social diversity.

Although the ecological constraints and benefits of philopatry hypotheses are said to be qualitatively similar (Emlen, 1994), one key difference seems to have been largely overlooked. Ecological constraint hypotheses primarily focus on mean reproductive success (Koenig and Pitelka, 1981; Emlen, 1982b; Stacey and Ligon, 1987; Koenig et al., 1992), emphasizing the average number of young produced when breeding alone or as part of a group. In contrast, the benefits of philopatry hypothesis were formulated on the idea of variance in reproductive success and emphasized variation in young produced on occupied territories through time in cooperative and noncooperative species (Stacey and Ligon, 1991). This key difference in fitness measures (i.e., mean vs. variance in fecundity) has important consequences for understanding how natural selection acts to promote cooperative behavior. Gillespie (1974, 1975, 1977) demonstrated that the fittest individuals are not always those that produce on average the highest number of offspring. Instead, he showed that in small populations, variance in fecundity can determine fitness as much as mean fecundity because the intensity of selection on reducing fecundity variance is inversely proportional to population size (Gillespie, 1974). Integrating these bet-hedging ideas into an inclusive fitness game theoretic framework, Lehmann and Balloux (2007) showed that helping behavior is selected for when fecundity variance is high. Thus, the simultaneous examination of mean offspring production and variance in offspring production [i.e., considering helping behavior and cooperative breeding as a bet-hedging strategy (Cockburn and Russell, 2011)] may shed light on the evolution of cooperative breeding behavior, particularly as it relates to spatiotemporal environmental variation.

To understand the role of fitness optimization in the evolution of cooperatively breeding behavioral phenotypes (i.e., breeding roles) in variable environments, we must consider the concept of bet-hedging, or risk aversion. Population geneticists have long understood that fluctuating selection resulting from environmental variability can favor the evolution of risk-averse strategies (Gillespie, 1974, 1975, 1977; Frank and Slatkin, 1990). Bet-hedging itself can be traced back more than 250 years to Bernoulli (1954; Stearns, 2000). In an evolutionary sense, bet-hedging strategies generally spread risk over multiple generations (i.e., years) by reducing variance in offspring production, which ultimately leads to an increase in the geometric mean lifetime reproductive success, but often a reduction in the arithmetic mean (Philippi and Seger, 1989). Although much rarer than these among-generation bet-hedging strategies, risk aversion can also operate within generations (Hopper et al., 2003). Within-generation bet-hedging spreads risk within a single generation and involves variability in the selection pressures to which a phenotype is exposed (Hopper et al., 2003). Importantly, within-generation bet-hedging encompasses any behavioral strategy that avoids having no or few offspring in any given generation, rather than maximizing the expected number of offspring (Sarhan and Kokko, 2007). Examples of within-generation bet-hedging are rarer than those of among-generation bet-hedging because within-generation bet-hedging only evolves under a much narrower set of demographic conditions (Hopper et al., 2003). Specifically, within-generation bet-hedging is only likely to evolve in small populations because the intensity of selection on reducing fecundity variance is inversely proportional to population size (Gillespie, 1974). For cooperatively breeding species in which populations are subdivided into kin-based social groups that are connected via dispersal, the conditions for within-generation bet-hedging to evolve are likely to exist (Lehmann and Balloux, 2007; Shpak, 2005). Such within-generation bet-hedging strategies could apply in any cooperatively breeding species in which, in addition to opportunities for helping, subordinates have options for direct reproduction, either by dispersing to breed independently outside of the group, becoming a breeder in the natal group, or gaining reproduction through extrapair paternity. Within-generation bet-hedging strategies to avoid having no or few offspring in any given generation may be most evident in cases of redirected helping in species like the long-tailed tit (*Aegithalos caudatus*), in which temporally variable ecological constraints drive individuals to switch from independent breeding to helping others (often relatives) later in the breeding season (MacColl and Hatchwell, 2002; Hatchwell and Sharp, 2006). Thus, when individuals have simultaneous opportunity to accrue fitness directly and indirectly within a given breeding season, environmentally induced selection to reduce fecundity variance can operate on risk-averse breeding strategies.

Although the conditions under which variance in fecundity can influence kin structure and cooperative breeding behavior have been modeled (Lehmann and Balloux, 2007), to my knowledge they have not been studied empirically. Here, I will study the effects of spatial and temporal environmental variation on fecundity variance in an avian cooperative breeder. Specifically, I will examine how mean and variance in group reproductive success change with increasing potential for cooperation, and how the potential for cooperation relates to environmental variation. I will test the hypothesis that cooperative breeding behavior is a risk-averse strategy to maximize fitness by reducing environmentally induced variance in fecundity. Such a within-generation bet-hedging hypothesis for social evolution predicts that (*i*) variance in reproductive success should be related to environmental variation in space and/

or time, (*ii*) variance in reproductive success should be related to the number of helpers in a group, and (*iii*) the numbers of helpers in the group should be related to environmental variation. Having helpers at the nest has been proposed to reduce the risk of complete clutch failure within a breeding season, either by preventing nestling starvation or depredation during harsh conditions, and/or by allowing for more clutches to be laid during benign conditions, resulting in an extended breeding season (Rubenstein and Lovette, 2007). Moreover, parental care that positively influences offspring survival may be favored in variable environments (Bonsall and Klug, 2011), and delayed reproduction strategies in general may be favored in unpredictable environments (Koons et al., 2008). Therefore, cooperative breeding itself may be a risk-averse strategy to maximize fitness by reducing variance in the number of offspring produced in a social group. Using data from a 10-year study of cooperatively breeding superb starlings (*Lamprotornis superbus*) living in a temporally and spatially variable savanna ecosystem (Rubenstein, 2009), I will explore how environmental variability in space (habitat heterogeneity) and time (climatic uncertainty) can directly influence fecundity variance. Moreover, I will examine whether fecundity variance relates to the potential for cooperation (number of helpers in the group), which may itself be related to environmental variability. Thus, this study will examine how spatiotemporal environmental variation influences fitness in cooperatively breeding birds living in unpredictable and heterogeneous environments, thereby providing insights into risk-averse social behavior and the evolution of complex animal societies.

RESULTS

To integrate spatial and temporal environmental variation into a framework for understanding the evolution of complex animal societies, I examined the environmental correlates of reproductive success in the cooperatively breeding superb starling using data from a 10-year field study representing 20 breeding seasons; birds typically breed twice a year during both the long and short rains. Superb starlings are endemic to the savanna of East Africa, which like most semiarid ecosystems is a spatially and temporally variable environment (Rubenstein, 2009). Superb starlings are obligate plural cooperative breeders, meaning that all groups have helpers and multiple breeding pairs that nest separately. They live in spatially subdivided populations (hereafter social groups) with high kin structure (Rubenstein, 2007c), thereby meeting the primary criteria to empirically examine within-generation bet-hedging strategies (Lehmann and Balloux, 2007). Because superb starlings live in such complex social groups with multiple breeding pairs, I quantified mean and variance in reproductive success at the level of the group as the proportion of eggs laid that fledged young in each nest for each group in each breeding season. This assumes that selection can operate on both the direct and indirect components of fitness, as has been shown theoretically in these types of kin-structured groups (Shpak, 2005; Lehmann and Balloux, 2007).

Rainfall in this region of Kenya is extremely variable from month to month and year to year. From 1998 through 2010, rainfall generally peaked three times per year; there were large peaks during both the short (November) and long (April–May) rainy seasons, as well as a small peak during July and August (Fig. 4.1*A*). The mean ± SD in annual rainfall was 529 ± 138 mm, which is characteristic of semiarid ecosystems worldwide (Austin et al., 2004). Annual rainfall ranged from 280 mm in 2000 to 822 mm in 2010. There was a negative relationship between mean monthly rainfall and the coefficient of variation in mean monthly rainfall (correlation: $F_{1,10} = 19.51$, $P = 0.0013$, $r = 0.81$; Fig. 4.1*B*), showing that the drier months during the prebreeding season were more unpredictable than the wetter months during the breeding season. Thus, there was both high within-year (i.e., seasonality) and among-year variation (i.e., temporal variability) in rainfall in this unpredictable environment.

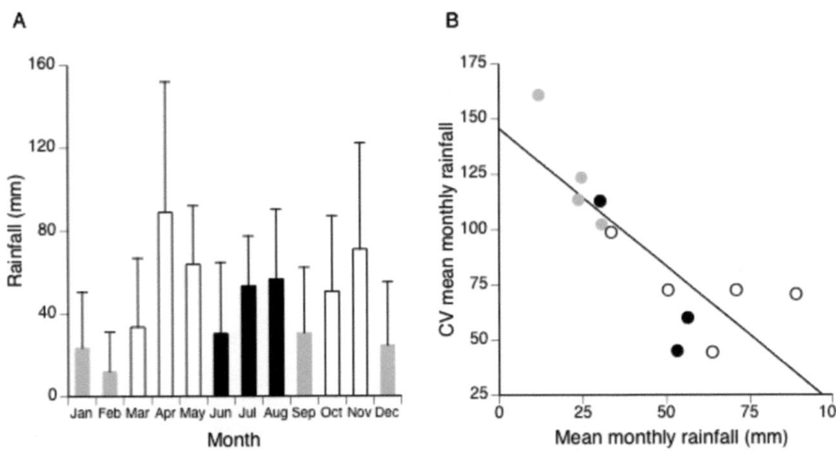

FIGURE 4.1 Rainfall patterns at the Mpala Research Centre, Laikipia, Kenya, from 1998 to 2010. (*A*) Mean ± SD in monthly rainfall is plotted. Peaks in rainfall were trimodal, highlighting the long and short rainy seasons, as well as a third peak in July and August. White bars indicate the primary breeding months, whereas graybars indicate the primary dry season months. (*B*) Mean monthly rainfall was negatively correlated with the coefficient of variation (CV) in mean monthly rainfall, showing that drier months were more variable than wetter months.

The high among- and within-year variation in rainfall influenced territory quality: There were significant differences among territories and among months in vegetation cover (generalized linear mixed model: territory, $F_{6,1622} = 22.85$, $P < 0.0001$; month, $F_{11,1563} = 4.24$; $P < 0.0001$). Vegetation cover is highly correlated with insect abundance and is thus a strong indicator of territory quality and a correlate of reproductive behavior (Rubenstein, 2007c). However, there was no effect of the interaction between territory and month on vegetation cover, suggesting that relative territory quality does not change much in this ecosystem (generalized linear mixed model: territory month, $F_{66,1623} = 0.86$; $P = 0.78$). In other words, although habitat quality differs among territories, high-quality territories remain better relative to low-quality territories in all months, seasons, and years.

Although the overall mean annual reproductive success was low in this population, as only 13% of all eggs laid fledged, there was significant variation in reproductive success among years (Wilcoxon test: $\chi^2 = 21.26$, df = 9, $P = 0.012$) and among territories (Wilcoxon test: $\chi^2 = 17.31$, df = 8, $P = 0.027$). However, mean fecundity was not related to variation in breeding conditions through time (i.e., climatic uncertainty or habitat heterogeneity); there was no relationship between mean reproductive success and breeding season rainfall (regression: $F_{1,7} = 0.073$, $P = 0.79$, $R^2 = 0.004$; Fig. 4.2*A*) or vegetation cover (regression: $F_{1,5} = 0.012$,

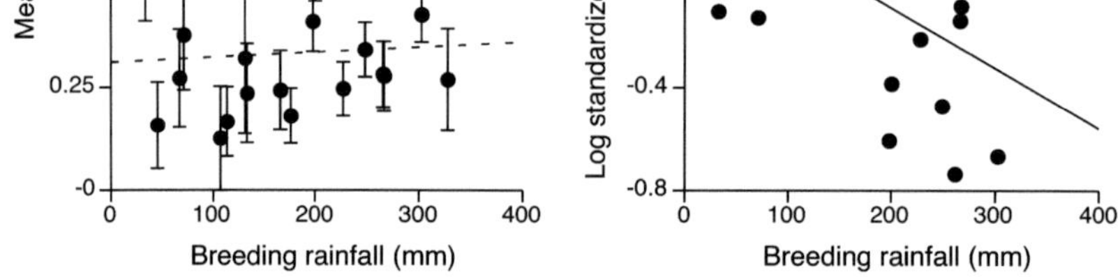

FIGURE 4.2 Reproductive success and climatic uncertainty. Reproductive success (RS) was estimated as the proportion of eggs fledged in each nest averaged for each group. Standardized variance in reproductive success was calculated as (variance in reproductive success)/(mean reproductive success)^2. Each point represents a breeding season (n = 19). (*A*) Mean ± SE reproductive success did not vary with breeding season rainfall, but (*B*) standardized variance in reproductive success was negatively related to breeding rainfall. Thus, fecundity variance decreased with increasing environmental quality or conditions.

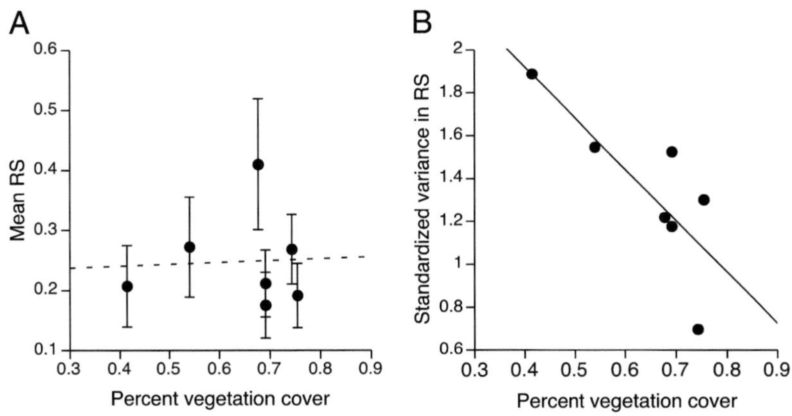

FIGURE 4.3 Reproductive success and habitat heterogeneity. Reproductive success (RS) was estimated as the proportion of eggs fledged in each nest averaged for each group. Standardized variance in reproductive success was calculated as (variance in reproductive success)/(mean reproductive success)^2. Each point represents a territory or group (n = 7). (*A*) Mean ± SE reproductive success did not vary with percentage vegetation cover, but (*B*) standardized variance in reproductive success was negatively related to percentage vegetation cover. Thus, fecundity variance decreased with increasing territory quality.

P = 0.92, R^2 = 0.002; Fig. 4.3*A*). In contrast, the variance in fecundity in time and space was related to environmental variation among territories and across years. There was a significant negative relationship between standardized variance in reproductive success and breeding season rainfall (regression: $F_{1,17}$ = 6.17, P = 0.024, R^2 = 0.27; Fig. 4.2*B*), suggesting that fecundity variance among territories declines with increasing environmental quality or conditions. There was also a negative relationship between standardized variance in reproductive success and vegetation cover (regression: $F_{1,5}$ = 8.42, P = 0.034, R^2 = 0.63; Fig. 4.3*B*), suggesting that fecundity variance among years declines with increasing territory quality. Together these results show that environmental variation in space and time had significant effects on variance in fecundity but not on mean fecundity.

Previous work in this system demonstrated that having helpers is critical, as nests with more helpers fledged more young (Rubenstein, 2007b). However, given the relationship between variance in reproductive success and environmental variability, does having more helpers actually increase mean reproductive success and reduce variance in reproductive success, and does helper number vary as a function of environmental variation? Using group size as an estimate of the potential

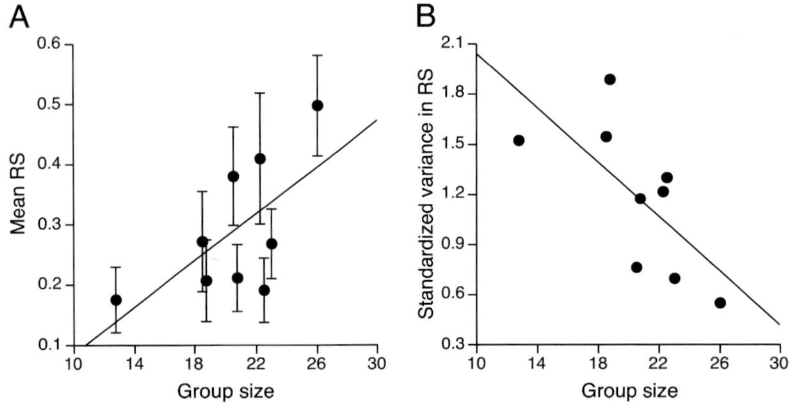

FIGURE 4.4 Reproductive success and helper number. Reproductive success (RS) was estimated as the proportion of eggs fledged in each nest averaged for each group. Standardized variance in reproductive success was calculated as (variance in reproductive success)/(mean reproductive success)^2. Because all superb starling groups have helpers, group size is a good estimate of the number of helpers in a group (Rubenstein, 2007b; Rubenstein and Shen, 2009). Each point represents a territory or group ($n = 9$) (*A*) Mean ± SE reproductive success showed a nonsignificant trend to increase with group size, whereas (*B*) standardized variance in reproductive success was negatively related to group size. Thus, mean fecundity tended to increase with increasing numbers of helpers, whereas fecundity variance decreased with increasing numbers of helpers.

number of helpers available in a group (Rubenstein, 2007b; Rubenstein and Shen, 2009), I found that mean reproductive success showed a trend to increase with increasing group size (regression: $F_{1,7} = 4.82$, $P = 0.064$, $R^2 = 0.41$; Fig. 4.4*A*), whereas standardized variance in reproductive success declined with increasing group size (regression: $F_{1,7} = 6.09$, $P = 0.043$, $R^2 = 0.47$; Fig. 4.4*B*). Additionally, group size was related to environmental variation in time but not in space. Group size, which was estimated during the long rains breeding season, was not influenced by vegetation cover (regression: $F_{1,5} = 0.48$, $P = 0.52$, $R^2 = 0.087$) or breeding rainfall (regression: $F_{1,2} = 1.15$, $P = 0.40$, $R^2 = 0.36$). However, just as many reproductive behaviors and components of superb starling physiology are influenced by rainfall in the prebreeding period leading up to the long rains breeding season (Rubenstein, 2007a,b; Rubenstein et al., 2008), so too is group size related to prebreeding rainfall (regression: $F_{1,2} = 27.32$, $P = 0.035$, $R^2 = 0.93$); groups were larger after relatively wetter dry seasons, suggesting that more helpers were available after favorable dry season conditions. Thus, fecundity (mean and variance) is related to the potential for cooperation, which is itself related to a different seasonal component of temporal environmental variation.

DISCUSSION

Having helpers at the nest is beneficial for superb starlings, as nests with more helpers fledge more young (Rubenstein, 2007b). Here, I further demonstrate that mean reproductive success increased with increasing group size [i.e., the number of available helpers (Rubenstein, 2007b; Rubenstein and Shen, 2009)] and that variance in reproductive success decreased with increasing group size. Group size was also directly related to prebreeding rainfall or temporal environmental variation in the dry season immediately before the primary breeding season. Rainfall during this period not only influences the potential for cooperation but also breeding behavior directly (Rubenstein, 2007a,b), as well as stress physiology (Rubenstein, 2007a) and immune function (Rubenstein et al., 2008). These results suggest that not only is having helpers beneficial but also that living in larger groups has added reproductive benefits. Thus, cooperative breeding and the formation of large, complex family groups in superb starlings may be related directly to environmental variation.

If cooperative breeding behavior within these large family groups is indeed a within-generation bet-hedging strategy to maximize fitness by reducing fecundity variance in spatially heterogeneous or temporally unpredictable environments, then variance in reproductive success should also be related to environmental variation in space and/or time. In support of this prediction, I found that reproductive success varied greatly among years and among territories, and that the variance in reproductive success was related to both climatic uncertainty and habitat heterogeneity. Variance in reproductive success among territories decreased with increasing environmental conditions across years (breeding season rainfall), whereas variance in reproductive success among years decreased with increasing territory quality (vegetation cover). Thus, mean reproductive success is similar in both good and bad times, as well as on high- and low-quality territories. As would be predicted from a bet-hedging hypothesis, constant levels of mean reproductive success are maintained across all environmental conditions, and importantly, reproductive success does not decrease as conditions deteriorate. Poor rainfall years, however, seem to exacerbate the differences between high- and low-quality territories, leading to greater variance in reproductive success among territories. Conversely, high rainfall can apparently mask the inherent differences in territory quality that drive patterns in reproductive success. Similarly, low-quality territories amplify the consequences of annual differences in breeding rainfall more than high-quality territories.

Together, these results are consistent with the hypothesis that cooperative breeding in starlings may be a risk-averse or within-generation bet-hedging strategy to maximize fitness by minimizing variance in fecundity in temporally and spatially variable environments. Additionally, these results are also consistent with both the ecological constraints and benefits of philopatry hypotheses. Ecological constraints clearly limit breeding opportunities in this species (Rubenstein, 2007a,c), but fecundity variance also seems to play an important role in shaping individual reproductive decisions. Additional studies in cooperatively breeding species in which subordinates have greater opportunities for independent breeding outside of the group will be needed to further disentangle these hypotheses.

If cooperative breeding is a strategy to reduce risk in variable environments, do spatially and temporally variable environments influence social behavior in similar ways, as has been proposed previously (Emlen, 1982a,b), or do the mechanisms underlying group formation differ in the different types of environments? Insider-outsider conflict theory (Giraldeau and Caraco, 1993; Higashi and Yamamura, 1993) provides a possible framework to explore how the tension over group membership between current group members (i.e., insiders) and potential joiners (i.e., outsiders) could differ in temporally and spatially variable environments. Insider-outsider conflict theory has recently been expanded to consider conflict resolution during group formation in cooperatively breeding species in which relatedness among group members can be high (Shen and Emlen, 2010). Instead of focusing on just offspring delayed dispersal, the theory emphasizes the importance of simultaneously considering both insider (parents) and outsider (joining mature offspring) interests to fully understand the evolution of cooperative breeding. In temporally variable environments where breeding conditions are unpredictable from year to year, insiders may be more likely to allow outsiders into the group to maintain a pool of available helpers as a form of insurance (Emlen, 1982a; Covas et al., 2004, 2008; Rubenstein, 2007a). However, outsiders may be less inclined to join groups in all but the poorest years without insider concessions (i.e., a share of reproduction) because outside breeding opportunities are likely to be available (Shen and Emlen, 2010). This environmentally induced conflict would not only lead to the formation of larger groups (i.e., a larger insurance pool) but also potentially to greater reproductive sharing within groups or lower reproductive skew. Once groups have formed, year-to-year environmental differences could still influence reproductive conflict and the degree of reproductive skew (Emlen, 1982a; Covas et al., 2004). Thus, temporally variable environments might favor large groups with multiple breeders and low reproductive skew (i.e., plural breeding), in which outsiders exert relatively greater control over group membership and reproductive conflict is high. In spatially variable, heterogeneous environments where territory quality varies across the landscape, the conditions on a given territory are more predictable from year to year than in temporally variable environments. Therefore, insiders might not only be less willing to accept outsiders into the group but also less inclined to share any reproduction with them. In contrast, outsiders may be more willing to join groups without concessions because outside breeding opportunities are likely to be limiting in all years. Thus, spatially variable environments might favor smaller groups with one breeding pair and high reproductive skew (i.e., singular breeding), in which insiders might exert relatively greater control over group membership and reproductive conflict is low. Overall, the mechanisms underlying group formation, the individuals that control group membership, and the types of social groups themselves (i.e., singular vs. plural) may differ in temporally and spatially variable environments.

Although the data presented here are consistent with the hypothesis that spatiotemporal environmental variation promotes cooperative breeding as a risk-averse behavioral coping strategy, it is not the only possible explanation. Kin-structured populations, or kin neighborhoods, resulting from environmental constraints could also influence avian cooperative breeding behavior (Hatchwell, 2009). Hamilton (1964b) was the first to realize that populations with limited dispersal, or population viscosity, will lead to greater opportunities for kin to interact. This idea, namely that reduced dispersal can lead to cooperation among relatives, forms the basis of most ecological constraints models of cooperative breeding (Emlen, 1982a; Koenig et al., 1992; Covas and Griesser, 2007), which are based largely on ideas of spatial variation in the environment. In general, habitat heterogeneity leads to reduced dispersal opportunities and therefore greater natal philopatry and the formation of kin neighborhoods that ultimately may give rise to kin groups. However, kin neighborhoods could also result from high temporal environmental variation, independent of processes like population viscosity. Variation in reproductive success, which could be driven largely by climatic uncertainty, is predicted to lead to an increase in the relatedness between group members because it decreases the number of effective relatives within a group (Lehmann and Balloux, 2007). Climatic uncertainty could therefore influence the formation of kin neighborhoods and ultimately kin groups, which is consistent with comparative results showing that cooperatively breeding species tend to live in temporally variable environments (Rubenstein and Lovette, 2007; Jetz and Rubenstein, 2011). Thus, spatial and temporal environmental variation can both influence demographic structure and the formation of kin neighborhoods, but for different reasons. Habitat heterogeneity could lead to kin-structured populations via spatial constraints on dispersal, whereas climatic uncertainty could lead to higher relatedness within groups because of decreased offspring production.

Whereas most theoretical and empirical studies examining the role of environmental constraints in the evolution of cooperative breeding have focused on mean reproductive success (Emlen, 1982b; Koenig et al., 1992), the results presented here suggest that we should also consider variance in reproductive success when studying social evolution. In superb starlings and other birds, cooperative breeding may be a risk-averse strategy to maximize fitness in a range of environmental conditions by reducing fecundity variance. In particular, when mean reproductive success does not differ between high- and low-quality territories, or between good and bad years, selection on variance may be important. Selection on variance is maintained when group sizes remain small and/or when dispersal rates are low (Lehmann and Balloux, 2007), both of which are hallmarks of cooperative breeders and especially likely to occur in temporally or spatially variable environments. However, spatial and temporal environment variation may influence cooperative breeding behavior in different ways. The mechanisms underlying group formation, the individuals that control group membership, and the types of social groups themselves may differ in temporally and spatially variable environments. Thus, to understand the adaptive value of cooperative breeding behavior in the heterogeneous and unpredictable environments where social species disproportionately occur (Jetz and Rubenstein, 2011), researchers must consider both the mean and environmentally induced variance in reproductive success. In addition to the within-generation bet-hedging hypothesis tested here, researchers should also consider social evolution in the context of among-generation bet-hedging, or risk spreading over multiple generations to maximize geometric mean lifetime reproductive success. Finally, the ideas presented here are not limited to birds: many species of social mammals (Solomon and French, 1997) and insects (Wilson, 1971; Costa, 2006) live in temporally variable environments or habitats where resources are distributed heterogeneously on the landscape. Ultimately, determining how spatiotemporal environmental variation drives patterns of and variation in fitness will provide important insights into the evolution of complex social behavior in a diversity of animal taxa.

MATERIALS AND METHODS

Study System and Species

A marked population of superb starlings was continuously monitored at the Mpala Research Centre, Laikipia, Kenya (0°17′ N, 37°52′ E) from

April 2001 through January 2011. Breeding activities of seven social groups were monitored over 10 long-rains and 10 short-rains breeding periods during this time. One additional group was added in January 2002, and another was added in January 2003; both were monitored through January 2011. Although birds have been recorded breeding during every month of the year, they typically only breed during both the long (April–May) and short (November) rains. Group size, which is a strong predictor of the number of available helpers (Rubenstein, 2007b; Rubenstein and Shen, 2009), was estimated annually for all groups from 2002 to 2005 during the long-rains breeding season. Active nests were checked every 1–3 days throughout the study during the hatching and nestling stages. Group reproductive success was quantified as the proportion of eggs laid that fledged young in each nest for each group in each breeding season. When nests were first encountered in the incubation stage and the number of eggs laid could not be determined, we used the mean clutch size of 3.5 eggs. When pairs had multiple clutches of eggs in a breeding period, the total number of eggs laid and fledged was summed before a proportion was calculated. Raw proportional data for each breeding pair were arcsine-square root transformed, and arithmetic means and SDs were used for all analyses, consistent with analyses of within-generation bet hedging (Gillespie, 1974, 1975; Crean and Marshall, 2009). Standardized variance in reproductive success was calculated as (variance in reproductive success)/(mean reproductive success)^2 (Weatherhead and Boag, 1997; Rubenstein, 2007b). All statistical tests were conducted in JMP v9 (SAS Institute, 2010). We used nonparametric Wilcoxon tests to examine differences in reproductive success among years and among territories. Regressions were used to examine the relationships between reproductive success (mean and variance) and climatic uncertainty (breeding rainfall) and habitat heterogeneity (vegetation cover). When necessary, summary data were logarithm transformed to meet assumptions of normality. Data from the 2006 short rains were excluded from some analyses because only 1 of 40 nests (1 of 124 eggs) fledged young, which greatly skewed the standardized variance estimates.

Rainfall

Daily rainfall data were collected continuously from 1998 through 2009 using an automated Hydrological Services TB3 Tipping Bucket Rain Gauge located at the Centre. In 2010, rainfall data were collected using a manual gauge located at the same place. Because the two datasets were highly correlated in previous years ($F_{1,138} = 2577.84$, $P < 0.0001$, $R^2 = 0.97$), we used the automated data from 2001 through 2009 supplemented with the 2009 manual data. Breeding rainfall was calculated as the total monthly rainfall for all months in which nests were initiated during a breeding season. Prebreeding rainfall was calculated as the total rainfall during December, January, and February each year (Rubenstein, 2007a,b; Rubenstein et al., 2008). This period represents the primary dry season and the 3 months with the greatest coefficients of variation in mean monthly rainfall (Fig. 4.1).

Habitat Quality

Territory quality was previously quantified using vegetation transect surveys conducted on seven territories at a single time point [see Methods in Rubenstein (2007c)]. Two groups were not surveyed because their territories encompassed areas at the Centre where constant human habitation and building, continuous livestock rearing, and seasonal grass and tree cutting altered the landscape and provided year-round food and water for birds. Briefly, seven 100-m transects starting at a randomly chosen point and compass direction were conducted at each site. For each 100-m transect, a metal pin was dropped every 2 m (50 points per transect, and 350 points per territory), and whether the pin was touching vegetation or bare earth was recorded (Holmes, 1974; Stewart et al., 2001). Territory quality was quantified for each transect as the percentage vegetation cover, or the proportion of pins that were touching a grass or forb species over the total number of pin drops per transect. Vegetation cover on each territory was positively correlated with the proportion of *Cynadon* grass, the dominant grass species in glades and an indicator of nutrient-rich soils (Augustine, 2003), and the abundance of grasshoppers, a primary food source for nestlings (Rubenstein, 2007c). Although previous work on this landscape showed that relative habitat quality does not change even though vegetation cover varies seasonally (Rubenstein, 2007c), monthly vegetation transects were conducted on all seven territories from February 2008 through January 2011 (36 months) using the same methods detailed above to confirm that relative habitat quality of superb starling territories does not change. Raw proportional data were arcsine-square root transformed and analyzed using a generalized linear mixed model with year and transect as random effects to account for repeated sampling of territories, as well as month, territory, and their interaction as fixed effects.

ACKNOWLEDGMENTS

I thank F. Ayala, J. Avise, D. Queller, and J. Strassmann for organizing the Sackler Colloquium and inviting me to participate; J. Ekiru, F. Lomojo, G. Rana, J. Ronjore, B. McCleery, T. Davis, and especially W. Watetu and G.

Manyaas for their help in the field; and J. Avise, S. Emlen, P. Sherman, and especially C. Botero, S.-F. Shen, and D. I. Rubenstein for providing discussion and/or comments on earlier versions of the manuscript. The Kenyan Ministry of Education, Science and Technology, the National Council on Science and Technology, the National Museums of Kenya Ornithology Section, the Kenya Wildlife Service, and the Mpala Research Centre enabled this work. D.R.R.'s long-term research was supported by Columbia University, the Howard Hughes Medical Institute, the Smithsonian Institution, Cornell University, the Miller Institute for Basic Research at the University of California at Berkeley, and the Museum of Vertebrate Zoology, as well as by grants from the National Science Foundation (IBN-407713), the American Museum of Natural History Chapman Fund, the American Ornithologists' Union, the Wilson Ornithological Society, the Society for Integrative and Comparative Biology, the Animal Behaviour Society, the Harvard Travellers Club, the Society of Sigma Xi, Cornell University, the Cornell Laboratory of Ornithology Benning Fund, and Cornell Sigma Xi.

Part II

COOPERATION WRIT SMALL: MICROBES

Perhaps no taxa are as promising for enhancing both our understanding of cooperation and our understanding of the organisms themselves as are microbes. Early work on microbes concentrated on purifying and isolating them for growth in pure culture. The postulates by Koch (1893) required this and were important for determining exactly which microbes caused which disease. But in nature microbes live in complex multispecies structured environments. Social interactions are profound, because microbes perform many functions (such as digestion) extracellularly that animals perform inside. One of the recent transformative elements of the study of microbes has been an appreciation of the importance of their social interactions. Many of the types of social interactions found in animals have their counterparts in microbes. Some cooperative interactions are much more easily studied in microbes, particularly if the goal is to illuminate the genetic basis of behavior or to use the power of experimental evolution.

Perhaps the best-studied social bacterium is *Myxococcus xanthus,* a species of δ-proteobacteria that spends its entire life in social groups (Velicer and Vos, 2009). It is a predatory bacterium that hunts other bacteria in social packs, dissolving its prey in pools of cooperatively produced enzymes before ingesting them. Movement usually is based on Type IV pili and is fundamentally social. When food is scarce, individual bacteria aggregate into a fruiting body. In this stalkless fruiting body, most or nearly all cells lyse, perhaps to the benefit of the remaining few, which form hardy spores. Experimental evolution has shown us much about the nature of sociality in *M. xanthus*. For example, when food was patchily distributed, the species evolved more efficient group hunting techniques (Hillesland et al., 2009). Under other circumstances, social cheaters can drive population crashes (Velicer and Vos, 2009; Fiegna and Velicer, 2005). In one fascinating case, a new cooperator evolved from the social cheater. But this work does not tell us how natural these events are; for that explanation we must turn to natural variation in wild fruiting bodies. In Chapter 5, Suzanne Kraemer and Gregory Velicer explore natural phenotypic variation in social traits of distinct clones within a fruiting body. They took 10 fruiting bodies from nature, and from them isolated 48 individual clones and examined their social phenotypes. These clones varied within fruiting bodies in swarming and in spore production, genetic traits likely to have arisen recently because the clones from the same fruiting body were nearly genetically identical. This fascinating work will shed light on the nature of sociality in the absence of a single cell bottleneck, where variations that benefit single clones within the group can spread, even at the cost of other group members.

One advantage to studying microbial social systems is that attributes that are strong but sometimes hard to measure in animals are easily examined in experimental systems. One such attribute can be called "restraint." It may not be easy to determine whether or not a cow in a herd is eating all it could or is holding back so that others may eat. If it were holding back, this would be a social trait that would benefit others, and thus would be expected to evolve under kin selection only if the genes for that trait are also present in others, and benefit accordingly. In an ingenious experiment described in Chapter 6, Joshua Nahum and colleagues examine the evolution of restraint in a nontransitive hierarchy often described by the rock-paper-scissors game in which no one type consistently dominates. They used *Escherichia coli* clones and the colicin system (Riley and Wertz, 2002). Colicins are costly to produce and resist, but sensitive strains are killed when producers release these substances. The researchers engineered double colicin producers and resisters so production and resistance would not be lost or gained in their system, and then, they asked how the three types of clone would fare under different migration schemes compared with how the resistor performed on its own. The authors found that the resistor strain exhibited the most restraint with restricted migration in the presence of all three strains, just the conditions where their models expect cooperation to evolve.

Cooperation among clonemates arises easily because the genes underlying cooperation are present in both partners. In microbes, cooperation often takes the form of extracellular secretions, including those used for quorum sensing, iron scavenging, and fruiting body formation. Therefore, a key question involves what favors the formation of clonal patches such that cooperation can be promoted. One answer involves the physical structure of the environment. For example, microorganisms growing on substrates are more likely to be in contact with clonemates than those living in a more fluid environment. Another possibility, and one investigated by Sara Mitri and colleagues in Chapter 7, is that other species can generate structure that favors within-species clonality. The authors use a modeling approach to understand how additional species can change interactions within species for the case of a growth-promoting secretion. This agent-based modeling approach uses one other species to stand in for all competing species. The authors' models indicate that other species can insulate secretors from selfish nonsecretors, even when the other species can use the secretions themselves. Other factors such as the role of dispersal and nutrient levels are also addressed in these models, which begin the important task of considering microbial sociality and ecology simultaneously, because these factors must influence how selection operates on these systems in nature.

5

Endemic Social Diversity Within Natural Kin Groups of a Cooperative Bacterium

SUSANNE A. KRAEMER*† AND GREGORY J. VELICER*

The spatial structure of genetic diversity underlying social variation is a critical determinant of how cooperation and conflict evolve. Here we investigated whether natural social groups of the cooperative soil bacterium *Myxococcus xanthus* harbor internal genetic and phenotypic variation and thus the potential for social conflict between interacting cells. Ten *M. xanthus* fruiting bodies isolated from soil were surveyed for variation in multiple social phenotypes and genetic loci, and patterns of diversity within and across fruiting body groups were examined. Eight of the 10 fruiting bodies were found to be internally diverse, with four exhibiting significant variation in social swarming phenotypes and five harboring large variation in the number of spores produced by member clones in pure culture. However, genetic variation within fruiting bodies was much lower than across fruiting bodies, suggesting that migration across even spatially proximate groups is limited relative to mutational generation of persisting endemic diversity. Our results simultaneously highlight the potential for social conflict within *Myxococcus* social groups and the possibility of social coevolution among diverse related lineages that are clustered in space and cotransmitted across generations.

---

*Department of Biology, Indiana University, Bloomington, IN 47405. †To whom correspondence should be addressed. E-mail: suskraem@indiana.edu.

Social evolution research seeks to explain the origin, maintenance, and diversification of both cooperative and competitive social traits. This goal requires understanding the character of social environments that mediate selection on these traits. The distribution of behavioral and genetic diversity within and across groups of social animals has received much attention (Krebs and Davies, 1997; Oxley et al., 2010; Waddington et al., 2010). In contrast, relatively little is known about the structure of diversity among natural groups of social microbes (Fortunato et al., 2003b; Vos and Velicer, 2006, 2008a; Gilbert et al., 2009; Köhler et al., 2009; Wilder et al., 2009; Wollenberg and Ruby, 2009). However, detailed knowledge of group composition is necessary for understanding the roles of mutation, migration, lateral gene transfer, genetic drift, and various forms of selection in shaping the evolution of social microbes in natural habitats.

Microbes engage in a wide range of social behaviors, both cooperative and antagonistic, that affect the evolutionary fitness of others (Velicer, 2003; West et al., 2007a; Nadell et al., 2009). Some of the most biologically complex forms of prokaryotic cooperation occur in the myxobacteria (order Myxococcales, δ-proteobacteria), which are best known for social development of multicellular, spore-bearing fruiting bodies in response to starvation (Shimkets et al., 2006). In particular, the predatory soil bacterium *Myxococcus xanthus* has become a model organism for the study of microbial sociality, including cooperative motility (Wu and Kaiser, 1995), social predation (Berleman and Kirby, 2009) and fruiting body formation (Shimkets et al., 2006), and its population biology (Velicer and Vos, 2009).

M. xanthus cells swarm in a coordinated manner through soil habitats in cohesive groups using two genetically distinct motility systems, one of which is obligately social [type IV pili-driven "S-motility" (Hodgkin and Kaiser, 1977; Wu and Kaiser, 1995)] and one of which allows individual cell movement ("A-motility") (Hodgkin, 1979; Sun et al., 2011). Swarms of *M. xanthus* in the soil kill and lyse prey cells of other microbial species with secreted antibiotics and lytic enzymes (Rosenberg and Varon, 1984). Upon starvation, swarming cells aggregate and develop into multicellular fruiting bodies (Shimkets et al., 2006). In these fruiting body aggregates a minority of cells convert to metabolically quiescent spores, whereas many other cells within the fruiting body lyse, possibly to the benefit of sporulating cells (Nariya and Inouye, 2008). The precise advantages of sporulation within fruiting bodies are unknown, although several hypotheses have been proposed, including enhanced dispersal, increased germination and/or growth rates in high-density groups, and protection from predation and/or environmental insults [summarized in greater detail in Velicer and Vos (2009)]. Here "fruiting body group" and "group" generally refer to either all cells that compose a particular fruiting body or a set of laboratory strains isolated from the same fruiting body.

Although bacterial growth by binary fission in structured habitats inherently generates clonal cell pockets (Nadell et al., 2010), cell groups forming *Myxococcus* fruiting bodies are not expected to be entirely clonal owing to mutation. The *M. xanthus* genome is large [>9 Mb (Goldman et al., 2006)] and fruiting bodies are thought to be constructed by ≈100,000 cells (Shimkets et al., 2006). If the mean *M. xanthus* mutation rate is roughly similar to that of *Escherichia coli* [$\approx 5.4 \times 10^{-10}$ per base pair per generation in one estimate (Drake et al., 1998)], any given fruiting body should contain at least dozens of mutational variants, even if the entire fruiting body group originated from a single cell. Although most mutations are deleterious (Eyre-Walker and Keightley, 2007) and are lost by selection or genetic drift, a small minority will persist and rise to high frequency either because they confer a selective advantage or nonadaptively by hitchhiking (Maynard Smith, 1991) or genetic drift (Wright, 1931). Persisting mutants might be socially defective cheaters that increase owing to a frequency-dependent advantage within groups (Velicer et al., 2000; Fiegna and Velicer, 2003; Ross-Gillespie et al., 2007; Sandoz et al., 2007). Alternatively, such mutants may be socially proficient strains that outcompete dominant genotypes owing to increased intrinsic fitness (Buttery et al., 2009) or the ability to socially exploit majority genotypes (Strassmann et al., 2000; Fiegna and Velicer, 2005; Vos and Velicer, 2009). Assessing the degree to which such persisting mutants migrate across social groups—within which cells interact during motility, predation, and development—is critical for understanding social evolution in the myxobacteria.

If intergroup migration is low, within-group diversity should derive primarily from endemic mutation and be lower than diversity across groups. In this scenario, relatedness values for social loci among interacting variants may often be high, thus promoting the maintenance of cooperation by kin selection (Hamilton, 1964a; Sachs et al., 2004; Foster et al., 2006). Under low migration, cotransmission of within-group social diversity across generations will be high (Sachs and Bull, 2005; Wade, 2007), and lineages that repeatedly and preferentially interact may coevolve to reduce within-group conflict. Even if spatially proximate, distinct lineage groups among which migration is low may diversify via differential trajectories of adaptation and drift.

Previous work has indicated that some *M. xanthus* genotypes disperse far, despite the overall differentiation of meter-scale populations isolated by distance across large spatial scales [e.g., >2,000 km (Vos and Velicer, 2008a)]. Across smaller scales (<300 km), local meter-scale populations were not differentiated at the genetic loci examined, and dispersal appears to be extensive (Vos and Velicer, 2008a). In another study, the spatial distribution of diverse multilocus genotypes among 78 centimeter-scale isolates did not appear to be significantly clustered, consistent with the possibility of extensive intergroup migration at this scale (Vos and Velicer, 2006). However, near the cellular (micrometer) scale, genetic variation in natural *M. xanthus* populations must be nonrandomly distributed due to the nature of bacterial colony growth by asexual binary fission in viscous environments (Nadell et al., 2010). Further work is required to better resolve patterns of genetic structure and degrees of dispersal and intergroup migration across a wide range of spatial scales.

A high level of phenotypic and genetic diversity has been documented among centimeter-scale *M. xanthus* isolates (Vos and Velicer, 2006, 2008b, 2009; Krug et al., 2008; Kraemer et al., 2010; Morgan et al., 2010), despite the fact that genetic diversity at this scale was found to be much lower than at only slightly larger sampling scales (Vos and Velicer, 2008a). For example, genetically similar centimeter-scale isolates were found to show extremely divergent competitive abilities during fruiting body development in forcibly mixed pairings (Vos and Velicer, 2009). However, the degree to which such diverse clones migrate across fruiting body–forming groups—either passively or by active motility—has remained unclear. In *Myxococcus,* cell–cell adhesion (Chang and Dworkin, 1994) and territorial kin discrimination (Vos and Velicer, 2009) may limit intergroup migration.

Using a new collection of natural isolates, here we have tested whether diverse social phenotypes coexist within the most discrete social unit of the *Myxococcus* life cycle, the fruiting body group. We then tested for group-level structure in genetic and phenotypic diversity to discriminate between scenarios of low vs. high migration among social groups residing in forest soils. Ten natural fruiting body groups were harvested from soil collected at three Indiana woodland locations separated by several kilometers. Fruiting bodies from a given kilometer-scale location originated from centimeter-scale (MC fruiting bodies; see *Methods)* or meter-scale (GH and KF fruiting bodies, see *Methods)* sites along sample transects. Forty-eight clones were independently isolated from each fruiting body and screened for diversity at several genetic loci and in several social phenotypes during group swarming and fruiting body development. Patterns of diversity within and across fruiting body groups were then analyzed. Detailed descriptions of the methods used can be found in *Methods.*

RESULTS

Variation in Swarming Phenotypes

Myxococcus group swarming on soft agar is a social trait that is driven by the type IV pili-based S-motility system in standard laboratory strains (Shi and Zusman, 1993). Natural fruiting body groups, each represented by 48 clones, varied significantly in their mean swarming rate on soft agar (Fig. 5.1*A*; Kruskal-Wallis test: $P < 0.001$). This result is consistent with previous work that documented extensive variation in soft-agar swarming among other natural isolates (Vos and Velicer, 2008b).

Seven fruiting bodies did not exhibit significant within-group variation in swarming rate according to Kruskal-Wallis tests ($P > 0.05$), but two of these groups included stark variation in another visual phenotype (colony color in fruiting body GH5.1.9 and degree of cell–cell adhesion in MC3.3.5). Moreover, although swarming rate variation within GH5.1.9 was not quite significant according to the Kruskal-Wallis test ($P = 0.08$), comparison of the mean swarming rates of the two color types revealed an extremely significant difference (yellow vs. orange, Wilcoxon rank-sum test, $P < 0.001$).

Three fruiting bodies (KF3.2.8, KF4.3.9, and MC3.5.9) harbored clones exhibiting significant within-group variation in soft-agar swarming rate according to Kruskal-Wallis tests (Figs. 5.1 *B–D* and 5.2; $P < 0.001$ in all cases). Among KF3.2.8 clones, *k*-means clustering into two groups reveals one majority rate phenotype [mean cluster swarming rate 5.63 mm/d; 95% confidence interval (CI) = 0.23] and a faster minority type (mean cluster swarming rate = 6.32 mm/d; 95% CI = 0.3) (Fig. 5.1*B*). Post hoc testing revealed a significant difference between the swarming rates of those phenotype clusters (Wilcoxon rank-sum test on cluster means, $P < 0.001$), suggesting that they represent two distinct motility genotypes.

Cluster analysis of the other two fruiting bodies with significant variation among clones (KF4.3.9 and MC3.5.9; Fig. 5.1*C* and *D*) suggested the existence of three swarming rate phenotype classes in each group. The KF4.3.9 fruiting body is characterized by a fast majority type (mean cluster swarming rate 5.54 mm/d, 95% CI = 0.37) and two slower minority variants (Fig. 5.1*C*) (mean cluster swarming rates of 4.89 and 2.85 mm/d, respectively; 95% CI = 0.54 and 0.16, respectively). In contrast, the MC3.5.9 clones grouped into a majority phenotype with an intermediate swarming rate (mean cluster swarming rate 4.15, 95% CI = 0.47) as well as faster and slower minority types (Fig. 5.1*D*) (mean cluster swarming rates of 7.14 and 2.94 mm/d, respectively; 95% CI = 0.49 and 0.31, respectively). Post hoc testing revealed that the mean swarming rates among the clusters within both KF4.3.9 and MC3.5.9 differ significantly (Wilcoxon rank-sum tests of all possible within-fruiting-body combina-

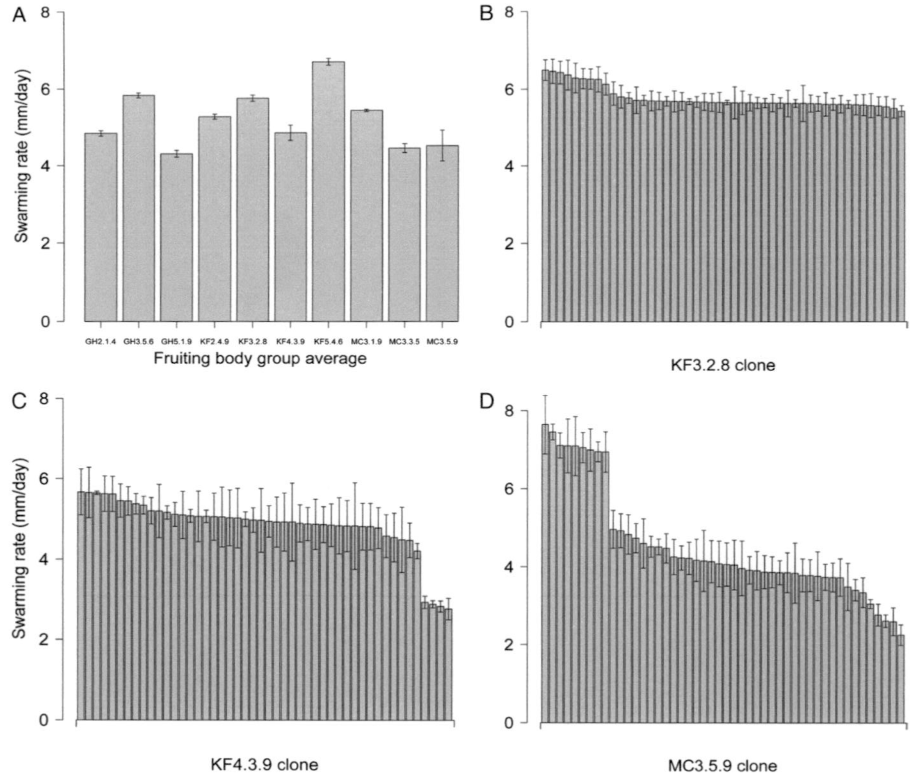

FIGURE 5.1 Swarming rates on soft agar. (*A*) Average swarming rates of all clones from each fruiting body group (n = 46–48 per fruiting body). (*B–D*) Average swarming rates for individual clones from fruiting bodies KF3.2.8, KF4.3.9, and MC3.5.9, respectively. Error bars represent 95% confidence intervals.

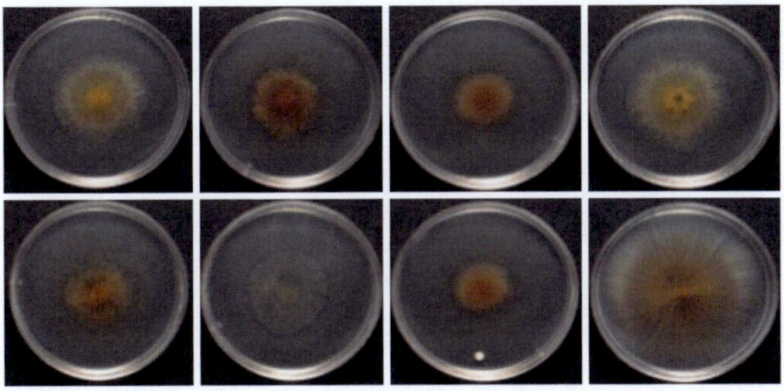

FIGURE 5.2 Swarming phenotypes of fruiting body MC3.5.9 isolates after 5 days of growth on soft agar plates. *Upper:* Left to right, c16, c19, c22, and c25; *Lower:* Left to right: c31, c35, c36, and c48.

tions of cluster means: $P < 0.001$ in all cases). In two cases, within-group variation in swarming rate was accompanied by variation in one or more additional phenotypes (colony color in KF4.3.9 and colony swarm pattern, color, opacity and degree of cell–cell adhesion in MC3.5.9; Fig. 5.2, Table 5.1).

Variation in Spore Production

Fruiting body groups, each represented by a subset of the 48 clones per fruiting body examined for motility, varied significantly in their mean levels of spore production (Fig. 5.3*A*; Kruskal-Wallis test: $P < 0.001$). This result is consistent with previous work that documented extensive variation in several developmental phenotypes among natural *M. xanthus* isolates (Fiegna and Velicer, 2005; Kadam and Velicer, 2006; Vos and Velicer, 2009; Kraemer et al., 2010).

Five of the 10 fruiting body groups examined were found to harbor significant within-group variation in spore production (Fig. 5.3*B–D*). In all five cases clones that sporulated poorly were in the minority. Spore production by the lowest sporulator within each of these five groups ranged from ≈10-fold below the level of the dominant phenotype cluster (MC3.3.5; Fig. 5.3*C*) to complete inability to produce viable spores (MC3.5.9; Fig. 5.3*D*). The isolates from fruiting bodies that harbored variation in spore production clustered into two distinct groups within each respective fruiting body by *k*-means cluster algorithms based on the criteria described in *Methods.* In two cases (GH5.1.9 and KF5.4.6), post hoc tests to compare cluster means were not possible because one cluster contained only a single clone. In the three remaining groups, the differ-

TABLE 5-1 Allelic and Phenotypic Variation Within and Across Fruiting Bodies

		Swarm	Spore	Colony	Gene					
Fruiting Body	Clone	Rate	#	Color	*pilA*	*0128*	*0176*	*0396*	*0533*	*4405*
GH2.1.4	9			y	1	1	1	1	1	1
	16			y						
	25			y						
	35			y						
	40			y	2					
	48			y						
GH3.5.6	2			y	2	2	2	2	1	2
	11			y						
	22			y	3					
	36			y	2					
	40			y						
GH5.1.9	17			y	1	3	3	1	2	2
	20			o						
	27			o						
	37			o						
	47		**Low**	y						
KF2.4.9	10			y	4	2	4	3	3	3
	17			y						
	21			y						
	38			y						
	42			y						
KF3.2.8	1			y	5	4	5	4	4	4
	11	**Fast**		y						
	12	**Fast**		y						
	16	**Fast**		y						
	20			y						
	26	**Fast**		y						
	28	**Fast**		y						

	29	**Fast**		y						
	30	**Fast**		y						
	35			y						
	37	**Fast**		y						
	39			y						
	45			y						
	48	**Fast**		y						
KF4.3.9	1	**Slow**	**Low**	t	1	2	6	1	5	5
	2	**Slow**	**Low**	t						
	3			y						
	19			y						
	23	**Slow**		y						
	28			y						
	30	**Slow**	**Low**	t						
	37			y						
	40	**Slow**		y	2					
	44			y	1					
KF5.4.6	4		**Low**	y	6	1	7	5	6	6
	11			y						
	28			y						
	29			y						
	36			y	1					
MC3.1.9	3			y	7	5	2	1	1	5
	16			y						
	25			y						
	28			y						
	47			y						
MC3.3.5	4		**Low**	t	2	6	7	6	1	2
	8		**Low**	y					2	
	9			y						
	16			y	1					
	21			y						

continued

TABLE 5-1 Continued

Fruiting body	Clone	Swarm rate	Spore #	Colony color	Gene *pilA*	*0128*	*0176*	*0396*	*0533*	*4405*
MC3.3.5	34			y						
(continued)	45			y						
MC3.5.9	2	**Fast**		y	1	6	7	6	2	2
	5			y						
	11	**Fast**		y						
	13	**Fast**		y						
	15	**Fast**		y						
	18			y						
	22	**Slow**	**Low**	t						
	23	**Fast**		y						
	25			y						
	27	**Fast**		y						
	29	**Slow**	**Low**	t						
	31	**Slow**		y						
	33			y						
	36	**Slow**	**Low**	t						
	39	**Slow**	**Low**	t						
	41	**Fast**		y						
	44			y						
	46	**Fast**		y						
	48	**Fast**		y						

Notes: Qualitative phenotype categories and allelic states at sequenced loci are shown for several clonal isolates from each fruiting body, including all clones that exhibited clear minority phenotypes for swarming and/or development. Distinct alleles at each locus are distinguished by a unique allele number (with numbers shown for only one representative of each allele). "Fast" and "Slow" designate individuals with minority swarming phenotypes in a given fruiting body, and "Low" designates individuals within low spore production clusters as identified by *k*-means cluster analysis (*Methods*). Colony color phenotypes are also indicated (y, yellow; o, orange; t, tan).

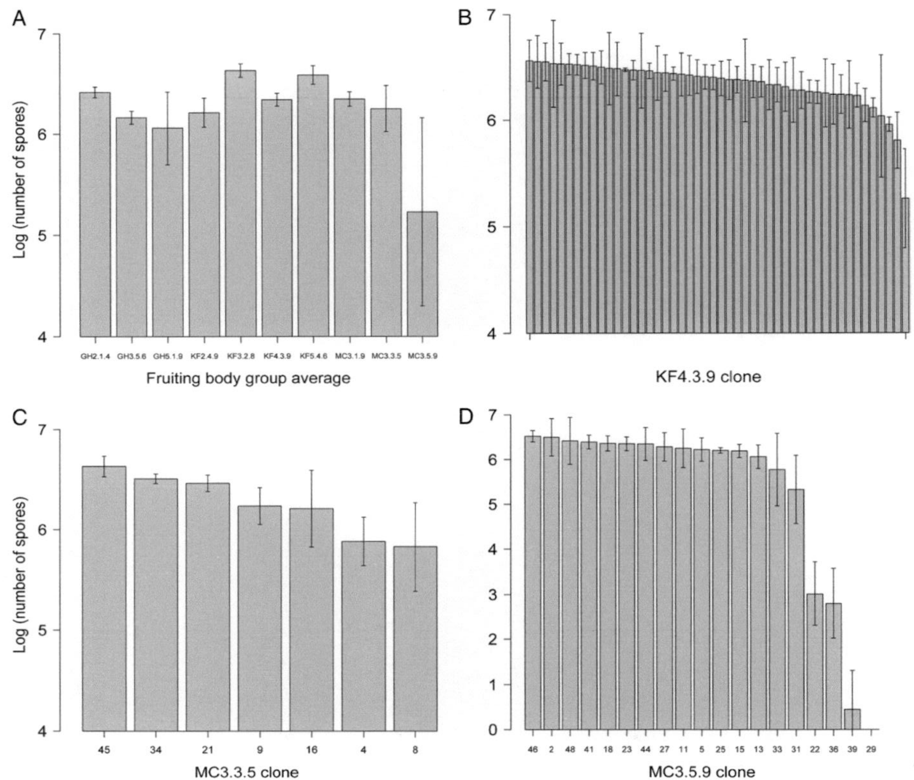

FIGURE 5.3 Spore production. (*A*) Average spore production of examined clones from each fruiting body group (n = 5–48 per fruiting body). (*B–D*) Average spore production by individual clones from fruiting bodies KF4.3.9, MC3.3.5, and MC3.5.9, respectively. Error bars represent 95% confidence intervals about means calculated from $\log_{10}$-transformed spore counts.

ences in spore production between the two clusters were either highly significant (KF4.3.9 and MC3.5.9; Wilcoxon rank-sum test, $P < 0.001$ in both cases) or marginally nonsignificant (MC3.3.5; Wilcoxon rank-sum test, $P = 0.09$).

Sporulation efficiency correlated significantly with swarming rate across all clones from all sampling locations (Spearman's rank correlation ρ: S = 3,758, $\rho = 0.56$, $P < 0.001$).

Genetic Structure of Fruiting Body Groups

All clones assayed for spore production were screened for genotypic variation within ≈500 base pair windows of six loci that contain above-average levels of variation among the laboratory strain DK1622 (Kaiser, 1979) and two *M. xanthus* isolates from Tübingen, Germany [A23 and A47 (Vos and Velicer, 2006)]. Genetic variation within fruiting body groups was found to be extremely low relative to variation across fruiting bodies. Seven *pilA* alleles were found across all clones (Table 5.1), but only 5 of the 10 fruiting bodies harbored *pilA* polymorphisms, and no more than two *pilA* alleles were present in any fruiting body group. Each of the five other loci was highly polymorphic across fruiting bodies, with either six or seven alleles detected at each locus (Table 5.1). However, only one of these loci was found to vary among clones from the same fruiting body, in which instance a minority allele of *Mxan_0533* was present in one clone of fruiting body MC3.3.5.

The vast majority of within-group phenotypic variation for swarming rate and spore production occurred between clones that are genetically identical at all (most cases) or most (a minority of cases) of the six loci examined (e.g., the swarming variants within fruiting body KF3.2.8 share the same alleles) (Table 5.1).

Phylogenetic Relationships

An unrooted maximum-likelihood tree and a Baysian inference tree were constructed with a sequence concatemer of the five loci other than *pilA* (Fig. 5.4). Both phylograms had similar topologies. Only one of the KF haplotypes (4.3.9) was found to group within the highly supported clade containing all of the GH and MC location haplotypes, with the KF5.4.6, KF2.4.9, and KF3.2.8 haplotypes branching more deeply. As reflected by this deep branching pattern, the KF3.2.8, KF2.4.9, and KF5.4.6 haplotypes were found to have only 0, 1, and 2 loci, respectively, that share an allele with one or more other fruiting body haplotypes (Table 5.1). KF3.2.8 is most similar to the laboratory strain DK1622 (Kaiser, 1979; Goldman et al., 2006).

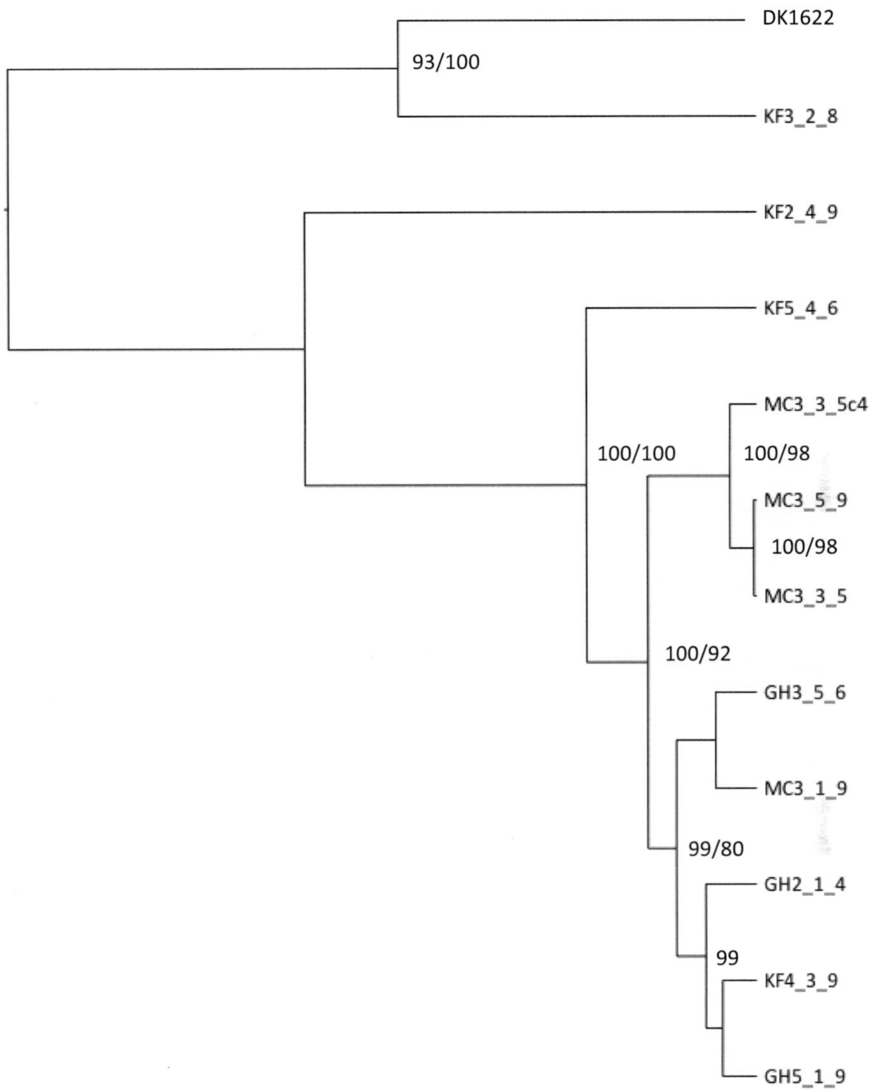

FIGURE 5.4 BI phylogram of 10 natural fruiting body groups based on a concatemer of the five loci (*Mxan_0128, Mxan_0176, Mxan_0396, Mxan_0533,* and *Mxan_4405*. BI and ML analyses produced similar topologies. Fruiting bodies harbored no internal variation at these loci with the exception of MC3.3.5, which contained a single minority variant MC3.3.5c4. Posterior probabilities >90 and bootstrap values >70 (based on 1,000 bootstrap replicates) are indicated at the nodes (posterior probabilities shown first).

Control for a Laboratory Origin of Minority Phenotypes

We tested whether the phenotypic variation demonstrated among isolated clones might have arisen during growth in the laboratory rather than in natural populations before soil collection. We examined four clones isolated from a fruiting body group (MC3.5.9) that exhibited a high degree of variation in both swarming rate and spore production. Specifically, we tested whether cultures from these four clones subjected to development and growth regimes similar to those experienced by the original fruiting body culture would generate an array of diverse phenotypes similar to that found among the 48 original MC3.5.9 clones. After cultures derived from these four clones had undergone development, heat, and sonication treatment and subsequent growth in CTT liquid, 48 clones were randomly selected from each culture after dilution plating and were screened for variation in swarming rate and fruiting body morphology. No significant variation was observed in either of these traits within any of the four clone sets, suggesting that the phenotypic variation documented here arose before fruiting body isolation.

DISCUSSION

The genetic and social diversity pervading natural *Myxococcus* populations is highly structured across local social groups within which cooperative—and likely antagonistic—interactions occur. This study and others have together shown that representatives of distinct but spatially proximate fruiting body groups vary starkly in social motility (Fig. 5.1*A*) (Vos and Velicer, 2008b) and several developmental phenotypes, including spore production (Fig. 5.3*A*), the rate of development, responsiveness to nutrient depletion in triggering development (Kraemer et al., 2010), and competitiveness in forced isolate pairings (Vos and Velicer, 2009). Here we have now also shown that pronounced social variation is present at high frequencies *within* many natural fruiting bodies—indeed within a majority of those fruiting bodies sampled here. Thus, diversity within fruiting bodies is high despite the fact that it was found to be much lower than diversity among fruiting bodies (Table 5.1 and Fig. 5.4). These results indicate that single clone isolates are likely to misrepresent the social phenotypes of other members of the groups from which they are isolated.

Possible Laboratory Effects

Control experiments strongly suggest that phenotypic variants isolated from the same fruiting body were already present at substantial frequencies in natural groups before sampling and did not arise in laboratory cultures. However, our data do not necessarily reflect accurately the frequencies of these variants in the soil at the time of sampling owing to possible variation in growth rate under laboratory conditions. Natural isolates of bacteria might vary in their degree of "preadaptation" to laboratory conditions (Velicer and Lenski, 1999), and some degree of phenotypic variation may be specific to laboratory settings. Nonetheless, it is plausible that the large trait differences documented here reflect heritable variation that is also manifested during motility and/or development in natural habitats.

Phase Variation

The patterns of variation among fruiting body groupmates documented here are not explicable by a previously documented form of "phase variation" in *M. xanthus*. In phase variation, bacterial cells switch between discrete phenotypic states at much higher rates than would be generated by the genomewide average mutation rate (Laue and Gill, 1994, 1995; Beaumont et al., 2009). In *M. xanthus* laboratory strain DK1622, color phase variation occurs in which ≈1% of cells derived from a yellow colony grow into tan colonies, whereas ≈25% of cells from a tan colony grow into yellow colonies (Laue and Gill, 1994). Although it has been suggested that tan and yellow cells may have different functional roles during development (Laue and Gill, 1995), the genetic basis and population-level effects of *M. xanthus* phase variation remain poorly understood.

Most minority variants in swarming rate and sporulation among our fruiting body isolates did not exhibit minority color phenotypes (Table 5.1). The one exception is that all seven clones identified as having both unusually slow swarming and unusually low sporulation within their two respective fruiting body groups (KF4.3.9 and MC3.5.9) grew as tan colonies rather than as the majority yellow phenotype (Table 5.1). However, groupmates among these seven clones varied significantly in both swarming rate (Fig. 5.1*C* and *D*) and spore production (Fig. 5.3*B* and *D*), indicating that even these variants do not represent simple dual-state phase variation.

We screened 48 colonies derived from each of four clones isolated from the most internally diverse fruiting body, MC3.5.9. Three of the four parental clones (c6, c16, and c24) were yellow, and one was tan (c29). None of the 48 colonies derived from the tan clone were yellow, as would be expected for ≈25% of colonies under DK1622-like phase variation. Among these four clones only one instance of apparent phase variation was observed. In that case, the biphasic diversity observed differed dramatically both from the patterns of diversity documented among our original fruiting body isolates and from DK1622 phase variation. Colonies derived from isolate MC3.5.9c6 showed two clearly distinguishable phenotypes of colony opacity during social swarming. Importantly, cultures derived from colonies of both opacity types form robust fruiting bodies and do not vary significantly in swarming rate. This limitation of variation among MC3.5.9c6 cells to colony opacity contrasts starkly with the variation observed among the original isolates from fruiting body MC3.5.9, which did not include similar variation in colony opacity but did include three distinct swarming-rate clusters and clones with severe developmental deficiencies. Colonies derived from the other three parental clones selected for the control experiments showed no variation for any visual phenotype. Thus, exhibition of phase variation is itself yet another phenotype that seems to vary among closely related groupmates within natural fruiting bodies.

Endemic Variation

Our results show that much of the detectable diversity within natural *Myxococcus* social groups derives from endemic mutation rather than intergroup migration. Here we consider migration to be the combination of dispersal to a new location and physical immigration into a new social group at that location. Five fruiting bodies, including the most phenotypically diverse one (MC3.5.9), contained only a single haplotype for all six loci sequenced here. Another four were polymorphic only at the *pilA* locus, with just two alleles present in each case. Only one fruiting body was polymorphic at more than one locus (MC3.3.5, polymorphic at *pilA* and *Mxan_0533*). In contrast, in almost all pairwise comparisons of fruiting body groups, the dominant six-locus haplotypes from the paired groups differed at most loci. The sole exceptional comparison is between MC3.3.5 and MC3.5.9, which share identical majority haplotypes. However, genetic variation seems to be structured even across these two fruiting bodies that were isolated at the centimeter scale because variants with similar phenotype profiles are represented by multiple clones within each fruiting body group but are absent from the other (Table 5.1).

In contrast to patterns revealing endemic variation, the occurrence of some alleles that are shared by clones from multiple fruiting bodies isolated from different locations is consistent with some degree of recombination across groups and populations, possibly mediated by phage transduction (Martin et al., 1978). Such patterns are not unexpected in light of previous evidence for horizontal gene transfer in *Myxococcus* populations (Vos and Velicer, 2006, 2008a; Vos and Didelot, 2009).

Migration

Our data suggest that the rate at which *Myxococcus* social variants arise by mutation and subsequently increase to detectable frequencies within their natal groups is high relative to the rate at which variants migrate into "foreign" groups and subsequently persist. The among-group migration rate will greatly affect the relative importance of within- vs. among-group selection (Wade, 1985) in determining the fate of new mutations in social genes. Low migration will promote spatiotemporal clustering of genetically similar lineages [i.e., high relatedness within groups (Foster et al., 2006; Gilbert et al., 2007, 2009)], high cotransmission of social diversity across generations (Wade, 2007), and the among-group component of selection in multilevel selection models of social evolution (Wade, 1985).

Biological traits that affect migration rate are thus likely to influence how cooperation is maintained in a metapopulation and the evolutionary forces causing socially proficient genotypes to diversify. In *Myxococcus,* most cells are highly cohesive owing to the production of cell-surface adhesins, which should hinder emigration away from natal kin groups. Moreover, kin discrimination mechanisms that hinder immigration (Travisano and Velicer, 2004) by members of neighboring groups appear to be pervasive in natural *M. xanthus* populations. This inference derives from experiments in which neighboring swarms of genetically very similar centimeter-scale isolates failed to merge on agar plates for most pairings (Vos and Velicer, 2009) (Fig. 5.5). Cooperation benefits buttressed by low migration may thus contribute to selection for cell–cell adhesion and territorial kin discrimination.

Limitation of Socially Defective Cheaters

Cheater strains with social defects in clonal groups that can exploit cooperative genotypes in mixed groups during *Myxococcus* development readily appear by mutation in laboratory populations (Velicer et al., 2000). When rare, these cheaters have a within-group advantage over cooperators but lose that advantage and impose cheating load (i.e., reduce group productivity) (Velicer, 2003) when they reach high frequencies (Velicer et al., 2000; Fiegna and Velicer, 2003). Many isolates described here exhibit low spore production (e.g., MC3.5.9c29; Fig. 5.3*D)* and/or slow swarming (e.g., KF4.3.9c1; Fig. 5.1C). These clones may represent cheaters of natural origin that defect from "fair" production levels for a social compound required for development or social motility but exploit others who produce more of that compound. Alternatively, socially deficient strains may be present owing to genetic drift or selection on some trait other than the socially defective one.

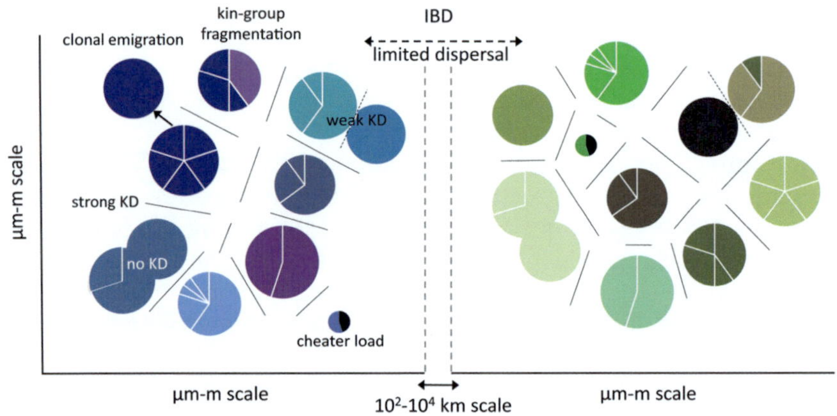

FIGURE 5.5 Simple hypothetical model of natural *Myxococcus* population biology. Circles represent social groups within which individuals directly interact. Sectors represent genetically distinct within-group variants. Distinctly shaded circles represent among-group genetic differentiation, with lines separating kin discrimination (KD) units (Vos and Velicer, 2009). Overlapping circles represent lack of KD between highly similar, but nonetheless genetically distinct, social groups. Multishaded circles represent kin-group fragmentation (see text). Small circles with a black sector represent cheater-infected groups burdened by cheater load (Velicer et al., 2000; Fiegna and Velicer, 2003; Velicer and Vos, 2009). *Left* vs. *Right* panels represent population differentiation across large spatial scales due to isolation by distance (IBD) (Vos and Velicer, 2008a). The arrow at *Left* represents establishment of a new clonal group by clonal emigration.

Only mutations creating socially defective cheaters that have an advantage within a group across an organism's entire life cycle will increase substantially within groups. The mutation rate to cheaters that have such a net within-group advantage (at least when rare) remains unknown. Pleiotropy may limit this mutation rate (Foster et al., 2004; Travisano and Velicer, 2004). Cheater mutations that are net beneficial within groups may nonetheless be net deleterious across groups in a larger metapopulation owing to among-group selection mediated by cheater load (Velicer et al., 2000; Fiegna and Velicer, 2003; Gilbert et al., 2007; Velicer and Vos, 2009) and promoted by limited migration (Fig. 5.5).

The mutation rate to socially defective cheating alleles that are net beneficial within groups might be lower than the rate at which such alleles are lost from a metapopulation owing to their net deleterious effect at the among-group level. In this scenario, the overall frequency of socially defective cheaters in a metapopulation should be determined by the relative magnitude of that mutation rate and the strength of among-group selection against the spectrum of cheaters that arise by mutation (Crow and Kimura, 1970; Van Dyken et al., 2011). Alternatively, if the combination of mutation rate to socially defective cheaters and intergroup cheater migration is sufficiently high relative to the rate of cheater loss via among-group selection, all social groups in a metapopulation could become infected by cheaters. Four of the 10 *Myxococcus* fruiting body groups examined here did not harbor variation in spore production or social swarming at frequencies above our detection limits, suggesting that these groups did not harbor cheaters at the time they were sampled. This result is consistent with (but not demonstrative of) the possibility that socially defective cheater frequencies in natural *Myxococcus* populations are largely determined by "kin selection–mutation balance" (Van Dyken et al., 2011), with kin selection in this scenario being mediated by selection among spatially structured kin groups.

Coevolution

High cotransmission of within-group diversity across generations aligns the evolutionary interests of clustered lineages (Sachs and Bull, 2005; Wade, 2007). Under low migration, diverse lineages that repeatedly and preferentially interact may coevolve to reduce conflict (Bouma and Lenski, 1988; Stewart et al., 2005; Weeks et al., 2007) and chimeric load (i.e., reduced group productivity caused by within-group diversity). Cheating load is one form of chimeric load. If coevolution within cheater-infected groups proceeds rapidly relative to the rate of cheater loss by among-group selection, immunity to socially defective cheaters and policing behaviors that suppress them may evolve, as has occurred in experimental populations (Fiegna et al., 2006; Zhang et al., 2009; Manhes and Velicer, 2011). Indeed, cheaters themselves may reevolve proficiency at cooperation by novel genetic routes (Manhes and Velicer, 2011) and thereby perhaps reach new cooperation fitness peaks (Wright, 1932) not accessible to noncheaters.

Clusters of socially cotransmitted lineages may also coevolve to reduce chimeric load generated by behavioral incongruities among interacting strains that are each socially proficient in clonal groups (Castillo et al., 2005; Fiegna and Velicer, 2005; Vos and Velicer, 2009). Cotransmitted lineages might even coevolve to perform distinct mutually beneficial functions that raise cooperative group productivity beyond that achievable by clonal groups. Unique trajectories of coevolution among clusters of cotransmitted lineages may promote diversification across kin groups.

Regeneration of Clonality

New clonal groups can be established by clonal emigration from an internally diverse group (Fig. 5.5) (Travisano and Velicer, 2004) or by local selective sweeps of adaptive mutants that purge variation from an existing kin group (Cohan, 2001). Alternatively, new kin discrimination alleles might arise by mutations that generate biological barriers to migration across clonal cell patches (Nadell et al., 2010) and thereby fragment a preexisting group into multiple kin discrimination units (Fig. 5.5). The rate at which clonal groups of cooperative genotypes are freshly established (or reestablished) is unknown but is an important parameter for understanding cheater–cooperator population dynamics.

What Maintains Diversity Within and Between Groups?

Natural variation in *M. xanthus* social traits documented here and previously (Krug et al., 2008; Vos and Velicer, 2008b; Kraemer et al., 2010; Morgan et al., 2010) may be nonadaptive and may have reached detectable frequencies by genetic drift or hitchhiking (Maynard Smith, 1991) or might reflect pleiotropic byproducts of evolutionary adaptation at some alternative trait. For example, even variation that causes large fitness differences during laboratory developmental competition experiments (Strassmann et al., 2000; Fiegna and Velicer, 2005; Vos and Velicer, 2009; Saxer et al., 2010) need not have been shaped by selection for within-group competitiveness during development. Indeed, the strongest developmental cheaters yet identified in *M. xanthus* originated in a selective regime in which evolving populations never underwent development (Velicer et al., 1998, 2000). Alternatively, selective sweeps may be driving some observed variants to fixation. Finally, *Myxococcus* social diversity may be maintained by various forms of balancing selection.

Within kin groups, frequency-dependent selection might maintain both cooperators and socially defective cheaters (Velicer et al., 2000) or multiple genotypes that mutually benefit one another owing to differential expression of cooperative traits (Manhes and Velicer, 2011). Within-group diversity might also be promoted by specialized performance among genotypes across variable environmental conditions [e.g., surface conditions (Shi and Zusman, 1993; Hillesland and Velicer, 2005; Vos and Velicer, 2008b), prey composition (Morgan et al., 2010), etc.]. Across kin groups, balancing selection might take the form of kin-group specialization to different microhabitats or nontransitive fitness relationships (Kerr et al., 2002) during competitive interactions, such as the production of anticompetitor compounds (Riley and Gordon, 1999) by adjacent kin groups.

The extensive diversity within natural *Myxococcus* social groups documented here suggests that within-group conflict is likely to play a major role in myxobacterial social evolution. Migration among kin groups seems to be low relative to the rate at which persisting variants arise by mutation and coevolution among socially cotransmitted *Myxococcus* lineages is likely to occur. The relative roles that the fundamental forces of evolution—mutation, distinct forms of selection, migration, genetic drift, and recombination—play in shaping natural social variation in the myxobacteria remain to be quantified. Doing so will require estimation of mutation rates, identification of loci and alleles responsible for observed social variation, screening for population genetic signatures of distinct evolutionary forces, and characterization of fitness relationships among social interactants under conditions relevant to natural habitats.

METHODS

Sample Collection and Strain Isolation

Soil samples were collected in a spatially nested design at three undisturbed woodland locations near Bloomington, Indiana [Old Meyers Road (GH) and Indiana University teaching and research preserves at Kent Farm (KF) and Moores Creek (MC)]. At each location, five sample sites were established at 10-m intervals along a line. At each of these meter-scale sample sites, five soil samples were collected at 2-cm intervals along the line. Samples were collected as described previously (Vos and Velicer, 2006) with a sterile 2-mL syringe from which the tip had been removed. Syringes were sealed with parafilm immediately after sampling to avoid cross contamination. After sampling, syringes were stored overnight at room temperature.

The day after sampling, ≈2 mm were removed from the ends of each soil core with a sterile scalpel, and the remaining core was crumbled onto selective agar medium [CTT medium with 1.5% agar (Hodgkin and Kaiser, 1977) containing the antibiotics and antifungals vancomycin (10 mg/L), nystatin (1,000 units/L), cyclohexamide (50 mg/L), and crystal violet (10 mg/L)]. Plates were incubated at 32 °C, 90% rH. After 2 weeks, plates were examined for the presence of fruiting bodies on soil particles. Ten spatially separated and individually discrete fruiting bodies were picked with a sterile toothpick from each plate. Each fruiting body was placed in a separate microcentrifuge tube containing 0.5 mL ddH_2O and heated at 50 °C for 120 minutes to kill nonspore cells. Samples were sonicated twice for 10 seconds to disperse spores and then transferred to CTT growth medium. Early samples (GH2.1.4, GH3.5.6, KF5.4.6, MC3.3.5) were transferred into CTT liquid and grown at 32 °C, 300 rpm, as were all liquid cultures described below. Cultures were grown until exponential phase (1–3 days) and then frozen with 20% glycerol at –80 °C (as were all frozen samples). Assuming a generation time of 4 hours [most likely a conservative underestimate (Velicer et al., 1998)], these cultures underwent no more than 18 generations of growth from the original group of fruiting body spores harvested directly from soil until frozen storage.

However, because of several instances in which contaminants that survived the heat and sonication treatments outgrew *Myxococcus* cells in liquid culture, subsequent samples (GH5.1.9, KF2.4.9, KF3.2.8, KF4.3.9, MC3.1.9, MC3.5.9) were transferred onto CTT hard (1.5%) agar after sonication and incubated at 32 °C, 90% rH (as were all agar-plate cultures described below). Plates were screened after 3 to 5 days for growth of *Myxococcus* cells, which grow into swarming colonies that are easily distinguished from contaminant colonies. When no contaminants were present, the entire *Myxococcus* population was harvested with a sterile scalpel and transferred into CTT liquid. If contaminant colonies were present, as much of the *Myxococcus* population as possible was harvested without touching contaminant colonies. Liquid cultures were incubated overnight and frozen. Cultures that underwent growth on agar plates likely underwent no more than 36 generations of growth from the original group of fruiting body spores harvested directly from soil until frozen storage.

Thawed samples (10 μL) from each of 10 frozen stocks derived from fruiting bodies isolated from the three sampling locations were diluted with CTT liquid into CTT soft (0.5%) agar (at 40 °C) at several dilution factors. Forty-eight spatially distinct colonies from each fruiting body culture were inoculated into separate flasks of CTT liquid, grown to high density, and frozen.

Fruiting body names reflect the sample location (GH, KF, or MC) and position (numbers). The first and second numbers in each name identify the meter- and centimeter-scale positions from which the respective soil sample was taken and the third number identifies the particular fruiting body taken from a given soil sample. Fruiting bodies from the GH and KF locations examined here were isolated from soil particles separated by meters in the sample plot, whereas the soil particles from which the MC fruiting bodies were isolated were from the same centimeter-scale plot because soil from other locations along the meter-scale transect did not yield fruiting bodies.

Swarming Motility Assays

Cells from all 48 clones representing each isolated fruiting body were inoculated from frozen stocks into 8 mL CTT liquid and incubated for 3 or 4 days. Cultures that reached exponential growth phase prior to others were diluted to avoid entry into stationary phase. The day prior to the swarming assay, cultures were diluted to 3 × 10^7 cells/mL. CTT soft (0.5%) agar plates were poured on the same day (25 mL in 9-cm-diameter petri dishes) and allowed to solidify uncovered for 15–20 min in a sterile laminar-flow hood before being covered and stored overnight at room temperature.

To initiate the swarming assays, 5 mL of each exponential-phase culture were centrifuged at 4,500*g* for 15 minutes and then resuspended with CTT liquid to 5 × 10^9 cells/mL. Ten microliters of each resuspended culture was then placed at the center of an agar plate and subsequently plates were incubated for 5 days. Swarm perimeters were marked after 1 and 5 days of incubation, and the distance swarmed between those time points for each replicate was measured as the average distance along four perpendicular vectors at a random orientation. More vectors were used for irregularly shaped swarms. All experiments (also those below) were performed in at least three temporarily independent replicate blocks.

Sporulation Assays

Five clones each were assayed for spore production from fruiting bodies that did not exhibit variation in swarming rate or other motility phenotypes. For fruiting bodies that did show variation, five clones of the majority swarming phenotype and minority-phenotype clones were assayed for spore production. For one fruiting body (KF4.3.9), all clones were included in the sporulation assay. Frozen samples were inoculated into CTT liquid, and resulting cultures were grown to visible turbidity but prevented from entering stationary phase by dilution if necessary. To initiate development, culture samples were centrifuged and resuspended to ≈5 × 10^9 cells/mL in TPM liquid (a buffered medium with no added carbon source) (Kroos et al., 1986). Ten microliters of each culture was spotted onto TPM hard (1.5%) agar plates and incubated. After 3 days, spores were harvested from the agar surface with a sterile scalpel, transferred into 1 mL ddH_2O and heated for 2 hours at 50 °C. After heat treatment spores were sonicated twice for 10 seconds and then diluted into CTT soft (0.5%) agar previously cooled to 40 °C. Plates were incubated 1 week, after which colonies were counted.

Control for Laboratory Origin of Minority Phenotypes

We tested the hypothesis that minority variants observed in our motility and sporulation assays might have originated by mutation during culture growth in the lab rather than in natural populations prior to soil collection. To do so, we tested for phenotypic variation within cultures derived from four randomly chosen clones isolated from fruiting body MC3.5.9, which exhibited a high degree of within-group variation in both motility and sporulation phenotypes. Cultures of the MC3.5.9 clones were subjected to growth in liquid medium (3–4 days), one cycle of development on TPM agar followed by heat and sonication treatments and subsequent growth again in CTT liquid prior to being diluted into CTT soft agar to allow isolation of clones for phenotypic analysis. Forty-eight clones from each initially clonal culture were isolated at random and examined for variation in swarming motility rate and phenotype as well as variation in fruiting body phenotypes after 10 μL of culture (5×10^9 cells/mL) were spotted onto CF hard (1.5%) agar (Hagen et al., 1978) and incubated for 5 days. Average swarming rates and photographs of fruiting bodies for all control clones are available upon request.

Statistics

All statistical analyses were performed with R software (R Development Core Team, 2009). Sporulation data were $\log_{10}$-transformed before analysis. Clones in populations harboring significant levels of variation were partitioned into phenotype clusters using *k*-means cluster algorithms. Specifically, optimal cluster values were selected by minimizing the within-cluster sum of squares (Everitt and Hothorn, 2009). If three clusters were present, nonindependent post hoc tests were performed with Bonferroni corrections.

DNA Sequencing and Phylogenetic Analysis

Five randomly selected clones representing the majority phenotype within each fruiting body were chosen for comparative DNA sequence analysis, as were (nonrandomly selected) clones that exhibited clearly distinct minority phenotypes in motility, sporulation, colony color, or degree of cell-cell adhesion. Approximately 500-bp fragments of six loci were sequenced for selected clones: *pilA* and five loci [loci *Mxan_0128, Mxan_0176, Mxan_0533, Mxan_0396,* and *Mxan_4405* (Goldman et al., 2006)] identified as being highly variable among the *M. xanthus* strains DK1622 (Kaiser, 1979; Goldman et al., 2006), A23 and A47 (Vos and Velicer, 2006) based on unpublished whole-genome sequence comparisons. Primer sequences and details of PCR and sequencing reactions are available upon request. All sequences were aligned with CodonCode Aligner Version 3.7.1 (CodonCode, Deadham, MA) and adjusted manually. Sequences are deposited at GenBank under the accession numbers JF819182–JF819591 and JF741968–JF742049.

Phylogenentic analysis was based on a 2,270-base concatemer of all loci sequences except *pilA,* which is more polymorphic than the other loci. Independent phylogenetic analyses were performed using maximum likelihood (ML) and Bayesian inference (BI) with Kimura-2 parameters of base substitution. We determined the ML phylogram using Mega Version 4.0 (Tamura et al., 2007) and assessed its support using 1,000 bootstrap replicates.

BI analyses were performed in BEAST version 1.6.1 (Drummond and Rambaut, 2007). Two MCMC runs with trees sampled every 1,000 generations were performed for 10 million generations and subsequently combined. Convergence was assured by visual inspection of parameter sample plots in Tracer version 1.4 (Drummond and Rambaut, 2007) and the first 10% of the analysis was discarded as burn-in. Bootstrap values >70% and posterior probabilities >95 were counted as high clade support.

ACKNOWLEDGMENTS

We thank H. Peitz for sharing genomic data; T. Platt for helpful suggestions regarding the control experiments; and E. Hanschen and M. Toups for helpful suggestions regarding statistical and phylogenetic analyses. This study was supported by National Institutes of Health Grant GM079690 (to G.J.V.). The sequences reported in this chapter are deposited in the GenBank database (Accession Nos. JF819182–JF819591 and JF741968–JF742049).

6

Evolution of Restraint in a Structured Rock–Paper–Scissors Community

JOSHUA R. NAHUM,* BRITTANY N. HARDING,* AND BENJAMIN KERR*†

It is not immediately clear how costly behavior that benefits others evolves by natural selection. By saving on inherent costs, individuals that do not contribute socially have a selective advantage over altruists if both types receive equal benefits. Restrained consumption of a common resource is a form of altruism. The cost of this kind of prudent behavior is that restrained individuals give up resources to less-restrained individuals. The benefit of restraint is that better resource management may prolong the persistence of the group. One way to dodge the problem of defection is for altruists to interact disproportionately with other altruists. With limited dispersal, restrained individuals persist because of interaction with like types, whereas it is the unrestrained individuals that must face the negative long-term consequences of their rapacity. Here, we study the evolution of restraint in a community of three competitors exhibiting a nontransitive (rock–paper–scissors) relationship. The nontransitivity ensures a form of negative feedback, whereby improvement in growth of one competitor has the counterintuitive consequence of lowering the density of that improved player. This negative feedback generates detrimental long-term consequences for unrestrained growth. Using both computer simulations and evolution experiments with a nontransitive community of *Escherichia coli,* we find that restrained growth can evolve under conditions of limited dispersal in which negative feedback is present. This research thus highlights a set of ecological conditions sufficient for the evolution of one form of altruism.

*Department of Biology and BEACON Center for the Study of Evolution in Action, University of Washington, Seattle, WA 98195. †To whom correspondence should be addressed. E-mail: kerrb@uw.edu.

Wisely and slow; they stumble that run fast.

William Shakespeare

The conflict between individual and group interests is a common element in many social dilemmas. Consider the rate at which an organism consumes shared resources. Prudent use of common resources promotes the longevity or fecundity of the group; however, any individual that exhibits restraint suffers in competition with those using resources rapidly. Rapacity is selectively favored and the displacement of prudent types by their unrestrained contemporaries occurs despite harmful consequences for the group (Maynard Smith, 1964; Williams, 1971). Restraint in the use of common resources is a form of altruism: behavior that is self-sacrificial and prosocial. Like other types of altruistic behavior, restraint faces a fundamental problem of subversion (Dawkins, 1976a; Okasha, 2008). How can restrained types persist in the midst of would-be cheaters—individuals that have a competitive edge because they are unrestrained? In this chapter, we address this question directly by outlining ecological conditions sufficient to favor the evolution of restraint.

One ingredient found in most explanations for the evolution of altruism, and thus relevant to the evolution of restraint, is positive assortment. Altruism stands a better chance when altruistic individuals disproportionately help those possessing the genes for altruism (Hamilton, 1975; Queller, 1992b; Pepper and Smuts, 2002; Fletcher and Doebeli, 2009; Godfrey-Smith and Kerr, 2009). One of the most obvious ways to achieve positive assortment is through interactions between genetic relatives (Hamilton, 1964a). In such a case, altruistic individuals disproportionately experience beneficial social environments (engineered by their kin), whereas selfish individuals tend to face a milieu lacking prosocial behavior (because their kin tend to be less altruistic). Interaction with kin can occur actively through the choice of relatives as social contacts or passively through the interaction with neighbors in a habitat with limited dispersal. There is now a large body of literature on the effect of active and passive assortment on the evolution of altruism (Matessi and Jayakar, 1976; Eshel and Cavalli-Sforza, 1982; Toro and Silio, 1986; Wilson et al., 1992; Queller, 1994; Pepper and Smuts, 2002; Lehmann and Keller, 2006; Pepper, 2007; Rankin and Taborsky, 2009). At a fundamental level, this research focuses on the distribution of interactions among altruistic and selfish individuals. However, in many systems, these individuals are also interacting with other members of their community (competing species, predators, prey, mutualists, etc.). It is less common to consider the role of broader ecological interactions on the evolution of various forms of altruism.

Here, we consider the evolution of restraint in communities where ecological interactions generate a type of negative feedback. One of the simplest communities with this property involves three members engaged in nontransitive competition. A simple scenario entails one player incurring a significant cost to harm a second player (e.g., through parasitism or allelopathy) and a third player possessing mildly costly resistance to the harm. Reminiscent of the children's game rock–paper–scissors, the harmer outcompetes the sensitive player, who outcompetes the resistant player; in turn, the resistant player outcompetes the harmer. Such nontransitivity has been reported in plant systems (Lankau and Strauss, 2007; D. D. Cameron et al., 2009) and as we see below, bacterial systems. More generally, in rock–paper–scissors games, each strategy beats one of the other two and is beaten by the third (e.g., paper covers rock but is cut by scissors). Imagine a nontransitive community in which, for convenience, we call the players Rock, Paper, and Scissors. Each type has a rate at which it displaces its victim (e.g., Rocks crush Scissors at some rate). Next, imagine a less-restrained variant of Rock, called Rock*, that displaces Scissors at a faster rate. In a Rock*–Paper–Scissors community, the abundance of Scissors decreases because of the increased prowess of Rock*. As a consequence, Scissors' victim (Paper) is liberated, which can displace Rock*. In an ironic twist, the improved Rock* decreases in abundance because of the expansion of its victim's victim. This form of negative feedback ensures that a higher displacement rate results in decreased abundance (Tainaka, 1993, 1995; Frean and Abraham, 2001; Marsland and Frank, 2001). Thus, more restrained players may be less prone to extinction, a phenomenon termed "survival of the weakest" (Frean and Abraham, 2001). A complication arises when considering a community with multiple variants present simultaneously (e.g., Rock and Rock* with Paper and Scissors). The same traits that allow Rock* to displace Scissors faster may render Rock* a better competitor against Rock. In this case, restraint has a selective disadvantage, despite its positive effects on abundance. How then can restraint evolve in a nontransitive community?

Spatial structure can play a critical role promoting restraint in nontransitive systems. Returning to our Rock–Paper–Scissors community, limitation of dispersal results in a patchwork of the three players. A patch of any one player chases its victim and is chased by its enemy (Durrett and Levin, 1997; Kerr et al., 2002). Within any patch, an unrestrained variant (Rock*) will replace its restrained counterpart (Rock). However, patches of unrestrained variants are more likely to go extinct. This difference in patch viability favors restraint. Limited dispersal ensures a type of positive assortment where restrained and unrestrained individuals tend to be surrounded by like types. This means that the long-term negative consequences of faster displacement are visited disproportionately on the less-restrained type. Consequently, restraint can be maintained evolutionarily in a structured nontransitive community. This outcome has been shown theoretically in nontransitive triplets and larger communities (Johnson and Seinen, 2002; Prado and Kerr, 2008), but there is little empirical work on this topic. This is despite the fact that nontransitive dynamics have been described in natural communities ranging from microbes to animals to plants (Jackson and Buss, 1975; Paquin and Adams, 1983; Sebens, 1986; Taylor and Aarssen, 1990; Sinervo and Lively, 1996; Clark et al., 2000; Birkhead et al., 2004; Lankau and Strauss, 2007; Sinervo et al., 2007; D. D. Cameron et al., 2009).

One well-studied nontransitive system involves strains of *Escherichia coli* that produce antimicrobial proteins termed colicins (James et al., 1996; Cascales et al., 2007). Colicin-producing cells possess a plasmid housing the colicin gene as well as a gene coding for a colicin-specific immunity protein. Cells that lack the plasmid, and thus lack immunity, are sensitive to the colicin. However, sensitive cells can experience mutations yielding resistance to colicins. Resistance is caused by alteration or loss of membrane proteins that bind or translocate the colicin. Because these same membrane components are involved in nutrient acquisition, resistance is often costly in the absence of colicins (measured by a reduced growth rate relative to sensitive cells) (Feldgarden and Riley, 1998, 1999). However, in some cases, the producer incurs even greater costs to carry the colicin plasmid and express immunity constitutively. Thus, these three players constitute a nontransitive community: the sensitive strain outgrows the resistant strain, the resistant strain outgrows the producer, and the producer kills the sensitive strain. Previous work with the three members of the colicin E2 system has shown nontransitivity both in vitro (Kerr et al., 2002) and in vivo (Kirkup and Riley, 2004). Nevertheless, there have been no experimental studies of the evolution of restraint in this system.

In this chapter, we describe experiments with bacteria that explore how positive assortment and negative ecological feedback influence the evolution of restraint. Of the three players (sensitive, resistant, and producer), we focus on the resistant strain. The mutations that define the resistant strain are costly, and there is evidence from numerous systems that secondary mutations can compensate for the initial costs of antimicrobial resistance (Schrag et al., 1997; Andersson and Levin, 1999; Reynolds, 2000; Nagaev et al., 2001; Andersson, 2006). Thus, we predict that this strain is the most likely to increase its growth rate, making it the most attractive candidate to study factors that would hinder such increase. We place the community in a metapopulation, structured into many subpopulations. We manipulate the pattern of migration within the metapopulation, which affects the degree of positive assortment. Migrations are either restricted to occur between neighboring subpopulations (Restricted treatment) or could occur between any subpopulations (Unrestricted treatment). The evolution of the resistant strain can be compared across migration treatments to gauge the effect of population structure on the evolution of restraint. To identify the role of negative feedback, the evolution of the resistant strain in the full community is compared with the evolution of the resistant strain evolving alone (Community and Alone treatments, respectively). By monitoring the resistant strain in three different types of metapopulations (Restricted Community, Unrestricted Community, and Restricted Alone), we assess the impact of both positive assortment and negative ecological feedback on the evolution of restraint.

RESULTS

Presence of Nontransitivity

As detailed in *Methods,* we constructed a strain that produced two colicins (Producer), a strain sensitive to both colicins (Sensitive), and a strain resistant to both colicins (Resistant). The double-colicin producer was used to decrease the likelihood of *de novo* resistance arising from the sensitive population during the evolution experiment. These three constructed strains are henceforth referred to as the ancestors. To confirm the nontransitive relationship, we performed pairwise competitions among the ancestral strains. Each competition was initiated with a ratio matching the proportions of two competitors when they first meet through migration within the metapopulation. The resistant ancestor was outcompeted by the sensitive ancestor (one-sample t-test; $t_5 = -5.78$, $P = 0.0022$). The producer ancestor was outgrown by the resistant ancestor (one-sample t-test; $t_5 = -3.62$, $P = 0.015$). The sensitive ancestor was always driven to extinction when mixed with the producer (giving a relative fitness of zero in all five replicates). Because each player was competitively inferior to the second player (but superior to the third player), these three strains form a nontransitive system (Fig. 6.1).

Ecological Dynamics

We propagated our bacteria as metapopulations using 96-well microtiter plates, where each well constituted a distinct subpopulation. We initialized the metapopulations with the nontransitive community (Community treatment) or the resistant strain alone (Alone treatment). Every 12

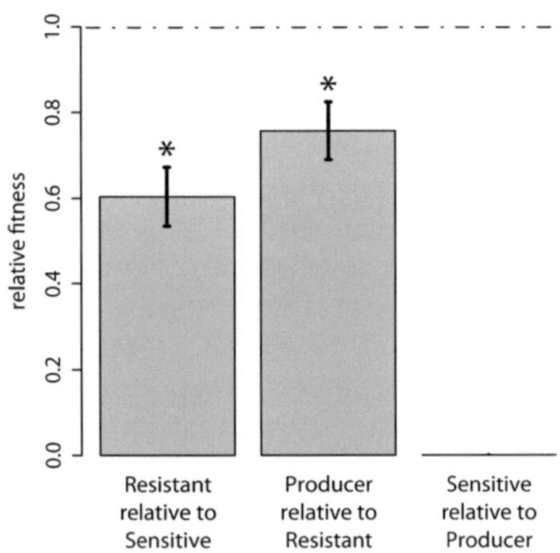

FIGURE 6.1 Pairwise competitions between the ancestral bacterial strains show nontransitivity. Asterisks signify that relative fitness is significantly less than one, and the error bars show the SEM. The resistant ancestor is dominated by the sensitive ancestor, and the ancestral producer is outgrown by the resistant ancestor. The sensitive strain is killed by the producer in all replicates, yielding a uniform relative fitness of zero. As each strain outcompetes one other strain but is outcompeted by the third strain, a nontransitive relation holds.

hours, each subpopulation was diluted into fresh growth medium, and migrations between subpopulations occurred. Within each metapopulation, migrations occurred between neighboring wells (Restricted treatment) or among any wells (Unrestricted treatment). We measured the abundances of all strains every six transfers. All three players were maintained in the Restricted Community and Unrestricted Community treatments for the duration of the experiment (Fig. 6.2*A* and *B*). The resistant strain persisted at a constant level in the Restricted Alone treatment for the length of the experiment (Fig. 6.2*C*).

Evolution of the Resistant Strain

We randomly sampled eight resistant isolates from the last transfer of the experiment. Each of these isolates was competed against a marked variant of the common resistant ancestor. To avoid pseudoreplication, we averaged relative fitness across isolates within each of five replicates of each treatment. We found that isolates from the Restricted Community treatment had the lowest competitive ability [single-factor ANOVA; $F_{2,12} = 9.36$, $P = 0.0036$, multiple comparisons by Tukey's Honestly Significant Difference (HSD)]. This is consistent with the evolution of a restrained

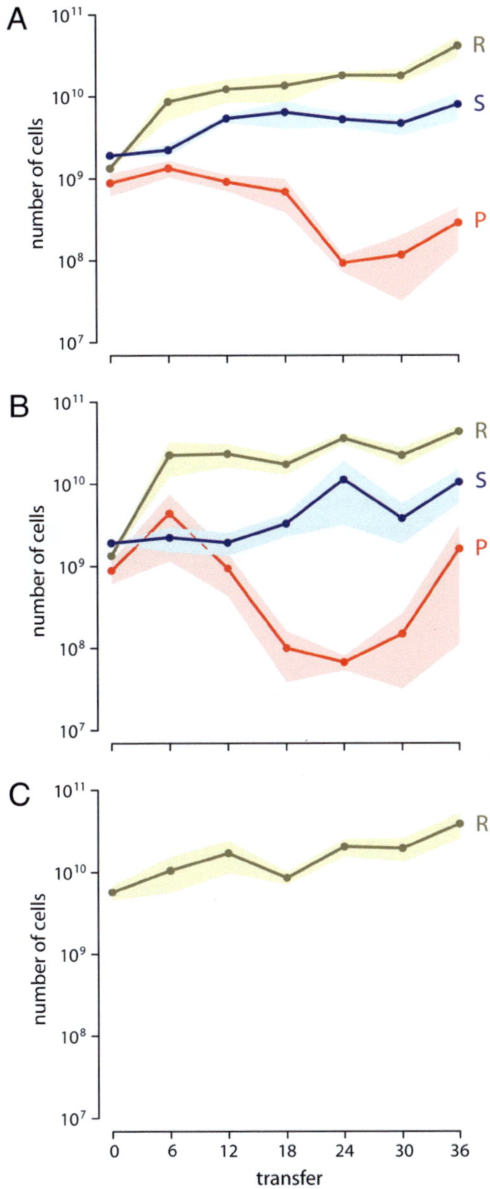

FIGURE 6.2 Bacterial abundance in (*A*) the Restricted Community treatment, (*B*) the Unrestricted Community treatment, and (C) the Restricted Alone treatment. Points represent mean abundance of the sensitive strain (S), resistant strain (R), and producer strain (P). Shading gives the SEM. All three players coexisted in the Community treatments for the duration of the experiment, and the density of the resistant strain was comparable across all three treatments.

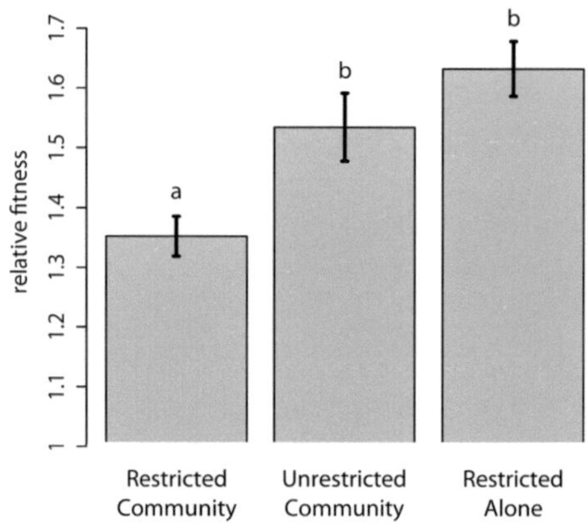

FIGURE 6.3 Fitness of evolved resistant isolates relative to their common ancestor. Mean relative fitness of each treatment is shown, and error bars give the SEM. The fitness of isolates from the Restricted Community treatment was significantly lower than the fitness of isolates from the other treatments. Letters distinguish treatments significantly different using post hoc comparisons. This pattern is consistent with the evolution of restrained growth in the Restricted Community treatment.

growth rate. Resistant cells in a full community evolved a significantly higher competitive ability under unrestricted migration than under restricted migration (Unrestricted Community vs. Restricted Community in Fig. 6.3).Resistant cells propagated alone evolved a significantly higher competitive ability than resistant cells in a nontransitive community (Restricted Alone vs. Restricted Community in Fig. 6.3). Thus, both population structure and the presence of the full community were important to the evolution of competitive restraint.

Simulation of Ecoevolutionary Dynamics

To better understand the evolutionary behavior of our system, we modeled the bacterial metapopulations using a lattice-based simulation (details in *Methods, SI Methods,* Table S1, and Figs. S1 and S2[†]). Each metapopulation was initialized with the three ancestral strains in a spatially clumped pattern. The basic algorithm consisted of a

---

[†]SI Methods, Table S1, and Figs. S1–S6 are available online as supporting information for the original PNAS article [108(Suppl 2):10831–10838] at www.pnas.org/lookup/suppl/doi:10.1073/pnas.110296108/-/DCSupplemental.

cycle of three stages: (*i*) growth/competition within wells, (*ii*) dilution of wells, and (*iii*) migration among wells. Thus, a simulated cycle corresponds to a transfer within our experiment. Every cycle, mutations to growth rate were permitted in resistant subpopulations. We simulated evolution within metapopulations in each of the three treatments described above (Restricted Community, Unrestricted Community, and Restricted Alone).

Although diversity was maintained in the Restricted Community treatment, the community tended to lose players in the Unrestricted Community treatment in the long run (e.g., after 100 transfers). Consequently, the Unrestricted Community treatment was excluded from analysis. The loss of diversity was robust to changes in several different parameters of the model and suggests that the Unrestricted Community treatment in the laboratory may have lost strains if it had been run for more transfers. This result is also consistent with previous work on the importance of limited dispersal to coexistence in this system (Durrett and Levin, 1997; Kerr et al., 2002). After evolving the metapopulations in each treatment, we determined the mean relative fitness of the resistant population. Consistent with our empirical results, we found the average growth rate of resistant strains from the Restricted Community treatment to be significantly lower than the average growth rate from the Restricted Alone treatment (Fig. S3).

To confirm the importance of positive assortment in the evolution of restraint, we ran an additional treatment: Restricted Community with Permutation. This treatment was identical to the Restricted Community treatment except that, at the beginning of each cycle, wells containing only resistant cells (ancestors or mutants) were randomly permuted. This operation allowed for mixing between the patches of resistant wells (capturing an element from the Unrestricted treatment). The average growth rate of resistant strains from the Restricted Community treatment was significantly lower than the average growth rate from the Restricted Community with Permutation treatment (Fig. 6.4).

The rate of displacement by fitter variants within any population will be slowed by population subdivision. We were curious if the lower growth rate of our Restricted Community treatment could be explained entirely by the fact that the evolving resistant population was divided into semi-isolated patches. To explore this possibility, we ran an additional simulation treatment: Restricted Alone with Shadowing. In this treatment, a Restricted Alone metapopulation evolved alongside a standard Restricted Community metapopulation, with the caveat that the Restricted Alone metapopulation's migrations and spatial distribution were forced to match the resistant portion of its paired Restricted Community metapopulation. In this way, the Restricted Alone shad-

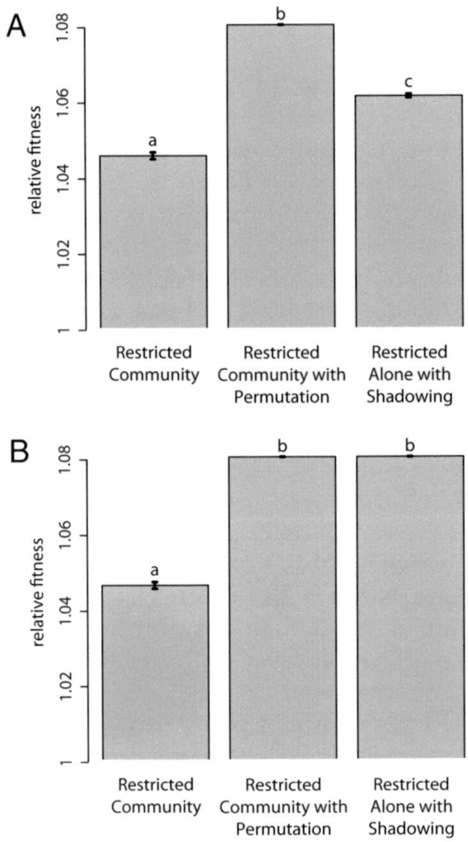

FIGURE 6.4 Mean resistant fitness relative to the resistant ancestor after simulated evolution in multiple treatments. Fitness values after (*A*) 100 and (*B*) 400 cycles are shown. Mean relative fitness of each treatment is shown, and error bars give the SEM. Letters distinguish significantly different treatments by post hoc comparisons. The fitness of resistant populations from the Restricted Community treatment was significantly lower than that of the other treatments at both time points. This pattern is consistent with the evolution of restrained growth in the Restricted Community treatment.

owed the Restricted Community. This meant that the Restricted Alone metapopulation was divided into patches. However, because mutation occurred independently in the Restricted Alone shadow and its Restricted Community master, mutations within a given patch in the shadow world had no effect on the survival of the patch in that world. We found that division into semi-isolated patches accounted for some but not all of the effect of lowering growth rate in the short term (Fig. 6.4*A*) (single-factor ANOVA; $F_{2,331} = 829.6$, $P < 0.001$, multiple comparisons by Tukey's HSD). However, simulations that ran for longer (Fig. 6.4*B*) show that the Shadowing treatment converges to the Permutation treatment (single-factor ANOVA; $F_{2,331} = 1{,}421$, $P < 0.001$, multiple comparisons by Tukey's HSD). We find the same patterns when we run simulations that exactly match the metapopulation size and number of transfers used in our experiment (Figs. S4 and S5). Thus, apparently, the connection between the presence of fast-growing variants within a patch and a greater probability of patch extinction was an important ingredient in explaining the evolution of restraint in the Restricted Community treatment.

DISCUSSION

For the resistant isolates considered here, the evolution of the lowest competitive ability occurred in the treatment in which migration was restricted and all three members of the nontransitive community were present (Fig. 6.3). If either migration was unrestricted or the resistant strain evolved alone, final competitive ability was significantly higher. The low competitive ability in the Restricted Community treatment presumably reflects a relatively low growth rate. There are a few possible explanations for this outcome. One explanation is that if the number of resistant cell divisions in the Restricted Community treatment was less than the number of divisions in the other treatments, isolates from the Restricted Community treatment might not have had enough opportunity to evolve a higher growth rate. However, we find no significant difference among the treatments in the total number of resistant cell divisions (*SI Methods* and Fig. S6). A second explanation is that restricted migration slows the spread of any advantageous mutant (Bolker et al., 2003). In this case, resistant mutants with a higher growth rate reach a lower frequency in the Restricted Community treatment than in the Unrestricted Community treatment by the end of the experiment. However, the resistant isolates with the highest growth rate came from Restricted Alone treatment; thus, a restriction to migration does not uniformly hinder the advent of fast-growing resistant mutants. A third explanation is that the presence of producers constrains the manner in which a resistant strain can compensate for the cost of resistance (e.g., reversion to sensitivity is not an option). This would limit the set of evolutionary options for resistant cells in the Restricted Community treatment relative to the Restricted Alone treatment. However, the growth rate of isolates from the treatment with the highest level of interaction between resistant cells and producers (Unrestricted Community) was similar to

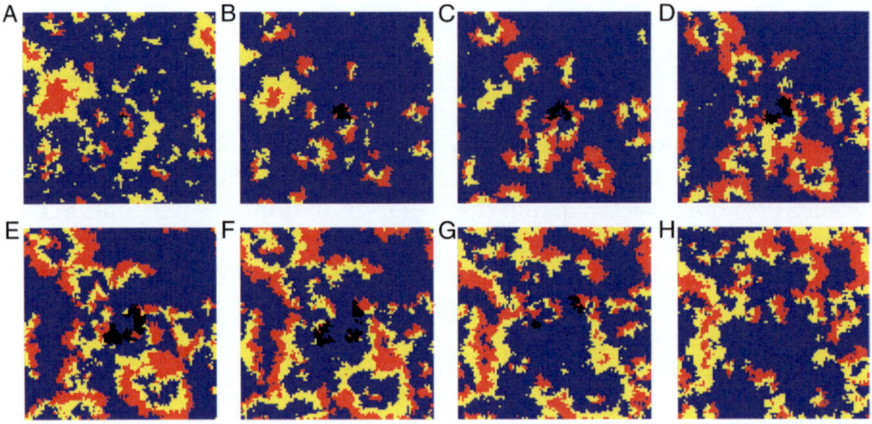

FIGURE 6.5 Snapshots of a metapopulation from an illustrative Restricted Community simulation recorded every 20 cycles (*A–H*). The metapopulation was initialized with the three bacterial strains sensitive (dark gray), resistant (white), and producer (light gray) in addition to a small patch of a mutant resistant strain (black) with an increased growth rate. The mutant initially outcompetes nearby ancestor patches (*A–E*) but is extinguished after outcompeting neighboring patches of the producer (*F–H*).

that of the treatment without producers (Restricted Alone). Additionally, not a single resistant isolate from any treatment reverted to sensitivity; thus, reversion did not explain competitive differences. Finally, the Restricted Community treatment's resistant population was divided into discontinuous regions by barriers consisting of the other strains (illustrated in Fig. 6.5), and such barriers would inhibit the spread of advantageous mutants. Our simulation-based treatment, Restricted Alone with Shadowing, where the resistant type was restricted to the patchy spatial distribution of Restricted Community evolved a lower growth rate, indicating that population subdivision may contribute to the low growth rate in the Restricted Community. Nonetheless, subdivision does not fully account for the restraint found in the Restricted Community treatment (Fig. 6.4 and Fig. S5). Thus, we do not find complete support for any of these explanations and, instead, favor the following alternative.

In the Restricted Community treatment, the nontransitivity of the full community provides a form of negative feedback, and the restricted migration ensures a form of positive assortment. We suggest that it is these two factors, negative feedback and positive assortment, that set the stage for the evolution of restraint. In the Restricted Community treatment, we have a set of patches chasing each other (Fig. 6.5). A faster-growing resistant mutant has a competitive advantage within a resistant patch, but a fast-growing resistant patch is more likely to burn through its victim (the producer) and consequently, face its enemy (the sensitive strain). This sequence of events is shown in Fig. 6.5 for a Restricted Community simulation in which wells with a faster-growing resistant mutant are labeled in black. Limited migration ensures that it is the unrestrained mutants that reap the negative long-term consequences (patch extinction) of their myopic strategy. When assortment is eradicated by shuffling the contents of multiple patches (as in the simulation-based treatment Restricted Community with Permutation), restraint is not maintained (Fig. 6.4). Without the negative feedback of the full community (e.g., in the Restricted Alone treatment) or the positive assortment resulting from limited migration (e.g., in the Unrestricted Community treatment), the evolution of restraint is not expected.

We have explored a model system under laboratory conditions, but our findings carry potential implications for other systems. In general, allelopathy permits nontransitivity, and allelopathic bacteriocins are widely distributed across bacterial taxa (Riley and Wertz, 2002). Nontransitive relationships have also been described in other ecological contexts. For instance, nontransitivity in male mating systems has been reported in common side-blotched lizards (Sinervo and Lively, 1996) and viviparous lizards (Sinervo et al., 2007), wherein the males exist in three color morphs: an aggressive morph can displace a less aggressive morph, which displaces a nonaggressive morph. The nonaggressive male is a female mimic, which disproportionately mates with females on the most aggressive male's territory. It has been argued that similar nontransitive mating systems are likely present in other animals, including some reptiles, fish, birds, and insects (Sinervo and Calsbeek, 2006; Sinervo et al., 2007), and nontransitive sperm competition has been reported in fruit flies (Clark et al., 2000) and domestic fowl (Birkhead et al., 2004). Another situation resulting in nontransitivity involves types differing in their colonization and competitive abilities. An overgrower (the best competitor) can displace a fugitive (the best colonizer), which displaces a preemptor (an intermediate colonizer that is resistant to overgrowth); then, the preemptor can displace the original overgrower (Edwards and Schreiber, 2010). This type of system was described for a rocky subtidal community (Sebens, 1986; Edwards and Schreiber, 2010), and nontransitivity in overgrowth patterns has also been reported in coral reef communities (Buss and Jackson, 1979). Another instance of nontransitivity involves a victim–exploiter relationship. This situation was reported in a grassland community in which grasses outcompete forb species but are disproportionately parasitized by a root hemiparasitic plant (D. D. Cameron et al., 2009). More broadly, many studies have proposed that nontransitive relations may be more prevalent than currently appreciated in systems with frequency-dependent selection or ecological tradeoffs (Gilpin, 1975; Sinervo and Calsbeek, 2006; Allesina and Levine, 2011).

Although the prevalence of nontransitivities in natural ecosystems remains to be determined (Verhoef and Morin, 2010), the ubiquity of spatial structure is widely recognized. Indeed, spatial structure is a component of many of the nontransitive systems described above. Structure may be most pronounced in sessile organisms (e.g., plants, some marine invertebrates, and microbes in biofilms); however, even populations of motile organisms can possess some degree of structure because of spatial limitations to dispersal and interaction. The spatial scale of ecological processes has been shown to be an important factor in the invasion of rare types (Chao and Levin, 1981; van Baalen and Rand, 1998), coexistence of multiple types (Kneitel and Chase, 2004), stability of communities (Morrison and Barbosa, 1987), and evolutionary trajectories of community members (Thrall and Burdon, 2002). We have shown that limited migration in a nontransitive community can promote the evolution of restraint. However, spatial structure can be important for the evolution of restraint in other types of communities as well.

As an example, limited dispersal can promote restraint within victim–exploiter communities (Boots and Mealor, 2007; Kerr et al., 2006). An inherent form of negative feedback exists when one species (e.g., predator, parasite, or herbivore) exploits another for critical resources (e.g., prey, host, or plant). To see this, consider a simple version of the Lotka-Volterra model, where the dynamics of exploiters (at density *E)* and victims (at density *V)* are described by (Eqs. 1)

$$\begin{aligned} \frac{dV}{dt} &= \beta V - \lambda VE, \\ \frac{dE}{dt} &= \lambda VE - \delta E, \end{aligned} \tag{1}$$

where β is the birth rate of victims, λ measures the attack rate of the exploiter, and δ is the death rate of the exploiter (we assume a conversion efficiency of unity). The nontrivial equilibrium for this community is $(\hat{V},\hat{E}) = (\delta/\lambda, \beta/\lambda)$. As the exploiter reduces its attack rate, its equilibrium abundance increases (as λ drops, $\hat{E} = \beta/\lambda$ grows). Nonetheless, an exploiter with a higher attack rate will displace a second exploiter exercising restraint (Korobeinikov and Wake, 1999). Selection for rapacious exploitation that results in community collapse constitutes an example of the tragedy of the commons (Hardin, 1968). Limited dispersal ensures that any tragedy of the commons that results from overexploitation primarily befalls the unrestrained exploiters. Several theoretical studies have explored the role of spatial structure in promoting restraint in victim–exploiter interactions (Mitteldorf et al., 2002; Killingback et al., 2006). There have also been experimental demonstrations that limited dispersal favors restraint in host–parasite communities in the form of reduced parasite virulence and/or infectivity (Kerr et al., 2006; Boots and Mealor, 2007; Eshelman et al., 2010).

A second example involves the role of structure in promoting restraint in hypercycle communities. A hypercycle is a series of self-replicative molecules cyclically linked, where each molecule catalyzes the replication of the next molecule in the cycle. Unstructured hypercycles are plagued by parasitic molecules, which receive greater catalytic activity from the previous molecule in the cycle while withholding catalytic support for the next molecule in the cycle. Boerlijst and Hogeweg (1991) showed theoretically that hypercycles in an incompletely mixed medium could keep parasitic molecules at bay. In a structured habitat, the hypercycle community organizes into a collage of rotating spirals. A parasitic molecule originating at the center of a spiral can lead to spiral demise and replacement by other spirals. Thus, short-term payoffs to the parasite (displacement within a spiral) can generate negative long-term consequences (spiral extinction) in a structured world. This favors the evolution of restrained molecules that avoid the immediate gains of parasitism.

Spatial structure and ecological feedback can also favor mutualistic behavior between species (Frank, 1994). Recently, Harcombe (2010) studied a case of bacterial cross-feeding. In lactose medium, *Salmonella enterica* consumes the acetate waste products of a mutant strain of *E. coli*. The *E. coli* mutant was a methionine auxotroph and could grow if *S. enterica* excreted methionine. Harcombe (2010) showed that, although methionine excretion was intrinsically costly, a mutant of *S. enterica* that exported an excess of methionine was able to displace WT *S. enterica* (which did not excrete methionine) when these types were grown on lactose plates with *E. coli*. The cooperative excretion by *S. enterica* was favored through a combination of ecological feedback (acetate was produced when *E. coli* obtained methionine) and spatial structure (ensuring that excreting cells had disproportionate access to acetate). When Harcombe (2010) destroyed either feedback (by growing the community on acetate plates so that *S. enterica* did not rely on *E. coli*) or structure (by growing the community in lactose flasks), the excreting *S. enterica* mutant was outcompeted by WT. This work shows that ecological feedback and positive assortment can be important ingredients in other forms of cooperation.

In all of the communities described above, a form of altruism exists. The elements that we have underlined as important to the evolution of restraint connect readily to prominent theoretical frameworks used to understand the evolution of altruism. In our nontransitive system, limited dispersal results in a preponderance of interaction between relatives. Kin selection arguments often focus on the coefficient of relatedness between interacting individuals (Eberhard, 1975; Griffin and West, 2002). In our system, limited dispersal results in higher coefficients of relatedness than in conditions of unlimited dispersal, a form of positive assortment (Pepper, 2000). The multilevel selection framework describes altruism as a behavior opposed by within-group selection but favored by between-group selection (Sober and Wilson, 1999; Wilson and Wilson, 2007). In the patchwork of a structured community, a restrained variant is at a local disadvantage (e.g., within its patch), but patches of restrained types may persist longer because of the negative feedback from rapid growth. We propose that multiple frameworks have relevance for understanding restraint in our system, because each framework focuses on (different) important elements underlying the evolution of altruism (Kerr, 2009).

Overall, we observe that a form of altruism can evolve in microbial metacommunities. With limited migration, similar types associate into patches that chase one another. The negative feedback resulting from the nontransitivity in our system means that patches filled with unrestrained variants are more prone to extinction. Thus, we see that altruistic restraint is favored precisely when those that run fast tend to stumble.

METHODS

Community Players

The bacterial community consisted of three players: a toxin-producing strain (P), a toxin-sensitive strain (S), and a toxin-resistant strain (R). P expressed two toxins (colicin E2 and colicin D). This strain was constructed by transforming the Col E2 and Col D plasmids sequentially into BK10 (*E. coli* K-12) cells followed by selecting for resistance to phage T5. S was constructed by transforming the pACYC184 plasmid encoding tetracycline (Tet) resistance into BK10 cells. R was constructed by a series of sequential selections on BK10—resistance to colicin E2, colicin D, and phage T6. Before marker additions (T5, Tet, and T6 resistance), these strains exhibited a rock–paper–scissors relationship. However, the growth inhibition of P and R by a low concentration of Tet (and the cost of T6 resistance in R) magnified the nontransitivity in our growth medium (LB + 0.25 μg/mL Tet).

Experimental Treatments

The evolution experiments involved propagating metapopulations of bacteria with two factors manipulated. The first experimental factor was the identity of the players in the metapopulation. Either the full community (S-R-P) was used or the resistant strain (R) was propagated alone (the Community or Alone treatments, respectively). In the Community treatments, each metapopulation consisted of two microtiter plates (192 wells with 200 μL growth medium each). In the Alone treatments, each metapopulation consisted of a single microtiter plate (96 wells with 200 μL growth medium each). The difference in the number of wells reflected our attempt to balance the total number of resistant cells across treatments (Fig. 6.2). The second factor manipulated was the pattern of migration within the metapopulations. Migration was either restricted to occur between wells directly bordering each other along cardinal directions or unrestricted (the Restricted or Unrestricted treatments, respectively). In both treatments, each well had a one-third probability of experiencing an immigration event from one random well in its neighborhood. In the Restricted treatment, this neighborhood included the wells directly north, east, south, or west of the focal well (using periodic boundaries to eliminate edge effects). In the Unrestricted treatment, the neighborhood included all wells minus the focal well. Migration events directly followed dilution of the entire metapopulation in fresh growth medium. Every 12 hours, 40-fold dilution was accomplished using a 96-slot pin multiblot replicator (5 μL in 200 μL*)*. Immediately after dilution, a BioRobot 8000 liquid-handling robot (Qiagen) executed the migrations, where each migration involved transferring 5 μL from the source well within the exhausted plate into the destination well within the fresh plate. Between transfers, plates were incubated (37 °C) and shaken (350 rpm using a microtiter shaker, Bellco Glass). For the Alone treatment, the metapopulation was initiated with the resistant strain in each well. For the Community treatment, the initial spatial arrangement of strains was obtained from the 100th transfer of a 192-point lattice-based simulation with a restricted neighborhood (*SI Methods*). Each metapopulation was propagated for a total of 36 transfers. The abundance of each strain was gauged every six transfers by selective plating (using Tet, T5, and T6). There were five replicates of each of three treatments: (*i*) Restricted Community, (*ii*) Unrestricted Community, and (*iii*) Restricted Alone.

Competition Assay

We picked eight random resistant isolates from the last transfer of each metapopulation (we denote any one of these strains as R_E). We marked our ancestral resistant strain (denoted R_A) with resistance to phage T5. Before the competition, R_E and R_A are grown separately in 200 μL growth medium for two 12-hour cycles (with 40-fold dilution at transfer). After this acclimation phase, we added 5 μL R_E and 5 μL R_A to a well containing 200 mL growth medium. The titer of each strain was assessed (by plating with and without phage T5) immediately after the competition was initiated and again after 12 hours. If $R_i(t)$ is the titer of strain R_i at time *t*, then the fitness of the evolved strain relative to its ancestor is given by (Eq. 2):

$$w\left(R_E, R_A\right) = \frac{\ln\left(R_E(12)/R_E(0)\right)}{\ln\left(R_A(12)/R_A(0)\right)}. \tag{2}$$

The same competitive assay was used to establish the nontransitive dynamic between the three ancestral players (simply with different selective plating schemes).

Simulation

We model the metapopulation as an $L \times W$ regular square lattice with periodic boundaries subjected to a cycle of three phases: (*i*) growth, (*ii*) dilution, and (*iii*) migration. Each lattice point *i* at time *t* is described by the vector (Eq. 3)

$$x_i(t) = \left\langle s_i(t), p_i(t), r_i^0(t), r_i^1(t), \ldots, r_i^K(t) \right\rangle, \tag{3}$$

where $s_i(t)$, $r_i^0(t)$, and $p_i(t)$ are the abundances of sensitive, resistant, and producer ancestors, respectively. The variables $r_i^1(t), r_i^2(t), \ldots r_i^k(t)$ are the abundances of each of *K* types of mutant resistant strains. These abundances are expressed in units of the limiting nutrient concentration (*SI Methods*).

During the growth phase, the dynamics of each strain (y) of each lattice point (i) are described by the following differential equation (*SI Methods*) (Eq. 4):

$$\dot{y}_i = y_i \frac{\mu_Y n_i}{\kappa_Y + n_i}, \tag{4}$$

where $n_i = 1 - \Sigma y_i$, μ_Y is the maximum growth rate, and κ_Y is the Monod constant (nutrient concentration yielding one-half maximum growth rate) of player *Y*. Each growth phase lasts *T* time units. Dilution at time *t* is given by (Eq. 5):

$$x_i(t') = \phi x_i(t), \tag{5}$$

where ϕ is the dilution factor and t' marks the postdilution state.

Migration happens with α uniform probability α. If a migration event occurs, a point within the focal point's neighborhood is chosen at random. For the Restricted treatment, the neighborhood is the four nearest lattice points (von Neumann neighborhood). For the Unrestricted treatment, the neighborhood is the entire lattice minus the focal point. In the case of migration, let the chosen neighbor of the focal point *i* be designated *j*. The state after migration (signified by t'') is given by (Eq. 6):

$$x_i(t'') = (1-\phi)x_i(t') + \phi x_i(t). \tag{6}$$

Removal occurs next. At point *i*, any player whose abundance is less than or equal to a critical value (a_{crit}) is removed. Also, the sensitive player is removed if the producer is present. In the simulation, the dilution, migration, and removal are assumed to be instantaneous and followed by a new growth phase. Lastly, mutation can occur with probability π. In the case of a mutational event, a fraction γ of the total abundance of the resistant players (ancestral and mutant) of a point is converted to a random resistant type.

We initialize lattice point *i* with the starting abundances of each ancestral player $[s_i(0), r_i^0(0), \text{ and } p_i(0)]$, and $[p_i(0)]$ using the same method as in the bacterial experiment (*SI Methods*). After *C* growth cycles, we measured the expected fitness of a randomly chosen resistant cell relative to the resistant ancestor. This mean fitness is (Eq. 7):

$$\overline{W} = \sum_{k=0}^{k} \left\{ \left(\frac{\sum_{i=0}^{L\times W} r_i^k(CT)}{\sum_{j=0}^{k} \sum_{i=0}^{L\times W} r_i^j(CT)} \right) \left(\frac{\ln\left(r^k(T)/r^k(0)\right)}{\ln\left(\mathrm{r}^0(T)/r^0(0)\right)} \right) \right\}. \tag{7}$$

In Table 6.1, we give values for all of the simulation parameters, which are tailored to our bacterial experiment or estimated from assays (*SI Methods*). For Figs. 6.4 and 6.5 and Fig. S3, we assume $L = 100$ and $W = 100$. For Figs. S4 and S5, we assume $L = 16$, $W = 12$, and $C = 36$, which are the values corresponding to our laboratory experiment.

ACKNOWLEDGMENTS

We thank P. Aulakh, C. Eshelman, H. Lindsey, S. Smith, and J. West for help in the laboratory; B. Bohannan, S. Forde, and M. Riley for discussion surrounding the early conception of this work; and M. Clifford, J. Cooper, S. Drescher, J. Gallie, C. Glenney, S. Heilmann, H. Lindsey, B. Miner, and J. West for comments on a previous draft of this manuscript. B.K. thanks J. Avise, F. Ayala, D. Queller, and J. Strassmann for the invitation to participate in the Sackler colloquium "In the Light of Evolution V: Cooperation." This material is based in part on work supported by National Institutes of Health Grant 07–004309 B-03 AM03, National Science Foundation under Cooperative Agreement Number DBI-0939454, a National Science Foundation Graduate Research Fellowship (to J.R.N.), and CAREER Award Grant DEB0952825 (to B.K.).

TABLE 6.1 Simulation Parameters

Parameter	Description (units)	Values(s)
L	Length of the lattice (points)	16 or 100
W	Width of the lattice (points)	12 or 100
T	Duration of growth phase (hours)	12
C	Number of growth cycles (unitless)	36, 100, or 400
K	Number of mutant resistant strains (unitless)	7
μ_Y $\left\{ K_s\ K_p\ K_{R^0}\ K_{R^1}\ K_{R^2}, K_{R^3}, K_{R^4}, K_{R^5}, K_{R^6}, K_{R^7} \right\}$	Maximum growth rate (abundance per hour)	0.61
	Monod constant of ancestral strains and resistant mutants (abundance)	{0.165, 0.93, 0.341, 0.27, 0.28, 0.29, 0.30, 0.31, 0.32, 0.33}
$\{r^k(0), r^0(0)\}$	Initial competition amount (abundance)	{1/40, 1/40}
ϕ	Dilution factor (unitless)	1/40
α	Probability of migration (unitless)	1/3
a_{crit}	Critical abundance for persistence (abundance)	0.00275
π	Mutation probability per transfer per well (unitless)	1/100
γ	Fraction of resistant subpopulation converted to a random mutant given a mutation event (unitless)	1/2

7

Social Evolution in Multispecies Biofilms

SARA MITRI,*†§ JOÃO B. XAVIER,‡ AND KEVIN R. FOSTER*†§

Microbial ecology is revealing the vast diversity of strains and species that coexist in many environments, ranging from free-living communities to the symbionts that compose the human microbiome. In parallel, there is growing evidence of the importance of cooperative phenotypes for the growth and behavior of microbial groups. Here we ask: How does the presence of multiple species affect the evolution of cooperative secretions? We use a computer simulation of spatially structured cellular groups that captures key features of their biology and physical environment. When nutrient competition is strong, we find that the addition of new species can inhibit cooperation by eradicating secreting strains before they can become established. When nutrients are abundant and many species mix in one environment, however, our model predicts that secretor strains of any one species will be surrounded by other species. This "social insulation" protects secretors from competition with nonsecretors of the same species and can improve the prospects of within-species cooperation. We also observe constraints on the evolution of mutualistic interactions among species, because it is difficult to find conditions that simultaneously favor both within- and among-species cooperation. Although relatively simple, our model reveals the richness of interactions between the ecology and social evolution of multispecies microbial groups, which can be critical for the evolution of cooperation.

*Department of Zoology, University of Oxford, Oxford OX1 3PS, United Kingdom; †Oxford Centre for Integrative Systems Biology, Oxford University, Oxford OX1 3QU, United Kingdom; and ‡Program in Computational Biology, Memorial Sloan-Kettering Cancer Center, New York, NY 10065. §To whom correspondence should be addressed. E-mail: sara.mitri@zoo.ox.ac.uk or kevin.foster@zoo.ox.ac.uk.

> It would seem justified to assert that, so far, no revision of the Darwinian paradigm has become necessary as a consequence of the spectacular discoveries of molecular biology. But there is something else that has indeed affected our understanding of the living world: that is its immense diversity.
>
> Ernst Mayr (2004)

DNA sequencing continues to reveal new species that could not be found with conventional methods. Nowhere is this more true than in the microbial world where the sequence-based estimates of species in a gram of soil commonly run into the thousands (Gans et al., 2005; Roesch et al., 2007). Only a fraction of these species would typically be identified by culture-based methods, revealing that the majority of microbial species will not grow in current laboratory conditions (Hugenholtz et al., 1998). This realization, along with the rapidly decreasing cost of DNA sequencing, has led to an impressive effort to identify and catalog microbial diversity across a wide range of environments. These environments include soil, which is often considered one of the most diverse environments, but also range out to marine environments including the open ocean (Yooseph et al., 2007), the massive microbial mats that form stromatolites (Baumgartner et al., 2009) (Fig. 7.1C), and hydrothermal vents where large numbers of rare species have been found (Sogin et al., 2006).

The survey of microbial life is also looking inward to the species that live in and on humans, as exemplified by the concept of the human microbiome. Numerous projects are under way to catalog genetic diversity in areas including the skin, the oral cavity, and the intestine (Dethlefsen et al., 2006; Ley et al., 2006a). Whereas intestinal communities have been found to be quite similar across humans as compared with other mammals (Ley et al., 2008), different people often carry different sets of microbial species, underlining the complexity of intestinal ecology (Guarner and Malagelada, 2003; Eckburg et al., 2005; Dethlefsen et al., 2006). The composition of the gut microbiota has also been found to have important implications for health and has been linked to a range of diseases including obesity, inflammatory bowel disease, and colonic cancer (Guarner and Malagelada, 2003; McGarr et al., 2005; Dethlefsen et al., 2006; Ley et al., 2006b; Manichanh et al., 2006).

Recognition of the vast diversity within microbial communities has occurred alongside another realization about microbial life: the importance of social interactions. It is now accepted that many phenotypes of one cell influence the ability of surrounding cells to divide and survive, which are social traits in an evolutionary sense (Crespi, 2001; West et

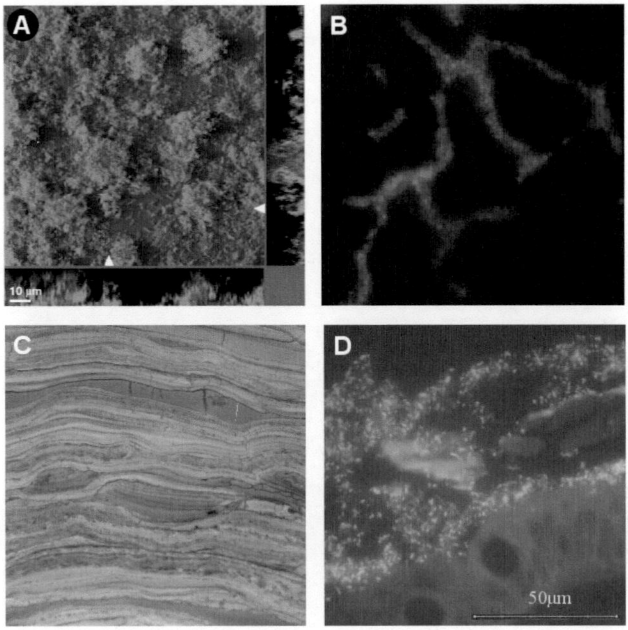

FIGURE 7.1 Microbial diversity: examples of natural microbial communities. (*A*) A two-species bacterial biofilm cultivated in the laboratory in which one strain evolves to increase its exploitation of the other. Adapted by permission from Macmillian Publishers Ltd: *Nature* (Hansen et al., 2007), copyright 2007. (*B*) A two-strain bacterial aggregate detected on a bean leaf surface (magnification 500×) [*Appl Environ Microbiol* (2005) 71(9):5484–5493, 10.1128/AEM.71.9.5484–5493.2005. Reproduced with permission from the American Society for Microbiology] (Monier and Lindow, 2005). (*C*) Stromatolite fossil that is ~2 billion years old. Modern stromatolites consist of multilayered sheets of microorganisms, and are a good example of very diverse, yet spatially structured microbial communities (copyright Merv Feick, http://www.Indiana9Fossils.com). (*D*) The detection of two of the species present in a bacterial biofilm covering the intestinal mucosae of a self-limiting colitis patient, imaged using triple-color fluorescence in situ hybridization [*J Clin Microbiol* (2005) 43(7):3380–3389, 10.1128/JCM.43.7.3380–3389.2005. Reproduced with permission from the American Society for Microbiology] (Swidsinski et al., 2005).

al., 2006; Foster, 2010). Social phenotypes in microbes include not only growth rate regulation (Kreft, 2004), which has the potential to affect the nutrients of surrounding cells, but also the widespread secretion of compounds that either promote or inhibit the growth of neighboring cells (Kerr et al., 2002; Griffin et al., 2004; Xavier et al., 2011).

In a bid to understand social phenotypes in microbes, theories of social evolution first developed for social animals have been used for the study of microbial groups (Brown, 1999; West et al., 2006; Nadell et al., 2010). A key prediction of this work is that the degree of mixing between different genotypes will be critical in determining the classes of social traits that evolve in microbial groups (Hamilton, 1964a). All else being equal, when cells of one genotype mix with many others, there is more potential for the evolution of competitive traits that harm neighboring cells than when strains are surrounded by clonemates. Clonal groups of cells are instead expected to display phenotypes that optimize the productivity of the group, like cells in a multicellular organism. For example, cells might display slow and efficient growth and secrete enzymes that harvest nutrients for all cells in the area. A growing body of empirical work has shown that genotypic mixing has the potential to limit cooperativity in a wide range of microbial traits (Greig and Travisano, 2004; Gore et al., 2009), including enzyme secretion (Griffin et al., 2004), iron scavenging (Diggle et al., 2007b), quorum sensing, and fruiting body formation (Foster et al., 2002; Buttery et al., 2009). Genetic mixing experiments also reveal the importance of the fitness costs and benefits for social phenotypes, with the potential for cooperation to be stabilized by either constraints on competitive traits (Foster et al., 2004; Harrison and Buckling 2009) or strategies that make cooperation carry little or no cost (Xavier et al., 2011).

Although our understanding of the evolution of social phenotypes in microbial populations in the laboratory is growing, we still understand little of how the theory and experiments relate to natural microbial communities (Little et al., 2008; Filoche et al., 2010; Foster, 2010). In particular, studies from social evolution typically consider well-mixed groups in liquid where local spatial structure is lacking (Griffin et al., 2004; Harrison et al., 2008). While shaking culture is an excellent technique with which to simplify and study interactions, microbes commonly form large surface-attached communities, known as biofilms. These biofilms carry spatial structure, and the potential for social interactions will typically be much greater than in liquid (O'Toole and Kolter, 1998; Hall-Stoodley et al., 2004; Monds and O'Toole, 2009; Nadell et al., 2009) (Fig. 7.1). In addition, the primary focus has been on mixing strains of a single species, which contrasts with the lesson from metagenomics that thousands of species are commonly present in any one environment. Natural microbial communities are thus often characterized by spatial structure and a multitude of species and environments, making it challenging to understand the links between social evolution and microbial ecology, not in the least because so many of the species involved cannot yet be cultured.

Our goal here is to develop models to explore the role of species diversity within biofilm-like microbial communities on the evolution of social phenotypes. In particular, we focus on the evolution of a growth-promoting secretion within a focal microbial species and ask: How does the presence and behavior of additional species affect the evolution of the growth-promoting secretion? This chapter is centered around a series of virtual experiments that use an individual-based simulation of microbial biofilms. The model captures many of the key biological and physical processes that affect cell groups, such as nutrient diffusion, secretion, cell division, and colony expansion. Although simulations are ultimately no substitute for experiments with real organisms, we can explore a much greater range of parameters than is possible with an empirical project. The analysis reveals a number of interdependencies between ecological competition among microbial species and the evolution of cooperation.

RESULTS

This chapter is centered on models of competition that investigate the evolutionary success of a strain that secretes a growth-promoting substance, such as an enzyme that diffuses outward and increases the availability of nutrients to all cells in proportion to its concentration. This focal strain is compared with a strain that does not secrete and by doing so saves energy that can be redirected into growth. The general question we ask is: What are the conditions that allow a cooperative secretor strain to outcompete the nonsecretor strain, or vice versa? Whereas we focus on a secretion phenotype, the general conclusions of the model should have relevance for any cooperative traits that affect the growth rate of neighboring cells (Kreft, 2004).

The framework used here is an agent-based model that employs mechanistic descriptions of solute diffusion and cell growth (Xavier et al., 2005; Xavier and Foster, 2007; Nadell et al., 2008) (*Materials and Methods*) and has been developed over the last decade for applications in the field of biochemical engineering. The underlying assumptions are described and justified in detail elsewhere (Matsushita and Fujikawa, 1990; Ben-Jacob et al., 1994; Kessler and Levine, 1998; Nadell et al., 2008), and empirical tests have demonstrated the framework's ability to make accurate predictions for real biological systems (Xavier et al., 2004, 2007).

Briefly, the simulations consider a two-dimensional surface on which a number of microbial cells (of the different phenotypes or species) attach, grow, and divide, resulting in a biofilm-like structure. Other geometries, including radial expansions from a point and three-dimensional simulations, can also be implemented but do not appear to affect evolutionary conclusions (Nadell et al., 2010). A constant concentration of nutrients is available at a fixed diffusion rate that cells take up, which leads to local gradients in nutrient concentrations. Cells may also secrete extracellular products, which become available to neighboring cells through diffusion. In the simulations presented, we assume that secretion carries an energetic cost of 30% of growth rate, in line with experimental results (Diggle et al., 2007b; Harcombe, 2010). However, we also investigate the effect of varying this cost (Figs. S1 and S6¶). In all experiments, cells are left to grow to a fixed total mass, at which point the fitness values of secretor and nonsecretor phenotypes (computed as the average number of cell divisions per unit time) (*Materials and Methods*) are compared to determine which of the two phenotypes would be expected to dominate in local competition. This cutoff point at which fitness is measured can be taken to model an environmental disturbance that occurs at a given frequency. The general effects of altering this parameter are discussed in Foster and Xavier (2007) and Brockhurst et al. (2007). Each cell is implemented as a circular agent, grows according to a Michaelis–Menten function of the substrate concentration in its local environment, and divides once it reaches a maximum radius (*Materials and Methods*). We do not consider active movement but cells can move passively due to the forces exerted between neighboring individuals as they grow and divide.

Single Species

In single-species simulations, Nadell et al. (2010) found that environmental nutrient concentration can determine whether a secretor or a nonsecretor strain is more evolutionarily successful. We begin this study by reproducing these results, which then serve as an experimental control with which to compare the effects of introducing additional species. In agreement with the previous study, our single-species simulations show that low nutrient concentrations result in tower-like clonal clusters of cells, whereas high nutrient concentrations result in the mixing of cell types as they grow (Fig. 7.2). Nadell et al. (2010) showed that this difference is due to changes in the depth of the growing front of the cell group, which depends on a multitude of factors in addition to nutrient concentration, such as the diffusion rates of nutrients into the cell groups, or on the growth rates of the cells. It should be kept in mind, therefore,

---

¶Figures S1 through S8 are available online as supporting information for the original PNAS article [108(Suppl 2):10839–10846] at www.pnas.org/lookup/suppl/doi:10.1073/pnas.1100292108/-/DCSupplemental.

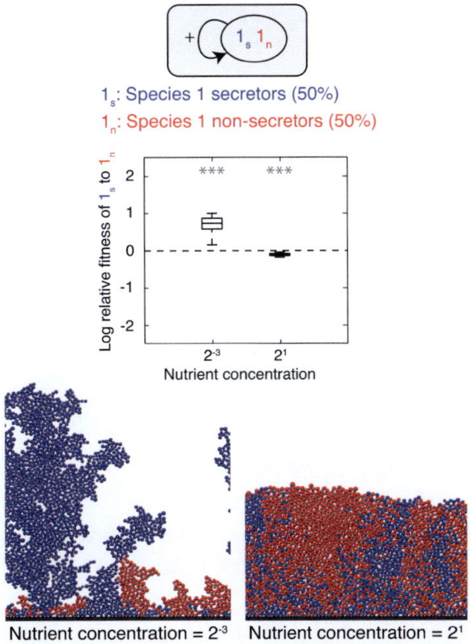

FIGURE 7.2 Secretors and nonsecretors of a single species. Equal proportions of two strains of the same species 1 are inoculated and left to grow to a fixed total biomass. Strain 1_s secretes a product that benefits both strains. Strain 1_n does not secrete the product. Product secretion incurs a cost of 30% of the cells' growth rate. Boxplots show log relative fitness (*Materials and Methods*) of secreting to nonsecreting cells [$\log(w(1_s){:}w(1_n))$] in 40 replicates with high and low nutrient concentrations. The dashed line shows the level at which the two phenotypes are equally fit. Asterisks indicate the significance of the difference between secretor and nonsecretor fitness, ***$P < 0.001$; **$P < 0.01$; *$P < 0.05$; ns, not significant. Below each boxplot is an image generated using the simulation from one of the 40 simulations that was closest to the median in the boxplot. It is shown that secretors can outcompete nonsecretors when the two phenotypes are well segregated, whereas they are at a disadvantage under conditions leading to high mixing.

that a change in nutrient levels in the simulations captures the effects of changing a number of factors.

When nutrients are low and growth results in clonal clusters, cells secreting a growth-promoting product (1_s) are more likely to be surrounded by others that also secrete the product. Consequently, the growth benefits of the product are preferentially directed toward clonal cells, whereas nonsecretors (1_n) rarely benefit from the secretions. In agreement with this logic, at a low nutrient concentration, secretor cells have a significantly higher fitness than nonsecretors, regardless of whether product secretion incurred a cost (30% of their growth rate) or not (Mann–Whitney test, df = 38, both $P < 0.001$, Figs. 7.2 and S1A). In contrast, increasing nutrient concentration leads to more mixing between the two cell types, such that the benefit of the secreted product is now equally distributed among both cell types. Secretors therefore grow as well as nonsecretors when secretion is free ($P = 0.39$, Fig. S1*A*), but have a significantly lower fitness when secretion incurs a cost [$P <$ 0.001, Fig. 7.2; see also Nadell et al. (2010)].

Ecological Competition Can Inhibit Cooperation

We next ask how the presence of additional species can affect the conclusions of the single-species model. We focus on how additional species will influence the competition between the two secretor phenotypes in our focal species. We do not analyze the competition playing out among the different species nor do we investigate mechanisms that can maintain species diversity in the face of the potential for competitive exclusion (Dethlefsen et al., 2006; Scheffer and van Nes, 2006; Gudelj et al., 2010; Robinson et al., 2010). Instead, we assume a simple model of species interaction that excludes the possibility for strong coevolutionary feedbacks among species (Box 7.1).

To investigate the effects of additional species, we introduce a new class of cells that can have different biological properties from the focal species. We assume that the second species uses the same nutrients as species 1 to grow so that it is an ecological competitor and can benefit from the secretions of species 1 (for results where species 2 does not benefit from the secreted product, see Fig. S2). As for the single-species model, we examine the outcome of competition between equal numbers of the secretor and nonsecretor phenotypes, with product secretion costing cells 30% of their growth rate. However, the total number of species 1 cells inoculated is now half that of the single-species case, with the other half being species 2 (results are qualitatively similar if density is doubled, Fig S.3).

The addition of species 2 led to a significant reduction in the relative fitness of secretors at low nutrient concentration (Mann–Whitney test, df = 38, $P < 0.001$), such that nonsecretors now have a significantly higher fitness than secretors ($P < 0.001$, Fig. 7.3*A*) and secretor cells are outcompeted. The poor performance of secretor cells is explained by an inability to compete when inoculated at low frequency (Fig. S4). In particular, adding a second species appears to interfere with the initial formation of

BOX 7.1 Species and Niche

In the various models presented throughout the chapter, the phenotype of the introduced species differs only minimally from the two phenotypes of the focal species. This similarity among the species poses the following question: How is the introduction of a second species different from simply increasing the frequency of one of the two phenotypes in the focal species? The key difference is that the focal species is mainly under selection in the focal environment, whereas the second species has its main selection component in different environments. A species in our model is thus functionally defined as a set of one or more phenotypes that share the same niche over evolutionary timescales. In this way, our "species 1" and "species 2" formally represent a dichotomous split between focal-niche and other-niche phenotypes. Each category could, in principle, contain multiple taxonomic species. In particular, species 2 is a proxy for multiple species that overlap only slightly (in space and/or time) with the focal species (see diagram). Whereas no one of these species interacts with the focal species enough for coevolution to be important, there is enough net overlap to influence selection on the focal species. Accordingly, we disregard changes in the fitness of the nonfocal species and concentrate solely on competition between the two focal phenotypes. A more realistic analysis would allow for a full range of niche overlaps rather than our binary division into complete niche overlap and minor overlap.

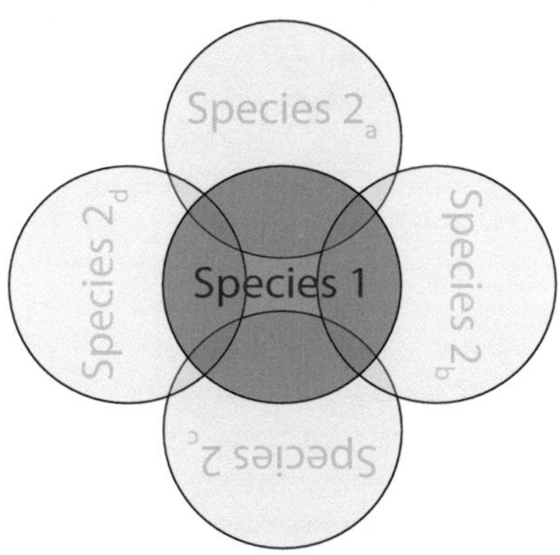

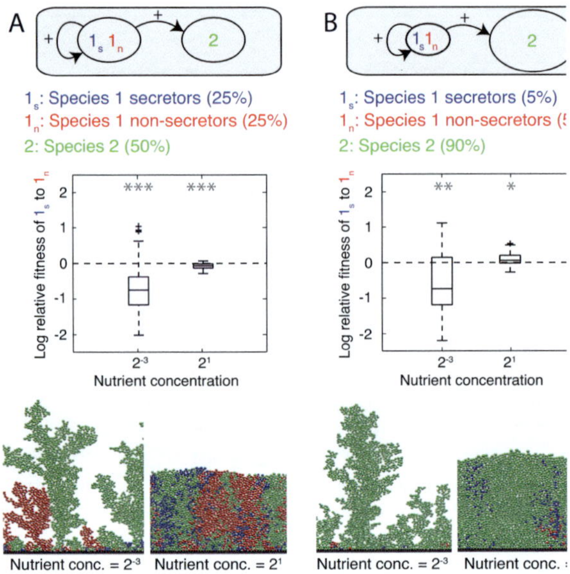

FIGURE 7.3 Ecological competition with a second species. A second species is added to the competition between secretors and nonsecretors (Fig. 7.2). This second species is intended to also approximate the effects of a mixture of many species (Box 7.1). Species 1 is equally divided into secretor and nonsecretor strains, whereas species 2 represents either (*A*) 50% or (*B*) 90% of the cells inoculated. All cells are then left to grow to a fixed total biomass. Strain 1_S secretes a product that benefits both strains of its own species, as well as species 2. Strain 1_n and species 2 do not secrete any products. Product secretion incurs a cost of 30% of the cells' growth rate. See Fig. 7.2 legend for explanations on data representation. It is shown that when cells are highly segregated, secretor cells lose their advantage (compared with Fig. 7.2, *Bottom Left*), independently of the two proportions of species 2. At high levels of mixing, however, secretors can outcompete nonsecretors when there is a high proportion of species 2 cells. The image (*B*, *Bottom Right*) shows the social insulation effect discussed in the text.

cooperative clumps of secretors such that secretors are more often overgrown. Consistent with the importance of ecological competition with species 2, we observe that the advantage of secretors over nonsecretors is significantly negatively correlated with the maximum growth rate of species 2 (Spearman's rank correlation test, $\rho = -0.51$, $P < 0.001$, Fig. S5A) and with the cost of secretions ($\rho = -0.67$, $P < 0.001$, Fig. S6A). In addition, decreasing the density of inoculated cells (by doubling the size of the growth area) results in a significant growth advantage for secretors over nonsecretors (Mann–Whitney test, df = 38, $P < 0.001$, Fig. S5C). The data thus far show that under low nutrient conditions, competition with a second species for nutrients and space can eliminate the advantage of cooperation.

Abundance of Additional Species Insulates Secretors from Nonsecretors

At high nutrient concentrations, we were surprised to find that the relative fitness of the secretor phenotype was significantly higher in the presence of species 2 than in its absence (Mann–Whitney test, df = 38, $P < 0.001$), although secretors still had a significantly lower fitness than nonsecretors ($P < 0.001$). To confirm that this result depended on the presence of species 2, we repeated the simulation, but instead of using equal proportions of both species, we started the simulation with 90% of the cells being of species 2 (Fig. 7.3*B*). Our focal species 1 is again divided equally among the two phenotypes, secretor and nonsecretor. This model is analogous to a conglomerate of multiple ecologically similar species (e.g., the model can be thought of as a mixture of 10 equally common species), where any focal species may often be in a minority (Box 7.1).

The higher initial proportion of cells of species 2 had no effect on the relative fitness of secretors and nonsecretors at low nutrient concentration compared with equal proportions ($P = 0.78$, Fig. 7.3*B*). The ecological competition effect still dominated and secretor strains fared poorly. However, at a high nutrient concentration, secretors now had a significantly higher fitness than nonsecretors ($P < 0.05$) and a higher relative fitness than when the initial number of cells of the two species was equal ($P < 0.001$). Overall, the proportion of species 2 inoculated together with species 1 correlated positively with the relative fitness of secretor cells (Spearman's rank correlation test, $\rho = 0.67$, $P < 0.001$, Fig. S7*D*). We hypothesized that this increase in the competitiveness of secretors was because species 2 was insulating the secretor strain 1_s from the nonsecretor strain 1_n, thereby reducing the access of the latter to the secretions of the former.

To examine this hypothesis further, we assessed the effect of species 2 on the spatiogenetic structure of species 1. In the high-nutrient case, increasing the proportion of inoculated cells of species 2 leads to an increase in the segregation index (*Materials and Methods*) between the two phenotypes of species 1 (Spearman's rank correlation test, $\rho = 0.94$, $P < 0.001$, Fig. S7*B*), suggesting a causal relationship between segregation and the increase in the relative fitness of secretors. This pattern was not observed at low nutrient concentrations (Fig. S7*A* and *C*).

Increasing the proportion of species 2 also decreases the number of cells of species 1 inoculated in the system, which might explain the increase in segregation among strains, independently of the presence of the additional species. To examine this idea, we repeated the simulation in the absence of species 2. This simulation was thus identical to that with species 1 alone (Fig. 7.2), except that the initial number of inoculated cells was 10 times lower. Secretors have a significantly lower fitness than nonsecretors (Mann–Whitney test, df = 38, $P < 0.05$) and the relative fitness of secretors is significantly lower than when species 2 was present ($P < 0.001$). However, the relative fitness of secretors is higher than in the original simulation with a higher number of inoculated cells ($P < 0.001$). This result suggests that reducing the inoculation density at high nutrient concentrations can increase segregation of the two phenotypes (see also Fig. S5*D*), but that species 2 was critical in acting as a social insulator that protects secretors from nonsecretors. Note that the insulation effect rests upon the assumption that the niches of the insulating species do not overlap perfectly with the focal species (Box 7.1). If the niches perfectly overlap, then the insulator species effectively become an excess of nonsecretors, which will tend to disfavor secretion (Fig. S4*B*).

Constraint on Multispecies Mutualism

We have explored the effect of competing species on the evolution of cooperative secretions in a focal species. Some species also exchange products or services that are mutually beneficial (Shimoyama et al., 2009). To investigate this possibility, we ran new simulations with an equal proportion of the two species in which the product secreted by strain 1_S provided benefits to both species 1 and 2, as in the simulations described above, but where species 2 additionally secreted a noncostly product that was beneficial to species 1. We assume that the trait of species 2 is not costly to focus upon the evolution of costly cooperation within species 1.

We found that the return benefit from species 2 slightly improves the prospects of the secretor cells of species 1. In particular, the fitness of nonsecretors is no longer significantly different from that of the secretors at low nutrient concentrations (Mann–Whitney test, df = 38, $P = 0.22$). However, this fitness improvement is rather small; that is, the relative fitness of secretors is not significantly different from the case where species 2 secretes nothing ($P = 0.12$). The beneficial product secreted by species 2 then does not strongly promote the fitness of secretor cells.

In the simulations described thus far, secreted products were always beneficial for cells of species 1. We next model a case involving the exchange of products that are only beneficial to the other species (Little et al., 2008) (Fig. 7.4*B*). Strikingly, secretors now have a significantly lower fitness than nonsecretors, independent of nutrient concentration (both $P < 0.001$). What explains the failure of secretors of species 1 to capitalize

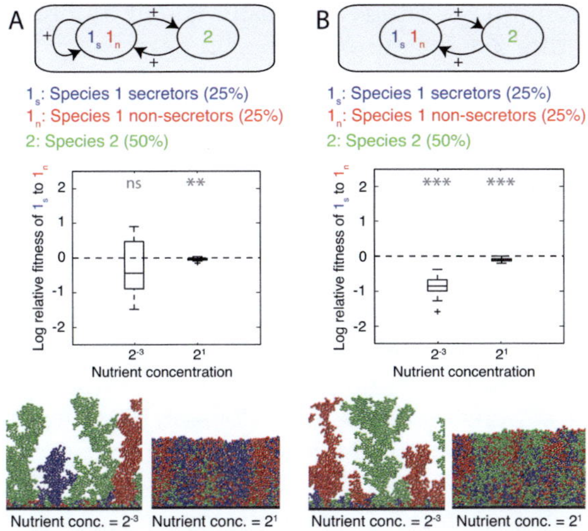

FIGURE 7.4 Multispecies mutualism. Species 2 now secretes a product that is beneficial to species 1, resulting in a mutualism between the two species. Species 1 is equally divided into secretor and nonsecretor strains, whereas species 1 and 2 are inoculated in equal proportions and left to grow to a fixed total biomass. Strain 1_S secretes a product that either benefits both strains of its own species, as well as species 2 (*A*) or species 2 only (*B*). Product secretion by 1_S incurs a cost of 30% of the cells' growth rate. In turn, species 2 secretes a cost-free product that benefits species 1. Strain 1_n does not secrete any products. See Fig. 7.2 legend for explanations on data representation. It is shown that secretor cells do not have a clear advantage over nonsecretors in any one of the four conditions considered here. This result is because mixing is important for the benefits of the two secreting strains to be shared, but is detrimental because it allows nonsecretors to grow faster than secretors, thereby undermining the mutualistic interaction.

on the return benefits from species 2? The answer is revealed by running simulations in which species 2 is mixed with either secretors or nonsecretors, but not both at the same time. At low nutrient concentration, nonsecretors perform better with species 2 than secretors with species 2 ($P < 0.001$, Fig. S8*A*) because spatial genetic segregation prevents secretors from interacting effectively with species 2. This explanation is further confirmed by the high-nutrient case where secretors with species 2 perform better than nonsecretors that are alone with species 2 ($P < 0.001$, Fig. S8*B*). However, the strain mixing that allows this positive effect is the same process in the full model that renders secretors vulnerable to competition from nonsecretors. The result is that in three-way competitions, the secretors always perform poorly.

Competition Among Microbial Groups

The results presented above predict the evolutionary trajectory within a group of microbes and form a good first step to understand the effect of additional species on cooperation within a microbial group. However, when the total productivity of microbial groups is important for their ability to colonize new patches ("hard" selection at the group level), there is the potential for higher-level evolutionary competition among different microbial groups. This competition can strongly affect the outcome of natural selection (Wilson, 1975; Rousset, 2004). In particular, it can favor genotypes that result in the most productive groups, even if those genotypes tend to do poorly within their groups.

To investigate the effects of competition among groups, we model the extreme case of maximum dispersal whereby after growth, cell groups disperse and mix with the cells of the same species in all other groups, before reforming groups containing two randomly chosen strains from the population. These groups then grow again before dispersing, and so on. Under this simple demography, we can estimate the potential for a rare secretor genotype to invade a population dominated by nonsecretors. This estimation is done by comparing the fitness of secretors in mixed groups with nonsecretors (a rare secretor genotype will tend to meet a nonsecretor genotype) to that of nonsecretor cells in the presence of other nonsecretors (nonsecretors are the common genotype and will tend to meet each other) (Xavier and Foster, 2007; Nadell et al., 2010). If the secretors tend to produce more cells per unit time in their groups, they will increase in frequency in the population: They are capable of evolutionary invasion. We can then reverse the problem and ask: Could a rare nonsecretor genotype invade a population of secretors? Under the assumptions of our model, we predict that if only one strain can invade, there will be a single strain at equilibrium. If both can invade, the prediction is that both can persist over evolutionary time.

As expected, competition among microbial groups increases the likelihood of the maintenance of cooperative secretions (Fig. 7.5). This result occurs because even though secretors often lose to nonsecretors in a group, the group they are in tends to do better and produces cells more rapidly than groups containing only nonsecretors (a phenomenon related to Simpson's paradox) (Sober and Wilson, 1999). Nevertheless, the overall effect of introducing a second species is similar to the within-group analyses of the previous sections: Under low-nutrient conditions, the addition of species 2 reduces the advantage of secretors. In the single-species invasion model, nonsecretors are unable to invade at low nutrient concentration (Fig. 7.5*A*, *Left*). By contrast, in two of four multispecies simulations where nutrients are low (Fig. 7.5*B–E*, *Left*), the nonsecretors can invade while excluding secretors. Under high-nutrient conditions, the effect of social insulation that promotes cooperation is again seen. With a majority of species 2 and high nutrients, secretors not only can invade but also can do so to the exclusion of nonsecretor cells (Fig. 7.5*C*, *Right*). The most significant deviation from the within-group results occurs for the two-species mutualism under high-nutrient conditions. Here, competition among microbial groups allows some secretors to be maintained, providing a way in which the constraint on mutualism discussed in the preceding section might be overcome.

DISCUSSION

Ecological Competition

Our model suggests that among-species interactions can strongly influence the potential for cooperation within a species in spatially structured microbial groups. We find that ecological competition with other species can preferentially harm secretor cells over nonsecretors. This result arises because investment in secretion can slow the growth of cell lineages at critical stages and lead to their overgrowth by another species. This initial investment leaves secretor cells vulnerable to being outcompeted by other lineages, particularly under low-nutrient conditions where resources are limiting and most lineages are eliminated through strong genetic bottlenecks (Fig. 7.3*A*, *Bottom Left*). The potential for such bottlenecks in growing microbial groups is empirically well documented (Gage, 2002; Hallatschek et al., 2007). Bottlenecks have been interpreted as being favorable for the evolution of cooperation because they promote genetic identity in the emerging clonal groups (Brockhurst, 2007; Nadell et al., 2010). Our study supports this interpretation in the single-species model (Fig. 7.2, *Bottom Left*), but suggests that this conclusion should be tempered by the fact that bottlenecks can also be indicative of strong ecological competition, which can eliminate cooperators before they have a chance to establish themselves.

The potential for ecological competition to preferentially harm cooperators was seen in a study that added *Staphylococcus aureus* to iron-limited cultures of *Pseudomonas aeruginosa* (Harrison et al., 2008). *P. aeruginosa* secretes iron-scavenging siderophores under iron-limiting conditions, and secreting strains are susceptible to the evolution of nonsecreting strains that use siderophores without producing them. The study is not a direct test of the results of our simulations, as it used shaking cultures

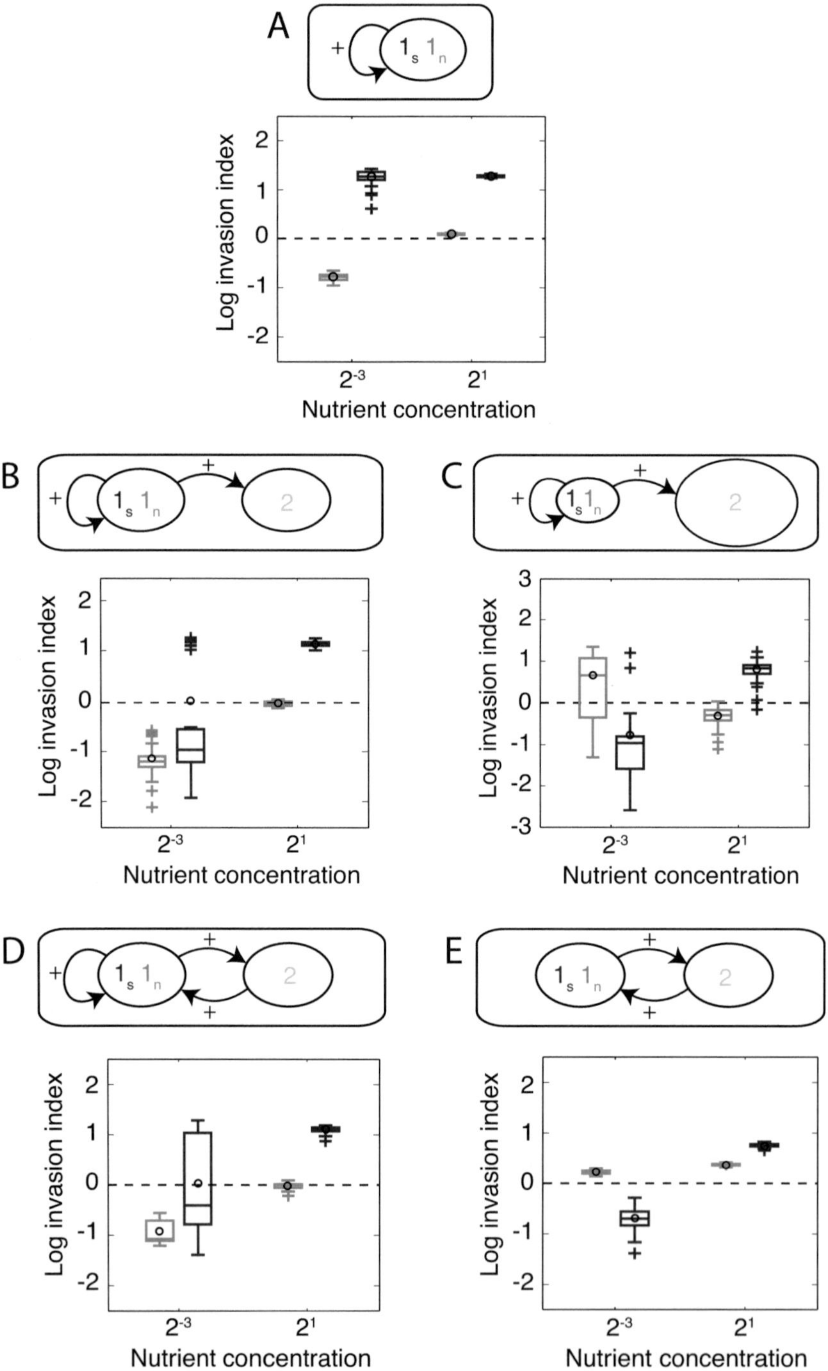
A
+
1_s 1_n
Log invasion index
2
1
0
-1
-2
2^{-3}
2^{1}
Nutrient concentration
B
+
1_s 1_n
+
2
Log invasion index
Nutrient concentration
C
+
1_s 1_n
+
2
3
-3
Log invasion index
Nutrient concentration
D
+
1_s 1_n
+
+
2
Log invasion index
Nutrient concentration
E
1_s 1_n
+
+
2
Log invasion index
Nutrient concentration

where spatial structure is lacking. Nevertheless, the addition of *S. aureus* promoted nonsecreting *P. aeruginosa* over secreting strains, thereby disfavoring cooperation. In contrast, other theory and experiments have highlighted the potential for ecological competition to favor cooperation within species. A model by Rankin et al. (2007) showed how ecological competition can strongly enrich for cooperation when noncooperative species compete poorly with other species. In support of this, a study on two termite species suggested that the species more affected by within-colony competition was more likely to be outcompeted by the other species (Korb and Foster, 2010).

What explains the difference between these results and our predictions? The key is whether within-species cooperation increases or decreases the ability to compete with other species. The model by Rankin et al. (2007) and the termite example concern competition among established social groups of each species where within-species cooperation improves the ability to compete with groups of the other species. By contrast, in our model, lone cooperator cells meet the other species before they have a chance to establish a clonal group, which can mean that cooperators are

FIGURE 7.5 Invasion analysis. The invasion index estimates the probability of a given minority phenotype to spread in a metapopulation consisting of many groups of the other phenotype (*Materials and Methods*). *A–E* correspond to Figs. 7.2–7.4. (*A*) Competition between secretors and nonsecretors of a single species (Fig. 7.2). (*B* and *C*) The invasion index of the two phenotypes is compared when an introduced species competes with the first species at (*B*) 1:1 inoculation (Fig. 7.3A) or (*C*) 1:9 inoculation densities of the two species (Fig. 7.3*B*). (*D* and *E*) The case of a mutualistic interaction with the second species (*D*) with self-benefit (Fig. 7.4*A*) or (*E*) without (Fig. 7.4*B*). See respective figure legends for details on simulations. Boxplots show log relative invasion index (*Materials and Methods*) of nonsecretor [$\log(I_{1_n \rightarrow 1_s})$, light gray] and secretor [$\log(I_{1_s \rightarrow 1_n})$, dark gray] cells separately in 40 replicates with high and low nutrient concentrations. Black circles show the mean of the distributions. The dashed line shows the level above which a phenotype can invade a metapopulation of the other. If the mean of only one of the two phenotypes is above the line, we predict that this phenotype would invade the other in a metapopulation. If both means are above the line, we expect the evolutionary equilibrium to consist of a mixture of both phenotypes. It is shown that under high nutrient conditions, secretors are expected to at least persist in the population, even though they were often at a disadvantage under local competition (Figs. 7.2–7.4). At low nutrient concentration, results are similar to the local competition simulations, where the presence of species 2 reduces the advantage of secretors over nonsecretors.

poor ecological competitors. It is interesting to speculate that this early-stage cost to cooperation may be important in both natural selection for quorum sensing regulation of secreted products (Diggle et al., 2007b) and the evolution of clumped dispersal (Gardner and West, 2006), which both limit the likelihood of being a solitary secreting cell. These mechanisms are not part of our simulations, but may help to restrict cooperation to established clonal groups in nature. If effective, there may be conditions under which microbial cooperation is favored rather than disfavored by ecological competition, as was seen in the termites (Korb and Foster, 2010).

Social Insulation

Under high-nutrient conditions, competitive effects are less severe and, accordingly, the impact of additional species upon within-species cooperation is reduced. Indeed, the model even predicts that interactions with other species can promote the evolution of secretor genotypes. Analysis of the spatiogenetic segregation in the simulations revealed that species 2 can act as a social insulator that keeps nonsecretor genotypes away from secretor genotypes. This insulation allows secretor cells to form patches in which they preferentially help their own genotype, in the same way that general spatial structuring can promote the evolution of cooperation (Hamilton, 1964a; Nowak and May, 1992; Rousset, 2004). Although the importance of social insulation effects in natural communities is not yet clear, our model suggests that it will be most important under relatively high nutrient conditions where many species meet and mix. One interesting candidate, therefore, is the human microbiome, and in particular the intestine, where cells can form dense biofilms containing multiple species (Fig. 7.1) (Macfarlane and Dillon, 2007).

Multispecies Mutualism

The conditions for the evolution of multispecies mutualism in the model were relatively restrictive. When the secretion of species 1 benefited cells of its own species as well as species 2 (Fig. 7.4*A*), the conditions for cooperation were similar to the case where there was no return benefit from species 2 (Fig. 7.3*A*). In the absence of within-species benefits, however, the evolution of costly secretions in the focal species was particularly unlikely (Fig. 7.4*B*). This result was due to an unexpected tension between the conditions that favor within- and among-species cooperation. One of the requirements for cooperation between groups of two species is cooperation within each species group (Foster and Wenseleers, 2006). However, within-species cooperation is favored by spatial segregation that keeps secretors away from nonsecretors, whereas among-species cooperation is favored by mixing that allows efficient exchange of mutual benefits. The tension between the requirements for the two forms of cooperation makes costly cooperative exchanges among microbial species relatively difficult to evolve. We know of no direct tests of this idea to date but some support comes from an example of cross-feeding among *Escherichia coli* strains where increasing spatial structure inhibited the benefits of the interaction among mutualists (Saxer et al., 2009). When within- and among-species mixing is coupled, however, our model suggests that mutualism is less likely to be favored.

This tension between within- and between-species cooperation has not been observed in previous theory on multispecies cooperation. For example, Doebeli and Knowlton (1998) performed an on-lattice simulation of two positively interacting species, which readily found conditions under which cooperation could be maintained. In addition, a simple model by Foster and Wenseleers (2006) predicts that among-species mutualism can be favored as long as within-species genetic assortment is high and there are reliable feedback benefits from the other species. The difference between these models and the current simulation is that both previous studies assume that within- and among-species assortment can be decoupled [Doebeli and Knowlton (1998) placed the two species on separate lattices]. This is a reasonable assumption for many mutualisms where interacting species have different ecologies. For example, genetic assortment within symbiont populations can be entirely independent of their degree of interaction with their host (Foster and Wenseleers, 2006): A bobtail squid can select for and interact with a near-clonal population of light-producing *Vibrio fischeri* (Visick et al., 2000). When within- and among-species mixing is coupled, however, our model suggests that mutualism is less likely to be favored.

On the basis of this argument, we predict that mechanisms that decouple mixing within and among species will promote the evolution of costly cooperation among microbial species. Species growing on different nutrient sources is one candidate mechanism suggested by our simulations. With different nutrient requirements, the segregation index within species is significantly higher than between species under low nutrient concentrations (Mann–Whitney test, df = 38, $P < 0.001$, Fig. 7.6, *Bottom Left*). Growing on different nutrients also reduces competition among species. Together with the effects on mixing that we observe, this result predicts that species with different metabolic lifestyles are most likely to be mutualists, which is broadly compatible with current data (Little et al., 2008). Mechanisms to select mutualistic partners may drive similar effects. One example is seen in the bacterium *Pelotomaculum thermopropionicum*, which uses its flagella to physically attach itself to the methanogenic archaeon *Methanothermobacter thermautotrophicus* with which it

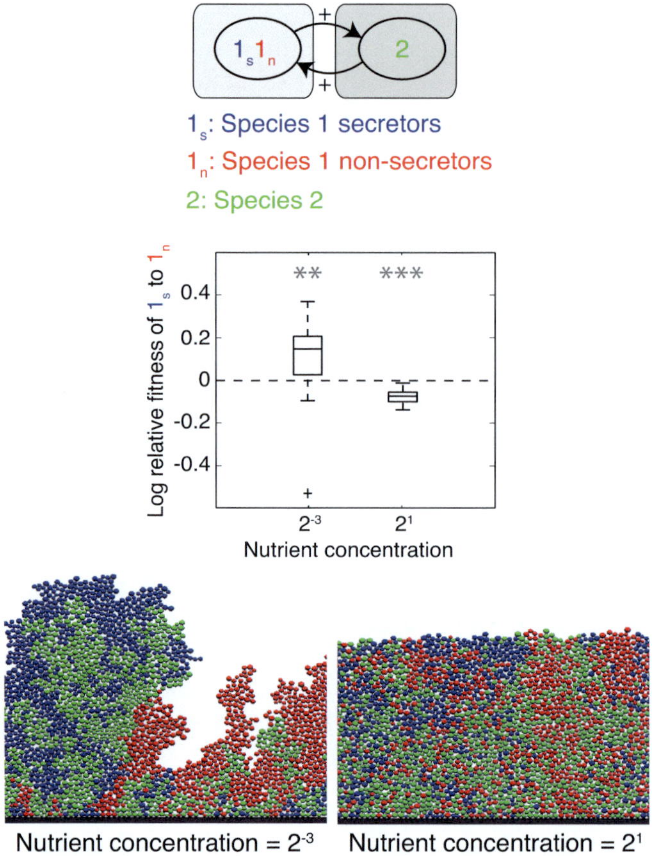

FIGURE 7.6 Mutualism when two species do not compete for nutrients: identical to Fig. 7.4*B*, but where species 1 and 2 consume different nutrients. See Fig. 7.4 legend for details on simulations and Fig. 7.2 legend for explanations on data representation. It is shown that reduced competition for nutrients between the two species can result in a significant advantage for mutualistic secretors under low nutrient conditions. It appears that this condition allows secretors of the two species to mix, while keeping the two phenotypes of species 1 separate. This result is not observed when nutrient concentration is high.

exchanges metabolic services (Shimoyama et al., 2009). Another candidate is chemotaxis by one species toward the secreting members of the other species, although additional mechanisms would presumably have to exist to ensure that the swimming species itself cooperates.

Mutualistic interactions have also been shown to persist between two engineered bacterial species under low inoculation density on agar plates (Harcombe, 2010). These experiments suggest that low colonization densities provide yet another mechanism to separate within- and between-species interactions by creating subpopulation structures, in which mutualist pairs can thrive in the absence of noncooperator strains. Selection for mutualism in this system was extremely strong, however, in that cells could hardly grow in the absence of secretors of the other species (Harcombe, 2010). More generally, the potential for within- and among-species cooperation in microbial communities will be promoted when there is higher-level competition among communities (invasion analyses, Fig. 7.5). The potential for such higher-level selection to shape microbial communities was seen in a large-scale simulation of microbial species growing and dispersing among a series of 10 interconnected flasks (Williams and Lenton, 2008). Over time, sets of species that limited harm to their local environment—a form of cooperation—were favored over more rapacious species that limited group productivity, something also seen in experiments that artificially selected for group productivity in real microbial communities (Swenson et al., 2000). Ultimately, the importance of local versus global competition will depend on the ecology of each species and the relative importance of within-community evolution compared with dispersal and colonization events.

Finally, we have deliberately focused on mutualisms where the investment in another species is an adaptation that carries an energetic cost. The majority of positive interactions among strains in nature may come about from cross-feeding by two species that comes at no energetic cost to the species involved. Such byproduct mutualisms are expected to often be evolutionarily stable (Connor, 1986; Foster and Wenseleers, 2006) and may be widespread in natural systems. However, our simulations suggest that even this form of mutualism can often be selectively neutral (Fig. S1*D* and *E*).

CONCLUSIONS

Although studies of microbial ecology and microbial sociality are progressing rapidly, we understand relatively little of the intersection between these disciplines. Our models indicate that this intersection can be important, owing to the interconnectedness of within- and between-species interactions in microbial groups. Contrary to typical social evolution predictions, we find that environmental conditions that promote genetic bottlenecks (and raise relatedness) can also increase ecological competition, thus disfavoring cooperation. Bottlenecks are also associated with segregation between species, which limits the potential for among-species mutualistic cooperation. The potential for social insulation by other species and the occurrence of higher-level competition among microbial groups, however, can counter these effects and favor cooperative phenotypes.

Ultimately, our simulations are simple and are able to capture only a small part of the complexity within real microbial communities. Nevertheless, we identify a number of familiar themes that can inform our understanding of microbial communities. A central theme is the importance of spatial structure for microbial interactions, which can simultaneously promote within-species cooperation and limit among-species interactions. Spatial structure in microbial groups can depend on a number of factors in addition to nutrient concentrations emphasized here. Motility will also influence spatial structures, where mixing may increase through undirected motility or decrease through chemotaxis. In addition, cell–cell adhesion can affect genetic mixing both within (Queller et al., 2003; Smukalla et al., 2008) and between species (Shimoyama et al., 2009). Related to this is the physical scale of social interactions. Strong spatial genetic structure may have no impact on the evolution of secretions that diffuse rapidly across strain and species boundaries.

The models also reemphasize the importance of the costs and benefits of social traits for the trajectories of their evolution (Hamilton, 1964a). The majority of the ecological barriers to cooperative evolution discussed here can be overcome by strategies that limit the cost of social traits, such as prudent regulation that produces a secretion only when it is cheap to do so (Xavier et al., 2011). The study of factors such as spatial structure and fitness costs promises a better understanding of when and why the members of microbial communities cooperate with one another.

MATERIALS AND METHODS

Model Framework

An individual-based model, described in detail previously (Xavier et al., 2005; Nadell et al., 2010), is used to simulate growing cell groups. Simulation parameter values (listed in Table 7.1) were taken from previous work (Nadell et al., 2010). At the beginning of each simulation, cells are placed at random positions on a surface and are left to grow to a fixed biomass. Most simulations are started with 120 cells but the effect of varying this number is discussed. Each cell grows according to the concentration of nutrients ($[G]$) and the concentration of extracellular secreted products ($[E]$) in its local microenvironment. The stoichiometry tables describing the metabolic model of cells (growth and secretion) can be found in Table 7.2. Cells whose radius exceeds a given value are divided into two new cells. Cells that overlap due to the growth and/or division process are moved to eliminate the overlap, causing the cell

TABLE 7.1 List of Parameters and the Values Used in Our Simulation Models

Symbol	Description	Dimension	Value
μ_{max_1}	Maximum cell growth rate of species 1	T^{-1}	1
μ_{max_2}	Maximum cell growth rate of species 2	T^{-1}	1
τ	Threshold for extracellular product concentration	$M_E L^{-3}$	4×10^{-3}
B_1	Growth factor increase of species 1 due to the presence of product secreted either by species 1 or by species 2 at or above threshold concentration τ	Dimensionless	3
B_2	Growth factor increase of species 2 due to the presence of product secreted by species 1 at or above threshold concentration τ	Dimensionless	3
C_{1_s}	Growth factor decrease in growth rate of strain 1_s due to the secretion of extracellular products	Dimensionless	0 or 0.3
C_2	Growth factor decrease in growth rate of species 2 due to the secretion of extracellular products	Dimensionless	0
D_G	Growth substrate (nutrient) diffusivity	L^2T^{-1}	4×10^4
D_E	Extracellular secreted product diffusivity	L^2T^{-1}	3×10^5
$[E_{1_s}]$	Local concentration of extracellular product secreted by strain 1_s	$M_E L^{-3}$	NA
$[E_2]$	Local concentration of extracellular product secreted by species 2	$M_E L^{-3}$	NA
G_{bulk}	Bulk concentration of growth substrate (nutrient)	$M_G L^{-3}$	2^{-3} or 2
$[G]$	Local concentration of growth substrate (nutrient)	$M_G L^{-3}$	NA
K_G	Half saturation constant for growth substrate concentration	$M_G L^{-3}$	3.5×10^{-5}
$N_{x,t}$	Number of cells of strain or species x in a cell group at time t	Dimensionless	NA
$R_{E_{1_s}}$	Rate of secretion of extracellular product by strain 1_s	$M_E M_X{}^{-1}T^{-1}$	1
R_{E_2}	Rate of secretion of extracellular product by species 2	$M_E M_X{}^{-1}T^{-1}$	0 or 1
w_x	Fitness of strain or species x	T^{-1}	NA
X_{1_s}	Concentration of biomass of strain 1_s (secretor cells)	$M_X L^{-3}$	NA
X_{1_n}	Concentration of biomass of strain 1_n (nonsecretor cells)	$M_X L^{-3}$	NA
X_2	Concentration of biomass of species 2	$M_X L^{-3}$	NA
Y_1	Yield of biomass of species 1 on substrate	$M_X M_G{}^{-1}$	0.5
Y_2	Yield of biomass of species 2 on substrate	$M_X M_G{}^{-1}$	0.5

Notes: M_E represents mass of extracellular enzyme, M_G represents mass of growth substrate, M_X represents cell biomass, L represents length, and T represents time. NA, not applicable.

TABLE 7.2 Stoichiometry of Cell Metabolism Used in the Simulation Models

	Solutes			Particulates			
Reaction	G	E_{1_s}	E_2	X_{1_s}	X_{1_n}	X_2	Rate expression
1_s growth	$-1/Y_1$			$1 + B_1 f([E_{1_s}]) + B_2 f([E_2])$			$\mu_{max_1} \frac{[G]}{[G] + K_G} X_{1_s}$
1_n growth	$-1/Y_1$				$1 + B_1 f([E_{1_s}]) + B_2 f([E_2])$		$\mu_{max_1} \frac{[G]}{[G] + K_G} X_{1_n}$
2 growth	$-1/Y_2$					$1 + B_1 f([E_{1_s}]) + B_2 f([E_2])$	$\mu_{max_2} \frac{[G]}{[G] + K_G} X_2$
Product secreted by 1_s		1		$-C_{1_s}$			$R_{E_{1_s}} \mu_{max_1} \frac{[G]}{[G] + K_G} X_{1_s}$
Product secreted by 2			1			$-C_2$	$R_{E_2} \mu_{max_2} \frac{[G]}{[G] + K_G} X_2$

$f([E_x])$ is a function of local product secreted by strain or species x, $f([E_x]) = \begin{cases} 0, & [E_x] < \tau \\ 1, & [E_x] > \tau. \end{cases}$
All other symbols are defined in Table S1.

group's front to advance where there are no physical barriers (e.g., the inoculation surface).

Nutrients come from above and the concentration of the nutrient source (bulk concentration, G_{bulk}) in the region above the upper boundary of growing cells is held constant (see Table 7.1 for values used) throughout the simulations. Under the assumption that reaction–diffusion is much faster than cell growth and division (Xavier et al., 2005), we update solute (nutrient and extracellular product) concentration fields after each cellular growth and division step. The new spatial concentration fields of all solutes *S* (*G* and *E*) are determined by solving the reaction–diffusion equation

$$\frac{\delta[S]}{\delta t} = D_s \nabla^2 [S] - \frac{1}{Y} r, \tag{1}$$

where *t* represents time, D_S is the diffusivity of solute *S*, $\nabla^2[S]$ is the Laplacian of the local solute concentration $[S]$, *r* is cell growth rate (computed using Table 7.2), and *Y* is the yield of biomass on substrate (see Table 7.1 for values used).

The individual-based simulation framework was written in the Java programming language. Numerical methods used in the model are detailed elsewhere (Xavier et al., 2005). Briefly, they include the Euler method to grow cells at each iteration, a hard-sphere collision detection method to identify pushing events between neighboring agents, and the FAS multigrid to solve reaction-diffusion equations to steady state (Rumbaugh et al., 2009; Press et al., 1997). All images were rendered using POV-Ray.

Calculating Fitness

Fitness *w* of a phenotype or species is calculated as the mean number of rounds of cell division per unit time that the cells of that phenotype achieve over the course of a simulation,

$$w_x = \frac{1}{t_{end}} \log_2 \frac{N_{x,t_{end}}}{N_{x,0}}, \tag{2}$$

where $N_{x,t}$ is the number of cells of phenotype or species *x* present within the cell group at time *t*, and t_{end} is the time that cells have taken to grow to the maximum total biomass. The log relative fitness of phenotype 1_s in local competition with phenotype 1_n (Figs. 7.2–7.4) is defined as $\log(w_{1_s}/w_{1_n})$.

Segregation Index

The segregation index used here is based on that used in previous work, with some minor differences (Nadell et al., 2010). To measure segregation in a population of M cells, we consider each cell c_i, $i = 1, \ldots, M$ in the population and identify all other individuals within a distance of 10 cell lengths. The N cells in this neighborhood are indexed by c_j, with $j = 1, \ldots, N$. In this case, we consider only cells of species 1. Cells of species 2 are treated as empty space. We define a phenotypic identity function, $p(c_i, c_j)$:

$$p(c_i, c_j) = \begin{cases} 0, c_j \text{ is not the same phenotype as } c_i \\ 1, c_j \text{ is the same phenotype as } c_i \end{cases} \tag{3}$$

Segregation with respect to a focal cell, $s(c_i)$, was calculated as the mean of the p function for every cell in its neighborhood:

$$s(c_i) = \frac{1}{N} \sum_{j=1}^{N} p(c_i, c_j). \tag{4}$$

Finally, we define the segregation index σ for the entire cell group as the mean value of $s(c_i)$ across the population of cells:

$$\sigma = \frac{1}{M} \sum_{i=1}^{M} s(c_i). \tag{5}$$

The segregation index measures the degree to which colocalized cells are clonally related to each other. The index is related to, and expected to correlate with, the relatedness coefficient from social evolution theory. However, the exact relation will depend on both the relative benefits of secretions to neighboring cells (Nadell et al., 2010) and the patterns of dispersal among different groups of cells (Rousset, 2004). Here, it is intended to illustrate only that genetic relatedness will tend to increase through the process of social insulation by other species.

Invasion Analysis

Analyses of relative fitness indicate which strains would be most likely to outnumber the other locally. The invasion analysis [based on that in Nadell et al. (2010)], on the other hand, is conducted to determine whether a rare mutant with a particular phenotype would survive in a metapopulation of cell groups where dispersal and colonization of new patches or hosts are common. We assume the existence of a very large number of cell groups where the great majority of groups are of a single dominant genotype and only a small minority will contain the mutant. Each group is seeded at random from the population with a particular number of strains. We focus here on groups seeded by two strains of species 1. We also assume that all subpopulations have identical conditions regarding the presence of the second species.

Under these conditions, a strain 1_x (rare mutant) can invade a metapopulation of strain 1_y (majority resident) if the fitness of 1_x in local competition with 1_y is greater than the average fitness of the whole metapopulation, denoted $\langle w_{1_y} \rangle$ (Maynard Smith, 1982). The fitness w_{1_y} was computed in 40 replicates of the simulations as in previous analyses in this chapter (with 1:1 inoculation frequencies of the two cell types). Because the great majority of cell groups in the metapopulation consist purely of the majority strain 1_y, $\langle w_{1_y} \rangle$ is approximated by the mean fitness of the majority strain, 1_y, when growing on its own (or with species 2). To calculate $\langle w_{1_y} \rangle$, the mean of w_{1_y} over 40 simulations is computed, where the cells of strain 1_x inoculated initially are replaced with 1_y cells (a total of 120 or 60 cells of strain 1_y together with 0 or 60 cells of species 2 are inoculated in the single- or multispecies simulations, respectively). The invasion index of a rare mutant 1_x into a metapopulation with majority strain 1_y was calculated for each of the 40 replicates as follows:

$$l_{1_x \to 1_y} = \frac{w_{1_x}}{\langle w_{1_y} \rangle}. \tag{6}$$

Under the assumptions of our model, we conclude that 1_x can invade in a population of 1_y when the mean of $l_{1x \to 1}1>1$.

Statistical Analysis

All simulations were repeated in 40 independent replicates. Boxplots are used to illustrate the distribution (medians, upper and lower quartiles, and outliers) of the 40 values. Because some of the data did not follow a normal distribution, nonparametric statistical methods were used to compare medians (Mann–Whitney tests) and to detect correlations (Spearman's rank correlation test). All statistical tests were conducted using Matlab.

ACKNOWLEDGMENTS

We thank Angus Buckling, Dan Cronforth, Laurent Keller, Wook Kim, Jan-Ulrich Kreft, Daniel Marbach, Carey Nadell, Nuno Oliveira, Jonas Schluter, and two anonymous reviewers for useful discussions and comments on the manuscript. S.M. is supported by the Swiss National Science Foundation; J.B.X. by a seed grant from the Lucille Castori Center for Microbes, Inflammation, and Cancer; and K.R.F. by European Research Council Grant 242670.

Part III

REAL SELFISH (AND COOPERATIVE) GENES

It is remarkable that a field founded on the concept of selfish genes (Dawkins, 1976b) got so far for so long without paying much attention to specifiable genes. That is probably because we learned how phenotypic strategies of cooperation and conflict could be understood as the results of genes maximizing inclusive fitness. However, studies at the genic level are now becoming common and should shed light both on the mechanisms and the manner in which social selection operates.

In Chapter 8, Brielle Fischman and colleagues review and extend what is known about the molecular genetic mechanisms of eusociality. Some of the information comes from studies of particular genes and pathways but much is now coming from evolutionary analyses of genome-scale data. To the seven sequenced genomes of social insects, the authors add their own transcriptome-based protein-coding sequences for 10 social and nonsocial bee species, representing three origins of sociality. Some of the patterns are idiosyncratic. For example, early results from the honeybee genome pointed to the importance of odorant receptors and immunity genes, but these do not hold up in the broader analyses. New findings include increased rates of evolution of brain-related genes in the primitively eusocial bees, conceivably because of the increased cognitive demands of their competitive social environment. Juvenile hormone and insulin are often important in caste. This is not surprising if caste is nutritionally based, although the effects of juvenile hormone are quite different than in nonsocial insects. There is also a rapid evolutionary change in proteins involved in fundamental carbohydrate metabolism. Again, this fits with a nutritional basis for caste, but it seems surprising that changes are common in such basic pathways. These issues should be clarified with additional genome sequences and functional studies of individual species.

In Chapter 9, Joan Strassmann and David Queller explore a microbial social system where it is possible to manipulate genes. In the social amoeba *Dictyostelium discoideum,* starved cells come together in large groups in which 20% of the cells sacrifice themselves to make a stalk that aids in dispersal of the others as spores (Kessin, 2001). Besides this impressive altruism, this species has been shown to have cheating, kin recognition, and even primitive farming of their bacterial food. Numerous genes of many functional types can be mutated to cheaters. Some cheaters could destroy cooperation, yet cooperation is maintained for a variety of reasons, one being the rather high genetic relatedness in the field, part of which is due to kin recognition mediated by highly polymorphic adhesion genes. Other controls on cheating that have been demonstrated include the evolution of resistor genes, power asymmetries, and lottery-like mechanisms. Studies of the *dimA* and *csaA* genes have shown that cheating can also be controlled by idiosyncratic pleiotropies of particular genes. The cheating allele would be favored by selection but other deleterious effects of the same allele keep it from spreading, suggesting that cheat-proof cooperation often may be built using elements that are essential for other reasons. Consistent with ongoing social conflicts and arms races, social genes evolve rapidly.

Dawkins (1976b) argued that all genes are selfish, but the ones that show the trait most distinctively are selfish genetic elements. These are the renegades of the genome, chunks of DNA that replicate in part at least via different pathways than most genes and thus can be selected to conflict with other loci. Transposons, for example, increase their representation by jumping from one place to another, often at some cost to the organism. Other examples include meiotic drive elements, various modification-rescue systems, imprinted genes, B chromosomes, and organellar genes. In Chapter 10, John Werren tackles the issues of the function and adaptation of these elements. He surveys the evidence, sometimes strong and sometimes suggestive, that such elements have had important functional consequences for their genomes. For example, parts of transposons sometimes evolve into regulatory regions, and defenses against selfish elements may have led to the eukaryotic intron-splicing apparatus. But contrary to some recent suggestions, Werren argues that there is as yet little evidence that these are the adaptive reasons for the maintenance of these elements. Instead, selfish genetic elements are maintained by their selfish behavior, but the new chunks of DNA that they sprinkle throughout genomes sometimes get co-opted, domesticated, or otherwise modified to cause some beneficial effect to the organism.

8

Molecular Evolutionary Analyses of Insect Societies

BRIELLE J. FISCHMAN,* S. HOLLIS WOODARD,* AND GENE E. ROBINSON*†‡§¶

The social insects live in extraordinarily complex and cohesive societies, where many individuals sacrifice their personal reproduction to become helpers in the colony. Identifying adaptive molecular changes involved in eusocial evolution in insects is important for understanding the mechanisms underlying transitions from solitary to social living, as well as the maintenance and elaboration of social life. Here, we review recent advances made in this area of research in several insect groups: the ants, bees, wasps, and termites. Drawing from whole-genome comparisons, candidate gene approaches, and a genome-scale comparative analysis of protein-coding sequence, we highlight novel insights gained for five major biological processes: chemical signaling, brain development and function, immunity, reproduction, and metabolism and nutrition. Lastly, we make comparisons across these diverse approaches and social insect lineages and discuss potential common themes of eusocial evolution, as well as challenges and prospects for future research in the field.

The social insects are exemplars of cooperative group living. Within their complex societies, there is a reproductive division of labor in which only a small number of individuals reproduce, whereas all other individuals belong to a functionally sterile worker caste that specializes in tasks important for colony growth and develop-

*Program in Ecology, Evolution, and Conservation Biology, †Department of Entomology, ‡Institute for Genomic Biology, and §Neuroscience Program, University of Illinois, Urbana, IL 61801. ¶To whom correspondence should be addressed. E-mail: generobi@illinois.edu.

ment (Wilson, 1971). Although there has been much theoretical research on the evolutionary forces that may select for eusociality (Strassmann and Queller, 2007; Nowak et al., 2010), less is known about the actual molecular mechanisms involved in transitions from solitary to social living and in the maintenance and elaboration of eusociality in insects (C. R. Smith et al., 2008).

The social insects provide a powerful comparative framework for investigating mechanisms involved in eusocial evolution. Eusociality has arisen independently at least 12 times in the insects (Cameron and Mardulyn, 2001; Brady et al., 2006; Hines et al., 2007; Cardinal et al., 2010), and eusocial insects have all converged on the following three characteristics: reproductive division of labor, cooperative brood care, and overlapping generations (Michener, 1974). Additionally, despite sharing this core set of traits, there are many differences among eusocial lifestyles, which may be related to ecological, phylogenetic, or other factors specific to particular eusocial lineages (Wilson, 1971). By comparing across social insect lineages, it is possible to both search for common mechanisms of eusocial evolution and explore how eusociality evolves under different conditions.

Analysis of adaptive evolution at the molecular level can yield great insights into the mechanisms underlying the evolution of complex phenotypes, such as eusociality. Genomic sequence provides a molecular record of how natural selection has shaped an organism's evolutionary history (Clark, 2006). Several methods have been developed for comparing genes and genomes to identify molecular signatures of adaptation. These methods were largely developed during the pregenomic era (Li, 1997) but gain enormous power when large genomic datasets are available, particularly for sets of closely related and phenotypically variable species (Clark et al., 2003; *Drosophila* 12 Genomes Consortium, 2007). For example, comparisons of primate genomes have identified adaptive genetic changes involved in the evolution of brain size in humans (Pollard et al., 2006), and comparisons of drosophilid genomes have shed light on the ecological pressures that shaped speciation in this group (Clark et al., 2003).

Here, we review some of the first contributions of molecular evolutionary research to our understanding of eusocial evolution in insects. This research has focused on the most well-studied social insects, which include several eusocial lineages within the order Hymenoptera, the ants, bees, and wasps, and the one eusocial lineage in the order Blattodea, the termites (Fig. 8.1). Some studies have performed targeted molecular evolutionary analyses of candidate genes that have been particularly valuable in species for which large amounts of genomic sequence are not yet available. Others have focused on comparative analyses of

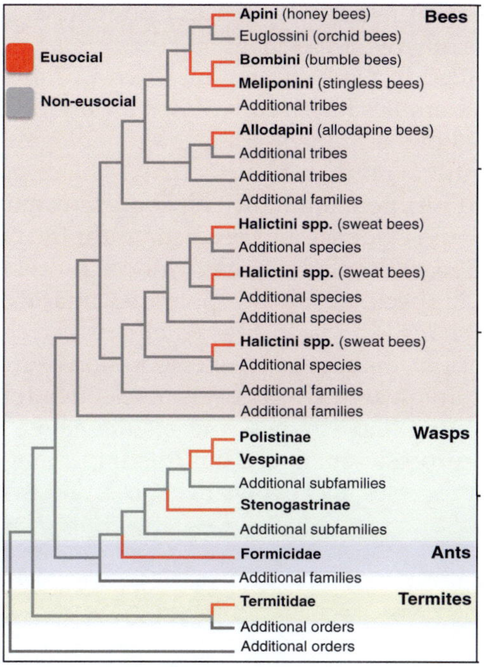

FIGURE 8.1 Cladogram showing the origins of eusociality in insects. Topology and reconstruction of evolutions of eusociality are based on multiple studies (Cameron and Mardulyn, 2001; Brady et al., 2006; Hines et al., 2007; Cardinal et al., 2010).

whole-genome sequence, which is currently available for six social insects, the honey bee, *Apis mellifera* (Honeybee Genome Sequencing Consortium, 2006), plus five ant species (Bonasio et al., 2010; C. D. Smith et al., 2011; C. R. Smith et al., 2011; Wurm et al., 2011), and for many solitary insects, including three solitary hymenopterans in the parasitoid jewel wasp genus, *Nasonia* (Werren et al., 2010).

We also draw heavily from our own recent genome-scale study of protein-coding sequence evolution in bees ("bee molecular evolution study"). This study analyzed ~3,600 genes from a set of 10 social and nonsocial bee transcriptomes; these species encompass three independent origins of eusociality (Woodard et al., 2011). Hundreds of genes were identified that exhibit a molecular signature of rapid evolution associated with sociality, defined as a higher ratio of nonsynonymous-to-synonymous nucleotide substitutions (d_N/d_S) in social relative to nonsocial bee lineages (Woodard et al., 2011). Throughout this review, evidence for rapid evolution is based on relative d_N/d_S, and positive selection is defined as $d_N/d_S > 1$, unless otherwise specified.

Genes identified in these studies are listed in Table 8.1. The insights gained from these studies have implications for understanding how evolutionary changes in the following five major biological processes might be involved in the evolution of eusociality: chemical signaling, brain development and function, immunity, reproduction, and metabolism and nutrition. We discuss evidence and predictions for the putative functional effects of identified molecular changes in these processes on social phenotypes. We also speculate on the potential adaptive significance of these molecular changes and consider whether these changes evolved in response to the origin, maintenance, or elaboration of eusociality, because each case likely involved a distinct set of selective forces. For the purposes of interpreting and synthesizing results across multiple studies, we present each process separately, but it is important to recognize that these biological processes may evolve in concert and that some molecular

TABLE 8.1 Genes Implicated in the Origin or Maintenance of Insect Society by Molecular Evolutionary Research

Gene	Function	Evidence	Type of Change[a]
Chemical signaling			
decapentaplegic	Gland development (Bradley et al., 2003; Harris et al., 2007)	Rapid evolution in eusocial bees (Woodard et al., 2011)	1
thickveins	Gland development (Bradley et al., 2003; Harris et al., 2007)	Rapid evolution in eusocial bees (Woodard et al., 2011)	1
PDGF- and VEGF-related factor 1	Gland development (Bradley et al., 2003; Harris et al., 2007)	Rapid evolution in eusocial bees (Woodard et al., 2011)	1
AmOr11	OR (Wanner et al., 2007)	Responds to main component of queen honey bee pheromone, 9-ODA (Wanner et al., 2007)	2
Neofem 2	β-Glycosidase-like (Korb et al., 2009; Weil et al., 2009)	Involved in signaling queen termite presence (Korb et al., 2009; Weil et al., 2009)	3

TABLE 8.1 Continued

Gene	Function	Evidence	Type of Change[a]
GP-9	Putative OBP (Keller and Ross, 1998; Krieger and Ross, 2005; Gotzek et al., 2007; Leal and Ishida, 2008; Gotzek and Ross, 2009)	Allelic variation associated with fire ant queen number (Keller and Ross, 1998; Krieger and Ross, 2005; Gotzek et al., 2007; Leal and Ishida, 2008; Gotzek and Ross, 2009)	1,2
Brain development and function			
dunce	cAMP/CREB signaling pathways (Silva et al., 1998)	Rapid evolution in primitively eusocial bees (Woodard et al., 2011)	1
nejire	CREB binding protein (Silva et al., 1998)	Rapid evolution in primitively eusocial bees (Woodard et al., 2011)	1
Immunity			
defensin	Antimicrobial protein (Viljakainen and Pamilo, 2008)	Positive selection in ants (Viljakainen and Pamilo, 2008)	1
termicin	Antimicrobial protein (Bulmer and Crozier, 2004; Bulmer et al., 2010)	Gene duplication, positive selection in termites (Bulmer and Crozier, 2004; Bulmer et al., 2010)	1,2
GNBP 1 and *2*	Pattern recognition receptors (Bulmer and Crozier, 2006)	Gene duplication, positive selection in termites (Bulmer and Crozier, 2006)	1,2
relish	Transcription factor, induces production of antimicrobial peptides (Bulmer and Crozier, 2006)	Positive selection in termites (Bulmer and Crozier, 2006)	1
Reproduction			
tudor	piRNA pathway (Siomi et al., 2010)	Rapid evolution in primitively eusocial bees (Woodard et al., 2011)	1

continued

TABLE 8.1 Continued

Gene	Function	Evidence	Type of Change[a]
capsuleen	piRNA pathway (Siomi et al., 2010)	Rapid evolution in primitively eusocial bees (Woodard et al., 2011)	1
vasa	piRNA pathway (Siomi et al., 2010)	Rapid evolution in primitively eusocial bees (Woodard et al., 2011)	1
csd	Sex determination (Beye et al., 2003; Hasselmann et al., 2008a,b)	Gene duplication, positive selection in honey bees (Beye et al., 2003; Hasselmann et al., 2008a,b)	1,2
Metabolism and nutrition			
MRJPs	Main components of royal jelly (Drapeau et al., 2006)	Gene family expansion, novel feeding-related functions in honey bees (Drapeau et al., 2006)	2
Hex-1 and Hex-2	Storage proteins (Zhou et al., 2006, 2007)	Unique insertions in termites (Zhou et al., 2006, 2007)	1
phosphofructokinase	Key regulator of glycolysis (Kunieda et al., 2006)	Rapid evolution in eusocial bees (Woodard et al., 2011)	1
hexokinase	Regulator of glycolytic flux (Kunieda et al., 2006)	Rapid evolution in eusocial bees (Woodard et al., 2011)	1
pyruvate kinase	Regulator of glycolytic flux (Kunieda et al., 2006)	Rapid evolution in eusocial bees (Woodard et al., 2011)	1

Note: Although many genes in this table are presumably involved in multiple biological processes, they are classified in one of five processes with known links to insect sociality: chemical signaling, brain development and function, immunity, reproduction, and metabolism and nutrition.

[a]Type of change: 1, protein coding sequence change; 2, novel gene; 3, change unknown.

changes could potentially affect multiple processes. We end with a discussion of future prospects and challenges for this young field.

CHEMICAL SIGNALING

Social insects use pheromones to coordinate the behavior and physiology of colony members, such as directing the foraging activity of nestmates, reinforcing dominance status, and inhibiting ovary development in workers (Le Conte and Hefetz, 2008). It is unknown whether chemical signaling was important during the origins of eusociality, because other mechanisms to mediate social interactions, such as physical interactions, serve similar functions in some social insect societies (Wilson, 1971). However, chemical signaling is certainly involved in the maintenance and elaboration of eusociality because it is crucial for the coordination and control of colony members. In humans, in whom vocalization is a major component of social communication, molecular signatures of adaptation have been detected in genes underlying both the production (Enard et al., 2002) and perception (Clark et al., 2003) of vocal signals. Early studies in social insects suggest that analogous changes have occurred in the molecular machinery underlying the production and perception of chemical signals.

Gland Development

Our bee molecular evolution study identified ~200 genes evolving more rapidly in social relative to nonsocial bee lineages (Woodard et al., 2011). Gene ontology enrichment analysis revealed that this set of genes was enriched for genes involved in gland development. This supports a role for these genes in chemical signaling, because glands are the primary organs involved in pheromone production in insects. Moreover, the evolution of complex chemical signaling in the social insects has been associated with the diversification of the gland repertoire (Wilson, 1971).

In other organisms, modular evolution, in which semiautonomous genetic pathways evolve as a functional unit and are reused in multiple contexts, appears to be a common evolutionary mechanism involved in morphological diversification (Wagner et al., 2007). The sequence changes identified in genes involved in gland development in social bees may have caused modular changes to the gland development program, resulting in functional changes to existing glands or the appearance of entirely new glands. This is supported by the evidence that several of these genes (*decapentaplegic, thickveins,* and *PDGF- and VEGF-related factor 1*) have specific roles in gland patterning during early development in *Drosophila* (Bradley et al., 2003; Harris et al., 2007).

Because diversification of gland function is a common characteristic shared by all social insects, it would be fruitful to investigate the sequence evolution and function of these genes in other social insect groups. It is possible that molecular changes in the same or similar genes were involved in gland evolution across other independent eusocial lineages.

Odorant Receptors

Given the diversity of chemical signals used by social insects, odorant receptor genes (ORs) have been predicted to be important targets of selection during eusocial evolution (Robertson and Wanner, 2006). Early support for this prediction was found in the genome of the honey bee, *A. mellifera,* which, at the time of its publication, contained the largest number of ORs yet found in an insect genome (Honeybee Genome Sequencing Consortium, 2006). However, as more insect genomes have been sequenced, it has been discovered that *A. mellifera* has an intermediate number of ORs, there is significant variation in OR number between the five ant genomes (Bonasio et al., 2010; C. D. Smith et al., 2011; C. R. Smith et al., 2011; Wurm et al., 2011), and several solitary insect genomes have among the most ORs found in insects so far (Engsontia et al., 2008; Robertson et al., 2010). Thus, the evidence no longer supports an association between sociality and expansion of the OR repertoire. Furthermore, studies in other organisms have revealed that ORs can function combinatorially and that bioinformatically predicted ORs may not all produce functional proteins, which, together, suggest that the number of ORs in a genome may not scale with the complexity of chemical communication in a species (Nei et al., 2008).

As a result of their functional specificity, ORs are particularly good targets for candidate gene studies, because the adaptive significance of OR evolution may be easier to interpret than for genes with broader functions (Nei et al., 2008). A functional genomics approach was used to identify a novel OR in the *A. mellifera* genome, *AmOr11,* which responds to the main component of the honey bee queen pheromone, (*E*)-9-oxo-2-decenoic acid (9-ODA) (Wanner et al., 2007). The queen pheromone attracts workers to the queen, partially inhibits worker ovary development, and acts as a sex pheromone, among other functions (Wanner et al., 2007). The specific molecular characteristics of *AmOr11* that are involved in the perception of 9-ODA are not yet known, but it appears that it arose early in *Apis* evolution (Plettner et al., 1997; Cruz-López et al., 2005; Urbanová et al., 2008).

Termite Queen Pheromone

Neofem2 is the first gene discovered in termites that is involved in signaling queen presence to workers. It was originally identified as being up-regulated in female neotenic "replacement" reproductives relative to other colony members in two species of *Cryptotermes* termites (Weil et al., 2009). Knocking down *Neofem2* in *Cryptotermes secundus* queens using RNAi caused an increase in aggressive behavior among workers, which is typically only exhibited under queenless conditions (Weil et al., 2009). Based on sequence similarity, *Neofem2* is most closely related to a β-glycosidase expressed in the salivary glands of the termite *Neotermes koshunensis* (Korb et al., 2009). β-glycosidases are enzymes that break down polysaccharides; in wood-dwelling termites, such as *N. koshunensis* and *C. secundus,* whose diet primarily consists of rotting bark, these enzymes are important for breaking down cellulose (Tokuda et al., 2002). It has thus been suggested that *Neofem2* evolved from a wood-digesting enzyme to pheromone (Korb et al., 2009). Supporting this speculation, β-glycosidases exhibit pheromonal activity in other insects, including the production of an egg recognition signal in another termite species (Korb et al., 2009). The specific molecular changes that have occurred in *Neofem2* as it evolved this new social function remain to be discovered. The story of *Neofem2* highlights the importance of considering the ecological context of social evolution in a given lineage, because the origin of a social pheromone from a wood-digesting enzyme is almost certainly a phenomenon specific to the wood-dwelling termites.

General protein-9 in Fire Ants

General protein-9 (*Gp-9*) alleles are strongly associated with variation in queen number in fire ants (genus *Solenopsis*). In monogynous (single queen) colonies, all females are homozygous for *B*-type alleles and will not tolerate the presence of multiple queens, whereas in polygynous (multiple queens) colonies, some individuals possess *b*-type alleles and do accept multiple queens but only if those queens also possess the *b*-type allele (Gotzek and Ross, 2009). *Gp-9* has been called a "greenbeard gene" (Keller and Ross, 1998), because workers carrying one allele favor queens that share the same allele. Molecular phylogenetic analyses of *Gp-9* both within and across *Solenopsis* species have revealed that the *b*-like alleles form a monophyletic clade, suggesting that monogyny was the ancestral condition in the genus and that polygyny arose once and has been maintained through multiple speciation events (Krieger and Ross, 2005; Gotzek et al., 2007).

At the protein sequence level, Gp-9 most closely resembles odorant-binding proteins (OBPs), which are expressed in chemosensory sensilla lymph and bind and transport soluble odorants (Gotzek and Ross, 2009). These results have led to the suggestion that *Gp-9* is an OBP that plays a role in pheromonal communication in fire ants (Gotzek and Ross, 2009). However, *Gp-9* is ubiquitously expressed in the hemolymph, suggesting it may be involved in functions that are unrelated to chemosensation (Leal and Ishida, 2008). In addition, *Gp-9* is found in a genomic region with a low recombination rate; therefore, other linked genes in the region may potentially have more influence on the regulation of queen number (Krieger and Ross, 2005; Gotzek and Ross, 2009). *Gp-9* alleles are also associated with variation in several life history traits in *Solenopsis* queens, including body fat and dispersal behavior (Gotzek et al., 2007), suggesting that *Gp-9* either acts pleiotropically or with other genes in the region.

Although the function of *Gp-9* is unresolved, molecular evolutionary analyses suggest that this gene is evolving adaptively, implying that *Gp-9* played an important role in fire ant evolution. A signature of positive selection was detected in the branch leading to the *b*-like allele clade (Krieger and Ross, 2005), suggesting that this allele had an adaptive benefit when it arose. In addition, all *b*-like alleles share the same amino acid residues at three diagnostic codon positions, and two of these positions show evidence of positive selection in *Solenopsis invicta,* the species in which it has been best studied (Gotzek et al., 2007).

BRAIN DEVELOPMENT AND FUNCTION

Some of the most striking differences between social and solitary insects are behavioral. Several social insect behaviors appear to be truly novel, such as symbolic dance communication in honey bees and slave making in ants (Wilson, 1971). Other behaviors exhibited by social insects appear to be modified forms of behaviors performed by solitary insects, for example, social foraging, which resembles nest provisioning in solitary insects. It is likely that molecular changes affecting nervous system development and function were important in the evolution of social insect behaviors, but very little is currently known.

Brain Evolution in Primitively Eusocial Bees

Our bee molecular evolution study detected a strong signal of rapid evolution in brain-related genes in primitively eusocial, but not highly eusocial lineages across two independent origins of each lifestyle (Woodard et al., 2011). Among these rapidly evolving genes were *dunce* and *nejire,* two genes that mediate learning and memory in invertebrates and vertebrates through cAMP/CREB signaling pathways (Silva et al., 1998).

The detection of molecular changes in brain-related genes exclusively in primitively eusocial bee lineages is perhaps surprising, given that this finding is not what may have been predicted by a prominent hypothesis about the relationship between sociality and brain evolution in vertebrates, the social brain hypothesis (SBH). Originally developed to explain the evolution of the enlarged neocortex in many social vertebrates, the SBH posits that the cognitive demands of social living are a strong selective force in brain evolution (Dunbar and Shultz, 2007). Given that highly eusocial bee societies have larger colony sizes, greater social complexity, and novel behaviors (i.e., dance communication in honey bees) relative to primitively eusocial bees, one might have assumed that the cognitive demands of social living are strongest in highly eusocial species and lead to stronger selection on brain-related genes.

Unique features of insect sociality and the primitively eusocial lifestyle may help to explain why selection on brain evolution appears to have been stronger in the primitively eusocial bees. First, unlike in vertebrate social evolution, where there has been an emphasis on increased individual cognitive abilities, there appears to have been an emphasis on increased connectedness among colony members in insect social evolution, often accompanied by a reduction of individual behavioral repertoires (Oster and Wilson, 1979; Gronenberg and Riveros, 2009). Therefore, individual cognitive abilities may not be correlated with group size in social insects, as has been found in vertebrates. There are also several distinguishing features of the primitively eusocial bee lifestyle that may have placed unique selective pressure on brain evolution in these lineages. Social structure in primitively eusocial bee colonies is typically maintained through fluid and dynamic dominance hierarchies (Michener, 1974; O'Donnell et al., 2007), which can be an especially cognitively challenging form of social interaction (O'Donnell et al., 2007; Salvador and Costa, 2009). In addition, a primitively eusocial bee queen is capable of behaving both solitarily, as she does during the colony-founding phase of her life cycle, and socially, as she does once she has reared her first brood of workers (Michener, 1974).

In both ants and wasps, which each evolved eusociality independent of bees, there are some species in which queens exhibit a similar "solitary-like" phase during colony founding and other species that found colonies in swarms, like highly eusocial bees do (Wilson, 1971). A comparison of brain-related genes and/or brain structure in ant and wasp species that do or do not establish colonies solitarily may provide clues as to whether this trait is a strong force in social insect brain evolution. One study in paper wasps reported brain region volume differences between swarm and independent-founding species, suggesting that these differences in colony founding can affect brain evolution (Molina et al., 2009).

IMMUNITY

Pathogens and parasites are thought to have been a strong selective force challenging the maintenance of sociality in a variety of organisms, including social insects (Wilson-Rich et al., 2009). Crowded living conditions, often with closely related individuals, facilitate pathogen transmission (Wilson-Rich et al., 2009). Social insects appear to have responded to this potentially dissolutive selective pressure in three main ways (Viljakainen and Pamilo, 2008). The first way is through "social immunity," which refers to group-level defenses, such as hygienic behaviors and the use of collected antimicrobial resins for lining nest cavities (Wilson-Rich et al., 2009). The second way is through increasing intracolonial genetic diversity via multiple mating by queens (Tarpy and Seeley, 2006) and high rates of genetic recombination (C. R. Smith et al., 2008) to enhance colony-level disease resistance. The third way is through adaptive evolution of immune genes (Viljakainen and Pamilo, 2008).

Molecular evolutionary analyses of immune genes have provided some of the best examples of positive selection acting in social insect genomes. This may be partly attributable to the fact that immune systems, in general, are often at the forefront of an ongoing evolutionary arms race with pathogens; thus, selection pressure on immune-related genes is typically quite strong (Lazzaro, 2008). In addition, many immune-related genes are functionally well characterized (Hoffmann, 2003), facilitating interpretations of the adaptive significance of sequence changes.

Immune Gene Evolution in Hymenoptera

When the first social insect genome was sequenced, that of *A. mellifera,* researchers were intrigued by the low number of immune genes found in *A. mellifera* relative to other fully sequenced insect genomes: those of the Diptera, *Drosophila melanogaster,* and *Anopheles gambiae* (Honeybee Genome Sequencing Consortium, 2006). Although the main components of canonical immune pathways are conserved, the *A. mellifera* genome contains smaller numbers of gene family members at all points along these pathways (Evans et al., 2006). It was hypothesized that the loss of immune genes was facilitated by novel forms of social immunity in social insects, resulting in relaxed constraint on immune genes (Evans et al., 2006). However, as more insect genomes have been sequenced, it has become apparent that sociality is not necessarily predictive of immune gene number. Rather, it seems that dipterans have unusually large immune gene repertoires, whereas the recently sequenced ant genomes (Bonasio et al., 2010; C. D. Smith et al., 2011; C. R. Smith et al., 2011; Wurm et al., 2011), the solitary wasps *Nasonia* (Werren et al., 2010), and the solitary pea aphid *Acyrthosiphon pisum* (International

Aphid Genomics Consortium, 2010) have similar numbers of immune genes as *A. mellifera* (Evans et al., 2006).

By contrast, molecular evolutionary analysis of individual immune genes in social Hymenoptera has provided evidence that sociality has driven immune gene sequence evolution. One study revealed that some immune genes are evolving more rapidly in species of honey bees, bumble bees, and ants relative to *Drosophila* (Viljakainen et al., 2009). This study also showed that immune genes are evolving more rapidly than non-immune genes in several honey bee species. Similarly, genes related to innate immunity and humoral immunity were among the fastest evolving (based on branch lengths in phylogenetic trees inferred from protein sequence) in *A. mellifera* in a comparison of over 3,000 genes among *A. mellifera, Nasonia,* and their common ancestor (Werren et al., 2010). Additionally, evidence for positive selection has been detected in the antimicrobial protein *defensin* in a study comparing the sequence of 27 ant species (Viljakainen and Pamilo, 2008). This study revealed that the signal and propeptide regions of *defensin,* which are cleaved off to activate the mature peptide, are evolving neutrally, whereas the active region of the peptide is under positive selection, including one amino acid site thought to mediate antimicrobial activity. Our bee molecular evolution study did not detect a strong signal of selection on immune genes, but that was likely because these classes of genes were underrepresented in our dataset (Woodard et al., 2011).

Immune Gene Evolution in Termites

A study of the termite *defensin*-like gene, *termicin,* in 11 *Nasutitermes* termite species revealed that this gene has duplicated repeatedly during *Nasutitermes* radiation and that positive selection has driven a divergence in the molecular charge of the gene copies (Bulmer and Crozier, 2004). Insect defensins are known to function by disrupting bacterial plasma membranes, and experimental evidence suggests that molecular charge may be a crucial component of this activity (Bulmer and Crozier, 2004). It was hypothesized that there is a selective advantage to having two termicins with different charge properties at specific sites (Bulmer and Crozier, 2004). In support of this hypothesis, results from this study suggest that ancestral termicins had relatively high positive charges and that in species in which there has been a gene duplication event, positive selection has driven a decrease in charge for one of the copies. Sequence analysis revealed a strong positive correlation between the strength of selection (d_N/d_S) and the change in molecular charge along different termicin lineages. Additionally, three amino acid sites that show a signature of positive selection have substitutions at these sites that contribute to a charge change, and they fall on the external surface of the predicted protein structure, suggesting that these sites may interact with a fungal membrane receptor (Bulmer and Crozier, 2004).

A different study of 13 *Nasutitermes* termite species also found evidence that gene duplication and positive selection are involved in termite immune gene evolution (Bulmer and Crozier, 2006). This study focused on genes encoding Gram-negative bacterial-binding protein 1 and 2 (GNBP1 and GNBP2), which are thought to have duplicated early in termite evolution, and the transcription factor *relish,* which induces production of antimicrobial peptides in *Drosophila.* All three genes show evidence of positive selection, with *relish* showing the strongest signal. Four of the five positively selected sites in *relish* are in a "spacer" region of the protein that is cleaved by the caspase Dredd. This cleavage is thought to activate *relish* by generating a DNA-binding Rel homology domain that translocates to the nucleus and binds to promoters of target genes (Stoven et al., 2003). Analysis of the *Drosophila simulans* ortholog also found positive selection in this spacer region (Bulmer and Crozier, 2006). It was hypothesized that microbial pathogens may be targeting this region of *relish* to prevent its activation, sparking an evolutionary arms race as *relish* evolves counterresponses to maintain its normal function (Bulmer and Crozier, 2006). Another study found evidence of positive selection in *termicin* but not in GNBP2 in two *Reticulitermes* termite species, a genus distantly related to the *Nasutitermes* genus (Bulmer et al., 2010). This study used a population genetics approach to analyze intraspecific polymorphism and interspecific divergence in coding sequence, and results indicated that *termicin* underwent a selective sweep driven by positive selection for beneficial amino acid changes.

REPRODUCTION

In many insect societies, queens are highly reproductive individuals, whereas workers perform almost no reproduction activity. Worker sterility is achieved through a variety of morphological, behavioral, and physiological mechanisms in social insects (Wilson, 1971). For example, in many social species, workers lack spermatheca for sperm storage. In addition, ovary development is tightly regulated by social cues, and queens and workers typically have grossly over- and underdeveloped ovaries, respectively, relative to solitary insects (Wilson, 1971). Sociality also has strong implications for reproductive behavior, particularly for mating frequency, which can affect genetic variation among colony members.

Ovary Development in Primitively Eusocial Bees

Our bee molecular evolution study identified some genes involved in ovary development evolving most rapidly in primitively eusocial bees (Woodard et al., 2011). Although both highly and primitively eusocial bee societies have a strong reproductive division of labor, the reproductive differences between queen and worker in primitively eusocial species are less extreme, and ovary development appears to be more sensitive to social cues in primitively eusocial species (Wilson, 1971). Perhaps the molecular changes in ovary development-related genes found only in the primitively eusocial lineages underlie some of the unique characteristics of the reproductive biology of this eusocial lifestyle.

Several genes (i.e., *tudor, capsuleen, vasa*) evolving rapidly in one or both of the primitively eusocial bee lineages interact together in the PIWI RNA (piRNA) pathway. The piRNA pathway is expressed only in gametic tissue, and it is involved in regulating gametic cell division and differentiation (Siomi et al., 2010). Functional PIWI genes have recently been discovered in *A. mellifera* (Liao et al., 2010), suggesting that the piRNA pathway is present and functional in bees. These genes are particularly good candidates for further study, because the tissue specificity of the piRNA pathway suggests that selection on these genes is specifically directed at changes related to reproductive processes, in contrast to genes with broader ranges of tissue expression, where the functional target of selection is harder to infer. Additional ovary development-related genes unrelated to the piRNA pathway also showed a signature of rapid evolution in these primitively eusocial bees (Woodard et al., 2011).

Sex Determination and *complementary sex determiner* in Honey Bees

More is known about the evolution of *complementary sex determiner* (*csd*) in honey bees than probably any other gene in the social insects. The story of *csd* involves the origin of entirely new genes and pathways, as well as a classic example of balancing selection. Sex determination in honey bees is based on genotype at the *csd* locus; individuals heterozygous at the *csd* locus develop into females, whereas hemizygous individuals develop into males (Beye et al., 2003). Sex in many Hymenoptera is probably determined through a similar single-locus system of complementary sex determination (Cook, 1993), but *csd* is the first and only locus that has been discovered thus far. The genomic region containing *csd* was first identified through mapping (Beye et al., 2003), and the function of the gene was confirmed by RNAi, which showed that reducing *csd* expression in genetically female eggs results in male-like development (Hasselmann et al., 2008a). Complementary sex determination not only regulates sex determination but influences many aspects of social insect biology that are influenced by kinship and degrees of relatedness, including kin selection and the genetic composition of colonies, which are important for division of labor and colony immunity (C. R. Smith et al., 2008).

The *csd* gene appears to be a honey bee–specific gene because it has been found in multiple *Apis* species (Hasselmann et al., 2008b) but not outside of the genus (Hasselmann et al., 2008a). The gene likely evolved through the duplication of an adjacent gene, *feminizer* (*fem*). The *csd* and *fem* genes are similar (>70%) in amino acid sequence, and both are serine/arginine-rich proteins, a class of proteins involved in RNA splicing (Hasselmann et al., 2008a). Both genes share two major domains, but *csd* has an additional hypervariable region located between these other domains (Hasselmann et al., 2008a). The *fem* gene has been found in several non-honey bee species and in *Nasonia* wasps, but not in any additional insect species, suggesting that it evolved sometime before the split between the hymenopteran superfamilies Apoidea and Chalcidoidia ~140 Mya but after the split from *Drosophila* ~300 Mya (Hasselmann et al., 2008a). The *fem* gene shares some functional and sequence similarities to *transformer* (*tra*), a gene involved in sex determination in *Drosophila,* and it perhaps evolved from an ancestral form of *tra* common to fly and bee lineages (Beye et al., 2003; Hasselmann et al., 2008a). RNAi experiments were used to show that *csd* acts upstream of *fem* in the sex determination pathway. Genetically female embryos treated with *fem* RNAi develop male heads, and RNAi knockdowns of *csd* cause male-specific *fem* splicing, suggesting that *csd* is involved in *fem* splicing (Hasselmann et al., 2008a).

The *csd* gene has been subject to rigorous population genetic analysis. Because homozygous males do not reproduce, it was predicted that there would be strong negative frequency-dependent selection at the *csd* locus (Hasselmann et al., 2008b). This prediction has been upheld, because at least 15 different *csd* alleles have been found in natural populations around the world in three different *Apis* species (Hasselmann et al., 2008b) and the gene has accumulated 10- to 13-fold more mutations than the rest of the genome (Hasselmann et al., 2008b). Pairwise nonsynonymous differences between alleles are highest in exons 6 and 7 (Hasselmann et al., 2008b), suggesting that this region is a target of positive selection, and is therefore presumably functionally important. Six fixed amino acid differences between *csd* and *fem* are located in the coiled-coil domain, which is important in protein binding (Hasselmann et al., 2008a). Strong positive selection was detected on the branch right after the split between the two genes, suggesting that positive selection played a role in their diversification (Hasselmann et al., 2008a).

METABOLISM AND NUTRITION

Transcriptomic analyses have shown that nutritional and metabolic pathways play an important role in queen-worker caste determination in every eusocial insect lineage thus far studied and also contribute to worker-worker division of labor in many species (C. R. Smith et al., 2008). Given these fundamental connections to eusociality, nutritional and metabolic pathways are well studied in social insects and several molecular evolutionary studies have identified changes associated with their function.

Major Royal Jelly Proteins

The evolution of the Major Royal Jelly Proteins (MRJPs) in honey bees is an excellent example of novel genes playing an integral role in the social biology of a species. In the honey bee, *A. mellifera,* the developmental fate of female larvae is determined by the amount of royal jelly they consume (Kamakura, 2011). Royal jelly is a protein- and lipid-rich substance secreted from the hypopharyngeal glands of brood-feeding "nurse" bees and fed to larvae, which triggers endocrine and epigenetic events that lead to the development of either a worker or a queen (Lyko et al., 2010; Kamakura, 2011). The main components of royal jelly are the MRJPs. The *A. mellifera* genome contains 10 *mrjp* genes, encoding 9 MRJPs (one *mrjp* is a pseudogene). These genes are arranged in tandem in the genome, have high sequence similarity (~60%) to one another, and have a conserved intron/exon structure, suggesting that they are a fairly young gene family (Drapeau et al., 2006). There is evidence that *mrjp* genes are also present in other *Apis* species (Drapeau et al., 2006; Yu et al., 2010).

The *mrjp* gene family in *A. mellifera* appears to have evolved via a gene duplication event from a member of the *yellow* gene family. The cluster of *mrjp* genes in the *A. mellifera* genome is flanked by members of the *yellow* gene family, and one of the flanking *yellow* genes, *yellow-e3,* shares the characteristic intron/exon structure of the *mrjp* genes, suggesting that it is their progenitor (Drapeau et al., 2006). Members of the *yellow* gene family are involved in pigmentation, reproductive physiology, and courtship behavior in insects (Ferguson et al., 2011).

The use of *mrjp* genes for larval feeding appears to be a derived social trait that is unique to honey bees. Although *mrjp*-like genes have been found in other social and nonsocial Hymenoptera species, evidence suggests that the *yellow* gene family is prone to duplication and that the *mrjp*-like genes in *non-Apis* species evolved independently of *Apis* (Werren et al., 2010; C. D. Smith et al., 2011). Furthermore, there is no evidence of a food-related role for any *mrjp*-like or *yellow*-like gene outside of *Apis* (Ferguson et al., 2011). Because many other social insect species manipulate larval nutrition for the purposes of caste determination without the use of specialized glandular secretions (Webster and Peng, 1988), the evolution of the *mrjp* genes in honey bees appears to be associated with the elaboration of eusociality and may have been correlated with or dependent on other evolutionary changes, such as changes in gland function.

Hexamerins

The work done on the termite hexamerins is another excellent example of linking genetic changes to protein function and social phenotype. In the lower termites, workers may develop into either reproductives or soldiers, depending on a number of social and environmental cues, and differentiation into the soldier caste is induced by high juvenile hormone (JH) titers (Zhou et al., 2007). RNAi studies in the termite *Reticulitermes flavipes* have shown that two hexamerin genes, *Hex-1* and *Hex-2*, are involved in the regulation of this caste determination (Zhou et al., 2006). In many insects, hexamerins act as storage proteins that sequester substances from the diet and release them when food is scarce or inaccessible, such as during early development (Zhou et al., 2007). It has been hypothesized that *Hex-1* and *Hex-2* work together to regulate caste differentiation in termites via direct interactions with JH (Zhou et al., 2006); however, elucidating the specific molecular mechanisms involved in JH action is a difficult challenge in insects in general (Riddiford, 2008).

Molecular evolutionary studies of *Hex-1* and *Hex-2* provide clues as to how these genes may interact with JH. Relative to 100+ known *Hex* genes in other insects, both termite *Hex* genes have distinctive insertions in their coding regions; the unique insertion in *Hex-1* contains a prenylation motif with a proposed function in JH binding, and the unique insertion in *Hex-2* shares sequence similarities to the well-characterized blowfly (*Calliphora vicina*) hexamerin receptor (Drapeau et al., 2006). Consistent with these predicted functions, follow-up experiments demonstrated that the Hex-1 protein has strong binding affinity for JH and the Hex-2 protein shows strong membrane affinity, as would be expected for a receptor protein (Zhou et al., 2006).

Hexamerins also exhibit novel social functions in other social insect species, suggesting that they may be particularly prone to social co-option. Evidence in honey bees (Martins et al., 2010) and *Polistes* wasps (J. H. Hunt et al., 2010) suggests that hexamerins may be important in caste determination in these social insect lineages, and in ants, hexamerins appear to be have been important in the evolution of elaborated life history characteristics (Wheeler and Buck, 1995).

JH, Insulin, and Vitellogenin Axis

In the highly eusocial honey bee, *A. mellifera*, the JH and insulin/insulin-like growth factor-1 (IIS) signaling pathways, as well as the yolk protein precursor vitellogenin (Vg), interact with one another and function in novel ways that are important in multiple social contexts. JH does not function as a gonadotropin in adult honey bees as it does in most insects; instead, it plays a strong role in caste determination and worker division of labor (Robinson and Vargo, 1997). The IIS signaling pathway interacts with JH and is also involved in worker division of labor. Foragers exhibit higher expression of genes in the IIS pathway in the brain relative to nurses, and down-regulating IIS signaling delays the age-related transition from nursing to foraging (Ament et al., 2008). This represents a reversal of the traditional positive relationship between high nutrition and IIS signaling, because foragers are nutritionally deprived relative to nurses (Ament et al., 2008). Vg also shows novel social functions in honey bees. It is highly expressed in some workers, although they are largely nonreproductive; it may be used by nurses in the synthesis of royal jelly (Amdam et al., 2003); and it functions as an antioxidant that may be involved in promoting longevity in queen bees (Corona et al., 2007).

The molecular changes underlying these novel functions of JH, IIS, and Vg are unknown, but insights from solitary insects may provide clues as to what these changes may be. The relationship between genetic variation and regulation of JH titers has been particularly well studied in crickets and butterflies (Zera et al., 2007), molecular evolution and function of the IIS pathway have been investigated across the complete genomes of 12 *Drosophila* species (Alvarez-Ponce et al., 2009; Grönke et al., 2010), and insect Vgs and their receptors are well characterized at the molecular level (Sappington and Raikhel, 1998).

Carbohydrate Metabolism

Several studies in bees suggest that the evolution of the highly eusocial lifestyle involved molecular changes in genes related to carbohydrate metabolism. Our bee molecular evolution study revealed that genes involved in carbohydrate metabolism are evolving more rapidly in eusocial relative to noneusocial bee lineages and are evolving most rapidly in highly eusocial lineages (Woodard et al., 2011). In particular, 15 genes encoding glycolytic enzymes showed evidence of rapid evolution in eusocial lineages, including enzymes that play a key regulatory role (e.g., *phosphofructokinase*) or are involved in glycolytic flux (e.g., *hexokinase, pyruvate kinase*) (Kunieda et al., 2006). Analysis of protein sequence evolution of genes with queen-biased brain gene expression in *A. mellifera* found that queen-biased genes involved in metabolism, including carbohydrate metabolism, were among the most rapidly evolving (based on branch lengths in phylogenetic trees inferred from protein sequence) relative to orthologs from several solitary insects (B. G. Hunt et al., 2010). Comparative analysis of the genome sequences of *A. mellifera, D. melanogaster,* and *A. gambiae* suggest that there may also have been bee-specific changes in gene copy number for carbohydrate-metabolizing genes (Kunieda et al., 2006). Given that carbohydrate metabolism is such a fundamental "housekeeping" process, it is not immediately clear why there has been unique selective pressure on these processes in highly eusocial bee lineages. Here, we offer three speculative hypotheses.

First, increases in the flight demands of highly eusocial bees may have placed strong selective pressure on increasing efficiency of glycolytic enzymes, because carbohydrates are the main fuel for flight in bees (Suarez et al., 2005). The individual foraging activity of highly eusocial bee workers appears to be higher than for solitary bees (Roubik, 1992), although, to the best of our knowledge, no direct comparisons of highly and primitively eusocial bee foraging activity have been performed.

Second, highly eusocial bees are unique in relying exclusively on a diet of modified stored sugars (i.e., honey) for long periods of time. Nest thermoregulation during winter months is completely reliant on honey stores as a fuel source to sustain workers, who shiver to produce metabolic heat to maintain optimal hive temperature (Southwick and Heldmaier, 1987). Perhaps these differences in diet have placed some novel selective pressure on glycolytic enzymes in highly eusocial lineages.

Third, perhaps the greatly extended life span of queens in highly eusocial species evolved through changes in metabolism-related genes, including those involved in carbohydrate metabolism. A connection between reduced metabolic rate and increased life span has been shown in many species (Finkel and Holbrook, 2000). In the honey bee, *A. mellifera,* queens exhibit an age-related reduction in IIS signaling (Corona et al., 2007) that regulates carbohydrate metabolism. If the molecular changes in carbohydrate metabolism genes in highly eusocial bees were attributable to selection for extended queen life span, it can be predicted that similar molecular changes may also be found in independent social insect lineages that also exhibit extended queen life spans (Wilson, 1971).

PROSPECTS AND CHALLENGES

Recent work on molecular evolutionary changes in social insects has identified specific genes, molecular pathways, and biological processes that appear to have been shaped by natural selection. Some of these changes can be plausibly associated with the origins, maintenance, or elaboration of eusociality, albeit speculatively.

Two insights emerge from this review. First, it appears that there have been unique genetic changes in different social insect lineages, suggesting that the multiple independent occurrences of eusociality have involved multiple molecular routes. These differences may reflect distinct ecological or other constraints for each lineage. For example, the evolution of a queen pheromone in termites from a wood-digesting enzyme seems fitting, given that many termite societies live in rotting wood (Korb et al., 2009).

Second, genetic changes also have occurred in similar biological functions across diverse species of social insects. This supports the concept of a genetic toolkit for eusociality (Toth and Robinson, 2009). This concept is reasonable, because despite the striking diversity among social insect species, they all have converged on a similar suite of traits, which are the defining characteristics of eusociality (Michener, 1974). Previous research suggesting components of a genetic toolkit for eusociality has focused on genes and molecular pathways that are associated both with solitary and related social behaviors in insects, for example, the *foraging* gene, which is involved in feeding behavior in *Drosophila* and a variety of other solitary organisms, and social foraging behavior in honey bees and ants (Toth and Robinson, 2009). Transcriptomic studies have also identified shared sets of genes whose expression patterns are associated with division of labor in independent social insect lineages (Toth et al., 2010).

The molecular evolutionary studies we reviewed identify biological processes and specific genes that may be excellent systems in which to investigate the concept of a genetic toolkit for eusociality further. Among the most promising are the following:

(*i*) *Hexamerins.* As discussed above, hexamerins have been shown to be involved in queen physiology and other social traits in a variety of social insects, and the work on *Hex-1* and *Hex-2* in termites demonstrates how hexamerin sequence evolution can be studied and linked to social traits.

(*ii*) *Gland development genes.* The rapidly evolving gland development genes identified in our bee molecular evolution study (Woodard et al., 2011) are also good candidates for further study, because the gene functions are relatively well characterized, and gland diversification is a universal phenomenon in social insect evolution.

(*iii*) *Brain-related genes.* The rapidly evolving brain-related genes identified in primitively eusocial lineages in our bee molecular evolution study (Woodard et al., 2011) are prime candidates for further study in primitively eusocial bees, as well as in ant and wasp species that share the primitively eusocial bee lifestyle feature of solitary nest-founding.

The molecular changes and biological processes highlighted in this review are currently the most well studied in social insects. There are almost certainly other equally important types of molecular changes and biological processes associated with social insect evolution that have not yet been discovered, perhaps because of the limited range of taxa subjected to these types of analyses thus far. This gap in our knowledge is largely attributable to a lack of genomic resources, especially for closely related social and nonsocial species. For example, some types of genetic changes, such as chromosomal rearrangements and patterns of DNA methylation, are not possible to study with only fragments of the genome. In addition, the identification of truly novel genes is limited by the small sample size of available genomes and less well-developed forward genetic analyses in social insects relative to model genetic organisms. As these limitations are overcome, it should be possible to search more broadly for different types of genetic changes associated with the evolution of eusocial traits. These analyses can be guided by several theoretical models that have been proposed to predict the types of genetic changes that are most important in social evolution (Nonacs and Kapheim, 2007; Linksvayer and Wade, 2009; Johnson and Linksvayer, 2010).

Whole-genome scans for molecular signatures of adaptive evolution specific to social insects will be particularly useful for generating new hypotheses and implicating new biological processes in social insect evolution. Candidate gene approaches across a broad sample of social and nonsocial insects will allow for greater accuracy in reconstructing the phylogenetic history of molecular changes and testing their associations with social evolution. Once specific sequence changes are identified, functional analyses are necessary to determine their effect on protein-, organismal-, and group-level phenotypes, as well as the adaptive significance of the phenotype change (Dean and Thornton, 2007).

This leads us to raise one important caveat for most molecular evolutionary studies in the social insects: the lack of species-specific information about gene function. As is often the case, gene function in this chapter is typically inferred from orthology to the fruit fly *D. melanogaster,* which shared a common ancestor with eusocial insect lineages over 300 Mya (Honeybee Genome Sequencing Consortium, 2006). Although gene function for molecular processes is generally highly conserved over evolutionary time, when interpreting findings, it is important to consider the possibility that a particular gene has evolved a novel function. Furthermore, many genes have multiple functions; thus, the target of selection can be difficult to infer solely from identifying molecular evolutionary changes.

Experimental approaches to determining gene function in social insects, via RNAi and transgenesis, will strengthen the interpretation of molecular evolutionary findings. Additional challenges arise in determining the adaptive or ecological significance of molecular changes, even when their functional significance is understood (Feder and Mitchell-Olds, 2003).

Despite these challenges, molecular evolutionary analysis of social insect societies holds promise for testing venerable theories of social evolution using genomic data. Multiple evolutionary scenarios have been proposed as potential routes to group living in insects. These include the composition of incipient social groups, such as associations between mothers and offspring (the "subsocial" route) or between related and unrelated individuals of the same generation ("semisocial" route) (Michener, 1974); mechanisms through which altruism is achieved, such as kin selection (Strassmann and Queller, 2007); parental manipulation of offspring or voluntary helpers at the nest (Charnov, 1978); and necessary preadaptations for social living, such as a monogamous mating system (Hughes et al., 2008) or progressive provisioning of offspring (Nowak et al., 2010). Wedding this rich theory with genome-scale molecular evolutionary analysis and functional experimentation holds the promise of finally answering the compelling question of how eusociality evolved in insects.

ACKNOWLEDGMENTS

We thank members of the G.E.R. laboratory for comments that improved this manuscript. Our bee molecular evolution study was supported by 454 Life Sciences (Roche Diagnostics Corporation) via the Roche 1GB contest, National Science Foundation Grant DEB07–43154 (to G.E.R., Primary Investigator), National Science Foundation Predoctoral Fellowship NSF DGE 07–15088 FLW (to B.J.F.), and National Institutes of Health–University of Illinois Sensory Neuroscience Training Grant PHS2T32DC006612 (to S.H.W. and A. S. Feng, Primary Investigator).

9

Evolution of Cooperation and Control of Cheating in a Social Microbe

JOAN E. STRASSMANN*† AND DAVID C. QUELLER*

Much of what we know about the evolution of altruism comes from animals. Here, we show that studying a microbe has yielded unique insights, particularly in understanding how social cheaters are controlled. The social stage of *Dictylostelium discoideum* occurs when the amoebae run out of their bacterial prey and aggregate into a multicellular, motile slug. This slug forms a fruiting body in which about a fifth of cells die to form a stalk that supports the remaining cells as they form hardy dispersal-ready spores. Because this social stage forms from aggregation, it is analogous to a social group, or a chimeric multicellular organism, and is vulnerable to internal conflict. Advances in cell labeling, microscopy, single-gene knockouts, and genomics, as well as the results of decades of study of *D. discoideum* as a model for development, allow us to explore the genetic basis of social contests and control of cheaters in unprecedented detail. Cheaters are limited from exploiting other clones by high relatedness, kin discrimination, pleiotropy, noble resistance, and lottery-like role assignment. The active nature of these limits is reflected in the elevated rates of change in social genes compared with nonsocial genes. Despite control of cheaters, some conflict is still expressed in chimeras, with slower movement of slugs, slightly decreased investment in stalk compared with spore cells, and differential contributions to stalk and spores. *D. discoideum* is rapidly becoming a model system of choice for molecular studies of social evolution.

*Department of Biology, Washington University, One Brookings Dr., Campus Box 1137, St. Louis, MO 63130. †To whom correspondence should be addressed. E-mail: strassmann@wustl.edu.

Natural selection favors cooperation when genes underlying it increase in frequency compared with their noncooperative counterparts (Hamilton, 1964a; Frank, 1998; West et al., 2007b). Evolutionary studies of cooperative interactions have focused on the selective advantages of cooperating, how cooperation is organized, whether cheating a cooperative system can occur, and how cheaters are controlled (Ratnieks, 1990; West et al., 2002c; Beekman and Ratnieks, 2003; Griffin et al., 2004; Sachs et al., 2004; Travisano and Velicer, 2004; Ratnieks et al., 2006; Wenseleers and Ratnieks, 2006b). These studies generally, but not always, focus on within-species interactions and have been behaviorally oriented. Social insects have been a major focus (Bourke and Franks, 1995; Robinson, 2002; Strassmann and Queller, 2007), with cooperative birds and mammals also getting considerable attention (Cockburn, 1998; Clutton-Brock et al., 2001; Cornwallis et al., 2010). The past few decades have seen phenomenal progress in understanding cooperation in these organisms by applying the powerful logic of kin selection (Queller, 1992a; Frank, 1998; West et al., 2007b).

Our advances in understanding the evolution of social behavior through kin selection have been very satisfying, but they have been isolated in some respects. This is because most organisms have not been seen to be particularly cooperative. They may come together briefly for mating but otherwise go about the business of securing nutrients, avoiding disease and predation, and producing progeny largely on their own.

COOPERATION IS WIDESPREAD

Behavioral ecologists have begun to study a wider selection of organisms and are finding cooperative interactions to be much more pervasive than previously appreciated. This is particularly true for microbes, wherein the structured environments necessary for cooperation have been discovered to be pervasive (Kerr et al., 2002; Griffin et al., 2004; Vos and Velicer, 2009). Microbes are particularly affected by the actions of their neighbors, because many functions that are internal in multicellular organisms are external in single-celled organisms. Secreted compounds involved in processes like iron sequestration or food digestion are vulnerable to exploitation by neighboring individuals (Travisano and Velicer, 2004; Buckling et al., 2007; West et al., 2007a). Microorganisms evaluate their numbers with quorum sensing, kill nonclonemates with bacteriocins, hunt in groups, and cooperatively swarm through their environment, to name just a few examples of their social attributes (Crespi, 2001; Riley and Wertz, 2002; Diggle et al., 2007a; West et al., 2007a). Sociality in nontraditional study organisms is only beginning to be understood, however.

COOPERATION, ORGANISMALITY, AND MAJOR TRANSITIONS IN EVOLUTION

The second reason for expanded interest in cooperation is a growing appreciation that it is important for how organisms came to be. Cooperative major transitions in life alter the raw material for natural selection in fundamental ways (Buss, 1987; Maynard Smith and Szathmáry, 1995). One of the earliest transitions brought molecules together into cells in which the fates of all were intertwined in a cooperative network. Eukaryotes themselves represent a major transition resulting from the capture of a bacterium that becomes the mitochondrion (Margulis, 1970). The level of cooperation between these partners is profound but not complete. Mitochondria are maternally inherited and do not go through meiosis, and thus will favor daughter production and have no interest in son production.

Another major transition resulted in multicellularity (Queller, 1997, 2000; Grosberg and Strathmann, 1998; Herron and Michod, 2008). Multicellularity has evolved multiple times in both bacterial and eukaryote lineages. Animals and plants have elaborated multicellularity into a plethora of diverse types. There are also a number of comparatively simple multicellular forms, like some single-species biofilms, the algal group Volvocales, or *Dictyostelium* (Herron and Michod, 2008; Strassmann and Queller, 2010). The transition to multicellularity is different from the transition to eukaryotes because the former involves an aggregate of like entities, whereas the latter binds different elements. The major transitions can thus be categorized as fraternal, with like cooperating with like, or egalitarian, where the cooperating units bring different things to the collaboration (Queller, 1997). Either kind of collaborative organism will usually retain conflicts, but these conflicts must be controlled if the partnership is to survive. How these controls operate is a major research topic under this view of life.

The selective factors that favored a past transition are not easy to study because they have already completed their work. There are living systems that could be considered to be more representative of transitional stages, however. These, we believe, may be the most productive for investigation into the advantages of cooperation and how conflict is controlled. We have argued elsewhere that organisms themselves can be defined as adapted bundles of cooperative elements, wherein actual conflict is at a minimum (Queller and Strassmann, 2009; Strassmann and Queller, 2010). In a 2D space, with one axis being cooperation and the other being conflict, organisms are those collaborative living units at the high end of cooperation and the low end of conflict. There is variation in the level of organismality, however, and those lacking complete cooperation and retaining conflict represent the best choices for studying the origins of cooperation.

LABORATORY-FRIENDLY, SOCIAL MODEL ORGANISMS

Kin selection has been very successful for generating predictions on the impact of queen number, mate number, and caste on sociality in social insects (Bourke and Franks, 1995; Bourke, 2011). Nevertheless, one would have to say that social insects fall short as an ideal model for studies of social evolution. They are long-lived, often do poorly in the laboratory (except ants), are not amenable to genetic experimentation, and have mostly already crossed the threshold to obligate sociality. Thus, social evolution research has not found its *Drosophila* here.

Another problem with the organisms currently favored for studies of cooperation is that the actual genes underlying cooperative behavior are elusive. This is particularly true for long-lived social insects and vertebrates, although the advances of genomics are slowly mitigating this (Robinson, 2002; Honeybee Genome Sequencing Consortium, 2006). Still, the twin powers of experimental evolution and single-gene knockouts are beyond the reach of most currently studied social organisms.

A social evolution *Drosophila* would need to address these issues; thus, it would probably be single-celled. In addition to being amenable for experimental evolution and single-gene knockouts, it should have full altruism, with some individuals dying to help others. This makes it easier to interpret the actions of different partners. Other attributes of the ideal social *Drosophila* include feasibility of study in a fairly natural environment, placement in a rich phylogeny with related species that vary in social traits, a sequenced set of genomes, and a collegial community of fellow investigators. Here, we make the case that the ideal model organism for social evolution has been found and is the social amoeba *Dictyostelium discoideum.* This choice is supported by the enormous progress in understanding social evolution that has been made with this organism in the past decade. In addition to *D. discoideum, Volvox* and its relatives are great for studying the origins of multicellularity in a clonal organism (Herron and Michod, 2008). *Myxococcus xanthus* offers all the advantages of a bacterial system (Velicer and Stredwick, 2002). There are also others, but we focus here on *D. discoideum* (Fig. 9.1).

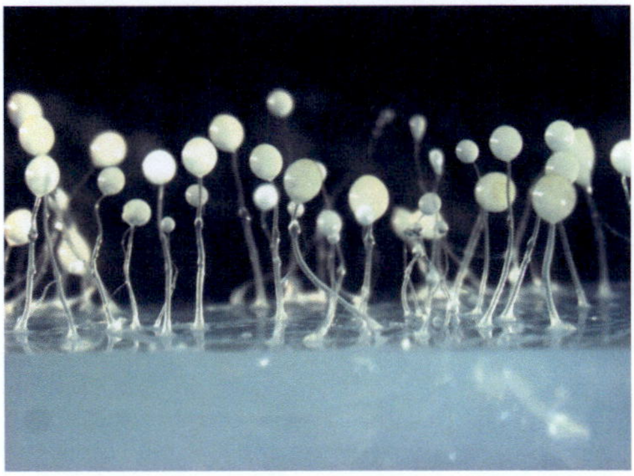

FIGURE 9.1 *D. discoideum* fruiting bodies on an agar plate.

DICTYOSTELIUM DISCOIDEUM AS A MODEL SYSTEM FOR COOPERATION

What Is a Social Amoeba?

Social amoebae are in the eukaryote kingdom Amoebozoa, sister to the Opisthokonts, or animals plus fungi (Baldauf et al., 2000). This kingdom is composed of solitary amoebae-like *Entamoeba* and *Acathamoeba*, the acellular slime molds such as *Physarum*, and the Dictyostelidae. There are over 100 species of *Dictyostelium*, divided into four major taxonomic groups (Raper, 1984; Schaap et al., 2006). *D. discoideum* is in group four and is the focal species here.

Individual amoebae of *D. discoideum* live in the upper layers of soil and leaf litter in the eastern Northern Hemisphere and in eastern Asia. The most intensely studied clone, NC4, and its derivatives such as Ax4, come from a temperate forest near Mount Mitchell in western North Carolina (Raper, 1984). *D. discoideum* amoebae are solitary predators on bacteria, which they consume by engulfment (Bonner, 1967). Although this is usually viewed as a solitary stage, they are always able to sense the density of nearby amoebae with a molecule called prestarvation factor (Kessin, 2001). Response to this factor is inhibited when bacteria are present (Kessin, 2001). When bacteria get scarce, and amoeba density is sufficient, they enter one of two stages, a sexual one, discussed later, or a social one (Fig. 9.2).

Social Cycle

The social stage, often called the developmental stage, occurs when *D. discoideum* amoebae begin to starve (Fig. 9.2). Amoebae have a quorum-

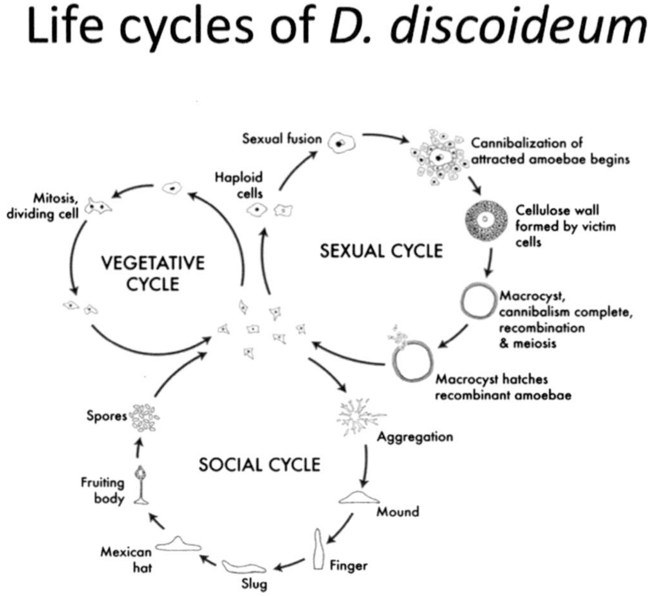

FIGURE 9.2 Colony cycles of *D. discoideum.* This study focuses on the social cycle, but the sexual cycle is a promising area for future study.

sensing mechanism; if there are enough other amoebae in the area, they begin to release cAMP and to make receptors to it, products of the *CAR* genes (Kessin, 2001; Alvarez-Curto et al., 2005). A signal relay system causes the amoebae to move up the cAMP gradient and form a mound of hundreds of thousands of cells. Differentiation begins in the mound stage, wherein some cells sort out toward the tip and express prestalk genes. The tip becomes the anterior of the slug and organizes forward movement. During movement, cells are lost from the slug posterior. At least some of these are capable of dedifferentiating and consuming any bacteria encountered (Kuzdzal-Fick et al., 2007). The slug itself will not fall apart on encountering bacteria. Some shed cells are former sentinel cells, full of toxins, and bacteria mopped up as they traverse through the slug (G. Chen et al., 2007).

The multicellular slug moves toward heat and light and away from ammonia (Kessin, 2001; Bonner, 2006). The cells at the tip then migrate down through the center of the aggregate and initiate stalk formation in a process called culmination. The stalk cells vacuolate and die, forming sturdy cellulose walls in the process that give them the strength to hold up the spherical ball of spores. The final fruiting body consists of about 20% stalk cells and 80% spore cells. Thus, the social stage is triggered by starvation and involves altruism, because the stalk cells die to support the spore cells (Kessin, 2001).

D. discoideum arrives at multicellularity not through development from a single cell but through aggregation of dispersed cells. Therefore, the social stage of *D. discoideum* is vulnerable to cheaters. This makes it fundamentally different from a metazoan that has gone through a single-cell bottleneck and had the interests of all cells in the organism reset to complete cooperation every generation (Maynard Smith, 1989b). This conflict, its control, and the resulting cooperation are what make *D. discoideum* such a great model for social evolution.

Why Have a Social Stage?

During the social process, three things happen, and we predict that all three are adaptive. First, spores are made. The adaptive value of a hardy spore is clear and has been demonstrated; it is not easily digested by predators and can withstand long periods of cold, heat, or drought (Raper, 1984). Second, the spores are only made atop a relatively long stalk composed of dead cells. These stalks can be anywhere from 1 to about 4 mm long, and their construction is the most vital part of the altruism story of *D. discoideum.* Why are spores made only atop stalks? It could be that cells are vulnerable during the transformation to spore, and doing so atop a stalk protects them from hazards in the soil. Another possibility is that dispersal is facilitated when the spores are lifted above the soil and that this is the main purpose of the stalk. In *D. discoideum,* spores are likely to be actively transported on small invertebrates, although the guts of vertebrates and stalks could increase the chance that they are contacted. The third advantage to grouping is slug movement; slugs move farther than amoebae, which could position them into a better place for dispersal. The complex orchestration of fruiting body formation could only have arisen through natural selection, but more work on the actual advantages is needed. In this review, we focus on the interactions of genetically different clones in this social process and not on the reasons why it is adaptive.

Chimerism and Cheating the Social Contract

Mixing of two or more genetically distinct clones is likely for social groups that form by aggregation. To see if this actually occurs, we collected tiny soil samples of 0.2 g at Mountain Lake Biological Station (Fortunato et al., 2003b). We reasoned that this was a reasonable scale over which social aggregation might occur. We found that our 0.2-g samples contained zero to five clones and that relatedness within the samples was about 0.52. These data support the view that chimerism is possible, at least in this population.

Later, we were able to find and genotype individual wild-fruiting bodies collected from the very rich resource of deer dung and nearby soil. This approach gave much higher relatednesses, between 0.86 and 0.98 depending on the sample and technique (Gilbert et al., 2007). Thus, relatedness is clearly high enough for kin selection under reasonable values of costs and benefits, and chimerism is common enough for social competition to be favored evolutionarily. Nevertheless, for cooperation to occur, there must be control of cheating. Here, we discuss what cheating is and then move on to evidence for it and its control in *D. discoideum*.

Complications with Defining Cheating

Cheating can only happen when one organism takes advantage of another; however, it is more than that. We would not say the lion cheated the gazelle out of its life with the lion's pounce and suffocating bite. This is because there is no expectation that the lion would behave in any other way. So, for an exploitative behavior to be considered cheating, there must be some expectation of cooperation that is not met. Cheating, therefore, is a fundamentally social action that takes place in the context of ordinarily cooperative acts, which the cheater somehow violates.

In *D. discoideum*, we talk of cheating in the context of cell allocation to the somatic, dead stalk and the living spores. The expected social contract is that the frequency of each clone among the spores will be the same as it was in the original mixture of aggregated cells. The same should be true in the stalk tissue. If this is not the case, we can say that the dominant clone cheated the minority clone by getting more than its fair share into spores, and cooperation can be put at risk when cheaters gain an advantage.

In many kinds of interactions, the starting and ending frequency may be viewed as enough information to determine if one partner is cheating the other. The formation of a fruiting body from an initial population of spores is a process that could vary for reasons other than social competition, however. Some clones may make longer or more robust stalks than others when they are entirely on their own. Some clones may migrate farther than others, losing cells in the process. Some clones may lose more cells from the slug than others even if they migrate the same distance. Variation is particularly expected in the highly variable environment of the soil. For example, a loose-grained soil may favor longer stalks for a given number of cells than a tighter-grained soil if the adapted trait is to rise above the surface. Selection on these traits can occur inde-

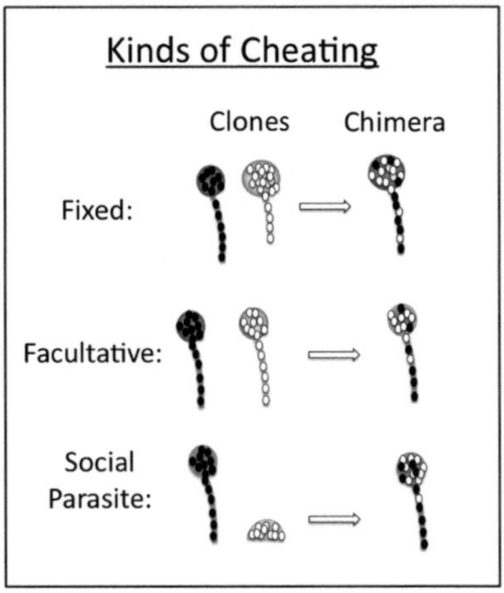

FIGURE 9.3 In the social stage, clones may take advantage of their partner in three different ways. They may allocate cells to spore and stalk in the same proportions as alone but allocate less to stalk than their partner, fixed cheating. They may modify their behavior in chimera to take advantage of their partner, facultative cheating. Third, a social parasite can only make fruiting bodies in chimera with a victim.

pendently of cheating but then have consequences in chimeras. If one clone in isolation allocates more to spore and continues to do so in the chimera with another clone that allocates less to spore, the first clone may then be viewed as a cheater, although it has behaved no differently in the chimera.

We will argue that even this case should be called cheating, because one clone does take advantage of the other. It might even have evolved for that purpose: Selection in chimeras could have favored variants that do suboptimal things on their own. We call this type of cheating "fixed," following Buttery et al. (2009). Cheating that results from behavior different from what they would do when clonal, in recognition that there is a partner to cheat, we then call "facultative" (Fig. 9.3). If the only information we have is how they behave in a chimera compared with starting frequencies, we cannot distinguish between these two and just call it "cheating."

It is probably worth pointing out that we are not implying any sort of conscious awareness to cheating in *D. discoideum*. In humans, cheating is value-based and assumes a certain awareness of the moral grounds of an act. This, of course, is impossible in an organism lacking a nervous system.

EVIDENCE FOR CHEATING IN *D. DISCOIDEUM*

Do Wild Clones Cheat?

When wild clones are mixed together, one clone often prevails over the other (Strassmann et al., 2000). Furthermore, there is a transitive hierarchy of cheaters (Fortunato et al., 2003a; Buttery et al., 2009). In all these cases, the clones are perfectly able to produce fruiting bodies with normal, although variable, spore/stalk ratios as pure clones. Buttery et al. (2009) found both fixed and facultative cheating among the clones.

Other evidence for social conflict among wild clones in the social stage comes from comparing chimeras with pure clones in their ability to migrate as slugs and to form tall fruiting bodies (Fig. 9.4). Chimeric slugs move less far than clonal slugs when cell number is controlled (Foster et al., 2002). This may be the result of increased competition to stay out of the front control region that becomes the sterile stalk. The other effect is that there are more spore cells in chimeric mixtures, presumably because there is less selective benefit to becoming a stalk to lift nonrelatives (Buttery et al., 2009).

Cheating by Single-Gene Knockouts

Nearly all the research by cell, developmental, and molecular biologists on *D. discoideum* has used a single clone, or descendants of that clone. This means that these studies could not reveal cheating even if it were common. The exception is that they could reveal circumstances under which a clone with a single gene that was knocked out cheated its immediate ancestor. Kessin and colleagues (Ennis et al., 2000) did just such a study. They made a large random collection of clones that each had a single gene disrupted by restriction enzyme-mediated integration (REMI), a process that randomly inserts a known sequence containing both restriction cut sites, and an antibiotic resistance gene (Kuspa and Loomis, 1992). Kessin and colleagues (Ennis et al., 2000) put a pool of REMI knockouts through 20 generations of selection in a well-mixed (low-relatedness) environment. At each round, they harvested the spores and began the next round from them; thus, any clone that

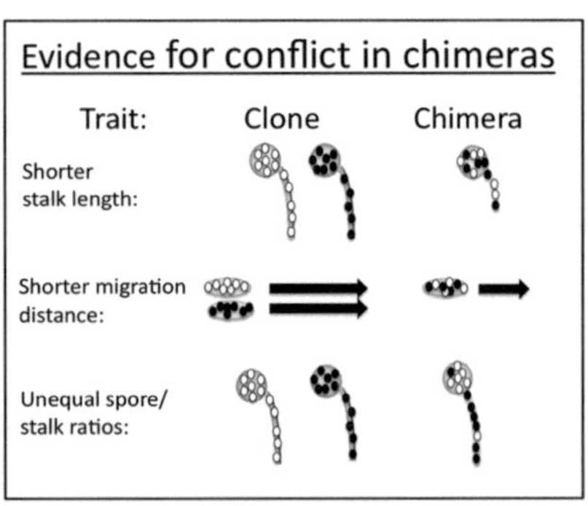

FIGURE 9.4 Conflict is manifested in chimeras in the form of shorter stalk lengths, shorter migration distances, and unequal spore/stalk ratios.

cheated the others increased in frequency over these rounds. They then characterized one mutant, *fbxA*. The *fbxA* knockout cheats its ancestor but cannot make spores on its own.

Pools of REMI mutants can also be screened to obtain cheaters that are able to make normal fruiting bodies on their own but cheat their ancestor in a chimera. A large study of this type used pools of REMI mutants and required that every mutant be able to fruit on its own (Santorelli et al., 2008). This approach identified over 100 different knockout mutants that cheated their ancestor. If knockout cheaters are so easy to generate and cheating is advantageous, one has to ask why these genes have not lost function in the wild. We discuss the answer to this question below in the section on the control of cheating.

CONTROL OF CHEATING

When wild clones come together in the social stage, cheating occurs between pairs of co-occurring wild clones. This could be the result of genetic or environmental factors. The work on single-gene knockouts suggests that at least some of the differences are genetic. Why are genes underlying victim status not eliminated from the population? We think the answer lies in the ways cheating is controlled. It can be controlled by high relatedness within social groups, which could result from kin discrimination. It can be controlled by positive pleiotropy, wherein a cooperation gene also has another essential function. Cheating can also be controlled if spore vs. stalk fate is the result of environmental rather than genetic factors. For example, spore fate could be the result of position in the mitotic cell cycle or it could be dependent on who starved first. Here, we take up these issues (Fig. 9.5).

Control of Cheating by High Relatedness

Cheaters can be controlled if relatedness within social groups is high enough. This is because the benefits of the sacrifice that stalk cells make will mostly go to relatives, and thus could be favored under kin selection. The importance of high relatedness can be seen in an experiment that used the knockout cheater *fbxA* (Gilbert et al., 2007). In this study, we showed that at low relatedness, the *fbxA* cheater knockout wins within groups at all mixture frequencies. This means that it should increase in frequency in the population. There is a tradeoff, however. The higher the frequency of the cheater in a group, the lower the spore production becomes, hurting the *fbxA* knockout and WT alike within that group. This means that the cheater knockout can only flourish at low related-

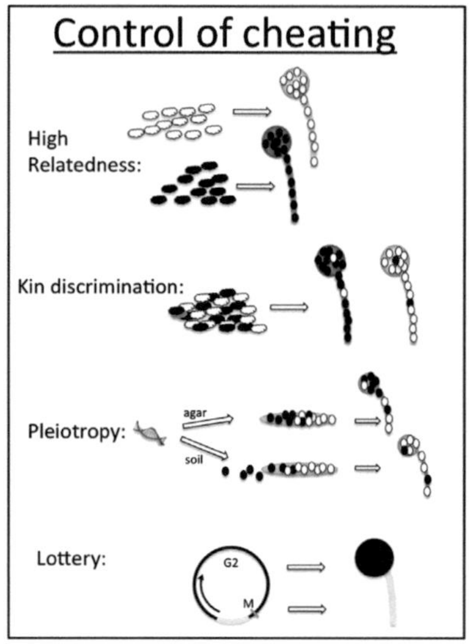

FIGURE 9.5 Cheating can be controlled in the social stage if fruiting bodies are clonal, as might happen if they arise from different patches. They may mix but then sort into nearly clonal fruiting bodies through kin discrimination. Pleiotropic effects may prevent cheating genes from spreading. Caste fate may be determined through a lottery, with cells in the M or S stage of the cell cycle becoming stalk and those in the G2 stage becoming spore. *D. discoideum* apparently has no G1 stage, although this is controversial.

ness because at high relatedness, it is selected against by its own compromised spore production.

We expect social parasites like this one to fail in nature because of the high relatedness within fruiting bodies found in the wild. If this is true, we should not find any clones within wild fruiting bodies that are unable to form fruiting bodies on their own. We tested this by plating cells from wild fruiting bodies clonally. Of 3,316 clonal isolates from 95 wild fruiting bodies, all were able to make completely normal fruiting bodies on their own. There was not a single social parasite like *fbxA*. Clearly, high relatedness within fruiting bodies is a powerful evolutionary deterrent to cheating. This does not mean cheater mutants that are competent on their own are equally controlled, however (Santorelli et al., 2008).

Control of Cheating by Kin Discrimination

One way of achieving high relatedness is to exclude nonkin from the group. This behavior could explain the difference in relatedness between small soil samples and fruiting bodies. Different clones might aggregate together to cAMP and then sort into genetically homogeneous slugs. Even different species coaggregate to cAMP and then separate (Jack et al., 2008); thus, it is not unreasonable to postulate a similar process within species.

Studies of chimerism between two clones of *D. discoideum* have generally found fairly homogeneous mixing, however (Strassmann et al., 2000; Fortunato et al., 2003a; Buttery et al., 2009). A couple of studies found some evidence for sorting, particularly between clones collected far apart (Ostrowski et al., 2008) or, in another study, particularly between clones found close together (Flowers et al., 2010). Neither approached the levels of sorting found in another species, *Dictylostelium purpureum* (Mehdiabadi et al., 2006).

At this point, we have a puzzle. Tiny soil samples have multiple clones of *D. discoideum,* but fruiting bodies are nearly clonal. Kin discrimination is weak as far as we can tell in laboratory mixtures of equal numbers of cells from two clones. The finding of an apparently selected molecular mechanism for sorting deepens the puzzle. Our supposition that sorting will occur in the aggregation stage or later means that cells are likely to discriminate when they are in direct contact with each other. This suggests that adhesion genes are likely candidates for recognition. To function as recognition genes, adhesion genes would have to be highly variable. The variability would provide an opportunity for discrimination that favors others carrying the same adhesion protein variant over others carrying different forms of the molecule. They should recognize self, with a homophilic binding site, or they should recognize a highly variable receptor.

There is excellent evidence that two cell adhesion genes, initially called *lagC* and *lagB* but now called *tgrC* and *tgrB,* are the kin discrimination genes in *D. discoideum* (Benabentos et al., 2009). These two genes are extremely variable and are part of a large gene family of generally much less variable genes. The protein produced by *tgrC* is hypothesized to adhere to the protein produced by *tgrB.* If one is knocked out, it causes development to fail at the aggregation stage. In that case, the amoebae aggregate begins to make a mound but then falls apart, as if a crucial component of recognition necessary for the subsequent altruistic steps were missing. The temporal coexpression, knockout behavior, high variability, and impact on sorting make these likely kin recognition genes. More work is clearly needed on this system to see if there are consequences of recognition other than sorting. It could be that it is advantageous to remain in the group but that the chimeric nature is recognized and responded to, causing reduced migration distances and shorter stalks, for example.

Control of Cheating by Pleiotropy

Pleiotropy means that a single gene has an impact on multiple phenotypic traits. It is therefore usually viewed as something that impedes selection on a specific trait, because any changes in the underlying genes will affect other traits as well. This conservative force in pleiotropic genes can have interesting consequences for social genes. If an altruistic trait is piggy-backed on an essential gene, a mutation that causes selfish behavior is unlikely to proliferate, because the essential function would also be lost.

Exactly how important this might be in social traits is unknown, because we know the genetic underpinnings for comparatively few traits. There are a couple of genes having an impact on altruism in *D. discoideum* that could be maintained through pleiotropy, however. They are from very different parts of the genetic landscape underlying altruism in *D. discoideum.* One is a cell adhesion gene, and the other is involved in the differentiation-inducing factor (DIF-1) signaling system.

Cell adhesion is an essential part of the social process because it is how the multicellular group stays together (Kessin, 2001). Variation in adhesion can have an impact on cell fate, because the cells at the front of the slug become stalk and the cells in the back three-quarters or so become spore (Bracco et al., 2000). One way of increasing the likelihood of becoming spore could therefore be to have reduced adhesion to the other cells and to slip back in the slug (Ponte et al., 1998; Queller et al., 2003). The knockout of the cell adhesion gene *csaA* has just this effect. When *csaA* is knocked out, adhesion is reduced. On agar, this has the impact of increasing the knockout's frequency in the spores, presumably because reduced adhesion allows it to slip out of the stalk-forming tip (Ponte et al., 1998; Queller et al., 2003). On the more natural substrate of soil, however, *csaA* knockouts apparently do not hold together enough to get into aggregations. It is therefore no surprise that the *csaA* gene continues to be expressed normally and that cheater knockouts have not prospered.

Another gene that could be a cheater were it not for pleiotropic effects is *dimA* (Foster et al., 2004). This gene was isolated in a screen of REMI mutants that are unresponsive to DIF-1, a small molecule that forces some cells to become stalk (more on this later). In chimeras with WT, *dimA* knockouts predominate in the prespore zone, presumably because they are insensitive to DIF (Thompson et al., 2004). Ultimately, however, they are in a minority in the actual spores. This could be true if they transdifferentiate from prespore cells to prestalk cells later in development, and this was shown to be the case (Foster et al., 2004). We interpreted this to be the result of another unknown function of *dimA,* an essential function that made the knockouts worse spore cells. This is another case in which pleiotropy inhibits the spread of a cheater.

Control of Cheating by Lottery

When two or more individuals take unequal roles in a social interaction, with one being the recipient and the other being the beneficiary, conflict can result. One way of controlling this conflict is if the partners do not know which role they will assume on entering the interaction. A human equivalent is called the veil of ignorance (Rawls, 1971), and it calls for resource allocation between partners by someone who does not know which lot he or she will get. A familiar example is the common family situation of dividing up a cake. If the child cutting the pieces does not get to decide which piece he or she gets, under the veil of ignorance model, he or she will be more likely to make the pieces equally sized. Cheating could be controlled in *D. discoideum* if there were a lottery to become spore based on the cell cycle.

The *D. discoideum* cell cycle has a very short G1 phase; thus, immediately after the mitosis (M) phase, cells enter the synthesis (S) phase and cytokinesis occurs during the S phase (Weijer et al., 1984). Therefore, in a population, the cells in S and early growth after synthesis (G2) phases tend to be the smallest cells with the fewest nutrient reserves. An experiment on a thin layer of cells not touching other cells, followed with videography, indicated that stalk cells were most likely to arise from cells that happened to be in the S or early G2 phase of the cell cycle at the time of starvation, whereas cells that happened to be in the late G2 phase became prespore (Gomer and Firtel, 1987). A variety of other experiments have also shown this (Araki et al., 1994; Azhar et al., 2001). If weaker cells are more likely to become stalk, this makes sense, because recently divided cells would have fewer nutrients. This cell cycle lottery system fits the veil of ignorance model. As cells encounter less and less food, however, it could be that those dividing earlier than others are selected against because these cells will be in the stage that sends them to stalk.

Another interesting result of this paper (Gomer and Firtel, 1987) suggests that delaying cell division may not be necessary for a cell to avoid becoming a stalk cell. This result was derived from careful observation of the fate of sister cells through videography. Every time a cell divided, one sister cell became prestalk or prespore according to the above musical chairs lottery mechanism, whereas the sister cell became a null cell, a third cell type that stained with neither prespore nor prestalk markers (Gomer and Firtel, 1987). The fate of these null cells is unclear. These null cells could become pstO, because that region of the slug also did not stain with prespore or prestalk markers. This region can be viewed as the most flexible area, with cells in that region remaining pstO on exposure to DIF, and perhaps becoming prespore otherwise. These interesting results remain controversial, however, and should be followed up on carefully (Shaulskyand Loomis, 1993; Jang and Gomer, 2011).

If a recently divided cell becomes stalk because it is smaller and weaker, cell division could be disfavored as starvation approached for this social reason. Under normal circumstances, however, amoebae will be selected to eat and proliferate as rapidly as possible. These two counterforces might achieve a compromise that could support altruism under a wide variety of conditions in *D. discoideum,* if one of two recently divided cells becomes stalk and the other becomes spore. This scenario is consistent with the data.

CONTROL OF CONFLICT BY POWER

We began the section on control of cheating with a discussion of social contracts and defined cheating as the violation of those contracts. In this case, we mean evolved contracts that favor the evolution of cooperation. One form of contract may be that the stronger individuals take the best roles. Here, we explore the evidence for this idea in *D. discoideum*.

First-Strike Power

One of the most common determinants of whether an individual in a social interaction becomes the altruist or the beneficiary is that individual's relative strength, or ability to prevail in a contest. Such contests under social and cooperative circumstances may look very similar to contests between nonsocial organisms for scarce resources such as good territories. The difference is that if the contest is between relatives, or mutually dependent individuals, after the contest is decided, the loser may acquiesce and go to work for the winner. Such contests can be valuable for all concerned, particularly if weaker individuals that lose contests are more effective in taking on the helping role than they would be with the winning, reproductive role.

How do we evaluate power in *D. discoideum* interactions? In some ways, all predictors of fate also involve power. The lottery system has a power element, because cells that recently divided may be weaker and go to stalk. If becoming a spore cell is competitive, the first amoebae to depart from growth and binary fission and enter the social stage may get a head start on preparing their weapons. Under this hypothesis, the first to starve would become spore. That this is the case has been very nicely demonstrated in both an experiment that manipulates timing of starvation in genetically identical cells and an experiment that uses an aggregation-initiation knockout. In the first experiment, cells were put into nutrient-free medium 4 hours apart. Those with the 4-hour head start in the social stage preferentially became spores (Kuzdzal-Fick et al., 2010). The other experiment used a knockout that was incapable of initiating aggregation but was capable of responding to the initial signal from others and relaying it (Huang et al., 1997). In this case, the single cell initiating aggregation became a spore.

Glucose Feeding, Condition, and Power

Power based on condition has also been studied directly by making chimeras of cells that were well fed with cells that were poorly fed. This was done by varying the amount of glucose in the medium of axenically grown cells. The cells fed with glucose were more likely to become spore than the glucose-starved cells (Leach et al., 1973). This effect holds with other metabolizable sugars and is absent with other sugars (Takeuchi et al., 1986). This is strong support for the hypothesis, but there could be something special about sugars; thus, we repeated this experiment with a glucose treatment and added another treatment to separate cells. In this treatment, we stressed the cells by growing them in a more acid pH than usual (Castillo et al., 2011). We affirmed the weakening effects of both treatments by documenting that they increased doubling times in the solitary stage. As expected, both acid-stressed and glucose-starved cells ended up preferentially in the stalk. Both treatments also made fewer spores when grown alone, however; thus, the chimera results are not attributable to competition alone (Castillo et al., 2011).

DIF-1 and Power

One of the delights in working with a microbial system is the accessibility of mechanisms. Whether a cell becomes spore or stalk is mediated by DIF-1, a small, secreted, chlorinated alkyl phenone (Kay, 1998). Stronger cells that are immune to its effects at biological levels produce DIF-1. Weaker cells can break it down but become stalk cells from its impact, mostly ending up in the lower cup or the basal disk, both of which are dead parts of the stalk (Thompson and Kay, 2000a,b). DIF-1 is unlikely simply to be a signal rather than a mediator of competition for several reasons. Signals are unlikely to include chlorine, something that is common for poisons. Levels of DIF-1 in the slug are about 62 nm (Kay, 1998), which is high, given that it can be lethal at concentrations as low as 200 nm (Masento et al., 1988). Signals have receptors and poisons do not, and no receptor has ever been found for DIF-1. Its small, toxic nature is just what might be expected of a poison (Atzmony et al., 1997). Unlike most morphogens, it is distributed evenly through the social stage and varies on its cell-specific impact (Kay and Thompson, 2009; Chattwood and Thompson, 2011; Parkinson et al., 2011). In some respects, it is a tame poison, incorporated into social life to mediate condition in a homogeneous mixture into different cell fates.

The condition variants resulting from position in the cell cycle or glucose feeding are tied to DIF-1 levels with weaker, more recently divided cells more vulnerable to DIF-1. There are single-gene knockouts with an impact on cell cycle and nutritional responses that further support the involvement of DIF-1, in a story nicely summarized by Chattwood and Thompson (2011). Cells that have *rtoA* knocked out lose the specificity toward stalk of the M and/or S cell cycle phase, producing fruiting bodies that are mostly stalk with tiny spore heads (Wood et al., 1996). This has been shown to be the result of high intracellular calcium, which has independently been shown to bias cell fate toward stalk (Baskar et al., 2000; Azhar et al., 2001; Chattwood and Thompson, 2011). Cells with high intracellular calcium are far more sensitive to DIF-1 (Schaap et al., 1996; Baskar et al., 2000). A similar story can be told with a gene that links nutritional status to cell fate, a *D. discoideum* homolog of the human retinoblastoma gene, *rblA* (MacWilliams et al., 2006; Chattwood and Thompson, 2011). Knockouts of *rblA* are hypersensitive to DIF-1 and preferentially become stalk.

Other work by Thompson and colleagues (Parkinson et al., 2011) has shown that the patterns linking DIF-1, or more generally stalk-inducing factors (StIFs), are also important in spore-stalk hierarchies of natural clones. These hierarchies are based on whether clones become spore or stalk when mixed pairwise with other clones (Fortunato et al., 2003a; Buttery et al., 2009). They separately evaluated response to and production of StIFs and found a threefold difference in production and a 15-fold difference in response; the latter was most powerful in explaining the hierarchy observed in natural clones (Parkinson et al., 2011). Thus, we know a satisfying amount about how power affects cell fate through DIF-1. There is more to learn, however, particularly because cheating can result from knocking out so many different genes (Santorelli et al., 2008). This led to another general approach to identifying resistance genes.

Genetic Control of Cheating by Noble Resistors

The evolution of resistance to cheater genes may limit their spread. To test this idea, we selected for resistors of cheater genes. We took one cheater, *chtC,* and exposed a pool of REMI mutants to it over successive rounds (Khare et al., 2009). We allowed selection of the REMI pool but not of the *chtC* knockout. We did this by removing the G418 resistance from the *chtC* clone so that we could kill it at each round, leaving the mutants we were selecting intact. We then simply added back in the naive *chtC* clone for the next round. This process resulted in a number of mutants that were resistant to *chtC* knockouts and could not be cheated by it. Interestingly, they were not cheaters of their ancestor; thus, we called them noble resistors (Khare et al., 2009).

SOCIAL GENES, ARMS RACES, AND THE RED QUEEN

Cheating and countering cheating are social processes that we predict will result in rapid evolution in the underlying genes. Our test of this hypothesis used the newly sequenced species *D. purpureum* and compared it with *D. discoideum* (Sucgang et al., 2011). Unfortunately, this is not an ideal pair of species because their proteins are as diverged as those of humans and fish. This means that silent amino acid changes (ds) are saturated, and thus are not useful in comparisons. Instead, we compared homologs; rates of amino acid change; and conservation scores, a measure of similarity that includes indels. We used two sets of social genes for comparisons. The first set was the 100 or so REMI mutants that cheated their ancestors when mixed equally with them (Santorelli et al., 2008). These genes did not show more rapid evolution, and thus failed to support our hypothesis that social genes evolve more rapidly.

The second set of genes we used was based on a social index, which was higher when a gene was more expressed in the social stage compared with the nonsocial stage. In this analysis, the more social genes had a lower probability of having homologs, an elevated rate of amino acid change, and a lower conservation score, supporting our hypothesis (Sucgang et al., 2011). The result could also be attributable to weaker purifying selection on social genes, however, and a better analysis would be between more closely related species.

OTHER ARENAS FOR COOPERATION: MUTUALISMS AND SEX

No review of social behavior of *D. discoideum* would be complete without mentioning two very exciting areas for future study. The sexual cycle is also a social cycle but has been studied very little. The other area is the discovery of a farming mutualism between *D. discoideum* and bacteria. This opens up the opportunity for studies of between-species symbioses.

Sexual Cycle Has Social Elements That Involve the Ultimate Sacrifice

The sexual cycle is triggered by starvation in the presence of sufficient numbers of other amoebae under wet, phosphate-poor conditions and begins with aggregation to cAMP (Bonner, 1967; Kessin, 2001). Two cells of different mating types fuse, forming a diploid zygote. The amoeba stage is ordinarily haploid and divides by mitosis; thus, no reduction division is necessary before sexual fusion. Aggregation does not cease with the formation of a diploid zygote (Urushihara, 1992; Ishida et al., 2005). Other amoebae continue to swarm in by the thousands, up the cAMP gradient. The zygote proceeds to consume the other cells by phagocytosis. The pace of consumption is slowed to a level that allows the waiting victim amoebae to construct an envelope around the aggregation, and this slowing is also regulated by cAMP. After a time, there is a firm wall around the zygote and its victims, and the latter are consumed and digested. Recombination and crossing over then happen, the zygote undergoes meiosis, and many recombinants are formed.

In a major recent advance, the sex-determining locus was identified and the presence of three mating types was confirmed, clearly establishing the genetic basis of sex (Bloomfield et al., 2010). The sexual cycle is somewhat of an enigma because it rarely leads to recombinant progeny under laboratory conditions (Kessin, 2001), but estimates of recombination rates of natural clones indicate they are very high, with a population ρ of 37.75 and baseline linkage disequilibrium achieved between 10 and 25 kb (Flowers et al., 2010). Getting the system to work in the laboratory would open up many interesting social questions to investigation. For example, we could select for social traits in sexually recombined pools and look for quantitative trait loci associated with social traits.

D. discoideum Farms Bacteria

There is another reason why *D. discoideum* is particularly good for studies of cooperation: mutualism. The standard view of the social stage of development is that all bacteria are purged from the aggregate (Kessin, 2001). There are known mechanisms for this that function at different stages, from mound, to slug, to final fruiting body. The sorus is considered to be sterile apart from the *D. discoideum* spores. Very recently, we discovered that this is not the case for about one-third of all clones (Brock et al., 2011). These clones carry bacteria with them through the social stage like a farmer might bring a flock of sheep to a different pasture. These bacteria are found within the fruiting body. When the spores hatch after favorable growing conditions have been encountered, they can feed on the proliferating population of the bacteria they brought. These farmed bacteria are better food than most wild bacteria. This farming mutualism is highly amenable for study, because all partners are microbial, advantages are clear, and the relationship is not obligate, at least at the species level. This discovery adds between-species cooperation to the things that can be studied about *D. discoideum.*

CONCLUSION

The ultimate advantage to an ideal model organism is what you can learn from it. In *D. discoideum,* we have shown that conflict exists in the form of shorter stalk lengths, reduced migration distances, and cheating to avoid the sterile caste. We have delineated cheating into fixed, facultative, and social parasite forms. We have shown that cheating can be controlled by high relatedness, kin discrimination, pleiotropy, or lotteries. We have shown that conflict can be controlled by conventions and power. The first cells to starve become spore, as do stronger cells. A small, toxic molecule called DIF-1 mediates social interactions. We and others have backed up much of this work with specific genes and knockouts. Further whole-genome outcomes are on the horizon, as is a much more detailed understanding of kin discrimination. Frontiers include the farming symbiosis and exploration of the sexual cycle. Clearly, this is a system that has yielded many important secrets about the cooperative side of major transitions.

ACKNOWLEDGMENTS

We thank John Avise and Francisco Ayala for inviting us to help organize the Arthur M. Sackler Colloquium on Cooperation. We thank many colleagues for helpful discussions of these points, especially Sandie Baldauf, Koos Boomsma, Kevin Foster, Richard Gomer, Ashleigh Griffin, Rob Kay, Richard Kessin, Adam Kuspa, Gene Robinson, Gadi Shaulsky, Chris Thompson, Greg Velicer, and Stuart West as well as our own laboratory group. Our research is supported by the U.S. National Science Foundation under Grants DEB 0816690 and DEB 0918931.

10

Selfish Genetic Elements, Genetic Conflict, and Evolutionary Innovation

JOHN H. WERREN

Genomes are vulnerable to selfish genetic elements (SGEs), which enhance their own transmission relative to the rest of an individual's genome but are neutral or harmful to the individual as a whole. As a result, genetic conflict occurs between SGEs and other genetic elements in the genome. There is growing evidence that SGEs, and the resulting genetic conflict, are an important motor for evolutionary change and innovation. In this review, the kinds of SGEs and their evolutionary consequences are described, including how these elements shape basic biological features, such as genome structure and gene regulation, evolution of new genes, origin of new species, and mechanisms of sex determination and development. The dynamics of SGEs are also considered, including possible "evolutionary functions" of SGEs.

The idea that some components of the genome can be "selfish" or "parasitic" has a long and controversial history. The first recognition that a gene could increase in frequency by imparting a drive relative to its homolog came with the description of X-chromosome meiotic drive dynamics in *Drosophila obscura* by Gershenson (1928). Later, Östergren (1945) investigated accumulation of supernumerary (extra nonvital) "B" chromosomes in plants and made the first explicit argument that some genetic material in an organism can be "parasitic."

Department of Biology, University of Rochester, Rochester, NY 14627. E-mail: werr@mail.rochester.edu.

These observations and their evolutionary implications were not widely known among biologists, however, in part because meiotic drive and supernumerary chromosomes were perceived as genetic peculiarities rather than important general phenomena. Three parallel threads then set the stage for more serious considerations of selfish genetic elements (SGEs) and genetic conflict ideas. First, empirical and conceptual developments in genetics and evolutionary biology led to wider acceptance of a gene-centric view of evolution (Dawkins, 1976c; Williams, 1996). Noteworthy in this regard was Dawkin's (1976) influential book entitled *The Selfish Gene,* which described genes as "selfish replicators" encoding phenotypes that increase their transmission to future generations and organisms fundamentally as "vehicles" for the transmission of genes. Second, rapid advances in molecular biology began to reveal that many eukaryotic genomes contain large amounts of repetitive DNA without any clear function, although their potential role within the genome was the subject of much speculation (Britten and Davidson, 1971). Seminal papers by Doolittle and Sapienza (1980) and Orgel and Crick (1980) first proposed that repetitive DNA could be considered parasitic or selfish replicators. Cosmides and Tooby (1981) explicitly introduced the concept of genetic conflict between nuclear and cytoplasmic (e.g., mitochondrial) elements over sex determination. The idea of SGEs and genetic conflict remained highly controversial, however, and a counterview was that such elements exist because they play important regulatory roles in cells and in evolution. Third, an increasing number of genetic studies began to uncover non-Mendelian and other elements within diverse organisms that appeared to have "self-promoting" features that cannot simply be explained as adaptations for the organism. These included discoveries of meiotic drive in diverse organisms; heritable elements, such as killer plasmids; and a genome-eliminating supernumerary chromosome that was an unequivocal example of a nonadaptive self-promoting replicator (Nur et al., 1988).

Werren et al. (1988) published the first general review of selfish or parasitic genes and defined an SGE as an element that has characteristics enhancing its own transmission relative to the rest of an individual's genome but neutral or detrimental to the organism as a whole. Examples include transposable elements (TEs), meiotic drivers, supernumerary B chromosomes, postsegregation killers, and heritable microbes and organelles that distort sex determination. In 1988, the idea that elements in the genome could be parasitic was still contrary to prevailing opinions of many molecular biologists, who viewed the cell and organism as a highly integrated machine, and therefore considered the idea that components of the cell could be maintained because of their selfish replication as a bizarre and foreign concept. In contrast, the SGE model is a more "ecological" view that considers the genome as a set of genetic elements with potentially different kinds of interactions, ranging from cooperative (mutualistic), to neutral (commensal), to selfish (parasitic) (Avise, 2001). According to this paradigm, genetic conflict can arise among components of the genome that have different transmission patterns (e.g., transposons, nuclear genes, cytoplasmic genes), and therefore conflicting genetic interests. The basic idea is as follows: When components of the genome have different transmission patterns, selection can act on an element to increase its transmission even if that is detrimental to the organism and/or other heritable components of the genome. Genetic conflict within the genome will then result, because enhanced transmission of an SGE decreases transmission of other genetic elements. An evolutionary "arms race" can then occur among different components of the genome over basic biological processes.

Werren et al. (1988) raised three basic questions about SGEs that are still the subject of study today: (*i*) What are their origins, (*ii*) how are SGEs maintained, and (*iii*) are SGEs important in evolution? Regarding this last question, they concluded that "selfish elements, and the 'intragenomic conflict' they create, may be an important force promoting evolutionary change. However, this possibility has not been demonstrated conclusively in any system" (Werren et al., 1988). They further observed that the pace of understanding of SGEs "is expected to accelerate with the application of molecular cloning techniques" (Werren et al., 1988). Subsequent advances have eclipsed this expectation. What has occurred in the intervening years is the genomics revolution, a veritable explosion of information and techniques that have begun to open the "black box" of genome structure, function, and evolution. Today, there are over 1,000 bacterial genomes and over 100 eukaryote genomes sequenced, with the numbers growing almost daily (http://www.ebi.ac.uk/genomes/). These data and advances in genetic techniques have helped reveal how genomes evolve and function. The story that is emerging increasingly supports a central role of SGEs in shaping structure and function of genomes and in playing an important role in such fundamental biological processes as gene regulation, development, evolution of genetic novelty, and evolution of new species.

Here, I describe the conceptual framework for SGEs and genetic conflict as well as their types, and I then discuss developments that reveal the role of SGEs in important biological processes. There are several common themes to the topic that are briefly listed here and elaborated on below:

(*i*) Antagonistic selection occurs between SGEs and other genome components, and this can lead to evolutionary change and novelty.

(*ii*) Sexual recombination and lateral movement between lineages is important to SGE maintenance and evolution.

(*iii*) Many genetic elements have mixed phenotypes with a combination of parasitic and beneficial features.

(*iv*) SGEs can also lead to an "evolved dependency" by the host, which has the appearance of mutualism but is not.

(*v*) SGEs can occupy "safe havens" within the genome, where their negative effects are mitigated or they are less likely to be excised or repressed.

(*vi*) SGEs can be "domesticated" or "co-opted" by genomes, resulting in the evolution of novel genes and functions.

I will end the paper with a discussion of "why" SGEs persist in nature and contrast the evidence and predictions for the view that SGEs persist because of their ability to replicate within genomes (the parasitic hypothesis) vs. the view that they persist because they promote the ability of populations to adapt and evolve (the "evolvability" hypothesis).

TYPES OF GENETIC CONFLICT

Genetic conflict occurs when different genetic elements (either within an individual or between individuals) have influence over the same phenotype, and an increase in transmission of one element by its phenotypic effects causes a decrease in transmission of the other. Not included in this definition are population changes in allele frequency at a locus, unless they result from antagonistic selection acting on the alternative alleles for the shared phenotype (Rice and Holland, 1997; Frank and Crespi, 2011). An example will illustrate the point. A meiotic drive allele reduces the nondriving allele among the gametes of heterozygous individuals; therefore, these alternative alleles (and tightly linked loci) experience antagonistic selective pressures over the phenotype (one is selected to drive and the other to suppress the drive).

Genetic conflicts historically have been divided into "intragenomic" conflict, which occurs within the genome of an individual, and "intergenomic" conflict, which occurs between individuals (e.g., male-female or sexual conflict, parent-offspring, social conflict) (Cosmides and Tooby, 1981; Werren et al., 1988; Hurst et al., 1992; Rice and Holland, 1997; Hurst and Werren, 2001; Burt and Trivers, 2006) (Fig. 10.1). Because the term "genome" is often used to include the sum of DNA across individuals within a species, however, less confusing terms to distinguish these levels may be "intraindividual" conflict and "interindividual" conflict, because these terms distinguish genetic conflicts within individual organisms (e.g., for transmission through gametes) as opposed to between individuals (e.g., male-female or parent-offspring conflict over reproductive effort). Fig. 10.1 shows several kinds of conflict that can occur within and between individuals. At one end of the spectrum are genetic conflicts between mobile elements (e.g., transposons), which are generally selected to transpose within a genome, and Mendelian components of the nuclear genome, which are selected to suppress transposition because of fitness costs to the individual. Evidence of this conflict includes diverse mechanisms that have evolved to restrain transposition (Johnson, 2007). Similarly, cytoplasmically inherited and nuclear inherited elements experience genetic conflict, primarily over sex determination, because of their differences in transmission through male sperm and female eggs (Cosmides and Tooby, 1981; Werren and Beukeboom, 1998). As indicated in Fig. 10.1, intercellular conflict can occur when heritable differences arise within the cell lineages of an organism by *de novo* mutation or unequal transmission of

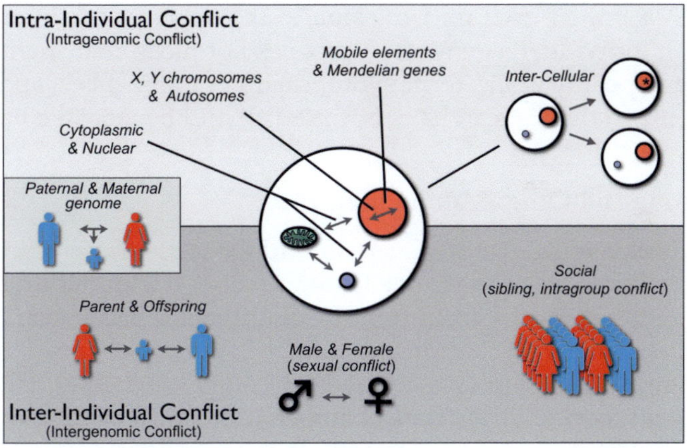

FIGURE 10.1 Types of genetic conflicts. Genetic conflicts can be categorized as intraindividual (or intragenomic) and interindividual (or intergenomic). Intraindividual conflicts occur among genetic elements with different inheritance patterns (e.g., cytoplasmic genes; nuclear genes; X, Y, and autosomally located genes; mobile elements). Intraindividual conflict also arises among cells within an organism that are genetically different because of de novo mutations or transpositions (*) or heteroplasmy attributable to unequal segregation. Genetic conflicts also occur between individuals, including parent-offspring, sexual, or social conflict. Paternal-maternal genome interactions within offspring have features of both intraindividual and interindividual conflict.

heritable organelles or microbes into daughter cells. It has been argued that development has been molded to minimize such conflicts by uniparental inheritance of organelles, metazoan development from single cells, and germline sequestering (Cosmides and Tooby, 1981; Maynard Smith and Szathmáry, 1995, p 360). At the other end of the spectrum are conflicts between individuals, such as parents, offspring, mates, or members of social groups, over phenotypes they jointly influence, such as resource use. Nature is not always as clean as our paradigms, and there is at least one category of genetic conflict that has features of both intraindividual and interindividual conflict, that is, paternal-maternal genome conflicts over resource allocation. This form of conflict is manifested as genomic imprinting of alleles during male and female gametogenesis, which differentially affects their expression in offspring (Haig, 2000b).

TYPES OF SGES AND THEIR CONSEQUENCES

Here, we have a rogue's gallery of the genome. SGEs can be placed into the following broad categories (Werren et al., 1988; Hurst and Werren, 2001; Burt and Trivers, 2006): TEs, biased gene converters, meiotic drivers, postsegregation drivers, and cytoplasmic drivers. These elements act to increase their own transmission to the detriment of other components of an individual's genome. This does not mean that such elements cannot have positive long-term evolutionary consequences, and some of the elements in this list can have both selfish and beneficial components.

Transposons and Other Mobile Elements

Mobile elements include plasmids, endogenous viruses, and TEs. TEs have the ability to copy and move to new locations within the genome; as a consequence, they can accumulate. Doolittle and Sapienza (1980) and Orgel and Crick (1980) first proposed that they can be considered SGEs, and this view is now widely accepted (although see below). TEs fall into two main categories: DNA transposons move via DNA copies, and retrotransposons use an RNA intermediate (Kidwell and Lisch, 2001). TEs can also be autonomous (encoding proteins that promote their transposition) or nonautonomous (not encoding proteins needed for transposition but using the cellular machinery or proteins provided by other TEs). An interesting category of mobile elements is group I and II self-splicing introns (Lambowitz and Zimmerly, 2004), which can be tolerated in the typically streamlined genomes of prokaryotes and organelles because self-splicing restores functional open reading frames in genes with the inserts, thus reducing negative fitness costs. Although group II introns are not found in eukaryotes, shared features with the spliceosome of eukaryotes have led to the proposal that the spliceosome machinery and eukaryotic introns evolved from group II introns (Lambowitz and Zimmerly, 2004). If correct, this hypothesis implies that acquisition of the spliceosome machinery to remove mobile group II introns set the stage for evolutionary expansion of introns in the genomes of higher eukaryotes.

In bacteria, the amount of mobile DNA ranges from 0% to 21% and varies with bacterial ecology rather than phylogeny (Newton and Bordenstein, 2011)—bacteria with greater exposure to other bacterial lineages show higher levels of mobile DNA. This suggests that opportunities for lateral acquisition, rather than benefits to the host, explain relative abundances of these elements. It is also clear, however, that mobile elements in bacteria (e.g., plasmids) can encode proteins that increase survival of their bacterial hosts, such as antibiotic resistance (Smillie et al., 2010). Evidence suggests that mobile elements in bacteria can be maintained by a combination of selfish features that promote their acquisition and retention in bacterial genomes and (in some cases) beneficial effects on their bacterial hosts. Very large plasmids tend to become immobile and carry important bacterial functions, indicating their evolution to mutualism (Smillie et al., 2010).

In eukaryote genomes, the abundance of TEs can vary widely (Biémont and Vieira, 2006). For example, ~40% of the human genome is composed of TEs, whereas only 3% of the pufferfish genome is (Blumenstiel, 2011). Plants vary similarly. As a result, TEs and other repetitive DNA can be major determinants of genome size within taxa (Bennetzen, 2005). TEs also vary considerably in the taxonomic breadth of their distribution (Feschotte and Pritham, 2007; Schaack et al., 2010a). Around 10 different DNA TE superfamilies are currently recognized (Feschotte and Pritham, 2007), and many show a broad host taxonomic distribution and signature of lateral transfer between taxa. There is evidence that poxviruses have vectored retroposons between reptiles and mammals, and members of four DNA TE families are found in both vertebrates and blood-sucking triatomid bugs, suggesting possible mechanisms for intertaxon transfers. A study in *Drosophila* genomes finds that approximately one-third of TE families originated from recent interspecies lateral transfers, with an estimated transfer rate of 0.04 events per family per million years (Bartolomé et al., 2009). In fact, lateral movement across host taxa is believed to be an important mechanism for long-term maintenance of TE families. The rationale is that evolutionary suppression of TEs by the host will lead to their eventual mutational degradation and loss, except for TEs that move laterally to "infect" and invade new hosts.

Although many active TEs have relatively short evolutionary associations with particular hosts, some can be maintained for long evolutionary time frames in a lineage because they occupy (or target) a safe haven within the genome. Safe havens are genome locations with either reduced fitness costs to the host or where hosts cannot readily remove the insert or evolve countermeasures. R1 and R2 retroelements appear to use a safe haven. They insert into highly conserved segments of the ribosomal RNA genes subject to strong selective constraint (Eickbush and Eickbush, 2007). Ribosomal RNA genes typically occur in large tandem arrays in eukaryotes; therefore, an insertion into any single copy has relative low fitness costs and new uninserted ribosomal RNA copies are continually produced by unequal chromatid exchange. These features probably explain why R1 and R2 elements have been maintained within lineages over long evolutionary timescales despite little evidence of fitness benefits or lateral element transfer (Eickbush and Eickbush, 2007).

TEs are known to induce harmful effects through various mechanisms, including insertions that disrupt coding sequences or *cis*-regulation regions, ectopic recombination between TE copies resulting in deletions and rearrangements, and the costs of transcription and translation of large numbers of TEs (Charlesworth et al., 1994; Kidwell and Lisch, 2001; O'Donnell and Burns, 2010). As a result, there has been strong selection on plant and animal genomes to evolve machinery to suppress TE activity, including DNA methylation suppression, repeat-induced point mutation in fungi, RNAi, and small RNA suppression pathways (Johnson, 2007; Blumenstiel, 2011). These mechanisms could have originally evolved for suppression of TEs and other exogenous DNA, and have subsequently acquired gene regulatory functions. In *Drosophila melanogaster,* the flamenco locus is a large genomic region containing TE insertions that is used by the piwi-interacting RNA (piRNA) pathway to suppress active TEs dispersed elsewhere in the genome. In plants, suppression of TEs can involve small noncoding RNAs that assist in transcriptional and posttranscriptional silencing and guide targeting of TEs for DNA methylation inactivation (Cantu et al., 2010). By increasing rates of CG-to-TA mutation, DNA methylation of TEs also accelerates their mutational degradation. *Neurospora* shows targeted repeat TE degradation by this mechanism. One side effect of their very efficient repeat elimination is that the maintenance of gene duplications is difficult in *Neurospora,* thus affecting its evolutionary trajectory (Johnson, 2007). DNA methylation suppression of TEs occurs in both animals and plants and has been invoked as a likely preadaptation for evolution of the placenta and genomic imprinting in mammals (Suzuki et al., 2007; Sekita et al., 2008).

Given the ubiquity and abundance of TEs, it is inevitable that some will be recruited by genomes for new cellular functions [reviewed in Feschotte and Pritham (2007), Feschotte (2008), and Sinzelle et al. (2009)].

This is variously referred to as "domestication," "co-option," or "exaptation." Classic examples include the utilization of TART and HetA TEs for telomeres in drosophilid flies and the likely origin of V[D]J recombination (used in vertebrates to generate immunoglobin diversity) from mariner-TC1 family TEs. New and exciting discoveries further indicate that domestication of TEs is important in the evolution of genomes, such as the evolution of new protein-coding genes (including regulatory DNA binding factors), *cis*-regulatory sequences, and regulatory small RNAs from TEs (Feschotte and Pritham, 2007; Feschotte, 2008; Sinzelle et al., 2009). Both the DNA binding and catalytic domains from the transposase genes of DNA TEs have been involved in domestication events in animals, plants, and fungi. The SETMAR gene in primates is a chimera derived from fusion of a mariner-like element with the SET domain from a histone methyltransferase gene 40–58 million years ago (Feschotte, 2008; Sinzelle et al., 2009). Its function is unknown but possibly involved in DNA repair. Widely distributed TEs have been involved in independent domestication events in diverse taxa, such as Pogo elements in both mammals and fission yeast (Sinzelle et al., 2009). Intriguingly, a number of regulatory DNA binding proteins appear to have evolved from the DNA binding domains of TEs, such as PAX6 (sensory development in metazoans), CENP-B (centromere function in vertebrates), and Bric-a-Brac (tissue development in insects). Feschotte (2008) reports that at least seven key DNA binding proteins probably evolved from TEs in taxa ranging from plants and fungi to metazoans. Retrogenes occur when host mRNA is reverse-transcribed and inserted into the genome. The process is dependent on RT proteins from retroelements. Retrogenes have been stripped of introns and usually degenerate as pseudogenes. They have also evolved into new functional genes (e.g., ~109 examples in the genome of *Populus*), however, and have acquired introns in some cases (Fablet et al., 2009; Zhu et al., 2009).

Although the vast majority of TEs that insert near or in protein-coding regions are deleterious, mounting evidence indicates that fragments of inserted TE DNA have also evolved *cis*-regulatory or posttranscriptional regulatory functions (Feschotte, 2008). Evidence for this includes conservation of TE-derived fragments in ~25% of human promoters and deeply conserved fragments in *cis*-regulatory modules of mammals, as well as evidence of a regulatory role in some cases. These regulatory elements have evolved from ancient insertions of TEs that are no longer active in the mammalian lineage. A basic interpretation is that TE insertions provide abundant sequence variation in regulatory regions on which selection can act.

In *Leishmania,* 3′ UTRs are important in posttranscriptional gene regulation. Bringaud et al. (2007) found evidence of accumulation of fragments from a now inactive family of retrotransposons in the UTRs of some predicted mRNAs, and further found that these genes showed lower-than-average mRNA levels. The pattern is suggestive of a possible evolution of a regulatory function. Caution is recommended in interpreting apparent overrepresentation of TEs near coding genes as an indicator of function, however. Elements inserted near protein-coding genes are less likely to be deleted because their removal increases the chance of harmful consequences to the adjacent gene. Therefore, inserts in these safe havens will persist longer, even if they are mildly deleterious. Studies that look for possible functional TE insertions based on distributions require null hypotheses that consider the mutational spectra (e.g., deletions) tolerated in regions of different distance from functional genes. Safe havens adjacent to genes also mean that such inserts have more time before deletion to evolve into functional *cis*-regulators, through mutational amelioration of their deleterious side effects and refinement of *cis*-regulatory effects.

Do TEs increase the rate of speciation in their hosts? Nonhomologous recombination among TEs can lead to chromosomal rearrangements that contribute to chromosomal-based speciation (Kidwell and Lisch, 2001). TE differences between related species may also contribute to reproductive isolation, however; arguing against this scenario is the rapidity by which TEs can jump species boundaries, as observed by P-elements in *Drosophila* (Kidwell and Lisch, 2001). Regarding extinction rates, a comparative study suggests that high TE loads increase the probability of extinction in plants, birds, and reptiles but not mammals (Vinogradov, 2004). Theoretical treatments indicate that TE activity can play a significant role in the extinction of parthenogenetic species through mutational load accumulation, particularly in small populations (Nuzhdin and Petrov, 2003; Dolgin and Charlesworth, 2006). In larger populations, clonal selection will lead to loss of active TEs, assuming their effects are mostly deleterious (Dolgin and Charlesworth, 2006). Zeh et al. (2009) propose an "epi-transposon" hypothesis that changing environments can lead to stress-induced breakdown of epigenetic suppression of TEs (e.g., methylation, piRNAs), with resulting extensive transposition providing new material for rapid adaptive shifts. They also note that such transposon release could lead to increased extinction rates. Alternatively, a changing environment could simply provide transient advantages for elevated mutation rates, which TEs can readily provide. Although intriguing, there is currently little direct support for these ideas.

Biased Gene Converters

Gene converters are a special class of SGEs that preferentially insert themselves into homologous uninserted sites in the genome. The most famous of these are the homing endonucleases (HEs) found in bacteria and eukaryotes (Stoddard, 2011). HEs are self-splicing group I introns coding for highly specific endonucleases that cut rare specific sites within the genome, typically the uninserted homologous sequence (hence the name "homing"). Cellular mechanisms use the inserted sequence and homologous flanking DNA as templates for repair, resulting in insertion of the HE into the previously unoccupied site. Other HEs actually splice out of proteins following translation ("inteins"). HEs are known to spread laterally between distant bacterial lineages and appear to maintain strong site fidelity. Their self-splicing ability ensures that the protein function is not disrupted. Using functionally important regions as insertion sites reduces the ability of hosts to evolve resistance to HEs, thus providing a "safe harbor."

Other forms of biased gene conversion have been found in recombination hotspots in humans through detailed analysis of the products of recombination (Jeffreys and Neumann, 2002). The extent to which such biased gene conversion can be considered selfish depends on whether conversion bias is dependent on the sequence of the putative SGE. There is growing evidence in eukaryotes of a general GC gene conversion bias in DNA repair of double-strand breaks, resulting in AT/GC heterozygotes producing more GC than AT gametes (Duret and Galtier, 2009). Can we therefore consider G and C to be our smallest selfish elements? Probably not, because conversion is likely attributable to a general bias in using G and C during double-stranded break repair rather than to a biased conversion attributable to a specific sequence motif. In any case, GC-biased conversion clearly has major consequences for genome composition and evolution (Duret and Galtier, 2009).

Meiotic Drivers

Meiosis results in a reduction of the diploid germ cells to haploid gametes. In general, meiosis is "fair," meaning that the two homologous chromosomes have an equal probability of ending up in functional gametes (sperm or eggs). Meiosis creates opportunities for SGEs that can increase their transmission relative to a nondriving homolog, however. Meiotic drivers are widespread in nature and include such examples as segregation distorter (SD) in *D. melanogaster,* X-chromosome drive in many animals, X and Y drive in plants, knob-containing chromosomes in maize, the *t*-locus in mice, supernumerary or B chromosomes in animals and plants, and centromere drive in different organisms (Lyttle, 1991; Camacho et al., 2000; Jaenike, 2001; Malik and Henikoff, 2009; Presgraves, 2009). Meiotic drive can take three basic forms. "True" meiotic drive (e.g., many B chromosomes, centromere drive) is accomplished by preferential segregation to the functional (egg or ovule) pole during gametogenesis. Germline overreplication occurs in some B chromosomes and results in their increased transmission to gametes. "Gamete killer" drive acts by selective elimination or functional disruption of gametes that do not carry the meiotic driving element. This latter form is typically found in males because they produce an excess of gametes that effectively compete for fertilization of eggs. Gamete killing results in increased fertilization of eggs by sperm with the driving chromosome from heterozygous males. SD illustrates the basic mechanism (Presgraves, 2009). SD is composed of two tightly linked loci near the centromere of chromosome 2 of *D. melanogaster.* The distorter locus *Sd* encodes a partial duplication of the gene *RanGAP*, and the *Rsp* responder locus contains variable numbers of a tandem repeat. WT chromosomes contain a normal copy of *RanGAP* and higher copy numbers of *Rsp*. In heterozygotes, WT sperm fail to develop properly because of interactions between the variant *RanGAP* (which mislocalizes to the nucleus) and *Rsp* repeats. A third linked locus enhances drive and a number of unlinked loci reduce drive, as predicted by conflict theory.

Centromere drive has been proposed as a mechanism for evolution of centromeric DNA (Malik and Henikoff, 2009). The basic idea is that competition for spindle binding favors expansion of centromere binding sequences to promote segregation to the function (egg) pole in meiosis. This process could have played an important role in the evolution of chromosome structure and will result in meiotic drive. B chromosomes are "extra" chromosomes that are not essential for viability. They are widespread in animals and plants, and many have mechanisms for increasing their transmission during gametogenesis, including overreplication in germ cells and/or preferential segregation to the egg nucleus rather than the polar body (Camacho et al., 2000). Most B chromosomes appear to be mildly parasitic, and their maintenance can be readily explained by drive mechanisms. Recent studies reveal that B chromosomes in some organisms can code for beneficial effects as well, however (Camacho et al., 2000). At the other end of the spectrum, the most extreme examples of SGEs are the B chromosomes found in haplodiploid insects that persist by destroying other chromosomes after fertilization of the egg (Werren and Stouthamer, 2003).

Sex chromosome drive is widespread (Jaenike, 2001). As a result of evolution of drive repression, however, X drive is often cryptic and only revealed in crosses between populations or species. For example, there are at least three cryptic X-drive systems in *Drosophila simulans* alone (Jaenike, 2001). Drive can also be difficult to detect if there is not an associated phenotype or linked genetic marker to detect deviation from Mendelian ratios. Genomic techniques are now opening new avenues for detecting drive. For example, a recent study in chickens using genomewide approaches revealed previously undetected drive around the centromere and telomeres of chromosome 1 (Axelsson et al., 2010), as predicted by the centric drive model. Such genomewide approaches are likely to uncover many more examples of drive in the near future.

The possible role of meiotic drive in speciation has a controversial and interesting history. Frank (1991) and Hurst and Pomiankowski (1991) first proposed that divergence in X and Y meiotic drive and suppression of drive could lead to abnormal gametogenesis and sterility in hybrids. They were, in part, attempting to explain Haldane's rule: the observation that when hybrid incompatibilities are asymmetrical, it is usually the heterogametic sex (XY males or ZW females) that show hybrid sterility. The drive model was vigorously dismissed by leading speciation researchers at the time (Coyne, 1992; Coyne and Orr, 1993) for two primary reasons: Meiotic drive was considered to be uncommon, and there was lack of direct empirical evidence for an association of drive and hybrid sterility. Today, there is mounting evidence in support of a significant role of meiotic drive in speciation (N. A. Johnson, 2010; McDermott and Noor, 2010; Presgraves, 2010). The change in landscape is attributable to the discovery that meiotic drive is often cryptic and much more common than previously thought and to detailed molecular and genetic studies of hybrid incompatibility genes that have revealed or implicated meiotic drive. Presgraves (2010) concludes that some form of genetic conflict is implicated in ~6 of the 14 hybrid incompatibility genes that have been relatively well characterized.

Postsegregation Distorters (PSDs)

A diverse array of PSDs exist, which, when present in an organism, act after progeny are produced to reduce the survival/fitness of progeny that have lost the driver. Although PSDs have arisen independently in many different organisms, a key feature of all PSDs is the involvement of a modification-rescue system, also called a toxin-antidote. I use the more general modification-rescue terminology because it does not assume a particular biochemical mechanism (cell toxicity) as the mode of action. A modification occurs in the "parent," and this modification must be rescued in the offspring. If the PSD element is not transmitted to the offspring, rescue cannot occur and "harm" will come to the progeny (often death).

Dramatic examples of PSDs are the killer plasmids of bacteria and yeast (Frank and Wolfe, 2009). In killer plasmids, the longer persistence of the encoded toxin relative to the antidote protein ensures that daughter cells die if the plasmid is lost. This acts to prevent both segregation loss of the plasmid and its displacement by a competitor plasmid. Restriction-modification (R-M) systems also have PSD properties (Kobayashi, 2001). They are widespread in bacteria and typically involve an enzyme that modifies DNA (e.g., by methylation) and a restriction enzyme that will cut DNA of a specific sequence that lacks the modification. These were originally believed to have evolved to provide protection to the cell against foreign DNA (e.g., bacteriophages). Kobayashi (2001) has convincingly shown that R-M systems also behave as postsegregation killers. R-M systems located on plasmids are like other PSDs in having a modification and rescue. The restriction enzyme protein persists longer in progeny cells than does the modification enzyme; therefore, if the plasmid is lost in a bacterial daughter cell, the result will be restriction of its DNA and death. It has been shown that R-M systems do indeed kill cells that lose them, either through segregation or displacement by a different plasmid, thus maintaining the selfish R-M system.

The dynamics of PSDs, such as killer and R-M system plasmids, can be complex and dependent on population structure (Kobayashi, 2001). In general, daughter cell killing only imparts an indirect and weak advantage to the progenitor cell lineage unless daughter cells compete with each other for resources, in which case the benefits are more direct. Therefore, the advantages occur most strongly in structured populations where related microbes co-occur. A second advantage of PSDs is prevention of displacement by competing plasmids. Some R-M systems are integrated into bacterial chromosomes, and this would seem to limit their advantage as postsegregation killers. Recent work shows that some integrated R-M systems are associated with TE-like structures that could facilitate their lateral movement between bacterial clones, however (Furuta et al., 2010). PSDs may have been co-opted by bacterial and yeast genomes for defense against competitors, viral protection, or other phenotypes. For example, killer plasmids share some features with bacteriocins in bacteria and diffusible killer toxins in yeast, which can act to kill competitor cells lacking the linked modification-rescue mechanism (Frank and Wolfe, 2009). Bacteriocins typically produce diffusible toxins, however, and therefore target different cell lineages. Nevertheless, some bacteriocin systems may have PSD features and may have originated from PSD plasmids in some cases. Frank and Wolfe (2009) have found the evolutionary addition of killer plasmid and killer virus DNA into yeast chromosomes, contributing to "killer chromosome" genotypes that produce diffusible substances targeting competitor cell lineages.

PSDs are also found in complex multicellular eukaryotes. A most intriguing example is the maternal effect dominant embryonic arrest (Medea) system in *Tribolium* beetles (Lorenzen et al., 2008), which is chromosomally integrated and was discovered in crosses between populations with and without the driver. When females carry the Medea element, their zygotes must also receive the element (either maternally or paternally) or the offspring perish. The maternal modification factor and zygotic responder are tightly linked. Recent work suggests that Medea may be caused by a TE insertion just upstream of a gene with maternal and zygotic function. It is a quandary why Medea-like elements have not been found more widely in eukaryotes, but one explanation could be that they can quickly go to fixation in populations and become hidden. Medea elements have the potential to drive desirable traits into host populations (e.g., vector resistance to pathogens). To explore this idea, an artificial Medea element was successfully constructed in *D. melanogaster* by coupling a gene for micro-RNA silencing of a maternally expressed gene required for embryogenesis with a zygotically expressing rescue (C. H. Chen et al., 2007).

An usual example of B chromosome-induced PSD occurs in *Nasonia* and *Trichogramma* wasps, which have haplodiploid sex determination; haploid males develop from unfertilized eggs and females from fertilized eggs (Werren and Stouthamer, 2003). The paternal sex ratio (*psr*) chromosome occurs in some males of these species. These males produce functional sperm, but *psr* induces improper condensation of the paternal chromosomes (except itself) in the fertilized egg, resulting in total loss of the normal paternal set. This converts the embryo into a haploid male that carries the supernumerary *psr* chromosome. That male's genome will be destroyed in the next generation, and *psr* will associate itself with yet another set of chromosomes destined for destruction. Because *psr* totally destroys the genome with which it becomes associated in each generation, it represents the most extreme example of an SGE in any organism. Haplodiploid sex determination promotes these extreme SGEs under certain population structures, and they have evolved independently in different wasp species.

Heritable Organelles and Microbes

It may seem odd to include vital organelles, such as mitochondria and chloroplasts, in a treatise on SGEs. These organelles can have genetic interests that diverge from that of the nuclear genome, however, resulting in genetic conflict (Fig. 10.1). Mitochondria and chloroplasts evolved from ancient bacterial symbionts, and each retains its own DNA despite extensive transfer of genetic material to the nucleus. Such symbioses are not simply events of the ancient past, because heritable microorganisms (those that are inherited during the reproduction of their hosts, often through the egg cytoplasm) are widespread in plants and animals and involve a diverse array of microbial taxa. Once it was assumed that these microbes must always be beneficial to their hosts because they are dependent on host reproduction for their transmission. Although many heritable microbes are mutualistic, others manipulate host reproduction in ways that enhance the microbes' transmission, and hence can be considered "reproductive parasites" (Werren et al., 2008).

The difference in inheritance patterns between heritable cytoplasmic elements and nuclear genes causes genetic conflict (Eberhard, 1980; Cosmides and Tooby, 1981; Werren and Beukeboom, 1998). With a few exceptions, inherited cytoplasmic elements are passed through the cytoplasm of the egg but not through sperm (in part, because of little cytoplasm in sperm). As a consequence, females transmit these elements, whereas males do not. In contrast, (autosomal) nuclear genes are typically inherited through both sexes. The result is cytonuclear conflict over sex determination and sex ratios, which, along with other forms of sex determination conflict, has likely played a role in sex determination evolution (Werren and Beukeboom, 1998). For example, mitochondrial variants induce pollen sterility in many plants, resulting in evolution of nuclear suppressor genotypes. Some of these systems are cryptic, as a result of evolutionary suppression, but are revealed in crosses between populations or related species.

Many different inherited microbes have evolved mechanisms to manipulate host reproduction because of their preferential transmission through females (Hurst et al., 1992; Werren and Beukeboom, 1998; Werren et al., 2008). Among the kinds of manipulations are conversion of males to functional females, induction of parthenogenetic reproduction in females, male-killing, and a form of sperm-egg incompatibility termed "cytoplasmic incompatibility" (CI). Noteworthy by its abundance in invertebrates and ability to perform all these manipulations is the α-proteobacterium *Wolbachia* (Werren et al., 2008). This bacterium is transmitted through eggs but also moves laterally between taxa. As a result, it is found in ~70% of terrestrial arthropods (Hilgenboecker et al., 2008). Many strains of *Wolbachia* induce a form of CI. These *Wolbachia* strains modify the sperm (by unknown biochemical mechanisms), such that the same strain of *Wolbachia* must be present in the egg to rescue the modification. If not, the sperm chromatin condenses improperly in the embryo, usually killing the offspring. By reducing the fitness of uninfected females, the infection can spread very rapidly in populations. Different strains of *Wolbachia* can be reciprocally incompatible because of differences in their modification-rescue system or host genetic interactions. An interesting possible case of evolved dependency occurs in the wasp *Asobara tabida* (Kremer et al., 2009). Removal of *Wolbachia* results in sterility as a result of elevated apoptosis in the developing female reproductive tract. Because closely related species do not require *Wolbachia* for ovarian development, this is likely either a case of PSD (killing of stem cells that have lost the bacteria) or an evolved host dependency on the presence of the parasite in reproductive tissues, which results in abnormal apoptosis in their absence. Such evolved dependencies to SGEs are likely to be common and are distinct from mutualisms, even though both will result in reduced fitness when the element is removed or inhibited.

Do inherited symbionts promote speciation in their hosts? Because CI can induce partial or complete reproductive incompatibility between diverging populations with different infections, it may promote reproductive isolation and speciation [reviewed in Werren (1998) and Bordenstein (2003)]. For example, in the *Nasonia* species complex, reciprocal CI between the species is a major contributor to hybrid reproductive incompatibilities and evolved early in the speciation process. A general role for *Wolbachia* in speciation has been criticized for several reasons, however, including the beliefs that (*i*) *Wolbachia* differences will not be stable between populations, (*ii*) CI is insufficient to maintain genetic divergence between populations, and (*iii*) *Wolbachia* infections are not common enough to be a major player. Nevertheless, supporting data continue to grow, including theoretical studies indicating that CI differences can be stable, maintain divergence, and select for premating isolation; empirical studies showing CI as contributing to reproductive isolation and reinforcement of mate discrimination between species (Jaenike et al., 2006); and the finding that *Wolbachia* infections are much more common than previously recognized (in ~70% of species rather than original estimates of 20%). Parthenogenetic species have also arisen courtesy of *Wolbachia* (Stouthamer et al., 2010). In haplodiploids, *Wolbachia* causes parthenogenesis by inducing diploidization of unfertilized haploid eggs, which leads to female development. Parthenogenesis causing *Wolbachia* often occurs as a polymorphism in sexual species, but host genetic changes favoring females that do not mate can lead to fully parthenogenetic species (Stouthamer et al., 2010). Subsequent loss of genes needed for sexual reproduction makes the process irreversible. Over 20 examples of *Wolbachia-induced* parthenogenetic species have been described in haplodiploids.

Wolbachia occurs within the germline in intimate proximity to the nucleus. An exciting recent discovery is that lateral gene transfers from *Wolbachia* to animals are common (Dunning Hotopp et al., 2007). Approximately one-third of invertebrate genomes show such transfers, sometimes involving large amounts of DNA (e.g., nearly the entire 1.2-MB genome of *Wolbachia* in *Drosophila anannassae*). Such transfers can either degrade over evolutionary time as a result of mutation accumulation or evolve novel functional genes, and there is evidence of the latter in some species (Dunning Hotopp et al., 2007; Werren et al., 2010).

Other Evolutionary Consequences of SGEs

SGEs have been invoked to play a role in many important biological phenomena, with variable levels of empirical and theoretical support. Examples include the evolution of sex, recombination, anisogamy, germline sequestration and uniparental inheritance of plastids, and alteration of mating systems (Hurst et al., 1992; Hurst and Werren, 2001; Burt and Trivers, 2006). Here, I will mention a few recent studies of interest. The scarcity of males caused by meiotic drivers and cytoplasmic sex ratio distorters has selected for changes in mating systems (Price et al., 2008). The germ granule, a key constituent in germ cell determination, contains proteins important in TE suppression (Lim and Kai, 2007), and SGEs may have promoted the evolution of a sequestered germline (Johnson, 2008). Suzuki et al. (2007) propose that DNA methylation suppression of TEs was a precursor for the evolution of genomic imprinting and the placenta in mammals. Such studies suggest that SGEs have an enormous potential range of evolutionary consequences.

Evolutionary Dynamics, Evolvability, and "Function" of SGEs

A longstanding debate concerns the evolutionary function of TEs, and this debate has been reinvigorated with recent discoveries of their evolutionary domestication into functional genes. Are TEs maintained over long evolutionary timescales because they induce beneficial mutations and innovations, thus allowing species to adapt (the evolvability hypothesis) or because they are self-replicating elements that are maintained by replicating at faster rates than they are lost (the parasite hypothesis)? Note that the parasite hypothesis does not preclude TE insertions evolving beneficial functions in the host (i.e., domestication) but argues that this is a consequence of TEs rather than the "reason" for the existence of TEs. There has been a recent resurgence of articles either implicitly or explicitly stating that an important evolutionary "function" of TEs is to promote genetic innovation and evolvability (Oliver and Greene, 2009; Aziz et al., 2010; Biémont, 2010; Britten, 2010). For example, Aziz et al. (2010) argue that the ubiquity of TEs is proof that they must exist to provide benefits in evolution. They state that "ubiquity is one of the indicators of essentiality" and further claim that these elements are "indispensible in every genome (elements of core genome) or every ecosystem (eco-essential genes)" (Aziz et al., 2010). The near-ubiquity of TEs can be readily explained by their ability to replicate within genomes and to move laterally between species, however. There is ample evidence that TEs can accumulate in genomes as a result of transposition and induce harmful mutations, that organisms have evolved many mechanisms to suppress TEs, that genomes are littered with fossil suppressed and degraded elements, and that TEs move laterally between species, often across large evolutionary distances. These observations are consistent with the parasitic hypothesis, and no additional conditions are needed to explain the persistence of TEs over evolutionary time or their widespread occurrence. In contrast, there is much less direct evidence supporting the argument that TEs persist in evolution because they enhance the evolvability of organisms or play some vital role in ecosystems (Aziz et al., 2010; Biémont, 2010).

To evaluate the concept of evolvability as it applies to SGEs, the questions need to be framed clearly. The evolvability concept can be applied to short-term adaptive evolution or long-term adaptation and innovation. Implicit to the short-term concept is that active TEs are maintained because of the beneficial mutations they induce (i.e., those insertions contribute substantially to the pool of active elements). If this is correct, there should be evidence of recent active TE insertions associated with adaptive evolutionary changes within species. TEs can be a significant source of standing genetic variation (O'Donnell and Burns, 2010), but the question remains of how often they induce beneficial as opposed to deleterious mutations. There is ample evidence of mutational costs of TE insertion events (Charlesworth et al., 1994; Kidwell and Lisch, 2001; O'Donnell and Burns, 2010). There has been relatively little evidence that young insertions lead to adaptive mutations, although some possible cases have emerged recently (González and Petrov, 2009). Examples include inserts associated with pesticide resistance in *D. melanogaster* and *D. simulans* as well as a genomewide study implicating TE insertions with temperature adaptation in two separate latitudinal clines in *D. melanogaster*.

González and Petrov (2009) also note that if adaptive TE insertions are frequent, we should observe fixation of TE insertions in species more often in high-recombination regions than are observed. They offer one possible explanation that beneficial insertions quickly evolve through mutation, and therefore are not readily recognizable as TEs. If true, this would suggest that they then contribute little to the generation of new active TEs. It is also possible that they are maintained as ancient polymorphisms in related species; however, again, we would expect them to degrade as active TEs. A third possible explanation is that TE insertions provide short-term adaptive mutations (e.g., attributable to habitat differences) but that they are eventually replaced by point mutations with fewer negative pleiotropic consequences. Although the data for frequent adaptive insertions are tantalizing, they are still largely indirect. Nevertheless, the findings suggest that a more nuanced evaluation of the mechanisms that maintain active TEs is needed, which includes models that incorporate deleterious, neutral, and beneficial insertions as well as mutational changes in insertions (Le Rouzic et al., 2007). The key empirical and theoretical question becomes "What portion of TE transposition comes from elements associated with beneficial vs. neutral or deleterious effects?" This is crucial for determining what maintains active TEs in a species.

A key observation argues against beneficial insertions being important in the maintenance of active TEs. If TEs are beneficial, we would expect to see specific TE types being maintained within lineages over evolutionary time (i.e., the phylogenies of TE elements should parallel that of the organisms in which they occur). With a few exceptions, this is not the case. Instead, the patterns of TE variation in most species indicate that they tend to invade a species, rapidly proliferate, and then are suppressed (Feschotte and Pritham, 2007; Feschotte, 2008).

The long-term evolvability argument is that TEs exist because of their long-term contributions to adaptation and innovation (e.g., new genes and gene regulation networks). In its naive form, the long-term evolvability argument is teleological and confuses cause and consequence. Simply, evolution is not anticipatory, and the observation that TE inserts can evolve into functional genes is not proof that they are maintained because they provide genetic material for long-term evolutionary innovation. The idea also suffers from the irreversibility problem. Once a TE insertion has evolved into a function gene or regulatory element, it is unlikely to evolve back into a TE, and thus does not contribute to the pool of active TEs supposedly being maintained for their long-term benefits. In other words, there is no evolutionary feedback to maintain active TEs for long-term benefits.

A more reasonable hypothesis for maintenance of TEs attributable to long-term benefits is based on clade selection (i.e., competition among lineages of species) (Oliver and Greene, 2009; Biémont, 2010). The clade selection hypothesis is that those clades (e.g., species, genera) with active TE elements are more likely to persist and radiate because of their ability to evolve to changing environments or to evolve innovations. This hypothesis is difficult to test because of the near-ubiquity of TEs. The parasitic vs. evolvability hypotheses do make clear contrasting predictions about what should happen in asexual vs. sexual species, however. The parasitic hypothesis predicts that active TEs will decline in asexual species because of the deleterious effects of transposition, whereas the evolvability hypothesis predicts that they will be maintained (or increase) because of their beneficial effects. Indeed, the benefits of TEs might be expected to be even greater in asexual species as a source of beneficial mutations to resist mutational decline and for adaptation to new environments. What few data currently exist support the parasitic hypothesis. Asexual Bdelloid rotifers have significantly reduced TE numbers compared with sexual relatives, including loss of retroelements (Arkhipova and Meselson, 2000), and asexual *Daphnia* species show reduced numbers of active TEs compared with sexual *Daphnia* species (Schaack et al., 2010b). More comparisons are needed between related asexual and sexual species to test the predictions of these alternative hypotheses.

A theme in some recent papers is that although TEs were "dismissed" over the past several decades as mere selfish DNA, new evidence now shows that they have evolved functions within the genome (Aziz et al., 2010; Biémont, 2010). Although perhaps a useful literary foil, these statements are not very accurate. Even the original paper by Doolittle and Sapienza (1980), which asserted that repetitive elements can be maintained because of their self-replicating properties, goes on to state that "we do not deny that [such elements] may have roles of immediate phenotypic benefit to the organism. Nor do we deny roles for these elements in the evolutionary process." This same theme has been maintained by advocates of the parasitic hypothesis for the past two decades (Werren et al., 1988; Hurst et al., 1992; Hurst and Werren, 2001; Burt and Trivers, 2006). Discoveries that TE insertions can evolve into function sequences and induce favorable mutations are not contrary to the SGE hypothesis for TE maintenance. What has been criticized by Doolittle and Sapienza (1980) onward are uncritical evolvability claims that TEs exist because they provide evolutionary benefits to organisms. As outlined above, evolvability arguments need to be precisely framed (and in nonteleological terms) to provide testable predictions, such as that beneficial mutations are required for maintenance of active TEs or that clade selection favors lineages with active TEs. This can then lead to more rigorous tests of the evolvability idea. To date, the data continue to support the view that TEs have important evolutionary consequences but are maintained because of their selfish replicative features.

CONCLUSIONS

Rapidly growing evidence emerging from genomics and advances in genetics indicate that SGEs are important motors for evolutionary change and innovation. Several general principles have reoccurred in the discussion above, and I will briefly revisit these themes here. The first is that SGEs lead to antagonistic co-evolution with other components of the genome. Important features of eukaryotic genomes (e.g., DNA methylation, RNAi, small RNA regulatory pathways, R-M systems) have evolved, at least in part, as defense mechanisms against SGEs. Many genetic elements have mixed phenotypes, with both selfish (parasitic) and "beneficial" ("mutualistic") features. The classic example is the mitochondrion, which is clearly beneficial but also shows selfish features (e.g., cytoplasmic male sterility) that reduce nuclear gene fitness, thus leading to genetic conflict. Evolutionary dependency can also evolve in hosts with ubiquitous SGEs, which can lead to irreversible dependence. Growing evidence supports a significant role of SGEs in eukaryotic development and speciation, and possibly also in extinction of species. Genome domestication of SGEs leads to evolutionary innovations, including acquisition of new genes and gene regulation from TEs, heritable microbes (e.g., *Wolbachia*), and selfish plasmids. Safe havens can promote longer associations of SGEs with host lineages and also may facilitate their domestication. Finally, distinctions are made between the evolutionary consequences of SGEs and the factors that maintain them over evolutionary time. Clear formulations of the idea of evolvability as a means for evolutionary maintenance of SGEs will facilitate rigorous testing of this idea. Nevertheless, current evidence strongly supports the view that SGEs are maintained by their transmission-enhancing phenotypes and that evolutionary innovations emerging from them are a consequence of their existence rather than the cause.

ACKNOWLEDGMENTS

I thank J. Strassmann, D. Queller, J. Avise, and F. Ayala for organizing *In the Light of Evolution V: Cooperation;* and A. Avery, R. Edwards, T. Eickbush, C. Feschotte, D. Loehlin, M. Clark, J. Jaenike, M. Lenoe, C. Landry, R. Minckley, P. Nagarajan, and D. Wheeler for assistance and comments on the manuscript. Because of space limitations, topic reviews are sometimes cited in lieu of the original research paper and the reader is encouraged to access the reviews for primary research sources. Support for this work came from National Science Foundation Grants DEB0821936 and NIH GM084917.

Part IV

SOCIALITY AND MEDICINE

Most biologists probably work in biomedical fields. If nothing in biology makes sense except in the light of evolution, then medicine should have much to learn from evolutionary reasoning. The rapidly growing field of Darwinian medicine (Williams and Neese, 1991) is based on this premise and seeks to provide insight on topics such as the evolution of virulence and diseases of altered evolutionary environments. A subfield recently called Hamiltonian medicine (Foster, 2005) investigates the impact of social evolution, cooperation, and conflict on disease.

In Chapter 11, Andrew Read and colleagues treat the vital problem of how to minimize the evolution of pathogen resistance and thereby extend the useful lives of our arsenal of antibiotic drugs. This involves a complex set of interacting causes, some of which have a social element and others do not. The authors challenge the dogma that we minimize the evolution of resistance by "radical pathogen cure": using enough of a drug to try to eliminate the pathogen from the patient's body. The reasonable rationale behind this practice is to lower the pathogen population size and minimize the occurrence of novel resistance mutations. But the authors argue that this ignores the selective phase, which may be more important in determining the time to drug impotence, particularly when resistance mutations arise with relative ease. In this selective phase, the radical pathogen cure provides the strongest possible selection for resistance. According to Read and colleagues, the social structure of the pathogen can powerfully augment this selection. When a host is infected by multiple strains of the pathogen (as is often true of malaria) and the total density of the pathogen is regulated, then wiping out susceptible strains with antibiotics can greatly increase the frequency of formerly rare resistant strains. This raises the possibility that the medical community is ignoring an important human social dilemma: that the best treatment for a patient may not be the best outcome for society as a whole.

Some human disorders can spring not from a failure of adaptation per se, but from disagreement and conflict over what is the correct adaptation. This is particularly so in the realm of human interpersonal relations, starting with fundamental conflicts between parent and offspring. Haig (1993) has argued that such conflicts can lead to pathologies in pregnancy when there is an upset in the precarious resolution of embryo-maternal conflict. Taking a radical step further, he has pointed out that the optimal strategy of an embryo's gene differs according to whether it came from the dam or the sire, with maternal loci being less selected to take resources from the mother. Remarkably, imprinted genes appear to behave in accord with this theory. In Chapter 12, David Haig extends this thinking in several directions. He notes that most of our kin belong to categories that have asymmetrical relatedness to our maternal and paternal genes, so that most of our psychological adaptations for dealing with kin, and perhaps pathologies, may reflect these kinds of conflicts. In particular, he shows how this perspective may illuminate unsolved problems surrounding the evolution of adolescence and the timing of sexual maturation in humans.

In Chapter 13, Steven Frank and Bernard Crespi extend and generalize the same theme: that conflict can lead to pathologies when opposing interests that are precariously balanced become unbalanced. These authors suggest that the conflict between maternal and paternal genes in offspring, through its demonstrated effects on the regulation and pathologies of growth, may be responsible for some cancers. They then discuss the exciting idea that this same balance is partly responsible for a wide spectrum of psychiatric disorders, such as autism that may result from an overexpression of paternal interests in offspring selfishness. Similarly, other disorders such as schizophrenia might result from an overexpression of genes underlying the maternal goal of greater social integration. Finally, the authors present a novel theory of conflict between autosomal and X chromosomes. The latter spend two-thirds of their time in females and therefore should be selected to give greater weight to female than to male adaptation. Autosomes should give equal weight. It will be fascinating to see if empirical tests support the authors' prediction that such conflict will underlie pathologies of expression along the male-female axis.

11

The Evolution of Drug Resistance and the Curious Orthodoxy of Aggressive Chemotherapy

ANDREW F. READ,[*†§] TROY DAY,[‡] AND SILVIE HUIJBEN[*]

The evolution of drug-resistant pathogens is a major challenge for 21st century medicine. Drug use practices vigorously advocated as resistance management tools by professional bodies, public health agencies, and medical schools represent some of humankind's largest attempts to manage evolution. It is our contention that these practices have poor theoretical and empirical justification for a broad spectrum of diseases. For instance, rapid elimination of pathogens can reduce the probability that *de novo* resistance mutations occur. This idea often motivates the medical orthodoxy that patients should complete drug courses even when they no longer feel sick. Yet "radical pathogen cure" maximizes the evolutionary advantage of any resistant pathogens that are present. It could promote the very evolution it is intended to retard. The guiding principle should be to impose no more selection than is absolutely necessary. We illustrate these arguments in the context of malaria; they likely apply to a wide range of infections as well as cancer and public health insecticides. Intuition is unreliable even in simple evolutionary contexts; in a social milieu where in-host competition can radically alter the fitness costs and benefits of resistance, expert opinion will be insufficient. An evidence-based approach to resistance management is required.

*Center for Infectious Disease Dynamics, Departments of Biology and Entomology, Pennsylvania State University, University Park, PA 16802; †Fogarty International Center, National Institutes of Health, Bethesda, MD 20892; and ‡Departments of Mathematics, Statistics, and Biology, Queen's University, Kingston, ON, Canada K7L 3N6. §To whom correspondence should be addressed. E-mail: a.read@psu.edu.

The evolution of drug-resistant pathogens significantly affects human well-being and health budgets. Consequently, existing and new antimicrobials should be viewed as precious resources in need of careful stewardship (Owens, 2008; Spellberg et al., 2008). An important aspiration is to maximize the therapeutically useful life span of a compound, the time a given antimicrobial yields clinical benefits before drug efficacy is undermined by resistance evolution. Attempting to do so is essentially an exercise in evolutionary management.

Various practices are widely thought to be effective resistance management strategies (American Academy of Microbiology, 2009; World Health Organization, 2010a; zur Wiesch et al., 2011). For instance, there is near-universal agreement that combination drug therapy, the coadministration of drugs with unrelated modes of action, prolongs the useful life of the component compounds for diseases as diverse as leprosy, HIV, malaria, and tuberculosis (TB). Another practice is the restriction of treatment to those patients who need it on clinical grounds, so as to reduce unnecessary selection for resistance. This philosophy underpins restrictions on the use of antibiotics in hospitals and in the community at large, and it has led to calls for reductions in drug use in animal feed.

A third practice thought to be an effective resistance management strategy is the use of drugs to clear all target pathogens from a patient as fast as possible. We hereafter refer to this practice as "radical pathogen cure." For a wide variety of infectious diseases, recommended drug doses, interdose intervals, and treatment durations (which together constitute "patient treatment regimens") are designed to achieve complete pathogen elimination as fast as possible. This is often the basis for physicians exhorting their patients to finish a drug course long after they feel better (long-course chemotherapy). Our claim is that aggressive chemotherapy cannot be assumed to be an effective resistance management strategy a priori. This is because radical pathogen cure necessarily confers the strongest possible evolutionary advantage on the very pathogens that cause drugs to fail.

At one level, our argument is simple. Elementary population genetics shows that, all else being equal, the stronger the strength of selection, the more rapid is the spread of a favored allele (Maynard Smith, 1989a). For drug use, the strength of selection is determined by how many people are being treated and, among the treated people, the treatment regimen. The more aggressive the regimen, the greater is the selection pressure in favor of resistance. Because overwhelming chemical force necessarily confers the strongest possible selective advantage on any pathogen capable of resisting it, radical pathogen cure can very effectively drive resistant pathogens through a population. As we will argue, this problem is especially important when there is genetic diversity among pathogens within an infected individual.

AIMS OF PATIENT TREATMENT

Ignoring economic considerations, patient treatment should seek to achieve the following:

(*i*) Make the patient healthy.
(*ii*) Prevent the patient from infecting others.
(*iii*) Prevent the spread of resistant pathogens to others.

The first aim concerns the health of the patient being treated. The second and third aims concern the effects of patient treatment on the health of others.

A single strategy cannot simultaneously best achieve all three aims. In the limit, zero treatment will usually be the best resistance management strategy. It is important to identify and justify compromises because this makes explicit problems in need of solution and is a prerequisite for evidence-based resistance management. There may come a time when resistance management strategies are required that put overall public health ahead of patient health (Foster and Grundmann, 2006). We do not think the problems of resistant pathogens are yet so dire as to require this. In our view, the current scientific challenge is to identify, among patient treatment regimens that are similarly effective at restoring health and preventing transmission, those regimens that best effect resistance management.

The aim of resistance management is to prevent clinical failures caused by high-level resistance. Resistance is often a continuous trait, and there can be varying degrees of intermediate resistance. Sometimes referred to as "tolerance," intermediate resistance confers the ability to survive concentrations of drug below those considered therapeutic (Fig. 11.1). We define high-level resistance as that which undermines patient health by causing therapeutic failure. It is the rate of spread of high-level resistance that needs to be managed because this determines the therapeutically useful life span of a drug.

The useful life span of a drug is determined by two processes. The first is the rate at which genetic events conferring high-level resistance on an individual pathogen actually occur. For simplicity, we refer to these events as *de novo* mutations, but we use this to include any heritable change that confers *de novo* high-level resistance on a pathogen individual. For example, in bacteria, this event can be the acquisition by lateral transfer of genetic material from another species. The second process affecting the

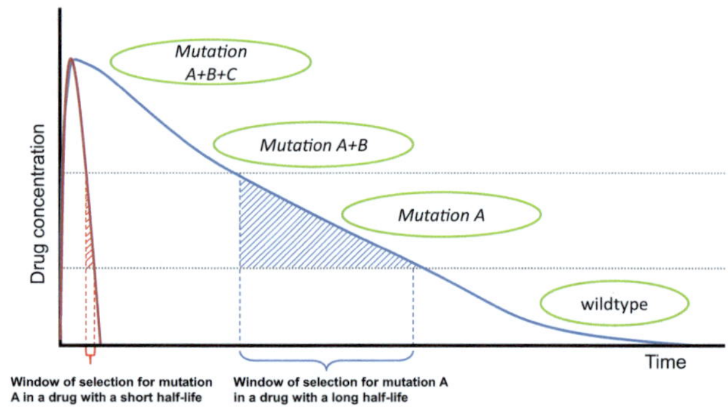

FIGURE 11.1 Hypothetical path to drug resistance. Solid curves show drug concentration in a treated patient for two drugs with very different half-lives; concentrations wane when treatment ceases. In this schematic, wild-type parasites can survive very low concentrations, with mutations A, B, and C conferring the ability to survive ("tolerate") successively higher drug concentrations. High-level resistance (full clinical resistance) is where treatment has a negligible direct impact on pathogens with all three mutations. The windows of selection for mutation A are shown. In those windows, parasites with mutation A have a selective advantage over wild-type parasites. Note that the duration of the window depends critically on the drug half-life, which for antimalarial drugs can vary from hours (e.g., artemisinin), to weeks (e.g., SP), to months (e.g., mefloquine).

rate of evolution is the strength of selection acting on this genetic change. Because both mutational and selection processes together determine the useful life span of a drug, resistance evolution can be retarded by managing mutations, selection, or, ideally, both. Our view is that conventional wisdom focuses too much on managing mutational events (genetic origins), often with the consequence that the selection pressures are ignored.

A REAL-WORLD CONTEXT

Our logic likely applies to a very wide range of pathogens, but, as we discuss further below, there will not be simple generalities. To make things more concrete, we base our discussion on malaria, a disease that typifies the clinical and financial problems posed by drug resistance.

Resistance has evolved to all classes of frontline antimalarial drugs (Hyde, 2005), and several have had to be withdrawn from use in many countries. The eventual failure of drugs in the face of parasite evolution is now accepted as inevitable by the World Health Organization (WHO) (Roll Back Malaria, 2008) and others (American Academy of Microbiology,

2009). A key component of the Global Malaria Action Plan is an explicit plan for a discovery pipeline to deliver replacement drugs continuously (Roll Back Malaria, 2008). This pipeline will cost more than U.S. $2.5 billion in research and development for the coming decade and, once the currently inadequate drug arsenal is rebuilt, U.S. $1.5 billion thereafter for every decade until malaria is eradicated (Roll Back Malaria, 2008). Even if we assume that an unlimited supply of drug classes can be discovered, more than money is at stake. Drugs can fail more rapidly than the time it takes to get them through modern regulatory processes, and the cost in terms of human suffering is high. National authorities switch their choice of first-line drug only when forced to by declining patient cure rates; thus, disease burdens are considerable. WHO currently recommends that a drug be withdrawn once treatment failure rates attributable to resistance reach 10% (World Health Organization, 2010a, p. 8). In practice, governments of poor countries do not have this luxury and often wait longer before drug withdrawal is implemented (World Health Organization, 2006, p. 15).

Severe (life-threatening) malaria involves the dysfunction of vital organs; for patients in this state, the sole aim of treatment is to prevent death. Uncomplicated malaria constitutes the bulk of treated cases and those that can drive transmission chains, and hence resistance evolution. The WHO *Guidelines for the Treatment of Malaria* (World Health Organization, 2010a, p. 6) state: "The objective of treating uncomplicated malaria is to cure the infection as rapidly as possible," with cure being defined as "the elimination from the body of the parasites that caused the illness." Patient treatment regimens recommended in the WHO guidelines are those designed to achieve rapid and full elimination.

It is clear that radical pathogen cure can, in the absence of resistance, achieve the first two aims of patient treatment (restore health and prevent disease transmission). The consensus view is that it can also achieve the third aim: "Resistance can be prevented, or its onset slowed considerably" by "ensuring very high cure rates through full adherence to correct dosing regimens" (World Health Organization, 2010a, p. 6). This is the orthodoxy that concerns us.

The strength of selection on resistance is primarily determined by the fate of resistant parasites in treated and untreated hosts. Resistant strains gain an advantage in treated hosts but often pay a cost in untreated hosts. In both types of host, the social milieu of strains within individual infections plays a very important role in mediating these costs and benefits. To explain why, we need to summarize some within-host ecology.

Genetic Diversity of Infections

Human malaria infections normally consist of more than one asexually proliferating parasite lineage ("clone"). Thus, the majority of *Plasmodium falciparum* clones in the world share their human hosts with at least one other lineage (Read and Taylor, 2001). Mixed infections arise from inoculations of genetically diverse parasites by a single mosquito or contemporaneous bites by multiple mosquitoes infected with different parasites. Consequently, the coexistence of drug-sensitive and drug-resistant parasites is common, and indeed may even be the rule (Day et al., 1992; Arnot, 1998; Babiker et al., 1999; Smith et al., 1999; Bruce et al., 2000; Jafari et al., 2004; Juliano et al., 2007, 2010; McCollum et al., 2008; Zhong et al., 2008; Owusu-Agyei et al., 2009).

A substantial body of epidemiological evidence is consistent with crowding effects within infections, whereby the population densities of individual genotypes are suppressed when other genotypes are present (Daubersies et al., 1996; Mercereau-Puijalon, 1996; Smith et al., 1999; Bruce et al., 2000; Hastings, 2003; Talisuna et al., 2003, 2006; Färnert, 2008; Harrington et al., 2009; Orjuela-Sánchez et al., 2009; Baliraine et al., 2010). For example, parasite densities are unrelated to the number of clones per host, and high turnover rates are observed in mixed-genotype infections.

Direct experimental evidence of crowding cannot be ethically obtained from human infections because formally demonstrating competition requires deliberate infection and/or the withholding of treatment (Read and Taylor, 2001). However, in a rodent malaria model, *P. chabaudi* in laboratory mice, we and others have experimentally demonstrated that densities of individual clones within an infection are severely suppressed when coinfecting clones are present (Jarra and Brown, 1985; Taylor et al., 1997a,b; Taylor and Read, 1998; de Roode et al., 2003, 2004a,b, 2005a,b; Raberg et al., 2006; Wargo et al., 2007; Huijben et al., 2010; Pollitt et al., 2011). This competitive suppression substantially reduces the density of transmission stages (Wargo et al., 2007; Huijben et al., 2010), and hence transmission of individual clones to mosquitoes (Taylor et al., 1997a; Taylor and Read, 1998; de Roode et al., 2004a). To date, there is no evidence of direct interference competition analogous to bacteriocin-mediated competition in bacteria (Riley and Gordon, 1999). Instead, the competition between coinfecting malaria parasites probably arises from competition for resources. Most likely, this competition is for access to red blood cells (Hellriegel, 1992; Yap and Stevenson, 1994; Hetzel and Anderson, 1996; Haydon et al., 2003; Antia et al., 2008; Mideo et al., 2008; Kochin et al., 2010; Miller et al., 2010; Pollitt et al., 2011), although other resources, such as glucose, may also be involved (de Roode et al., 2003). Immune-mediated apparent competition, wherein the immune response provoked by one strain suppresses the population densities of a coinfecting strain (Read and Taylor, 2001), likely also plays a major role (Mota et al., 2001; Raberg et al., 2006).

This in-host competition has profound effects on the evolution of drug resistance because it affects the fitness costs and benefits of resistance. We take these in turn.

Costs of Resistance

It is generally assumed that resistant pathogens are less fit than their wild-type ancestors in the absence of drug treatment and that this is the main force slowing the evolution of resistance. In malaria, there is good evidence of this (Hastings and Donnelly, 2005; Walliker et al., 2005; Babiker et al., 2009; World Health Organization, 2010a). One consequence of the social ecology within a host is that it acts as a serious multiplier of these costs of resistance. Costs of resistance arise from metabolic inefficiencies associated with efflux or detoxification mechanisms, which can include negative pleiotropic effects on other cellular and biochemical processes or reduced biochemical efficiencies associated with target site mutations (Hastings and Donnelly, 2005). These reductions in performance can be quite small (e.g., a few percent), but small differences can be greatly magnified by competition between clones. For example, in mice, the social context of the infection can translate modest differences in performance into differences well in excess of 90% (Fig. 11.2). The social context within which resistant strains are circulating is thus a potent determinant of the fitness costs of resistance, the main brake on the spread of drug resistance.

Benefits of Resistance

The flip side of this ecology is that the fitness advantages resistant parasites experience in treated hosts are greatly magnified in mixed-clone infections. Consider the consequences of radical pathogen cure where competition is occurring. Aggressive chemotherapy will kill all sensitive or tolerant parasites. This will result in competitive release and enhanced transmission of any highly resistant strains that are present. In rodent models, this is precisely what happens (de Roode et al., 2004a; Wargo et al., 2007; Huijben et al., 2010) (Fig. 11.3). Thus, radical parasitological cure enhances the transmission of the resistant strains. The impact of this competitive release on the rate of spread of resistance can be very substantial, as was first recognized by Hastings and colleagues (Hastings, 1997; Mackinnon and Hastings, 1998; Hastings and D'Alessandro, 2000). Where multiclone infections dominate, this within-host ecology

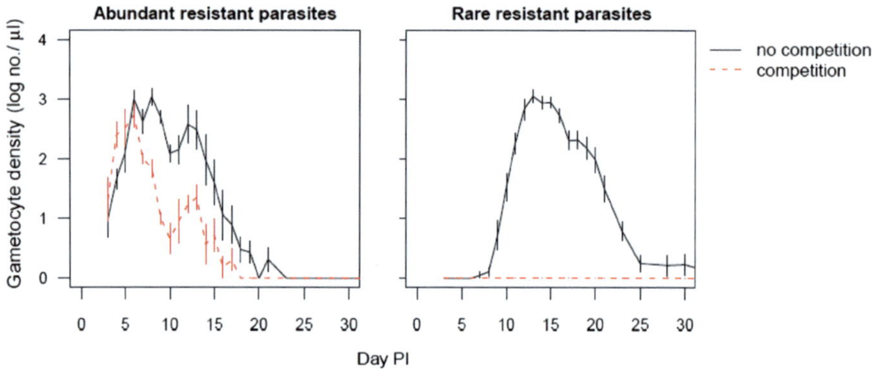

FIGURE 11.2 Costs of resistance are greatly affected by competition. Transmission stage densities of the resistant *P. chabaudi* clone in laboratory mice in the absence of drug treatment are shown. Infections were initiated with 10^6 (*Left*) or 10^1 (*Right*) resistant parasites and either no sensitive parasites (no competition, solid lines) or 10^6 sensitive parasites (competition, dashed lines). Performance of the resistant clone alone includes any physiological costs to resistance. When the resistant clone shares a host with a sensitive clone, performance is greatly reduced, and is effectively zero when rare in the inoculum (*Right*). Thus, the costs of resistance depend critically on whether competitors are present and the frequency of resistant parasites in an infection. PI, post-infection. Plotted points are the mean (±SEM) densities in peripheral blood from 5 to 10 mice per group, estimated by quantitative PCR using protocols described elsewhere (Huijben et al., 2010).

can be the primary determinant of the speed at which resistance spreads, and a far more important selective force than the simple survival advantage conferred by resistance (Hastings, 1997, 2003, 2006; Mackinnon and Hastings, 1998; Hastings and D'Alessandro, 2000; Mackinnon, 2005; Talisuna et al., 2006).

For instance, in an infection composed of two equally represented clones, aggressive treatment can effectively double the absolute fitness of the resistant strain if that strain can fully exploit the "infection-space" created by the removal of its competitor. If the resistant clone was rare before treatment, the effect can be substantially greater (Fig. 11.3). In nature, there is wide variation in the number of clones per person. Next-generation sequencing techniques are already discovering patients with more than 15 *P. falciparum* clones (Juliano et al., 2010), some of which are represented at frequencies significantly less than 1%. Were those rare clones drug-resistant, aggressive chemotherapy could increase transmission success of resistant parasites >100-fold.

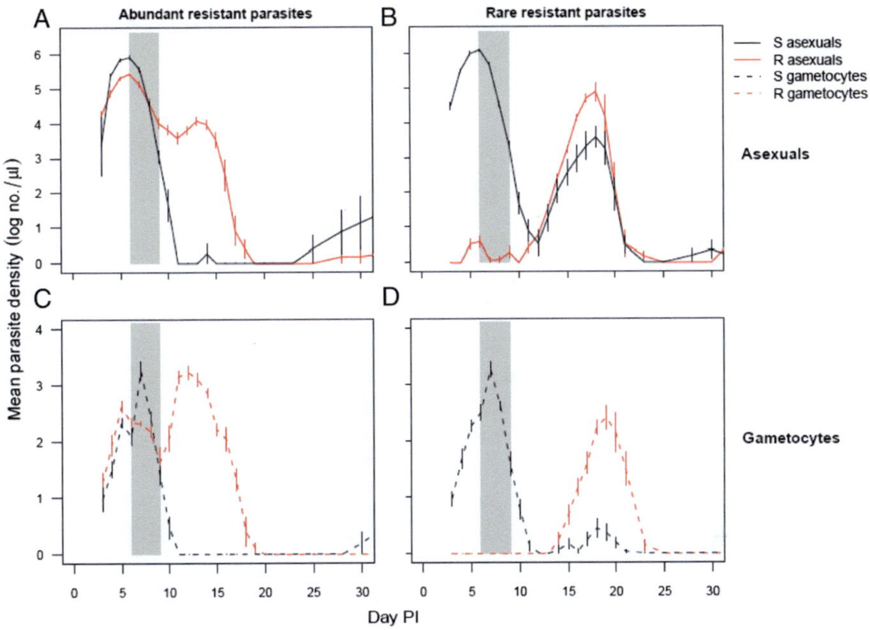

FIGURE 11.3 Competitive release of drug resistance. Infections of *P. chabaudi* were initiated in laboratory mice with 10^6 sensitive parasites (dark lines) and either 10^6 (*A* and C) or 10^1 (*B* and *D*) resistant parasites (gray lines). Panels *A* and B, densities of asexual parasites (within-host replicative stages). Panels C and *D*, densities of gametocytes (transmission stages). Gray bars indicate period of drug treatment (four daily doses of 8 mg/kg of pyrimethamine). R, resistant; S, sensitive; PI, post-infection. Drug treatment rapidly suppresses sensitive parasites, allowing resistant parasites to dominate post-treatment populations; the expansion following competitive release is especially marked when the resistant clone is rare. In untreated mice, resistant parasite densities are markedly lower than sensitive parasite densities throughout the infections, particularly when they were rare initially (compare with Fig. 11.2, which details the transmission stage densities of resistant parasites in the untreated mice in the same experiment). Plotted points are the mean (±SEM) densities in peripheral blood from 5 to 10 mice per group, estimated by quantitative PCR using protocols described elsewhere (Huijben et al., 2010).

Putting this slightly more formally, highly resistant parasites have a relative fitness advantage in treated hosts simply because drug treatment reduces the fitness of susceptible parasites. This advantage plays out even if all infections in a population consist of just a single clone. When hosts are infected with multiple lineages, however, the removal of competitors by drug treatment also leads to absolute fitness gains if resistant clones are able to capitalize on the newly emptied niche space in the host. These absolute fitness gains can be very, very large when resistant parasites are otherwise kept at very low numbers by competitive suppression.

Whence Conventional Wisdom?

Thus, radical parasite cure, by rapidly eliminating sensitive competitor strains, confers very strong selection in favor of resistance. Despite this, radical parasite cure is frequently advocated as a resistance management strategy. This conventional wisdom is based on two arguments. Both have to do with managing the initial mutational inputs into the system, essentially trying to prolong the time until high-level resistance appears in the first place. The first argument is that aggressive chemotherapy maximally reduces parasite numbers, and thus the probability that resistance mutations will occur in a treated patient [e.g., White (2004) and World Health Organization (2010a, p. 129)]. This clearly has to be true.

The second argument is essentially a subtle variation of the first. The idea is that when multiple independent mutations are required to confer high-level resistance, it is essential to try to minimize positive selection in favor of any partially resistant mutant because these partially resistant mutants can be important mutational stepping stones toward full (high-level) resistance (Hastings and Watkins, 2006). Partially resistant parasites only have an evolutionary advantage at lower drug concentrations; thus, from a resistance management perspective, it is important to minimize the probability that such parasites encounter those lower concentrations. Low drug concentrations in a patient can arise in several ways, not least after a course of chemotherapy has finished and the drug is being metabolized or excreted from the body (Fig. 11.1). During some of that time, there is a period [the "selection window" (Stepniewska and White, 2008)] when parasites that are able to survive low drug doses have a selective advantage. The aim of aggressive chemotherapy is to ensure that no parasites from the treated infection remain alive during the selection window, thus reducing the number of parasites in the overall population experiencing that source of selection for low-level resistance.

DOUBLE-EDGED SWORD

Thus, aggressive chemotherapy is a double-edged sword for resistance management. It can reduce the chances of high-level resistance arising *de novo* in an infection. But when an infection does contain resistant parasites, either from *de novo* mutation or acquired by transmission from other hosts, it gives those parasites the greatest possible evolutionary advantage both within individual hosts and in the population as a whole. How do the opposing evolutionary pressures generated by radical cure combine in different circumstances to determine the useful life span of a drug? There will be circumstances when overwhelming chemical force retards evolution and other times when it drives things very rapidly. We contend that for no infectious disease do we have sufficient theory and empiricism to determine which outcome is more important. It seems unlikely that any general rule will apply even for a single disease, let alone across disease systems.

Consider again the case of malaria. There will be many cases where the resistance management gains of radical pathogen cure (reduced mutational inputs) will not outweigh its costs (maximal selection for high-level resistance). For instance, where high-level resistance is conferred by a single point mutation [e.g., atovaquone (White, 2004)], the mutational stepping stone argument is clearly irrelevant. Moreover, there are about 10^{12} parasites in an infection at the time radical cure commences (White, 2004), so that every point mutation in the genome can potentially occur in a single infection. There are at least one-quarter of a billion symptomatic cases of malaria each year (World Health Organization, 2010b), so that at least 10^{20} parasites could see a new drug each year. Among these 10^{20} parasites, it is quite plausible that there already exists at least a single parasite completely resistant to most yet-to-be invented drugs. Aggressive chemotherapy can reduce the chances of *de novo* resistance mutations occurring in treated patients, but it can make no impact on the probability that such mutations occurred before treatment. Aggressive use of a new drug will very effectively find these resistant "needles in the haystack."

Even when we can be confident that mutational inputs in patients receiving treatment do limit the rate of evolutionary change (something that is extremely hard to know, especially for new drugs), there is an important quantitative argument to be had about the advantage of managing mutational inputs by aggressive chemotherapy. This is because aggressive treatment regimens increase the probability that any high-level resistance that has arisen *de novo* will avoid stochastic loss and reach transmissible frequencies. It is extremely challenging for a very rare resistant mutant to replicate to transmissible densities in a host [e.g., Mackinnon (2005), Pongtavornpinyo et al. (2009), and Hastings (2011a)], not least because it will likely compete with the ancestral strain from which it arose. The performance of the mutant can be especially poor if *de novo* resistance is associated with large fitness costs. Large costs can erode as compensatory mutations accumulate (Levin et al., 2000; zur Wiesch et al., 2011), but this requires persistence and large population sizes, both of which are countered by competition. Thus, even when aggressive chemotherapy reduces the probability that *de novo* mutations occur, it can, by eliminating competitors, increase the population-wide probability that *de novo* mutations survive to transmit from hosts, and hence escape stochastic loss.

Moreover, there are ways to manage mutational inputs that do not have the unfortunate consequence of simultaneously maximizing selection for the very mutations they are trying to prevent. Combination therapy is an example. As WHO puts it (World Health Organization, 2010a), if resistance to one drug has a per parasite probability of 10^{-12} of spontaneously arising, the probability of resistance to two drugs with independent modes of action arising spontaneously in the same parasite is 10^{-24}, a vanishingly small probability. The duration of the selection window (Fig. 11.1) depends critically on the half-life of the particular drug. The window can be weeks long in some cases [sulfadoxine-pyrimethamine (SP)] or just a few hours in others (artemisinin and its derivatives). Judicious choice of a drug or drug combination can thus affect the likelihood of stepping stones to high-level resistance.

EVIDENCE-BASED RESISTANCE MANAGEMENT

The foregoing suggests to us that radical parasite cure is not a priori the best way to manage resistance and that it could even promote the very evolution it is intended to retard. The scientific challenge is to determine how the contrasting evolutionary consequences of aggressive chemotherapy determine the rate of resistance evolution and whether, among the vast array of possible regimens, there are other ways of treating patients that would better delay resistance.

It might be, of course, that the other aims of patient treatment (restore health and prevent infectiousness) can be achieved only by radical parasite cure (Hastings, 2011b). If radical parasite cure is indeed critical for clinical management, an empirical question, we might be stuck with evolutionary mismanagement as an unavoidable side effect. If so, it is important to recognize this. Claims that resistance evolution is retarded by aggressive treatment regimens might be obscuring a serious evolutionary problem in need of solution.

Rational development of treatment regimens that deliver effective resistance management requires a sound knowledge base (Read and Huijben, 2009; Goncalves and Paul, 2011; zur Wiesch et al., 2011), and there is considerable scope for investigating the evolutionary consequences of different treatment regimens for a wide range of diseases. Ideally, these would involve quantitative comparisons of how contrasting regimens affect each of the aims of patient treatment: health, infectiousness, and resistance management. In principle, such studies can be done on animal models [e.g., de Roode et al. (2004a), Wargo et al. (2007), and Huijben et al. (2010)] and, in a more limited way, on humans [e.g., Harrington et al. (2009)]. It is possible to measure the evolutionary consequences of competing resistance management strategies in hospitals (Brown and Nathwani, 2005; Martínez et al., 2006; R. L. Smith et al., 2008), and it might even be possible in human communities. Penilla et al. (2007) randomly allocated 24 villages in Mexico to one of four different methods of applying public health insecticides and compared the rate of rise of resistant mosquitoes over several years. None of the putative resistance management strategies slowed the spread of phenotypic resistance. Empirical assessments of evolutionary outcomes are problematic for a drug against which resistance has yet to arise, but once high-level resistance has arisen, there is an ethical imperative to do such studies.

Mathematical models have much to offer, but the challenges are formidable even *in silico.* Consider malaria. As we argued above, the strength and direction of selection are critically affected by the interactions between competing pathogen lineages within a patient and how drug treatment affects this ecology. Treatment determines what is transmitted, and changes in the force of infection will, in turn, affect the genetic diversity within an infection, and hence the ecology. Such feedbacks defy standard population genetics approaches, which track gene frequencies without explicit population dynamics (Mackinnon, 2005). Evolutionary-epidemiological models [e.g., Gandon and Day (2009)] are computationally intensive, and we are unaware of any real-world context in which resistance evolution is adequately modeled. Unfortunately, the complexity of the situation does not make it go away. Quantitative predictions of the impact of different treatment regimens on the useful life of a drug have to involve this social ecology. Such modeling efforts would also evaluate the resistance management consequences of reductions in disease transmission by other measures, such as mass drug administration or transmission-blocking interventions [e.g., World Health Organization (2011)]. These too will reduce force of infection, and hence alter the in-host ecology. Reductions in force of infection might reduce the benefits of resistance by reducing the multiplicity of infection, and hence the levels of competitive release; however, as argued above, the costs of resistance will also be lower if there is less competition.

The difficulty of adequately capturing the relevant features in a mathematical model points to an important bottom line: Intuition (expert opinion), a very poor guide to evolutionary trajectories at the best of times, is really going to struggle in this context.

HOW TO TREAT PATIENTS?

A corollary of our observation that radical pathogen cure can very seriously promote the evolution of resistance is that less aggressive drug treatment could prolong the useful life span of a drug. Because even small changes in relative fitness can alter the useful therapeutical life span of a drug by decades (Hastings and Donnelly, 2005), there is a strong case for investigating the clinical consequences of lighter touch chemotherapy.

Drug treatment is often continued after patient health is restored; this is a major reason why patients fail to complete prescribed drug courses. Could there be room to harness the in-host ecology to reduce the fitness advantages of resistance, in effect retaining some drug-sensitive pathogens to suppress resistance (Wargo et al., 2007; Read and Huijben, 2009; Huijben et al., 2010)? Critically, patient health does not necessarily require immediate parasite elimination by drugs. To affect clinical recovery, the immune system often just needs to battle fewer parasites or have a longer time period over which to ramp up. It may be, for instance, that only minimal intervention with drugs is required before immunity controls and clears disease-causing pathogens. This could involve a very short course of treatment with a rapidly clearing drug (or drug combination), perhaps repeated at well-spaced intervals. Given a bit of help, immunity can deal very effectively with resistant parasites without imposing any selection for resistance (Cravo et al., 2001; Rice, 2008a,b; Taubes, 2008). Some currently heretical rules, such as "stop taking drugs when you feel better, and take them again if you get sick," bear examination in such contexts. Critical questions are how best to combine dose and duration, how much it is necessary to have an impact on pathogen densities at first treatment, and how far apart pulses of treatment should be.

A general principle that should guide the rational development of patient treatment guidelines is to impose no more selection for resistance than is absolutely necessary. There might be cases where rules like "hit hard and hit early" (Ehrlich, 1913) or "ensure very high cure rates" (World Health Organization, 2010a) are consistent with this, but we doubt that they apply across a wide swath of diseases. For instance, *de novo* resistance mutants are a major threat to the health of patients infected with highly mutable pathogens like HIV. In such a case, it probably is wise to use aggressive chemotherapy to reduce pathogen biomass, and hence the probability of *de novo* mutations. For many diseases, however, patients are at far higher risk of acquiring resistance from other patients. In TB, for example, up to 99% of cases of drug-resistant infections are acquired from the community (Luciani et al., 2009). In these circumstances, the merits of managing *de novo* mutations with aggressive chemotherapy are less clear. Chloroquine became ineffective against malaria because the highly resistant progeny of a single parasite in Asia spread across the entire African continent (Wootton et al., 2002; Talisuna et al., 2004). SP, another inexpensive and initially highly effective antimalarial, was similarly undermined by vast epidemics derived from very few genetic events (Roper et al., 2004). Those resistant parasites enjoyed maximum evolutionary advantage in patients who adhered to regimens effecting radical cure of susceptible parasites.

More broadly, resistance management strategies will probably have to be tailored to particular drug-bug combinations and epidemiological circumstances. For instance, where single-clone infections dominate (acute childhood diseases or malaria where force of infection is low), the relative fitness of resistant and sensitive strains will be quite different from situations where most infections have a high multiplicity of infection. Where there is lateral transfer of resistance genes from the environment (many bacteria), persistent subpopulations [e.g., *Escherichia coli* (Levin and Rozen, 2006)], or infection sites that are difficult to treat [e.g., TB (Dye, 2009)], or where treated stages are diploid [e.g., helminths (Prichard and Tait, 2001)], things could again be different. Where the social interactions between coinfecting strains differ from those we have described for malaria [e.g., West et al. (2006)], things could be different again.

It might also be that patient treatment regimens need to be modified as resistance evolution proceeds. Perhaps, for instance, aggressive chemotherapy can reduce the probability that mutations to high-level resistance will occur. If so, it could be worth moving to less aggressive regimens as soon as high-level resistance is detected in a region. Regimens involving lower doses or shorter treatments will impose weaker selection on that new resistance. Such a switch may be difficult in practice. Health messaging may require constancy, or it may be that by the time unambiguous evidence of high-level resistance has been obtained and policy changed, it is already too late.

CODA

Arguments somewhat analogous to ours have also been made for bacterial diseases (Lipsitch and Samore, 2002; Rice, 2008a,b). Aggressive chemotherapy could be particularly problematic in the case of many bacterial infections, where exhortations for patients to adhere to long-course regimens probably generate sustained selection on gut commensals to harbor resistance genes. These can be readily passed to any disease-causing bacteria that subsequently invade. Our discussion also has strong parallels with the management of *Clostridium difficile* in hospitals, where aggressive use of broad-spectrum antibiotics is responsible for the competitive release of the more virulent *C. difficile* (Vonberg et al., 2008).

An analogous situation also occurs in cancer therapy, where cell lineages within a tumor compete for access to space and nutrients. There, the argument has recently been made that less aggressive chemotherapy might sustain life better than overwhelming drug treatment, which simply removes the competitively more able susceptible cell lineages, allowing drug-resistant lineages to kill the host (Gatenby, 2009; Gatenby et al., 2009). Mouse experiments support this: Conventionally treated mice died of drug-resistant tumors, but less aggressively treated mice survived (Gatenby et al., 2009). Elsewhere, we and others have also argued that by concentrating on malaria control rather than vector control, selection for insecticide-resistant mosquitoes can be managed and even eliminated, obviating the need for an insecticide discovery pipeline (Koella et al., 2009; Read et al., 2009; Gourley et al., 2011). In all this, the key issue is to impose only the selection needed to achieve health gains and no more.

There is widespread agreement that stewardship of antimicrobials means restricting their use to only those patients who need them. We suggest that a similar default philosophy of sparing use should apply at the within-host level to patient treatment regimens. Overwhelming chemical force may at times be required, but we need to be very clear about when and why that is. Aggressive chemotherapy will, under a wide range of circumstances, spread resistance.

ACKNOWLEDGMENTS

Our arguments benefited from discussion with V. Barclay, C. Bergstrom, S. Bonhoeffer, N. Colegrave, M. Ferdig, A. Griffin, J. Hansen, J. Juliano, M. Laufer, B. Levin, J. Lloyd Smith, M. Mackinnon, S. Meshnick, N. Mideo, S. Nee, R. Paul, J. de Roode, P. Schneider, F. Taddei, and A. Wargo, not all of whom agree with our conclusions. We thank J. Antonovics, A. Bell, K. Foster, M. Greischar, S. Reece, and an anonymous referee for comments on the manuscript and members of the Research and Policy in Infectious Disease Dynamics Program of the Science and Technology Directorate, Department of Homeland Security, and the Fogarty International Center, National Institutes of Health, for stimulating discussion. Award R01GM089932 from the National Institute of General Medical Sciences supported the empirical work reported here; work under Awards R01AI089819 and U19AI089676 from the National Institute of Allergy and Infectious Diseases contributed to conceptual development. This work greatly benefited from a Fellowship to A.F.R. at the Wissenschaftskolleg zu Berlin.

12

Genomic Imprinting and the Evolutionary Psychology of Human Kinship

DAVID HAIG

Genomic imprinting is predicted to influence behaviors that affect individuals to whom an actor has different degrees of matrilineal and patrilineal kinship (asymmetric kin). Effects of imprinted genes are not predicted in interactions with nonrelatives or with individuals who are equally related to the actor's maternally and paternally derived genes (unless a gene also has pleiotropic effects on fitness of asymmetric kin). Long-term mating bonds are common in most human populations, but dissolution of marriage has always affected a significant proportion of mated pairs. Children born in a new union are asymmetric kin of children born in a previous union. Therefore, the innate dispositions of children toward parents and sibs are expected to be sensitive to cues of marital stability, and these dispositions may be subject to effects of imprinted genes.

"The burden of making and the duty of exacting compensation ran on the mother's side as well as the father's. A father and son, or two half-brothers, would for the purposes of the blood-feud have some of their kindred in common, but by no means all."

Pollock and Maitland (1895)

Department of Organismic and Evolutionary Biology, Harvard University, Cambridge, MA 02138. E-mail: dhaig@oeb.harvard.edu.

The opening quotation comes from a discussion of Anglo-Saxon law. An individual could be liable to pay wergeld for the slaying of his mother's kinsman by his father's kinsman and be entitled to receive wergeld for the same slaying, because each individual combined two lines of descent. The individual is divisible. Just as his loyalties can be divided by obligations to the two sides of his family, so too can his genome be divided between genes he shares with his mother and genes he shares with his father. Blood is thicker than water, and blood does not mix (in the sense that genes do not blend).

Genetically determined behaviors that benefit the father's side of the family may be favored by natural selection when a gene has been transmitted by a sperm but not when the same gene has been transmitted by an egg. Conversely, a behavior that benefits the mother's side of the family may be favored when a gene has been transmitted by an egg but not when the same gene has been transmitted by a sperm. In such circumstances, imprinted alleles, genes that are differently expressed when inherited via eggs and via sperm, can supplant unimprinted alleles that are expressed independent of parental origin (Haig, 1997, 2000b).

Imprinted genes have been considered prime candidates for involvement in disorders of human social interaction, such as autism and schizophrenia, because of their predicted role in interactions among kin (Badcock and Crespi, 2006; Isles et al., 2006; Crespi, 2008; Goos and Ragsdale, 2008; Úbeda and Gardner, 2010, 2011). Not all social interactions promote imprinted gene expression, however. The principal purpose of this paper is to clarify the rather specific conditions that favor stable maintenance of imprinted gene expression, but this task requires a broader understanding of how humans innately categorize kin. These questions will be addressed with a particular focus on effects of partner change and on internal genetic conflicts during sexual maturation and adolescence.

ASYMMETRIES OF RELATEDNESS

Consider an imprinted locus at which the established allele is silent when paternally derived but expressed at level $x > 0$ when maternally derived. This pattern of expression is an evolutionarily stable strategy (ESS) when two conditions are met:

$$\sum_i p_i \frac{dw_i}{dx} < 0 \tag{1A}$$

$$\sum_i m_i \frac{dw_i}{dx} = 0, \sum_i m_i \frac{d^2 w_i}{dx^2} < 0 \tag{1B}$$

where dw_i/dx is the effect of a change in x on the fitness of individual i, and p_i and m_i are coefficients of patrilineal and matrilineal relatedness of the category to which individual i belongs. Inequality (1A) specifies that extra x reduces patrilineal inclusive fitness. This condition maintains silence of paternally derived alleles. Eq. (1B) specifies that x is a local maximum of matrilineal inclusive fitness (Haig, 1997). These conditions are equivalent to

$$\sum_s m_s \frac{dw_s}{dx} < -\sum_a p_a \frac{dw_a}{dx} \tag{2A}$$

$$\sum_s m_s \frac{dw_s}{dx} = -\sum_a m_a \frac{dw_a}{dx} \tag{2B}$$

where s indexes symmetric kin (individuals for whom $m_s = p_s$) and a indexes asymmetric kin (individuals for whom $m_a \neq p_a$). An individual's symmetric kin include herself, her offspring, and her grandoffspring, but most other categories of kin are asymmetric, including "fullsibs," because of uncertainty of paternity. Thus, the right-hand sides of 1A and Eq. (1B) can be considered to represent the marginal effect of x on the individual's own survival and reproduction (individual fitness).

Eq. (2B) describes a tradeoff in the maximization of matrilineal inclusive fitness. At the ESS, the marginal effect of x on individual fitness is balanced by a marginal effect of opposite sign on indirect fitness obtained via asymmetric kin. If the value of Eq. (2B) is negative, then extra x increases individual fitness at a cost to matrilineal asymmetric kin. If the value of Eq. (2B) is zero, then x simultaneously maximizes both components of inclusive fitness (most plausible if x has no effects on matrilineal asymmetric kin). If the value of Eq. (2B) is positive, then extra x increases the fitness of matrilineal asymmetric kin at a cost to individual fitness.

Substitution of the right-hand side of Eq. (2B) for the left-hand side of 2A yields

$$\sum_a m_a \frac{dw_a}{dx} > \sum_a p_a \frac{dw_a}{dx}, \tag{3A}$$

which can be rearranged to give

$$\sum_j (m_j - p_j) \frac{dw_j}{dx} > \sum_k (p_k - m_k) \frac{dw_k}{dx}, \tag{3B}$$

where j indexes matrikin (individuals for whom $m_j > p_j$) and k indexes patrikin (individuals for whom $m_k < p_k$). This partition allows kin to be assigned to three mutually exclusive classes: symmetric kin ($m_i = p_i$), matrikin ($m_i > p_i$), and patrikin ($m_i < p_i$). Inequality (3B) states that inactivation of the silent paternal allele is maintained when the summed effects of extra x on fitness are worse for patrikin than for matrikin, where fitness effects are weighted by the asymmetries of relatedness (terms in parentheses).

The ESS for a locus at which the established allele is silent when maternally derived but expressed at level $z > 0$ when paternally derived is obtained by substitution of z for x and reciprocal substitution of m for p throughout the above analysis.

If maternal-specific expression of x has effects on two individuals only, (1A) and Eq. (1B) become

$$p_1 \frac{dw_1}{dx} + p_2 \frac{dw_2}{dx} < 0 \tag{4A}$$

$$m_1 \frac{dw_1}{dx} + m_2 \frac{dw_2}{dx} = 0. \tag{4B}$$

These conditions describe a tradeoff in which the two individuals' fitnesses are differently weighted for genes of maternal and paternal origin. Condition (4A) can be expressed in a convenient form using a substitution from Eq. (4B):

$$\left\{ \frac{p_2}{p_1} - \frac{m_2}{m_1} \right\} \frac{dw_2}{dx} < 0 \tag{4C}$$

Condition (4C) shows that the maintenance of paternal silence depends on a difference in the ratios of matrilineal and patrilineal relatedness for the two individuals affected.

If one of these individuals is the actor within whom x is expressed ($p_1 = m_1 = 1$), then (4C) and Eq. (4B) become

$$\{p_2 - m_2\} \frac{dw_2}{dx} < 0 \tag{5A}$$

$$\frac{dw_1}{dx} = -m_2 \frac{dw_2}{dx}. \tag{5B}$$

Maternal-specific expression of x either benefits self at a cost to an individual who is patrikin ($dw_1/dx > 0$, $dw_2/dx < 0$, $p_2 > m_2$) or benefits an individual who is matrikin at a cost to self ($dw_1/dx < 0$, $dw_2/dx > 0$, $p_2 < m_2$). The ESS at a maternally silent locus, with paternal expression level z, is obtained by swapping m_2 for p_2 and z for x:

$$\{m_2 - p_2\}\frac{dw_2}{dz} < 0 \tag{6A}$$

$$\frac{dw_1}{dz} = -p_2\frac{dw_2}{dz}. \tag{6B}$$

Paternal-specific expression of z either benefits self at a cost to an individual who is matrikin ($dw_1/dx > 0$, $dw_2/dx < 0$, $p_2 < m_2$) or benefits an individual who is patrikin at a cost to self ($dw_1/dx < 0$, $dw_2/dx > 0$, $p_2 > m_2$).

KINSHIP CATEGORIES

Other individuals evoke different innate dispositions in ego: some are sexual rivals, and others are potential mates; some are parents, and others are offspring; some are friends, and others are strangers. These dispositions constitute an implicit categorization of others that represents the way natural selection has parsed social interactions in a particular evolutionary lineage. The dispositions define the categories: all individuals who evoke a disposition belong to a category defined by the disposition. If the members of a category are, on average, related to ego, then the disposition will evolve, in part, shaped by its effects on the fitness of kin. An individual who evokes the disposition can be considered to be treated as a kinsman and the disposition can be considered a kin-directed behavior.

An actor can treat another individual as belonging to an innate category without recognizing that a category exists or recognizing particular individuals as members of the category. As a simple example, hormones secreted into the maternal circulation by a fetus affect another individual who necessarily carries copies of the maternally derived alleles of the fetus (Haig, 1996). However, if a behavior is to be preferentially directed toward a particular category of kin within a larger group of similar individuals, then the actor must discriminate among individuals and the individuals that belong to a category must be learned by social context. As a classic example, goslings have an innate disposition to follow "mother," but the individual that is recognized as "mother" by a particular gosling is learnt through a process of imprinting (in an earlier sense of the word). Similarly, human children may possess innate dispositions in their interactions with "mother," "father," "brother," or "sister," but the particular individuals who evoke these dispositions must be learnt from social context.

We probably possess more-or-less discrete instinctive categories for primary kin, such as "self," "mother," "sister," and "daughter," and perhaps for some secondary kin, such as "sister's daughter" or "daughter's daughter." However, for more distant kin, I suspect we possess a vague sense of some individuals as closer kin than others, with behavioral dispositions that vary with degree of perceived kinship. Where we place a particular individual on this continuum will be determined by things we have been told, and how often, and in what contexts we have interacted with them, particularly during childhood.

A disposition evolves according to the average relatedness of the individuals who evoke the disposition, not according to the relatedness of any particular genealogical category. Thus, instinctive categories should be distinguished from the categories that would be determined by an omniscient geneticist. Suppose, for example, that a disposition is evoked by females, born within a few years of the actor, who live in close association with the actor's "mother" during the actor's own childhood. Such a disposition will often have been evoked by genealogical sisters, and the innate category can be labeled, for convenience, as "sister," even though it may sometimes have been evoked by individuals who were not offspring of the actors' mothers.

Hamilton's second principle of the genetical evolution of social behavior was that "The situations in which a species discriminates in its social behaviour tend to evolve and multiply in such a way that the coefficients of relationship involved in each situation become more nearly determinate" (Hamilton, 1964b). In other words, natural selection will tend to favor actors who are able to subdivide beneficiaries into categories with a lower variance of genealogical relatedness. By this process, innate categories would more nearly come to approximate genealogical categories. However, this conclusion should be qualified by the observation that an individual may benefit from being classified by an altruistic actor as a member of a category with a coefficient of relatedness greater than the individual's "true" relatedness. Thus, natural selection on actors to make ever finer discriminations of kinship may be opposed by natural selection on a subset of beneficiaries to confound such discrimination (Haig, 2000a).

Instinctive categories should also be distinguished from the cultural classification of kin (Feinberg and Ottenheimer, 2001). Cultural evolution exploits our innate dispositions for various cultural and rational ends. For example, by defining another individual as a brother, a cultural tradition or a political innovator attempts to evoke dispositions appropriate to innate brotherhood in interactions with that individual (Johnson, 1987; Salmon, 1998). To the extent that this evocation is successful, culture thereby changes the coefficients of relatedness associated with an innate category in ongoing natural selection. By this means, culture can shape the innate taxonomy of kin.

Symmetric Kin

"Self" and "offspring" are symmetric kin. These innate categories are evolutionarily ancient and undoubtedly have accrued a rich set of innate dispositions. "Grandoffspring" are also symmetric kin. Genes of maternal and paternal origin favor the same outcomes when fitness tradeoffs affect symmetric kin alone. Therefore, significant effects of imprinted genes on symmetric kin are predicted only if a gene's expression also affects asymmetric kin. For example, gene expression might mediate a direct tradeoff between the fitness of symmetric kin (e.g., "self") and asymmetric kin (e.g., "mother").

Perfect symmetry of matrilineal and patrilineal relatedness is an ideal that is probably rarely realized, although selection favoring imprinted expression will be weak when asymmetries of relatedness are small. For example, fullsibs, considered as a genealogical category, are symmetrically related to ego. However, "fullsibs," considered as an innate category, are ego's matrikin because of the possibility of undetected cuckoldry. The asymmetry of relatedness associated with "fullsibs" will be small, however, whenever the probability of shared paternity is high. As another example, ego's offspring will be asymmetrically related to ego when ego's spouse is asymmetrically related to ego, as occurs under some forms of inbreeding (Haig, 1999; Wilkins and Haig, 2003a), but the asymmetries of relatedness will be small, except under close inbreeding.

Mother

"Mother" is the most important category of matrikin. Strong effects of imprinted genes are expected in an offspring's relations with its mother, both prenatally and postnatally, because mothers have large effects on the fitness of offspring and are associated with an extreme asymmetry of relatedness from an outbred offspring's genetic perspective ($m - p = 1$). Paternally expressed genes are predicted to increase the demands offspring impose on mothers, whereas maternally expressed genes are predicted to reduce these demands (Wilkins and Haig, 2003b; Haig, 2004). The strength of these effects will be attenuated when mothers establish stable breeding bonds with a particular male because paternally derived genes of an offspring then have an interest in a mother's continued reproduction. Her future offspring are potentially also the offspring's father's future offspring. However, when a mother changes partner, her continued reproduction expends limited maternal investment on maternal halfsibs, who are unrelated rivals from the perspective of paternally derived genes of existing offspring.

Effects of imprinted genes during fetal development are broadly consistent with theoretical predictions that paternally expressed genes should enhance growth and maternally expressed genes should restrain overgrowth (Eggermann et al., 2008; Haig, 2010). Evolutionary speculation about postnatal effects of imprinted genes on a child's relations with his or her mother has focused on the phenotypes of Prader-Willi syndrome (PWS) and Angelman syndrome (AS), with the former caused by loss of paternally expressed genes at 15q11–13 and the latter by loss of maternally expressed genes from the same region (Buiting, 2010). Therefore, PWS is predicted to exhibit absence, or weak development, of behaviors that elicit resources from mothers, whereas AS is predicted to exhibit an overdevelopment of such behaviors (Haig and Wharton, 2003; Úbeda, 2008).

Infants with PWS exhibit poor suck, weak cry, and excessive sleepiness, suggesting paternally expressed genes from 15q11–13 promote suckling, strength of cry, and wakefulness (all phenotypes that are expected to enhance maternal costs). From about the age of natural weaning, children with PWS develop an insatiable appetite associated with "foraging" behaviors. These phenotypes have been interpreted as a pathological expression of "weaning conflicts" that occurred when our ancestors were transitioning from predominant reliance on the breast to reliance on supplemental foods (Haig and Wharton, 2003; Úbeda, 2008).

The happy affect and smiling demeanor of children with AS contrast with the less effusive personality of children with PWS. Children with AS are proposed to express strongly behaviors that normally function to elicit maternal care, attention, and attachment (Isles et al., 2006; Brown and Consedine, 2004). However, the overtly social personality of children with AS is combined with a profound deficit in communication (Haig, 2008). Speech and gesture are largely absent. Therefore, maternally expressed genes at 15q11–13 appear necessary for the normal development of language. Badcock and Crespi (2006) have suggested that genes of maternal origin have been selected to act in the language centers of the child's brain to promote attentiveness to maternal instruction and maternal example, coordinating maternal and child needs for the benefit of the matriline.

Father

Fathers are patrikin of their offspring. Therefore, maternally expressed genes in offspring are predicted to favor increased demands on fathers relative to the effects of paternally expressed genes. By contrast, paternally expressed genes are predicted to show greater solicitude to the needs of fathers.

Most mammals probably lack an innate category of "father." The evolution of more-or-less stable mating bonds between men and women has allowed fathers to recognize their offspring and offspring to recognize their fathers (often with a fair degree of confidence). Human fathers recognize offspring as babies born to women with whom they are involved in a more-or-less exclusive sexual relationship. Human offspring recognize fathers as adult males closely associated with their mother during infancy and early childhood (Chapais, 2008).

Recognition of fathers means fathers can be avoided as mates. From the genetic perspective of a daughter, mating with her father is associated with a direct cost of producing inbred, rather than outbred, offspring but an indirect benefit of an extra, albeit inbred, paternal half-sib (with the daughter herself as the mother). The direct cost is experienced equally by the daughter's maternal and paternal alleles, but the indirect benefit is experienced solely by the daughter's paternal alleles. Therefore, maternally expressed genes are predicted to promote strong aversions to sexual relations with fathers (Haig, 1999).

Sibs

Uterine sibs sometimes have different fathers and are thereby matrikin. Paternal-specific expression of imprinted genes is expected to benefit self at the expense of uterine sibs, whereas maternal-specific expression is expected to benefit uterine sibs at a cost to self. Two factors in human evolution have probably had opposing effects on the intensity of conflict between genes of maternal and paternal origin over relations with uterine sibs. On the one hand, the evolution of persistent pair-bonds increased the proportion of uterine sibs that were fullsibs, thus reducing asymmetries of relatedness and mitigating potential conflicts. On the other hand, the evolution of shorter interbirth intervals and prolonged childhoods increased opportunities for competition among sibs (Kennedy, 2005; Sellen, 2007; Humphrey, 2010).

Sibs are both sharers of common genes and competitors for common resources; hence, the characteristic admixture of affection and aggression in many sibling relations. Innate dispositions toward sibs are expected to be sensitive to relative age, with rivalry more intense among sibs who are closer in age. An older sib usually has greater power to help or harm a younger sib than the younger has to help or harm the older (although younger sibs will often attempt to recruit the even greater power of a parent on their behalf). The age-related asymmetry in power between a pair of sibs is expected to lessen as they grow older because they become closer together in relative age.

"Younger uterine sib" is likely to have been an evolutionarily salient category of matrikin because the arrival of a new baby will often have been accompanied by a reapportionment of maternal care away from older sibs. Consider two scenarios. In the first, a child grows up with his or her mother and a "father" who disappears and is replaced by an unfamiliar adult male, after which the mother has a new baby. From the perspective of the older child, the new baby is associated with a large asymmetry of relatedness ($m - p = 0.5$). In the second scenario, the "father" and mother remain together for the birth of a new baby. In this scenario, the new baby is associated with a much smaller asymmetry of relatedness because he or she is likely (although not certain) to be a fullsib of the older child.

New babies evoke a single instinctive category if the innate dispositions of older sibs are the same in the two scenarios. In this case, gene expression will have evolved according to a gene's average relatedness to babies in the different scenarios, weighted by the long-term average frequencies of each scenario. If the two scenarios evoke different innate dispositions, then the babies belong to different instinctive categories and gene expression will have evolved according to scenario-specific coefficients of relatedness.

Innate dispositions of younger children toward "older uterine sibs" are probably less responsive to a mother's change of partner than dispositions of older children to "younger uterine sibs" because a newborn child has not herself or himself experienced the change of partner and has few direct cues about the paternity of older sibs. The latter are more powerful and better informed than younger sibs, and thus may often set the tone of sibling relations. Infants and toddlers may, at first, express behaviors designed to ingratiate themselves to older sibs in an attempt to elicit help and avoid harm, with the overt expression of rivalry intensifying as disparities of power lessen with age.

Paternal halfsibs are patrikin, but it is unclear whether we have evolved innate dispositions that are specific for this category of kin. Relations with paternal halfsibs are usually less intimate than relations with maternal halfsibs because a father's contact with his offspring becomes attenuated once his sexual relations with their mother ends, especially if he has offspring with another woman. Interactions with paternal halfsibs are more intense in polygynous households in which the offspring of two or more women compete for family resources (Jankowiak and Diderich, 2000), but this situation has probably been less frequent than living with maternal halfsibs.

Extended Kinship

Asymmetries of matrilineal and patrilineal relatedness are created whenever individuals of one sex disperse to reproduce, whereas individuals of the other sex remain in their natal group. If the variance of reproductive success is similar in the two sexes, then random pairs of individuals are more likely to share genes of maternal origin than genes of paternal origin in matrilocal groups with male-biased dispersal, but the reverse is true in patrilocal groups with female-biased dispersal (Haig, 2000a, 2010; Brandvain, 2010; Úbeda and Gardner, 2010, 2011; Van Cleve et al., 2010). Thus, differential dispersal of the sexes can result in genes having effects that discriminate between matrikin and patrikin without other individuals being explicitly recognized as belonging to the mother's family or father's family.

Whether human reproductive dispersal has been female-biased or male-biased, on average, is controversial (Ember, 1978; Alvarez, 2004). What is not controversial is that human groups exhibit a flexibility of social organization such that ties of matrilineal and patrilineal kinship predominate in different populations, with strong ties to both sides of the family maintained in many groups (Rodseth et al., 1991; Marlowe, 2004; Chapais, 2008). Most, if not all, cultures distinguish between matrilineal and patrilineal kin. A key unanswered question is whether this cultural distinction is reinforced by innate dispositions that distinguish "mother's kin" from "father's kin," or whether the two kinds of kin are lumped together in a single instinctive category with asymmetries of relatedness determined by social context.

The recognition of particular individuals as belonging to particular categories of kin enables discrimination among members of social groups on the basis of degree of relatedness (nepotism). The evolutionarily oldest and strongest ties are between mothers and their offspring, and among uterine sibs. If adults maintain associations with their mothers and uterine sibs, then second-order ties are facilitated between children and their mother's mother and mother's sibs (Chapais, 2008). Strong ties of patrilineal kinship are more tenuous because recognition of fathers is less certain and (evolutionarily) more recent. Perhaps for these reasons, ties of patrilineal kinship are often reinforced by strong patriarchal ideologies.

KITH AND KIN

Queller (2011) distinguishes social effects mediated via kin from those mediated via kith or kind. Kin selection involves fitness effects on individuals who share genes via genealogical descent. Kind selection involves fitness effects mediated by identity by state rather than by descent. Kith selection involves an actor's effects on other individuals that feed back to the actor's own individual or inclusive fitness. I will not discuss kind selection except to draw attention to Queller's perceptive discussion of the relation between "phenotype matching" and green-beard effects (Queller, Chaper 1, this volume).

Each individual has two parents who may be genetically unrelated but have a common interest in the survival and reproduction of their mutual offspring. The parents are each other's kith, and their relation engenders a complex intertwining of kith and kin effects because the affines of the father and mother are, respectively, matrikin and patrikin of the offspring. Put another way, an individual's matrikin are kith from the perspective of paternally derived genes, whereas an individual's patrikin are kith from the perspective of maternally derived genes. Further entanglement of kith and kin occurs when parents are themselves kin because of consanguineous matings.

A husband may benefit from investment in the health and well-being of his wife because this feeds back to increased fitness of his children. By extension, a husband is kith of his wife's family, who are matrikin of his offspring. His investment in relations with his wife's parents, and their investment in their son-in-law, may feed back to increased fitness of his children and their grandchildren. By further extension, a mother is kith of the paternally derived genes of her own offspring. These genes have an interest in her well-being to the extent that the offspring's individual fitness depends on continued investment by a healthy mother. Moreover, the offspring's patrilineal inclusive fitness may benefit from maternal investment in fullsibs.

Kith relations are contingent in ways that kin relations are not. The love of a child is more robust to bad behavior by the child than is love of a spouse to bad behavior by the spouse. The sharing of genes by descent is a brute fact that is unchanged by changes in the personal relations of kin, but spousal fitnesses are decoupled when either partner pursues other reproductive opportunities.

PARTNER CHANGE

In preindustrial societies, it was a lucky child who reached maturity living in a household with both biological parents because of high rates of parental death and divorce (Hewlett, 1991; Marlowe, 2005). Some of our ancestors undoubtedly grew up in families with both parents present, but others grew up in families in which one or both parents were absent. Behaviors that best promoted inclusive fitness are likely to have differed between intact and disrupted families because parents differed in their ability (or willingness) to invest in offspring and divorce was associated with predictable changes in relatedness for the children of former marriages.

As long as a couple remains together, their mutual offspring are fullsibs and symmetric kin of existing offspring (ignoring, for the moment, children conceived by extrapair copulations). However, once parents change partners, subsequent offspring of the mother and father are, respectively, maternal halfsibs (matrikin) and paternal halfsibs (patrikin) of the parents' mutual offspring. Therefore, a child's innate dispositions toward parents and younger sibs should be sensitive to whether or not his or her parents remain together, and these dispositions may be particularly sensitive to influences of imprinted genes after divorce or parental death.

Conflict in a child's relations with his or her parents is expected to intensify after parental separation, especially after parents acquire new partners (Emlen, 1995, 1997b), because genes of maternal origin in the child have no direct interest in the father's continued reproduction, whereas genes of paternal origin have no direct interest in the mother's continued reproduction. Therefore, genes of paternal origin are expected to promote reduced cooperation with mothers after divorce, either expressed as increased demands for maternal resources, increased competition with maternal halfsibs, or reduced expression of helpful behaviors. Genes of maternal origin are expected to have opposing effects. As a result, conflicts within the child's genome are predicted to intensify after divorce.

When marriages dissolve, children usually remain with their mothers and contact with their fathers declines; social interactions with maternal halfsibs tend to be stronger than with paternal halfsibs; and ties to the mother's extended family strengthen, whereas ties to the father's family weaken (Furstenberg and Cherlin, 1991). If similar biases were present in our evolutionary past, then the dissolution of pairbonds would have been associated with a statistical shift toward greater interaction with matrikin and a concomitant shift in the selective forces acting on imprinted genes in children. Paternally derived genes of children would therefore favor a greater emphasis on self-beneficial behaviors and a reduced emphasis on kin-beneficial behaviors after parental divorce.

SEXUAL MATURATION

Age at sexual maturity is a pivotal life-history variable (Stearns, 1992). In standard life-history theory, risk of death is the primary factor favoring earlier reproduction because individuals who delay maturation may not survive to reproduce or may not remain alive long enough to raise their offspring. Other things being equal, higher mortality of young adults favors earlier reproduction. Thus, high risks of subadult mortality have been proposed to explain early reproduction at small size in human pygmies (Walker et al., 2006; Migliano et al., 2007). Early maturation, in this case, is assumed to reflect a genetic change in the pygmy gene pool. Facultative responses are also possible. Thus, early reproduction by poor African-American women has been interpreted as a rational response to low life expectancy (Geronimus, 1997).

Theoretical discussions have focused on effects of pubertal timing on individual fitness with indirect effects on the fitness of relatives, for the most part, neglected. In this section, I will focus on indirect effects. My motivation is that a number of imprinted regions of the human genome influence pubertal progression and timing. This suggests that variation in age at maturity has affected the fitness of asymmetric kin, as well as individual fitness (Haig, 2010). I do not address the relative importance of direct and indirect effects. The selective forces acting on pubertal timing are undoubtedly complex, and a comprehensive review is beyond the scope of this paper.

The timing of ego's transition to adulthood would have had varied consequences for the fitness of ego's kin depending on ecological conditions: whether ego remained in his or her natal group or moved to another group, how much ego contributed to communal goods, and the extent to which ego's offspring competed for limited resources with other group members (Haig, 2010). Rather than attempt a global analysis that sums fitness effects across all categories of kin, I will consider a simple model in which the level of x (expressed from one locus) accelerates ego's pubertal development, whereas the level of z (expressed from another locus) decelerates pubertal development, and consider two ways in which ego's age at maturity could affect the fitness of a younger uterine sib.

In the first scenario (another-mouth-to-feed), ego ($m_1 = p_1 = 1$) and a younger sib ($m_2 = 0.5 > p_2$) compete for limited maternal investment until ego leaves the parental home. Earlier puberty reduces ego's fitness ($\partial w_1/\partial x < 0$, $\partial w_1/\partial z > 0$) at a benefit to the younger sib ($\partial w_2/\partial x > 0$, $\partial w_2/\partial z < 0$), who acquires more resources because of reduced competition with ego. Ego's genes of paternal origin have less of an interest in the younger sib's welfare than ego's genes of maternal origin. Therefore, this scenario predicts maternal-specific expression of accelerators of puberty and paternal-specific expression of decelerators of puberty. From Eq. (5B) and Eq. (6B), the joint ESS is characterized by

$$-\frac{\partial w_1}{\partial x} = m_2 \frac{\partial w_2}{\partial x}, \quad \frac{\partial w_1}{\partial z} = -p_2 \frac{\partial w_2}{\partial z}. \tag{7}$$

Ego is predicted to undergo puberty at a younger age than is optimal for his or her individual fitness. Production of x is "altruistic" because it benefits the younger sib at a cost to self, whereas production of z is "selfish" because it benefits self at a cost to the younger sib.

In the second scenario (helper-at-the-nest), ego helps raise the younger sib by providing child care and contributing food to the household pot or otherwise reducing maternal workload (Turke, 1988; Bereczkei and Dunbar, 2002; Kramer, 2002; Hrdy, 2009), but this help is withdrawn when ego begins to reproduce on his or her own. Earlier puberty enhances ego's individual fitness ($\partial w_1/\partial x > 0$, $\partial w_1/\partial z < 0$) at a cost to the fitness of the younger sib ($\partial w_2/\partial x < 0$, $\partial w_2/\partial z > 0$). Ego's genes of paternal origin have less interest in the fitness of the younger sib than ego's genes of maternal origin. Therefore, this scenario predicts maternal-specific expression of decelerators of puberty and paternal-specific expression of accelerators of puberty. The joint ESS is characterized by

$$\frac{\partial w_1}{\partial x} = -p_2 \frac{\partial w_2}{\partial x}, \quad -\frac{\partial w_1}{\partial z} = m_2 \frac{\partial w_2}{\partial z}. \tag{8}$$

Ego is predicted to undergo puberty at an older age than is optimal for his or her individual fitness. Production of x is "selfish" because it benefits self at a cost to the younger sib, whereas production of z is "altruistic" because it benefits the younger sib at a cost to self.

Human sexual maturation is delayed relative to the other great apes. The two scenarios make different predictions about the reason for delayed maturation in the human lineage. In the another-mouth-to-feed scenario, delayed maturation is favored because it allows ego to accumulate more embodied capital and become a better parent (Gurven et al., 2006). In the helper-at-the-nest scenario, delayed maturation is favored because ego obtains greater returns from indirect investment in a younger sib than from direct investment in his or her own offspring (Reiches et al., 2009; Kramer and Ellison, 2010). The two scenarios are, of course, not mutually exclusive: A child can both compete with his or her sibs for limited resources and provide help to his or her parents. Moreover, experience gained in care of younger sibs is a form of embodied capital when a child has offspring of his or her own.

What would one expect if age at maturity were contingent on whether ego's mother and father stayed together for the birth of the younger sib? Partner change causes a decrease in p_2, thus discounting the cost of competition with the younger sib, or discounting the benefit of helping raise the younger sib, for genes of paternal origin. In the another-mouth-to-feed scenario, reduced relatedness to younger sibs is predicted to favor increased production of z from alleles of paternal origin to slow the onset of puberty. Partner change, by itself, does not promote a change in x because m_2 is unchanged. However, the acceleration of puberty due to increased z may favor enhanced production of x from alleles of maternal origin as a countermeasure. By contrast, in the helper-at-the-nest scenario, reduced relatedness to younger sibs directly favors increased production of x from alleles of paternal origin to hasten the onset of puberty.

Blended vs. Unblended Relatedness

What coefficient (or coefficients) of relatedness should be associated with an innate kinship category in models of inclusive fitness? Inclusive-fitness theory usually employs a coefficient that averages relatedness for alleles of maternal and paternal origin, as if maternal and paternal alleles were blended together in offspring rather than retaining their separate identities. By contrast, the present paper uses parent-specific coefficients. There has been surprisingly little discussion of if, and when, blending is appropriate given that the two approaches make different kinds of predictions about what should be observed in nature. I will use the helper-at-the-nest scenario to illustrate the difference of approach and predictions.

A "conventional" model of pubertal timing would use age at puberty, y, as the variable for direct optimization. Larger values of y would be associated with a benefit to the younger sib (dw_2/dy) at a cost to self ($-dw_1/dy$). At the optimal age of puberty,

$$-\frac{dw_1}{dy} = r_2 \frac{dw_2}{dy}, \tag{9}$$

where $r_2 = (m_2 + p_2)/2$ is a coefficient that blends matrilineal and patrilineal relatedness. By contrast, an "imprinting" model would treat y as a function of the level of expression of one or more genes [e.g., $y = \phi(x, z)$, where x is a pubertal accelerator and z is a pubertal decelerator] [Eq. (8)]. The model then makes statements about levels of gene expression at evolutionary equilibrium.

This comparison immediately identifies the attraction of the conventional approach. The use of blended relatedness allows statements to be made about phenotype y, whereas the use of parent-specific relatedness views y as an issue in dispute between opposing parties. The latter models usually do not predict how the dispute will be resolved at the level of outward phenotype. The differences should not be overstated. When considering the effects of husband replacement on optimal age at puberty, r_2 (the blended coefficient) changes in the same direction as p_2 (the patrilineal coefficient), with m_2 (the matrilineal coefficient) unchanged. Therefore, predictions about the direction of change in phenotype will be similar for the two kinds of model, absent the possibility that genes of paternal origin have no influence. Nevertheless, I would argue that parent-specific coefficients are more appropriate than blended coefficients except in the special case where effects of imprinted alleles can be excluded. Even in the latter case, my preference is to use parent-specific coefficients and let the "blending" of relatedness emerge from the mathematics of the model.

Effects of Father Absence

Girls are reported to enter puberty and begin reproduction at younger ages when raised in households in which biological fathers are absent (Surbey, 1990; Maestripieri et al., 2004). Previous attempts to explain this association have focused on the role of father absence as a predictor of the daughter's expected fitness when "choosing" among alternative reproductive strategies. Thus, absence of her father during a girl's early childhood has been proposed to predict lower paternal investment by the potential fathers of her own offspring. Her poor prospects of finding a reliable spouse have been conjectured to favor an earlier onset of reproduction (Belsky et al., 1991). A somewhat simpler hypothesis is that daughters themselves expect less parental investment from delaying maturation in a family with a single parent than in a family in which both biological parents are present (Ellis, 2004). These hypotheses interpret the association between father absence and early maturity as the outcome of a conditional strategy of a common genotype. An alternative interpretation is that early maturation of daughters and absence of fathers are genetically correlated (Mendle et al., 2006, 2009).

The helper-at-the-nest scenario predicts earlier menarche in disrupted families, whereas the another-mouth-to-feed scenario predicts a delay in menarche. Therefore, the association of early menarche with father absence is consistent with older daughters having been selected to delay reproduction to help mothers raise younger sibs when these are likely to be fullsibs but not when these are likely to be halfsibs. The assumptions of the model are simplistic, however. In particular, partner change is assumed to change patrilineal relatedness to the younger sib but not to affect the form of the fitness functions. Thus, the model does not consider direct effects of father absence on the daughter's expected fitness or the purported value of father absence as a cue to the quality of the mating market.

The helper-at-the-nest hypothesis is compatible with the effect of father absence being either the expression of a conditional strategy or the result of a genetic correlation. In the first instance, the presence of a girl's father would be used as a cue to delay maturation. In the second instance, genes that predispose men to short-term relationships would become statistically associated with genes of paternal origin that predispose daughters to mature early, and thus avoid sacrificing personal reproduction for the benefit of maternal halfsibs. Conversely, genes that predispose men to long-term relationships would become associated with genes that predispose daughters to help parents raise fullsibs.

Early reproduction by elder daughters maximizes the potential for reproductive overlap between mothers and daughters. Either could help the other raise offspring at the expense of personal reproduction. What determines who becomes the helper at whose nest? Cant and Johnstone (2008) have argued that when a young woman moves into the extended family of her husband, her mother-in-law is predisposed to become the helper at the younger woman's nest because the older women is related to the younger's offspring (and therefore has a genetic incentive to help), whereas the younger woman is unrelated to the older's offspring (and therefore has no genetic incentive to help). Perhaps a similar dynamic can play out after divorce between a younger woman and her own mother. In this case, the older woman is symmetrically related to the younger woman's offspring (she is their maternal grandmother), but genes of paternal origin in the younger woman are unrelated to potential offspring of the older woman. This would create a bias in favor of the older woman helping the younger.

Effects of Birth Order

Elder daughters probably provide most effective help for sibs several years younger than themselves. Therefore, the helper-at-the-nest hypothesis predicts earlier menarche for daughters with fewer younger sibs and later menarche for elder daughters in larger families. A British study found correlations broadly consistent with these predictions: Menarche was delayed in girls from larger families, but girls born later in a family of a given size had earlier menarche (Dann and Roberts, 1993). However, in a Spanish study, first-born daughters had earlier menarche than second-born and third-born daughters but later menarche than fourth-born or higher-born daughters (Apraiz, 1999). Effects of birth order on age at menarche may be complex and highly contingent. The greater competence of elder daughters to provide help is accompanied by a greater ability to compete for limited resources. Elder daughters have also spent early childhood in a smaller family than the family experienced by their younger sibs at the same age.

Expectations about the effects of family size are complicated because larger families contain more competitors for limited resources but also more opportunities for help. Analyses that consider effects of birth order commonly assign lowest birth order to oldest offspring, regardless of family size, and assign higher birth orders to younger sibs. In this formulation, birth order is linearly related to ego's number of older sibs but provides no information about number of younger sibs (potential beneficiaries of help). From an evolutionary perspective, it might be more informative to perform these analyses with numbers of older and younger sibs as independent predictors.

Effects of Imprinted Genes

Previous sections have explored the hypothesis that mothers (and maternally derived genes of daughters) benefit from delayed maturation because of help provided to mothers by older daughters. The effects of father absence on timing of menarche were construed as supportive of this hypothesis. However, evidence from the effects of imprinted genes on pubertal progression is not readily compatible with the helper-at-the-nest hypothesis.

PWS and Silver-Russell syndrome are caused by the absence of paternally expressed genes or increased dosage of maternally expressed genes. Both syndromes are associated with reduced linear growth in childhood and a weak (or absent) pubertal growth spurt (Davies et al., 1988; Wollmann et al., 1995; Hauffa et al., 2000). These phenotypes suggest matrikin benefited from slower childhood growth. Moreover, individuals who receive both copies of chromosome 14 from their mother experience precocious puberty (Kotzot, 2004). This phenotype suggests that earlier puberty benefited matrikin, perhaps via reduced competition for resources among uterine sibs. Thus, the effects of imprinted genes are more easily reconciled with the another-mouth-to-feed scenario than with the helper-at-the nest scenario.

Given the centrality of age of first reproduction to life-history theory, it is perhaps surprising how little we understand about the fitness tradeoffs that influence variation within and among human populations in age at puberty in either sex. A detailed study of the effects of imprinted genes on pubertal timing and progression promises to provide important clues about the evolution of the distinctive human life history.

Adolescence

Adolescence has been defined as the period from onset of puberty to independence from parents (Casey et al., 2010). The duration of adolescence, by this definition, is highly variable within and among human populations. Popular opinion views adolescence as a time of heightened conflict between parents and offspring and of internal turmoil within the adolescent psyche. Adolescence is both a period of reorganization of neural circuits within the brain (Casey et al., 2010) and a period in which decisions are made about where to live and whom to marry that may have divergent effects on matrilineal and patrilineal inclusive fitness of the child and of his or her parents.

Humans exhibit variation, even within sibships, in the degree to which adults maintain close ties with parents, siblings, and more distant kin. Relations with parents during adolescence are often perceived by young persons as a conflict between their desire for autonomy and parental attempts to control their choices (Surbey, 1998). Parents often perceive the adolescent as self-absorbed and as neglecting responsibilities to family. Adolescence is a life-history transition in which the expression of imprinted genes may have significant effects within the brain. The maternal and paternal genomes of the adolescent agree about individual fitness but may disagree over how much individual fitness should be killed for investment in indirect fitness via kin.

A long tradition in anthropology has emphasized the role of marriage as a form of exchange between patriarchal groups with young women as the commodity of exchange (Tylor, 1889; Durkheim, 1963; Lévi-Strauss, 1980). This was sometimes a direct exchange of daughters between groups, and it sometimes involved a transfer of family wealth, either a payment for a bride or a payment to place a daughter in a favorable situation. The freedom of young people to choose their own partners was curtailed. Conflicts between parents and offspring over the choice of marriage partners are the stuff of legend and literature. Parents usually believe they are acting in their child's best interests (they believe they have more experience than their child in identifying a suitable spouse), but evolutionary theory recognizes that the genetic interests of parent and offspring may diverge.

Material benefits that a spouse brings to a marriage can be transmitted to affinal kin of the spouse, but genetic benefits are transmitted only to offspring of the marriage. Offspring are therefore expected to place a greater emphasis than parents on the genetic qualities, rather than material resources, provided by mates (Trivers, 1974; Apostolou, 2007a,b; Buunk et al., 2008). Mother and father may disagree over the relative value of material and genetic benefits provided by potential spouses of their child if material benefits flow unequally to matrikin and patrikin. For the same reason, maternal and paternal genomes of the child may disagree about the optimal attributes of a spouse.

The choice of where and with whom to reside may also have important fitness consequences for a young couple. Families can both provide support for personal reproduction and demand support for kin. The expression of imprinted genes within the brain raises the possibility that some of these conflicts, over where to live and who to marry, may be internalized within the adolescent psyche.

DISCUSSION

Life-history theory is concerned with tradeoffs in fitness: between the benefits of muscle and fat, between immune function and reproductive effort, between quantity and quality of offspring, and between reproduction now and reproduction later. Inclusive-fitness tradeoffs may involve the personal fitness of different individuals. Social tradeoffs exist between eating food now or bringing it back to the camp to be shared, between being a dad or a cad, between sponging on mother's kin or father's kin, and between helping one's mother raise sibs or having a child of one's own. Psychology is concerned with tradeoffs in mental function: between immediate and delayed gratification, between empathizing and systemizing, between focused and diffuse attention, and between impulsiveness and executive control. A key challenge for a synthesis of these fields will be to understand how psychological tradeoffs mediate life-history tradeoffs.

Our species' innate taxonomy of kin is defined by evolved dispositions that are directed toward some individuals but not others based on environmental cues that are correlated with degree of relatedness. An innate disposition defines the membership of a category, and the membership defines the coefficient of relatedness associated with the category. All individuals who evoke a disposition belong to the category, and all members of the category determine the relatedness associated with fitness consequences of the disposition. Thus, innate kin categories need not correspond exactly to genealogical categories, and, given enough time and genetic variation, the cultural categorization of kin can shape our innate dispositions.

An unresolved issue is the richness of our innate categorization of kin both in terms of the number of different kinds of kin for whom we have distinct dispositions and in terms of the complexity of dispositions toward each particular category. No one would seriously argue that innate structure is absent in our interactions with mothers, but there is no similar consensus over whether we innately distinguish fullsibs from halfsibs, let alone mother's brother's daughters from father's sister's daughters.

When tradeoffs exist between the individual fitnesses of relatives, inclusive fitness assigns relative values to effects on different categories of kin based on each category's degree of relatedness to an actor. For most categories of kin, relatedness differs for genes of maternal and paternal origin. The inclusive fitness of maternal and paternal alleles will be maximized by different allocations of fitness among kin, creating the potential for conflicting goals within individual organisms and a deep-seated biological ambivalence in relations among kin.

Maternal and paternal genes have a common interest in the effective functioning of the individual actor, but phenotypes that are determined by agents with different fitness functions are not expected to show the degree of integration and physiological efficiency one would expect of a phenotype determined by agents with identical interests. Perhaps such internal conflicts can partially account for inefficiencies of mental function and a high frequency of pathology in human social interactions.

ACKNOWLEDGMENTS

The paper has benefited from the comments of Bernard Crespi, Edgar Dueñez-Guzman, Sarah Hrdy, Karen Kramer, David Queller, Stephen Stearns, Robert Trivers, and two anonymous reviewers. The work was supported by a Collaborative Innovation Award from the Howard Hughes Medical Institute to Catherine Dulac.

13

Pathology from Evolutionary Conflict, with a Theory of X Chromosome Versus Autosome Conflict over Sexually Antagonistic Traits

STEVEN A. FRANK*‡ AND BERNARD J. CRESPI†

Evolutionary conflicts cause opponents to push increasingly hard and in opposite directions on the regulation of traits. One can see only the intermediate outcome from the balance of the exaggerated and opposed forces. Intermediate expression hides the underlying conflict, potentially misleading one to conclude that trait regulation is designed to achieve efficient and robust expression, rather than arising by the precarious resolution of conflict. Perturbation often reveals the underlying nature of evolutionary conflict. Upon mutation or knockout of one side in the conflict, the other previously hidden and exaggerated push on the trait may cause extreme, pathological expression. In this regard, pathology reveals hidden evolutionary design. We first review several evolutionary conflicts between males and females, including conflicts over mating, fertilization, and the growth rate of offspring. Perturbations of these conflicts lead to infertility, misregulated growth, cancer, behavioral abnormalities, and psychiatric diseases. We then turn to antagonism between the sexes over traits present in both males and females. For many traits, the different sexes favor different phenotypic values, and constraints prevent completely distinct expression in the sexes. In this case of sexual antagonism, we present a theory of conflict between X-linked genes and autosomal genes. We suggest that dysregulation of the exaggerated conflicting forces between the X chromosome and the autosomes may be associated with various pathologies caused by extreme expression along the male–female

*Department of Ecology and Evolutionary Biology, University of California, Irvine, CA 92697-2525; and †Department of Biosciences, Simon Fraser University, Burnaby, BC, Canada V5A 1S6. ‡To whom correspondence should be addressed. E-mail: safrank@uci.edu.

axis. Rapid evolution of conflicting X-linked and autosomal genes may cause divergence between populations and speciation.

Pathologies often arise from perturbations of evolutionary conflict. In conflict between different components of the genome, the opposing genes push in opposite directions on a particular trait, such as sex ratio or offspring growth rate (Burt and Trivers, 2006). The regulation of such traits under conflict becomes dominated by a balance of opposing forces. This precarious regulatory balance contrasts with the typically supposed design of regulation to achieve efficient and robust expression (Foster, 2011; Werren, Chapter 10, this volume). Mutation or knockout of one side in the conflict leads to the other side dominating expression, often pushing the trait to an extreme in the absence of the opposing force. Extreme expression typically causes pathology.

In this chapter, we develop the idea of pathology arising from perturbations to evolutionary conflicts. We discuss several examples of evolutionary conflicts, the ways in which conflict may lead to exaggerated opposition of forces on a trait, and the occasional breakdown in the normal balance of opposing forces that leads to pathology. We also present a theory of evolutionary conflict between X-linked and autosomal genes over traits that differ in their consequences for male and female fitness. Perturbations to the X–autosome conflict may lead to pathologies of extreme expression along a male–female continuum in trait expression.

The first section develops the general concept of pathology arising from evolutionary conflict. Although the evolutionary dynamics and mechanistic constraints vary greatly between cases, pathology seems likely to increase with the difference between the optimal phenotypic values favored by the conflicting parties. The difference in conflicting fitnesses sets the potential instability of regulatory control built from opposing forces. The degree of pathology in particular cases also increases with the rarity of pathological expression, because rarity reduces the intensity of selection. Weaker selection allows greater exaggeration of opposing forces between conflicting parties, creating greater instability and pathology when the uneasy balance between strongly opposing forces does break down.

The second section analyzes the pathology of mammals derived from growth-related conflicts between paternal and maternal components of the genome (Haig, 2010). Several regulatory control networks of growth do appear to be a conflict between exaggerated paternal enhancers of growth and opposing maternal brakes on growth rate. We consider pathologies arising from imbalances between these strongly opposing forces (Úbeda and Wilkins, 2008). Overly aggressive growth may lead to cancer.

The third section extends our discussion of growth-related pathologies in mammals by considering morphological and behavioral pathologies. Overexpression of normally paternally expressed factors in humans associates with characters such as a protruding tongue, a wide mouth, and excessive feeding solicitation behavior by offspring. By contrast, overexpression of normally maternally expressed factors associates with characters such as growth hormone deficiency, low birth weight, lack of appetite, and poor sucking ability (Eggermann et al., 2008). We also discuss psychiatric pathologies that associate the paternally expressed tendencies with autism and the maternally expressed tendencies with psychosis (Crespi and Badcock, 2008).

The fourth section reviews antagonism between the sexes (Rice and Holland, 1997). Distinct male and female characters interact in mating and fertilization. The sexes often conflict because, in a mating, males push to increase the chance of fertilization success, to increase current female investment in the male's offspring, and to reduce future female mating. Females may push back by resisting male control over fertilization, future mating, and patterns of maternal resource investment in different offspring. Perturbations to these conflicts may lead to infertility.

A different sort of antagonism between the sexes occurs when the same trait is expressed in both males and females, such as aspects of metabolism, physiology, or structure (van Doorn, 2009). Often, males and females are favored to express this common trait in different ways. To the extent that the trait cannot be modulated completely to different expression in the two sexes, natural selection favors a balanced expression of the trait that averages the best trait value in each sex. In some cases, there is no conflict, but rather an intermediate outcome between the divergent characters favored in males and females.

The fifth section presents our theory of X versus autosome conflict. For a trait expressed in both sexes, the autosomes typically favor an intermediate expression that weights equally the best trait expression in males and females. By contrast, the X chromosome favors an intermediate value that weights the trait expression favored by females twice as much as the trait expression favored by males. This conflict between the X chromosome and the autosomes can lead to exaggeration of the opposing forces and to pathology when perturbations disrupt the conflict.

We conclude by reiterating the importance of pathology in the study of conflict. Normally, one cannot see the strongly opposed forces in a conflict, because the observed trait typically reflects an intermediate balance that might be expected in the absence of conflict. Perturbation of the conflict often leads to extreme expression and pathology (Burt and Trivers, 2006), revealing the hidden nature of evolutionary design.

MODEL OF OPPOSING FORCES

In this section, we summarize conclusions from a model of conflict. The model describes how a particular balance of opposing forces leads to a particular level of pathology when the balance is perturbed. We give the conclusions here and present the details of the model in *Appendix A*.

Fig. 13.1 shows the main concepts. Two parties, *A* and *B*, are in conflict, each with different optima for some character. The observed character value arises as an outcome of the opposing forces: *B* pushing for higher values, and *A* pushing for lower values. The opposing forces may become exaggerated as each side pushes harder against the other, with little net change in the outcome. As long as the opposing forces continue to balance, one often cannot see the underlying opposition that leads to a particular character value, such as a particular growth rate. However, when the force imposed by one party is knocked out, for example, by mutation, then the exaggerated force imposed by the other party may push the character value beyond its own optima. Such exaggerated expression, now revealed by the lack of opposition, may lead to a pathological character that is so extreme that it is disadvantageous to all parties.

The model in *Appendix A* develops these ideas of exaggeration and pathology in a simple way. The conclusions from the model are as follows: (*i*) Between conflicting parties, the greater the divergence of favored trait values is, the greater the tendency for a trait to be the outcome of a precarious balance between strongly opposed forces. (*ii*) The less frequently perturbations occur, the weaker the penalty against the pathologies that result from perturbation. A weaker penalty allows evolution of more extreme exaggeration for the conflicting forces and thus greater pathology when the balance is perturbed. (*iii*) The weaker the fitness consequence is for perturbation to a particular opposition of conflicting forces, the greater the opposition of forces becomes. The opposing forces diverge toward an ever more precarious balance until the consequences of pathology or other costs of exaggeration outweigh the tendency for opponents to push oppositely on the trait.

GROWTH PATHOLOGIES: CANCER

The paternally derived genes of a mammal may do better by enhancing early childhood growth at the expense of maternal survival. The paternal push for growth arises because the fathers of particular offspring

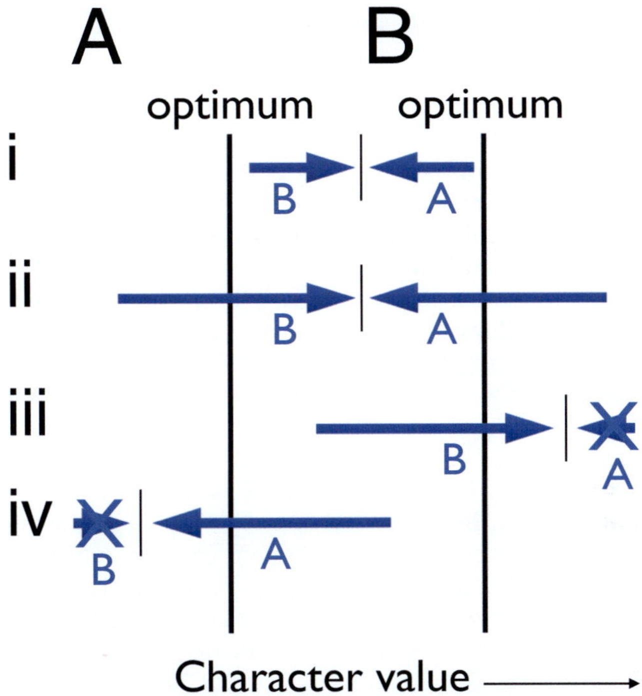

FIGURE 13.1 Pathology from evolutionary conflict. The conflict arises between two parties, *A* and *B*, which have distinct optima for some character. For example, *A* may be a mother and *B* a father, and the character value may be the growth rate of their child. In this case, the father favors a higher growth rate for the child than does the mother. (*i*) Party *B* pushes for higher character value, and party *A* opposes by pushing for lower character value. An observer often can see only the resolution measured as the character value that results from the hidden opposing forces. (*ii*) The resolution in *i* is not at either optimum, so *B* may push harder for an increase in character, which is then opposed by a stronger push by *A* in the other direction. This exaggeration of forces may be difficult to see, because the observed character value may be nearly unchanged under the stronger opposing forces that continue to balance at essentially the same level. (*iii*) The force imposed by *A* is knocked out. *B*'s force, now unopposed, may push the character value to a high level beyond B's own optimum, causing a pathological outcome that is disadvantageous to all. (*iv*) A knockout of *B*, causing *A*'s unopposed force to push the character value too low, leading to pathology that is disadvantageous to both parties.

are frequently unrelated to other offspring produced by the same mother. By contrast, maternally derived genes may do better by slowing childhood growth to balance current offspring success against future maternal reproduction (Haig, 2010).

The opposition of parental interests can influence the regulatory networks that control growth. Several paternally derived genes exaggerate childhood growth rate; several maternally derived genes compensate by slowing growth (Haig, 2010). The net growth rate depends in part on how the conflict is resolved.

Epigenetic imprints of several growth-regulating genes appear to mediate the parental conflict over offspring growth (Fowden et al., 2011). The paternally derived allele may carry an imprint that silences expression, causing only the maternal allele to be expressed. Or the maternal allele may be imprinted and silenced, so that only the paternal allele is expressed.

The insulin growth factor gene *IGF2* is maternally imprinted and paternally expressed. In mice, this gene is perhaps the most important stimulator of fetal growth and determinant of offspring size. The paternally imprinted and maternally expressed gene *H19* produces a noncoding RNA associated with reduced expression of *IGF2* and a lower rate of growth (Gabory et al., 2009). There appears to be a broad network of imprinted genes influencing growth in mice, in which the maternally expressed *H19* acts to repress many growth-promoting components of the imprinted network (Gabory et al., 2009; Fowden et al., 2011). Several other imprinted loci affect growth. There is a tendency for growth stimulation to be associated with paternally expressed loci and growth repression to be associated with maternally expressed loci (Fowden et al., 2011).

The opposition of parental forces can lead evolutionarily to repeated enhancement of paternal pushing toward faster growth and repeated counterresponses of maternal pushing toward slower growth (Haig, 2010; Wilkins, 2011). To the extent that such opposition escalates over evolutionary history, the growth regulatory network becomes a precarious balance between strongly opposing forces that may be easily perturbed (Fig. 13.1). Such perturbations may lead to pathology (Úbeda and Wilkins, 2008; Haig, 2010).

Cancer is excessive growth. Thus, cancer may be a common pathology arising from perturbations to a precarious balance between strongly opposing growth promoters and growth repressors. Some evidence does connect perturbations of imprinted growth regulators to early stages of cancer progression (Lim and Maher, 2010; Monk, 2010).

Higher expression than normal of maternally silenced *IGF2* or lower expression than normal of paternally silenced *H19* or *CDKN1C* leads to a broad spectrum of overly rapid growth pathologies known as

Beckwith–Weidemann syndrome. The risk of certain childhood cancers, such as Wilms' tumor and hepatoblastoma, is increased >100-fold in individuals with this rapid growth syndrome (DeBaun and Tucker, 1998). Other childhood cancers are also significantly increased in frequency (Rump et al., 2005), with a tissue distribution that closely matches that of typical sporadic childhood cancers. These excess, widely distributed cancers are consistent with the interpretation that an overly active *IGF2* pathway exerts its growth effects broadly by stimulating cell replication in many tissues.

An indirect link between imprinting and childhood cancer comes from the association between higher birth weight, accelerated fetal growth, and higher rates of most of the major childhood cancers (Troisi et al., 2006; Milne et al., 2007; Laurvick et al., 2008; Callan and Milne, 2009; Samuelsen et al., 2009). To the extent that perturbations to imprinting can lead to misregulated growth, this association between growth and cancer may also link misregulated imprinting to cancer.

An inherited loss of the maternal *IGF2* imprint causes a fivefold increase in human colorectal tumor risk (Cui et al., 2003). In a mouse study, knockout of the normal maternal *IGF2* imprint led to expression of the maternal allele, increased *IGF2* dosage, and higher sensitivity of the insulin growth factor signaling pathway (Kaneda et al., 2007). These growth-stimulatory changes in the *IGF2* network may increase the number of intestinal progenitor cells at risk for progression or enhance the effects of other growth-promoting mutations (Kaneda and Feinberg, 2005). Somatic loss of imprinting for growth-promoting genes such as *IGF2* has been associated with early stages in cancer progression (Feinberg et al., 2006).

The key question remains: How much of cancer pathology arises from perturbations to maternally and paternally opposed growth regulation? At present, the strongest hints come from the *IGF2* network and from the fact that some other key cancer-related loci, such as *RB1* associated with retinoblastoma and *WT1* associated with Wilms' tumor, are imprinted and are involved in growth (Dallosso et al., 2004; Buiting et al., 2010). These hints suggest that some fraction of cancer pathology may indeed come from growth-related conflicts. However, on the basis of the current evidence, the total cancer risk from growth conflict remains unclear.

The open problem concerns how deeply growth conflict and imprinting influence broad aspects of cellular proliferation. On the negative side, we have only a small number of known genes that fit. On the positive side, the number of genes that fit has increased steadily as data accumulate. It has been technically difficult to identify imprinted genes, leaving open the possibility that the known imprinted genes are just a small fraction of the total amount of imprinting.

With respect to the problem of identifying imprinted genes, Gregg et al.'s (2010) recent study is interesting. In their analysis of mouse brains, they estimated that >1,300 loci have the kind of parent-of-origin effects typical of imprinting. If widespread imprinting does in fact occur, then the conflicting interests of mothers and fathers over offspring growth may indeed lead to a growth regulation system precariously poised between strongly opposing forces. The pathologies from perturbations to a conflict-influenced regulatory design might contribute significantly to cancer risk.

GROWTH PATHOLOGIES: MORPHOLOGY AND BEHAVIOR

The previous section discussed how the mother–father conflict over offspring growth rate may lead to tissue-level pathologies and cancer. In this section, we follow the same conflict in relation to two syndrome pairs. We begin with the syndromes' morphological and feeding-related pathologies. We then turn to psychiatric pathologies, which are more complex.

Morphology and Feeding-Related Behavior

The Beckwith–Weidemann syndrome (BWS) often associates with over-expression of the normally maternally silenced and paternally expressed *IGF2* (Cohen, 2005). The opposing Silver–Russell syndrome (SRS) often arises by repression of *IGF2* (Eggermann, 2010). Not all cases have a known direct association to *IGF2*. It is not clear whether those other cases derive primarily from different growth-related pathways or from unknown connections to regulation of *IGF2* (Eggermann et al., 2008).

BWS individuals often have an enlarged tongue and high birth weight and height (Cohen, 2005). Other abnormalities, such as enlarged kidneys, may follow from a general tendency for rapid growth. Excess placental inclusions associated with rapid fetal growth occur. BWS individuals typically become adults of normal size and proportion, suggesting that the growth abnormalities are concentrated in the preweaning period associated with the primary demands on maternal resources. SRS individuals are small at birth and remain small through development, have significantly reduced subcutaneous fat, and have poor muscle tone (Eggermann, 2010). SRS babies typically lack interest in feeding and may have difficulty taking more than a small amount of food (Blissett et al., 2001). Growth hormone therapy is often an effective treatment.

The second pair of imprinted gene pathologies opposes Angelman syndrome (AS) and Prader–Willi syndrome (PWS). These syndromes associate with imprinted loci on the long arm of chromosome 15, although other causes may be involved. AS typically associates with loss of the normally maternally expressed gene *UBE3A* of the ubiquitin pathway (Johnstone et al., 2006), whereas PWS individuals usually lose function of normally paternally expressed factors in the same chromosomal region (Haig and Wharton, 2003).

AS individuals often have a protruding tongue, a wide mouth, and excessive mouthing behavior (Dan, 2009). PWS individuals tend to have growth hormone deficiency and low birth weight (Haig and Wharton, 2003). Before the typical age of weaning at 2 or 3 years, they also lack appetite and have poor sucking ability, a weak cry, and a low activity level. After typical weaning age, they tend to overeat, perhaps associated with growth compensation derived from low size and weight at weaning age.

Overall, the two syndromes that are biased toward paternal expression, BWS and AS, have preweaning attributes associated with obtaining excess maternal resources. By contrast, the two syndromes that are biased toward maternal expression, SRS and PWS, have preweaning attributes associated with reduced acquisition of maternal resources.

The growth and feeding behavior of the two syndrome pairs fit well with the maternal–paternal conflict theory (Haig, 2010). By this theory, the design of regulatory control arises from opposition of forces rather than maximizing efficiency or enhancing robustness against perturbations. These syndromes may be the extreme expressions among numerous opposing forces in the regulation of preweaning growth and feeding behavior. If so, there may be a variety of potential perturbations leading to varying degrees of deviation from normal. Also, the breakdown of the normal paternal and maternal opposition of forces may lead to other pathologies besides mother–child resource transfer.

Psychiatric Pathologies

Crespi and Badcock (2008) suggested a continuum of psychiatric pathologies arising from the precarious balance between opposed maternal and paternal interests over maternal investment in each offspring. This theory of psychiatric pathology is more speculative than the growth-related pathologies, because complex mental aberrations are harder to quantify and are perhaps influenced by a broader spectrum of causes. In addition, severe pathologies can be difficult to relate to simple theories such as the interests of opposing parties in a conflict, because pathologies are by definition abnormal and maladaptive, favoring no clear interests with respect to design. Failure is always harder to parse than coherent design, because the logic that explains failure arises only from a full understanding of the forces that create normal design. In other words, explaining the causes of pathology is hard. However, it is worth trying, because the causes of pathology lead back to the nature of design. And understanding cause is likely to be helpful in treatment.

To repeat: It is important to keep in mind that pathologies are abnormal and maladaptive. To give a simple example on the basis of the concepts illustrated in Fig. 13.1, suppose mother favors a trait associated with the quantity 10, and father favors 20. The mother might push toward the low end with a contribution that, by itself, causes a value of −15, and the father may respond with a push that, by itself, causes a value of 30. The opposing forces combine additively to a precarious compromise of 15, between the two favored values. However, a loss of the push by either side leads to a pathologically extreme outcome that is maladaptive for both parties.

Clearly, psychiatric pathologies do not sit along a single line of numbers. However, it is worthwhile to ask how much of pathology can be arrayed along an axis between the opposing forces of behavioral regulation favored by maternal and paternal interests.

The Crespi–Badcock (2008) theory defines a psychiatric pathology axis with autism at one end and psychotic disorders such as schizophrenia at the other end. By their theory, normal behavior arises from a balance between opposing forces. The balance arises mechanistically from the relative dominance between the "selfish" limbic and the "social" neocortical brain systems.

Paternally expressed genes tend to push for greater growth and enhanced demand on maternal resources associated with enhancement of placentation, growth factors, suckling, tongue, orofacial muscles, and engagement with mother in infancy. The paternally expressed push for relatively greater development may lead to excess limbic control, which motivates behavior underlying solicitation for food in infancy and, more generally, behaviors that may be regarded as primarily selfish or self-centered. Many paternally expressed genes influence the hypothalamus, a core component of the limbic system.

A paternal bias in imprinted gene expression most commonly arises from reduced expression of normally maternally expressed genes, as in AS (Dan, 2009). Paternal bias associates with relative dominance of limbic versus neocortical function, possibly causing overdevelopment of limbic self-centered behavior and underdevelopment of neocortical social aspects of behavior. Excess self-centered and reduced social behaviors associate with autistic spectrum pathologies. In addition, low IQ may arise because IQ develops in part from neocortical functions, which are relatively reduced when a paternal bias enhances limbic relative to neocortical control. Both AS and BWS associate with excess relative expression of certain paternally expressed genes and an increased risk of characteristics associated with autistic behavior (Bonati et al., 2007; Kent et al., 2008).

A bias toward maternally expressed genes, as in PWS, may associate with increased dominance of the neocortex, enhancing social aspects of behavior sometimes to the extremes of pathology (Badcock, 2010). The definitions and delineations of those behaviors that are social or pathological remain somewhat vague at present, leading to difficulties of interpretation and controversy. According to Crespi and Badcock (2008), social hyperexpression associates with hyperdevelopment of language leading to auditory hallucinations, hyperdevelopment of self in a social context leading to megalomania, hyperdeveloped theory of mind leading to paranoia, amplification of social emotions of elation or depression, and other behaviors sometimes associated with psychosis, schizophrenia, bipolar disorder, and depression.

The example of PWS illustrates the connections between growth, offspring demand on maternal resources, and the mechanistic bases of psychiatric pathologies. In PWS, there is a great reduction in numbers of oxytocin-secreting neurons in the hypothalamus (Swaab et al., 1995; Muscatelli et al., 2000), apparently associated with reduced relative effects on brain development from paternal gene expression and greater relative effects from maternal gene expression. In adults, oxytocin has been called a natural "antipsychotic" (Caldwell et al., 2009) because it appears to connect people socially (Rosenfeld et al., 2010). PWS children do not bond normally with their mothers, and they are complacent and undemanding (Crespi, 2011). Mechanistically, the hypothesis is that a relative bias toward maternal gene expression caused by reduced paternal gene expression associates with lower oxytocin, weak attachment, relatively reduced limbic compared with neocortical functions, and dysregulation of social interactions and bonding. PWS associates with a greatly increased risk of psychosis, especially in cases caused by inheriting two copies of maternally derived chromosome 15 (Webb et al., 2008), presumably creating a maternal expression bias.

CONFLICT BETWEEN THE SEXES

The previous sections discussed conflict over offspring growth rate. In that case, the conflict occurs between maternally and paternally derived genes over the expression of traits within the offspring. In this section, we introduce two other types of conflict between the sexes, each type with its own structure of competing interests and expression of traits. This introduction reviews prior work on sexual conflict.

In the following section, we extend prior work with our own theory of conflict between the sexes. Our theory develops a conflict in which X-linked and autosomal genes are favored to push in opposite directions on traits with different effects on male and female fitness.

Sexual Conflict: Sex-Limited Traits

Many traits arise from male–female interaction. Examples include the timing and frequency of mating and the processes of fertilization. These traits typically depend on the interaction between male and female characters, such as male courtship and female response to courtship. Each character involved in sexual interaction is often expressed only in one sex. Different male and female characters may be in conflict (Rice, 1984; Chapman, 2006).

For example, males express proteins in their seminal fluid that manipulate their mates' reproductive physiology. A male can gain by pushing his mate to invest more in immediate reproduction associated with fertilization by that male's sperm or by pushing his mate to reduce copulation frequency in the future with other males. Females may, in turn, gain by pushing against these male manipulations. That type of sexual conflict matches the structure of Fig. 13.1. The opposing male and female pushes on traits influencing mating and reproduction may become exaggerated. Pathology may occur when a perturbation blocks or alters expression by one of the opposing parties.

Numerous male and female characters conflict over mating and fertilization (Eberhard, 1996; Rice and Holland, 1997; Holland and Rice, 1998; Lew et al., 2006; Chapman, 2008). Infertility is perhaps the most likely type of pathology, arising from abnormalities in fertilization or mating (Lew et al., 2006). However, few studies have directly analyzed the role of conflicting, exaggerated sexual characters in pathology.

Sexual Antagonism: Traits Expressed in both Sexes

Many traits are expressed in both sexes, such as structural components and basic aspects of metabolism, physiology, and morphology. Although both sexes often express the same gene that influences a basic biochemical or structural function, the male and female optima for that trait will sometimes differ. Males may, for example, gain from diverting more resources to muscle growth; females may gain from diverting more resources to fat deposition.

Different male and female optima favor modulation of the trait separately in each sex, leading to sex-limited expression that moves the trait toward its distinct optimum in each sex (Lande, 1980; Rice, 1984). However, various constraints may prevent complete uncoupling of the trait expression between the sexes, or evolutionary dynamics may take a long time to produce pure sex-limited expression. The degree of coupling, or correlation, between the sexes in the expression of the trait determines the degree of potential sexual antagonism (van Doorn, 2009; Rice and Chippindale, 2001).

In the case of a trait expressed in both sexes, current theory suggests that no conflict of interest occurs. Instead, for each individual gene influencing the trait, natural selection favors an averaging of the separate optima in males and females. This averaging of distinct optima is often called *intralocus* antagonism, to emphasize that the divergent selective pressures of male and female optima act simultaneously on the same locus (Rice and Chippindate, 2001; van Doorn, 2009).

From the perspective of a single locus, this averaging of distinct optima is like a situation in which individuals express the same trait in two distinct habitats. The favored trait value is an average of the trait values favored in each habitat. In the case of sexual antagonism, the gene lives alternately in the two distinct habitats of male and female bodies.

SEXUAL ANTAGONISM: A THEORY OF X VERSUS AUTOSOME CONFLICT

The previous section reviewed the theory of sexual antagonism for a trait that is expressed in both sexes. In that case, each gene favors a trait that averages the distinct male and female optima. All genes on the autosomes favor an equal weighting of the male and female optima, because the reproductive value of those autosomal genes is the same in both sexes. By contrast, genes on the X chromosome favor weighting the female optimum twice as strongly as the male optimum, because X-linked genes in females have twice the reproductive value of X-linked genes in males.

The different weightings of male and female optima by autosomal and X-linked genes create a conflict of interest. Haig (2006a,b) briefly mentioned this conflict, but did not develop the consequences. To understand the consequences, consider that X-linked genes are selected to push more strongly toward the female optimum than are autosomal genes, and autosomal genes are selected to push more strongly toward the male optimum than are X-linked genes. With conflict, there is the potential for exaggeration, in which the conflicting parties push oppositely and increasingly hard on the trait. The resulting precarious balance may lead to pathology (*Appendix B,* notes on X inactivation and inbreeding).

More generally, conflicts driven by the different weightings of male and female fitness arise between various genomic subsets: Mitochondria favor the female optimum, X chromosomes favor a weighting of two-thirds of the female optimum and one-third of the male optimum, autosomes favor equal weighting of the optima, and Y chromosomes favor the male optimum. Here, we develop the X–autosome conflict, but note that other genomic conflicts of this sort may also be important. For example, mitochondria push metabolic traits toward the female optimum and may therefore be opposed by other genomic components that push the regulation of metabolic traits toward the male optimum. Exaggeration and the potential for pathology may follow.

X versus autosome conflict has been discussed in a variety of situations, such as meiotic drive (Burt and Trivers, 2006). However, apart from Haig's (2006a,b) brief comments, we could not find in the literature mention of the conflict between different genomic subsets, such as the X and the autosomes, over divergent male–female optima. Given the very simple logic of the conflict, it is not clear why the extensive discussions of sexual antagonism have not emphasized this particular aspect of X versus autosome conflict.

The evolutionary dynamics of sexual antagonism for a trait expressed in both sexes may explain the lack of discussion about X versus autosome conflict. The stable outcome, with the highest fitness, would be modulation of the trait to express differently in the two sexes. With sex-limited expression, each sex if favored to match the trait to its own optimum, and the conflict disappears.

The literature discusses extensively the evolutionary path to pure sex-limited expression and complete sexual dimorphism (Lande, 1980; Rice, 1984; van Doorn, 2009; Connallon and Clark, 2010). However, the data suggest that a significant correlation between the sexes remains for traits with divergent optima between the sexes (Chenoweth et al., 2008; Bonduriansky and Chenoweth, 2009; Poissant and Coltman, 2009; van Doorn, 2009; Poissant et al., 2010; Stewart et al., 2010). Such correlation can arise because constraints of regulation and expression prevent tuning of the traits separately in each sex. Alternatively, the constraints may slow the evolutionary path toward sex limitation sufficiently to maintain a balance between the rate at which sex-limited expression is enhanced and the rate at which new antagonisms arise. In any case, given the observed correlation between the sexes in traits for which sexual antagonism occurs, there is wide scope for X-linked versus autosomal conflict.

Any behavioral, metabolic, physiological, or structural trait with divergent male and female fitness will be subject to X–autosome conflict whenever traits are not completely tuned in each sex to achieve perfect sex-limited expression. To the extent that the conflict induces exaggerated and opposing forces by the X chromosome and autosomes, subsequent evolutionary change to enhance sex-limited expression may become more difficult to achieve. Thus, the conflict, once established, may tend to be maintained because of the complexities in trait regulation induced by the conflict.

Observations have not previously been interpreted in light of this particular kind of X chromosome versus autosome conflict. The most obvious prediction is widespread interaction between X–linked and autosomal genes over sexually antagonistic traits, with the X–linked genes pushing toward the female optimum and the autosomal genes pushing toward the male optimum. However, it may be difficult to see those sorts of interactions in a particular population. If, for example, a particular pair of X-linked and autosomal genes interact as predicted, but lack polymorphism, their interaction would be hidden from observation.

Loss-of-function mutations or chromosomal duplications provide one type of perturbation that can lead to pathology and provide a window into the underlying genetic architecture of trait regulation. Our theory predicts a simple directionality along the male–female axis. X chromosomes push traits toward expression favored by females. Knockout of X-linked genes therefore tends to cause excess expression in the direction favored by males. Similarly, autosomes push traits toward expression favored by males. Knockout of autosomal genes therefore tends to cause excess expression in the direction favored by females.

The most interesting, and controversial, discussion of a male–female axis in the recent literature concerns differences in behavior. By that theory, extreme maleness associates with autistic characteristics (Baron-Cohen, 2009) and extreme femaleness associates with psychotic characteristics (Crespi and Badcock, 2008; M. Brosnan et al., 2010). Our theory predicts that X knockouts associate with extreme maleness. Thus, by the theory of a male–female behavioral axis, one would expect X-linked knockouts to be associated with autistic characteristics. To evaluate this hypothesis fully, one would have to estimate the relative number of genes influencing autism on the X chromosome and the autosomes and then show that the X carries a disproportionate share. Not enough data exist at present. Some intriguing hints of X-linked effects associated with autistic tendencies have been reported (Marco and Skuse, 2006). Other extremes along a male–female axis may also be evaluated with regard to our predictions about the alternative directions of pathology associated with X-linked and autosomal genes.

Hybridization between populations or species provides another sort of perturbation that can reveal the underlying genetic architecture of traits. Genes in conflict may tend to diverge relatively rapidly between populations (Frank, 1991; Hurst and Pomiankowski, 1991; Werren, Chapter 10, this volume). Upon hybridization, mismatched X-linked and autosomal genes may cause pathological expression of traits. Such pathologies in crosses between populations are referred to as hybrid incompatibilities.

Our theory predicts hybrid incompatibilities between X-linked and autosomal loci. These X–autosome incompatibilities may be dispersed widely throughout the genome, because many traits may be subject to sexually antagonistic selection. Many observations suggest relatively rapid divergence of X chromosomes or widespread X–autosome incompatibilities in hybrids (Coyne and Orr, 2004; Carneiro et al., 2010; Lu et al., 2010).

CONCLUSIONS

Some traits are regulated by the opposition of conflicting forces. For example, early offspring growth in mammals balances the powerful opposing pushes of paternal enhancement and maternal slowing. These opposing forces appear to have become exaggerated by the conflict. Nonetheless, the typical outcome remains intermediate and apparently normal because the opposing forces come to a precarious balance. When a mutation or other block to one of the exaggerated forces occurs, the unopposed push in the opposite direction often causes a pathologically disrupted growth trajectory.

Increasing evidence supports this conflict interpretation for the regulation of early offspring growth in mammals. The interesting question is: How often is the evolutionary design of regulatory control dominated by the precarious balance of conflicting and exaggerated forces rather than by the efficiency and robustness of control? We do not know the answer to that question. In this paper, we reviewed theory for sexual conflicts that suggests opposing forces may be important for many characters. We also gave some examples of particular traits that may be regulated by conflict. Although those examples are preliminary with regard to empirical support, they do show the wide range of organismal characters and associated pathologies that may ultimately have to be understood in the light of evolutionary conflict.

From previous studies, conflicts have been invoked to explain childhood growth, excessive male-like or female-like characteristics, infertility from exaggeration of mating or fertilization traits, and psychiatric disorders of misregulated social behavior. Sexual differences are often the first kind of trait that can be studied with regard to strong contrasts, because male–female dimorphism can appear binary and relatively easy to identify. How many other traits follow the evolutionary path of exaggerated conflict and occasional pathology? Again, we do not know. However, it would certainly be worthwhile to consider the wide range of genomic conflicts and social conflicts that may be associated with pathologically disrupted genetic or social regulation. The normal and apparently cooperative working of genomes, insect societies, and other groups may be regulated in part by precariously balanced opposing forces.

How does conflict influence the design of regulatory control? Scant research has focused on that interesting question (Foster, 2011). Speculating briefly, genes that share common interests may be more cooperative when opposed by a group of genes with conflicting interests. For example, the paternally imprinted and maternally expressed genes *TP73*, *RB1*, and *CDKN1C* are all in the same regulatory pathway influencing the cell cycle (Boominathan, 2007; Buiting et al., 2010). In general, do genes with common interests often segregate into common pathways? And do genes with opposing interests tend to segregate into different pathways with opposing effects? Or, as with *IGF2* versus *IGF2R* (Haig and Graham, 1991), do conflicting genes frequently interact directly within the same pathway, perhaps causing opposing tendencies in regulatory control?

In this paper, we also added to the theory of conflict. Previously, a variety of male–female conflicts were identified. For example, we reviewed the maternal–paternal conflict over offspring growth rate and the male–female conflicts over mating and fertilization. Our theory focused on the conflict between X chromosomes and autosomes. When a trait has different consequences for males and females, natural selection favors the sexes to express the trait differently. However, many traits of metabolism, physiology, and structure arise from a common genetic basis in the two sexes. Those traits may be difficult to tune perfectly to different expression in the sexes.

To the extent that expression is constrained to be correlated between the sexes, genes tend to favor an averaging of the best trait values in males in females. Our theory of conflict arises because autosomal genes tend to weight the sexes equally, whereas X-linked genes tend to weight females about twice as much as males. Once this sort of conflict occurs, the autosomal and X-linked genes may push in opposite directions on the trait, with the opposing forces becoming exaggerated. Once exaggerated, all of the tendencies for pathology and consequences of regulatory control arise that we have emphasized throughout. The X versus autosome conflict may be particularly important, because it applies to any trait with different optima in males and females. By contrast, the other sexual conflicts that we reviewed are usually confined to a particular type of trait, such as growth or mating. Thus, the X versus autosome conflict may be particularly associated with widely dispersed genetic interactions throughout the genome, providing another hypothesis for rapid evolution and hybrid incompatibilities between species involving the X chromosome.

In all cases of disrupted conflict, the particular disease pathologies are interesting in themselves. The ordering of the different human childhood overgrowth and undergrowth pathologies is the most obvious example. More speculatively, the ordering of psychosocial pathologies such as autism and various psychoses may turn out to be an interesting component of psychiatric disease.

Beyond the explanation of particular diseases, pathologies are interesting because they reveal the underlying evolutionary design. In most individuals, the opposing forces precariously balance. One cannot see the underlying conflict. The conflict becomes apparent only upon perturbation and the observation of pathology. Once one recognizes the axis of conflict, it may be possible to order apparently different pathologies along that axis. The extreme pathologies at the opposite ends of the axis of conflict reflect the exaggerated pushes in opposing directions. Once we recognize the paired extremes and the underlying structure of normal regulation, we may begin to understand many graduations in the traits along the conflict axis. Pathology reveals design.

APPENDIX A: CONFLICT BETWEEN TWO INDIVIDUALS OVER A TRAIT THAT INFLUENCES THE FITNESS OF BOTH PARTIES

We consider two parties in conflict over a trait (Fig. 13.1). To present the simplest case, suppose the final trait, x, is the sum of the contributions from the two parties, $x = x_A + x_B$. The first party has optimal trait value, m_A, and the second party has optimal trait value, m_B. The expected fitness of each party is given by

$$w_i = K - a(x_A + x_B - m_i)^2 - pc(x_B - x_A)^2,$$

where $i = A$ or B, allowing this single equation to describe the fitnesses of the two opposing parties.

The first two terms of the fitness equation describe a typical stabilizing selection function, in which the final trait x is favored to converge to the optimum m_i, with fitness falling off quadratically from the optimum.

The last term of the fitness equation quantifies the penalty for opposition of forces acting on the trait. The penalty rises with the distance between the contributions of the two parties. That distance is weighted by a cost parameter, c, that scales the penalty for perturbation in relation to distance, and a probability parameter, p, that describes the probability that a perturbation occurs.

A perturbation may, for example, be the knockout of the contribution by one party, leaving the other party's contribution as the sole determinant of the trait. Such a knockout affects fitness by moving the trait in relation to the optimum, m_i, and by invoking the penalty that depends on the distance between the parties and the scaling, c.

Assuming no constraints on the traits, the optimum is

$$x_i^* = m/2 + b(m_i - m_j),$$

where $m = (m_A + m_B)/2$ is the midpoint between the opposing optima, j is the opposing party to i such that if $i = A$, then $j = B$, and vice versa, and $b = a/4pc$. The conclusions given in the main text follow.

APPENDIX B: CONFLICT BETWEEN X-LINKED AND AUTOSOMAL GENES OVER A TRAIT WITH DIFFERENT FITNESS CONSEQUENCES IN MALES AND FEMALES

Suppose, for a particular trait, that the fitness of a female is maximized at F^*, and the fitness of a male is maximized at M^*. Optimally, each sex would separately express its own maximal trait value in a sex-limited way. However, a certain fraction of trait expression may arise from genes that influence the trait in the same way in both sexes, creating a genetic correlation between trait values in males and females. If so, then divergent selection on these jointly expressed genes will pull in different directions in the two sexes. For the phenotypic contribution to trait expression shared by the two sexes and encoded by autosomal genes, natural selection typically favors the average of the optimal values in the two sexes. The simple averaging arises because the total reproductive value of autosomal genes is the same in the two sexes. In this case, there is no conflict of interest, because each autosomal gene weights the two sexes equally.

If both X-linked and autosomal genes influence the part of trait expression shared by the sexes, a conflict of interest occurs between the different components of the genome. The reproductive value of X-linked genes is twice as great in females as in males, compared with the equal reproductive value weighting of the two sexes by autosomal genes. Thus, X-linked genes pull toward the female optimum and, relative to the X, autosomal genes pull in the other direction toward the male optimum. Here, we present a simple model to illustrate this X versus autosome conflict. To keep things simple, we do not consider a full genetical model, but instead use a phenotypic model with reproductive value weightings. We also mention some interesting extensions with regard to X inactivation and inbreeding.

Let X and A be the contributions of X-linked and autosomal genes to the trait value. We assume that X makes the same contribution to male and female trait values independently of the fact that females have two X chromosomes and males have one. The ploidy normalization for the sexes may happen in various ways, and the particular mechanisms can have interesting consequences. Our initial description ignores those ploidy issues. Our discussion of X inactivation and inbreeding at the end of this section hints at some important extensions to the theory that need to be studied further.

We start by writing the trait expressed in females as

$$T_f = \delta F^* + (1-\delta)(X+A),$$

where $1 - \delta$ is the fraction of the trait that is not sex limited in expression and is controlled by a combination of X-linked and autosomal genes. The fraction that is sex limited, δ, is at the female optimum, F^*. The distance between the actual trait expressed and the optimum is $T_f - F^*$.

Quadratic Fitness

We write the fitness of a female as

$$\begin{aligned} w_F &= 1-\alpha(T_f - F^*)^2 - b(X-A)^2 \\ &= 1-a(X+A-F^*)^2 - b(X-A)^2, \end{aligned}$$

where $a = \alpha(1-\delta)^2$, with α as the weighting of the fitness penalty for distance from the optimum trait value. The last term is a penalty for divergent contributions of the X and autosomal genes, as in *Appendix A*. The expression for male fitness, w_M, is the same, replacing the female optimum F^* by the male optimum M^*.

The fitness of an autosomal gene is the average of the fitnesses of the females and males, $w_A = (1/2)w_F + (1/2)w_M$, whereas the fitness of an X-linked gene is weighted twice as strongly toward females as males, $w_X = (2/3)w_F + (1/3)w_M$.

We assume that the contributions of X and A are normalized with respect to ploidy differences, as mentioned above. With that assumption, we can find the evolutionarily stable strategy (ESS) values, X^* and A^*, by jointly maximizing the X-linked and autosomal fitnesses and solving for the ESS values. Without loss of generality, we can set $F^* = 0$ and define $d = a/5b$ and $M' = (5/24)M^*$, allowing us to write the ESS values as

$$\begin{aligned} A^* &= M'(1+d) \\ X^* &= M'(1-d). \end{aligned}$$

These solutions show that the X-linked genes push toward the female optimum at $F^* = 0$ and the autosomal genes push toward the male optimum at M^*. As the relative cost of pushing on the trait, b, becomes small, d increases, causing exaggeration of the opposing forces.

Consequences of X Inactivation

If there is X inactivation of one X allele in females, then the situation is more complex. About 15% of genes on the human X chromosome escape inactivation, and another 10% of X-linked loci are variably expressed on inactive X chromosomes (Carrel and Willard, 2005). Thus, a significant number of X-linked loci may be expressed from both copies and may conflict with autosomes. Occasional diploid expression on the X is sufficient to create the conflict.

Among loci with complete X inactivation, different cells may inactivate different copies of the X. Thus, each cell may express only one of the X copies, but each individual female may express both copies. The consequences of inactivation for a particular phenotype depend on the particular tissue that controls the phenotype and the relative fraction of each X chromosome inactivated in that tissue. If there is sufficient mixture of expression of the two copies in the focal tissue, then the phenotypic consequences may in some cases be equivalent to diploid expression.

In certain cases, most of the focal tissue may express only one particular copy, or the phenotype may be dominated by one particular X copy. If so, we would need to account for three types of fitness classes for an X-linked gene: the copy of the gene in males, the expressed copy of the gene in females, and the silent copy of the gene in females. We have not done the full analysis of this model. Here are a few conjectures based on concepts from class-structured models (Taylor and Frank, 1996; Frank, 1998).

With no inbreeding, the conflict between X-linked and autosomal genes disappears with X inactivation, because, for each copy of an X linked gene, the probability that it is expressed in males or females is equal in each generation. In particular, there is a one-third chance of being in males and expressed, a one-third chance of being in females and expressed, and a one-third chance of being in females and unexpressed. With no inbreeding, an unexpressed allele has average fitness and so does not contribute to evolutionary change. (It is more accurate to say that the reproductive values of alleles in the two sexes are equal for autosomal loci and the reproductive value of alleles in females is twice that in males for X-linked loci, as above.)

If there is inbreeding, there will be a correlation between the expressed and latent trait values of the two X-linked copies in females. That correlation causes an unexpressed (inactivated or imprinted) X-linked copy to have its fitness associated with its own latent trait value, adding a further push toward the female optimum and creating once again a conflict between X-linked and autosomal genes, including X-linked loci subject to X inactivation.

Gaussian Fitness and Genetics

Many aspects of this preliminary phenotypic model deserve further study. We mention just two. First, the simple quadratic fitness function used here is a special case of a Gaussian fitness function, which becomes quadratic when the selective intensity is weak. For example, if we focus only on selection on the X chromosome by setting $A = b = 0$, and we rescale so that $F^* = 0$ and $M^* = 1$, then the expressions for Gaussian fitness functions are

$$w_{\mathrm{F}} = e^{-a_{\mathrm{f}}X^2}$$

$$w_{\mathrm{M}} = e^{-a_{\mathrm{m}}(1-X)^2},$$

where a_f and a_m are the selective intensities on females and males for deviations from each respective optimum. The ESS phenotype favored by the X chromosome maximizes $w = (1/3)w_{\mathrm{M}} + (2/3)w_{\mathrm{F}}$, which can be obtained by solving for X in

$$a_{\mathrm{m}}(1-X)e^{-a_{\mathrm{m}}(1-X)^2} = 2a_{\mathrm{f}}Xe^{-a_{\mathrm{f}}X^2}.$$

Similarly, the ESS phenotype favored by autosomes in the absence of contribution from the X chromosome maximizes $w = (1/2)w_{\mathrm{M}} + (1/2)w_{\mathrm{F}}$, which can be obtained by solving for A in

$$a_{\mathrm{m}}(1-A)e^{-a_{\mathrm{m}}(1-A)^2} = a_{\mathrm{f}}Ae^{-a_{\mathrm{f}}A^2}.$$

Typically, the X chromosome favors a phenotype relatively closer to the female optimum than that favored by the autosomes.

The second issue concerns the range of underlying genetic assumptions for which the ESS phenotypic model correctly expresses the key evolutionary forces. Such phenotypic models are generally accurate for alleles that contribute additively to phenotype, under the assumption of a continuous spectrum of mutational effects and when accounting for the possibility of alternative equilibria (Frank, 1998). By contrast, many genetic models of sexually antagonistic traits find significant complexities with respect to the dominance interaction patterns among alleles (Rice, 1984; Patten and Haig, 2009; Fry, 2010). Those genetical models did not analyze the X versus autosome conflict. So it remains an open question how the genetic complexities of dominance and polymorphism would play out in a model of interactions between X-linked and autosomal loci. Often, if one studies a polygenic model and allows a spectrum of allelic effect sizes and parameters of dominance and epistasis, the ESS phenotypic model captures reasonably well the long-term evolutionary forces of the polygenic model. However, the particular problem of X versus autosome conflict remains to be studied in full genetical detail.

ACKNOWLEDGMENTS

S.A.F.'s research is supported by National Science Foundation Grant EF-0822399, National Institute of General Medical Sciences Models of Infectious Disease Agent Study Program Grant U01-GM-76499, and a grant from the James S. McDonnell Foundation. B.J.C.'s research is supported by the Natural Sciences and Engineering Research Council.

Part V

ARE HUMANS DIFFERENT?

Evolutionary principles for cooperation that have been developed from studies of diverse social organisms should apply to humans. The more immediate roots to human cooperation and conflict also may be seen in primates. However, there are challenges in studying humans and their close relatives. Objectivity is essential. There are many possibilities for study techniques (such as the questionnaire or survey) in humans, but these also offer many opportunities for confusion. One powerful approach to studying human cooperation is to look at what humans do and what the outcomes are, just as one might do for other social animals. This technique can be particularly informative when the human group lives in ways consistent with humans over most of their evolutionary past. The Dogon people of Mali, reported on by Beverly Strassmann in Chapter 14, are millet-and-onion-farming agriculturalists who do not use contraception, adhere largely to indigenous religions, practice polygyny, and have high mortality rates. In a 25-year-longitudinal study, Strassmann has investigated the hypothesis that the Dogon are cooperative breeders, where some individuals help rear nondescendent kin rather than their own progeny. She does not find that the data support this hypothesis. First, neither women nor men delay reproduction in order to raise siblings. Although parents force daughters to care for extra siblings, this is better viewed as parental manipulation because the presence of siblings reduces survivorship. Similarly, grandmothers do not appear to be effective alloparents. Rather than increasing survivorship, the presence of paternal grandmothers does the opposite, doubling the hazard of death for a child. What matters most for survival is the presence of the mother, and other relatives are not adequate replacements. Task cooperation occurs within the groups that work and eat together, but conflict is always present in ways that Strassmann carefully explains.

In an overview of vertebrate interactions, Dorothy Cheney demonstrates in Chapter 15 that animals ranging from chickadees to chimpanzees are aware of their own status, and that of their companions, and behave accordingly. Eavesdropping on how individuals interact with others can change behaviors. Relatednesses are often known and impact interactions. In vervet monkeys, for example, an individual who has been attacked may turn and subsequently attack a relative of her opponent. Dominance hierarchies also impact such interactions. But some animal interactions are more subtle. Ravens are more likely to cache food in hidden sites when competitors are present, for example. However, the calculations of gain, cost, and punishment necessary for reciprocal altruism (here called contingent altruism) seem largely lacking outside of humans. Instead, there is a great deal of tolerance in interactions and a lack of direct payback among close relatives and long-time partners. Yet it is in these relationships where cooperation overwhelmingly occurs. A common feature of cooperative acts is that they are not necessarily transitive. Some individuals consistently take on the risky jobs, be it male chimps patrolling their territorial edges or female lions leading the hunt. This is also true in organisms (such as wasps) with much simpler brains, where cooperation flows from workers to the queen.

Observations of humans and primates in natural situations can teach us much about behavior, but environmental complexity can make causation difficult to discern. An alternative is to examine choices made under highly regulated circumstances. To address social acts such as generosity, trust, fairness, and punishment, many purportedly relevant games have been applied to humans, one simple example being the Dictator Game that allows a subject to decide whether to share a quantifiable resource with an unseen other. [This game typically yields donations of 20–30% of the resource.] Although such games have weaknesses, they seem to indicate that humans are willing to donate but only at levels indicating they consistently value themselves most highly. These and other experiments further indicate that humans favor relatives, long-term partners, and group members over outsiders, and they will suffer costs to punish cheaters. As described by Joan Silk and Bailey House in Chapter 16, versions of social games involving food or tools that likewise have been used with primates produce complex results. Cooperation clearly occurs and tracks levels of sociality in the groups, but some results are controversial and remain open to alternative interpretations.

In the modern world, most of a person's material possessions are items that no individual could possibly make by herself. Instead they were produced with the learned and specialized expertise of others. In Chapter 17, Robert Boyd and colleagues argue that learning from others (and not intelligence alone) is the key to human success, the characteristic that has made us so adaptable. Initially in human history, most adaptations involved direct climatic protection, food acquisition, and food storage. Thus, the sharing and acquiring of information from others is a particular kind of intelligence. Boyd and his coauthors argue that cultural learners have an advantage because they can grasp the best from the past even if they innovate personally only occasionally. Tools and customs certainly make life for humans easier or possible.

The study of cooperation and conflict has come a very long way from the time, almost 50 years ago, when Hamilton (1964a,b) first pondered how to explain the evolution of worker behavior in social insects with a strange genetic system. Such analyses have spread out taxonomically, extending even to microbes. They have deepened mechanistically as we probe the molecular and genetic basis of cooperative phenomena. The findings are also beginning to show practical applications, as in medicine, and they have proven essential for understanding the structure of life, from cells to multicellular organisms to societies. Not least, study of the complex mix of cooperation and conflict helps us to understand what makes the human animal both ordinary and remarkable.

14

Cooperation and Competition in a Cliff-Dwelling People

BEVERLY I. STRASSMANN

In animals that breed cooperatively, adult individuals will sometimes delay reproduction to act as helpers at the nest who raise young that are not their genetic offspring. It has been proposed that humans are also a cooperatively breeding species because older daughters, grandmothers, and other kin and nonkin may provide significant childcare. Through a prospective cohort study of children's (n = 1,700) growth and survival in the Dogon of Mali, I show that cooperative breeding theory is a poor fit to the family dynamics of this population. Rather than helping each other, siblings competed for resources, producing a tradeoff between the number of maternal siblings and growth and survival. It did not take a village to raise a child; children fared the same in nuclear as in extended families. Of critical importance was the degree of polygyny, which created conflicts associated with asymmetries in genetic relatedness. The risk of death was higher and the rate of growth was slower in polygynous than monogamous families. The hazard of death for Dogon children was twofold higher if the resident paternal grandmother was alive rather than dead. This finding may reflect the frailty of elderly grandmothers who become net consumers rather than net producers in this resource-poor society. Mothers were of overwhelming importance for child survival and could not be substituted by any category of kin or nonkin. The idea of cooperative breeding taken from animal studies is a poor fit to the complexity and diversity of kin interactions in humans.

Department of Anthropology and Research Center for Group Dynamics, Institute for Social Research, University of Michigan, Ann Arbor, MI 48109. E-mail: bis@umich.edu.

It has been reported that humans are a cooperatively breeding species that depends on individuals other than the mother and father for the successful rearing of offspring (Hrdy, 2005a, 2009). These individuals are known as alloparents and may be siblings, aunts, uncles, grandparents, and other kin or nonkin. One of the most thorough studies took place among the Maya of the Yucatan, Mexico. Hard-working girl-farmers paid back to their parents 94% of their cumulative consumption costs before leaving home to marry, and boys repaid 80%, enabling mothers to have more closely spaced births (Kramer, 2005a,b). In other species, alloparents seem to enhance their inclusive fitness by helping (Emlen, 1997a), but it is not yet clear whether this is also the case for Mayan children. In a study on the fishing atoll of Ifaluk in Micronesia, women ($n = 7$) whose two firstborn children were daughters had a mean of nine surviving offspring, and women ($n = 11$) who bore two sons first had a mean of five surviving offspring (Turke, 1988). This difference of four offspring was attributed to the role of elder daughters as helpers at the nest and was an enormous effect size for such a small sample ($n = 18$). Given that the analysis was bivariate, there is a strong possibility that the results reflect confounding variables.

In 19th century Finland, the survival of the maternal grandmother was correlated with improved grandoffspring survival (Lahdenperä et al., 2004). This study controlled for occupation (e.g., farmer, priest, or landless laborer) but did not distinguish among farmers by the size of their holdings; hence, the phenotypic correlation between grandmaternal survival and grandoffspring survival may be caused by variation in wealth. A recent review (Sear and Mace, 2008) concluded that cooperative breeding was prevalent in traditional farming populations that had high fertility and high mortality, but only 6 of 17 studies controlled for wealth. A meta-analysis showed that the positive association between grandparental and grandchild survival was found only for the maternal and not the paternal side, although in these farming populations, the children were more likely to live with their paternal grandparents (Strassmann and Garrard, 2011).

Even in forager populations, the data on grandparental investment are mixed. In the Hadza of Tanzania, foraging grandmothers worked longer hours gathering tubers than they did in their prime, and children who received food from a grandmother or great aunt had higher nutritional status than children who were without alloparents (Hawkes et al., 1997, 1998). In the !Kung, having four surviving grandparents was not associated with improved grandchild survival or nutritional status (Draper and Howell, 2005).

Here, I use evolutionary social theory to explore the family dynamics of the Dogon of Mali, West Africa. My underlying premise is that cooperative breeding is a facultative response to particular socioecological environments and that there is no strong evolutionary or empirical argument for viewing cooperative breeding as the species typical or evolved pattern in humans. In exploring Dogon social dynamics, I will focus on (*i*) siblings, (*ii*) extended families, and (*iii*) grandparents and grandchildren to test predictions from the hypothesis that humans are cooperative breeders against predictions from the following evolutionary hypotheses: kin selection (Hamilton, 1964a), life history theory (Stearns, 1992), parent–offspring conflict (Trivers, 1974), and local resource competition (Clark, 1978). My research design is a prospective cohort study of the growth and survival of 1,700 children (*Methods*).

ETHNOGRAPHIC BACKGROUND

The Dogon are traditional agriculturalists whose staple crop is pearl millet (*Pennisetum glaucum*). They have been the subject of a 25-year longitudinal study of human evolutionary biology that I initiated in 1986 and that is presently ongoing. The Dogon retain many of the features that have characterized humans over our evolutionary past, including the absence of contraception, polygyny, preservation of the indigenous religion, a subsistence economy, life in a tight-knit web of close and more distant kin, and high mortality levels (Strassmann, 1992, 2000). Marital residence is patrilocal (wives take up residence with the husband's family), and descent is patrilineal (father to son). Women who are married to the same man are never sisters, and closely related women are not allowed to marry into the same patrilineage, a custom that disfavors female kinship bonds (Strassmann, 2003).

The Dogon are one of the most traditional people of Africa, and their cliffside villages have been designated a United Nations Educational, Scientific, and Cultural Organization (UNESCO) World Heritage site. On a seasonal basis, the area attracts tourists, but the revenues that are generated do not trickle down to the population. Infrastructure is sparse (electricity is absent and latrines are rare) in this area of Mali, which is a country that had a per capita income of $470 in 2006 (US Department of State, 2010). The United States Agency for International Development made condoms and low-cost hormonal contraceptives available in 2010; however, the population remains pronatalist, and the demand for these products in rural areas is nonexistent. The modal fertility per woman per lifetime is 10 live births (Strassmann and Gillespie, 2002).

RESULTS AND DISCUSSION

Siblings

The cooperative breeding hypothesis predicts that (*i*) older siblings help to rear younger siblings and (*ii*) the productive labor of children enables mothers to have more closely spaced births (Kramer, 2005a,b). It points to the beneficial impact that siblings have on each other during the juvenile period. By contrast, life history theory emphasizes the finite nature of parental resources and the tradeoff between offspring number and offspring quality. Kin selection theory (Hamilton, 1964a) predicts that siblings will help each other when $rB > C$ in Hamilton's rule. The theory of parent–offspring conflict (Trivers, 1974) and parental manipulation (Alexander, 1974) or social dominance theory (Emlen, 1997a) point to the divergent genetic interests of parents and offspring and the ability of parents, who are older and more powerful, to manipulate offspring, making them serve the genetic interests of parents.

How do the Dogon data stack up against these theoretical expectations? In a previous study of 176 children who were followed for 8 years, the addition of one extra child (age 0–10 years) to the extended family increased the odds of child death by 26% (Strassmann, 2000). In the present study (*Methods*), child growth (measured annually) decreased linearly as the number of maternal siblings increased (Fig. 14.1). Thus,

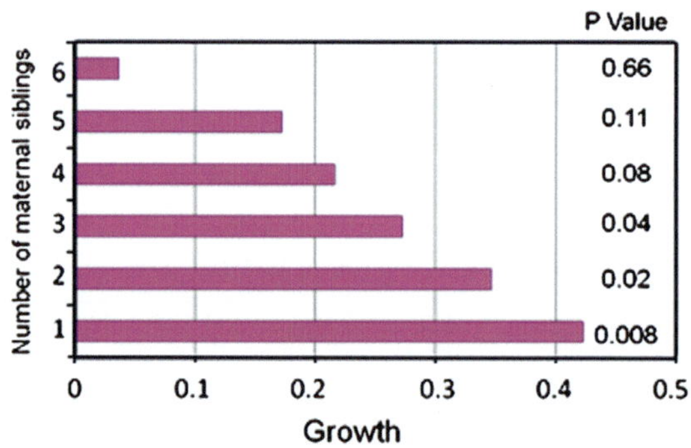

FIGURE 14.1 Child growth by number of maternal siblings. Growth is measured as the annual change in *z* score for height for age. The reference is seven maternal siblings. The other variables controlled include child's age, year of study, sex, age by sex interaction, village of residence, survival of the paternal grandparents, sex by survival of the father's mother interaction, sex by survival of the mother's mother interaction, birth order (mother's offspring), standardized wealth rank, and mother's marital rank ($n = 572$ and the ages of the children were 41–98 months).

children were stunted in direct proportion to maternal family size. Sibling competition can also be seen in data on female lifetime reproductive success, which showed a tradeoff between the number of offspring born and the proportion that was successfully reared (Strassmann and Gillespie, 2002). Reproductive success reached a maximum of 4.1 offspring who survived to age 10 years at 10.5 maternal live births, and the 95% confidence limits around the maximum were 3.4–4.8 live births (Strassmann and Gillespie, 2002). Because of child mortality, the reproductive success of women who had 10 or 11 live births was not significantly greater than that of women who had 6 live births (Strassmann and Gillespie, 2002). In support of life history theory, the prevailing effect of siblings on each other was competition and not cooperation.

Behavioral scan data on childcare and work performed in the agricultural fields and the village show differences by sex and age (Fig. 14.2). Girls aged 5–9 years were observed doing childcare in the fields as often as adult women of reproductive age (Fig. 14.2*A*). Boys did relatively little childcare (Fig. 14.2*A* and *B*). Children of both sexes also performed non-childcare chores in the fields and the village (Fig. 14.2*C* and *D*). Children might prefer to play, but they are constrained by social norms, especially parental expectations and occasional scolding. Apparently, reprimands are also needed in the Maya (Kramer, 2005b). In the Efe, the services of fostered orphan boys help to offset the costs of their sustenance (Ivey, 2000; Ivey et al., 2005). Dogon children depend on parents for nutrition and survival in a harsh environment; their labor is obligatory, and I propose that it reflects parental manipulation (Alexander, 1974). Children's work is also cooperative; however, it is not directly analogous to cooperative breeding in birds and other mammals because the latter refers to the postponement of personal reproduction caused by constraints on independent breeding opportunities (Emlen, 1997a). Human children who work and tend siblings are constrained not only by their economic insufficiency but also by their subordinance to parents as well as their reproductive immaturity.

In the context of parental coercion, Dogon children can become alloparents who habitually carry heavy infants on their backs. Sometimes, a Dogon allomother is an orphan, and the infant that she carries is the younger of two twins whose prospects for survival are poor. The obligation to take care of siblings, especially after the birth of twins, was cited by children as a reason for being unable to stay in school. If there is parent–offspring conflict over allomothering, then the reproductive success of adults who served as alloparents in childhood should be lower than that of individuals who were comparatively free of such responsibilities. Future research on cooperative breeding in humans should test this prediction to assess the role played by parental coercion.

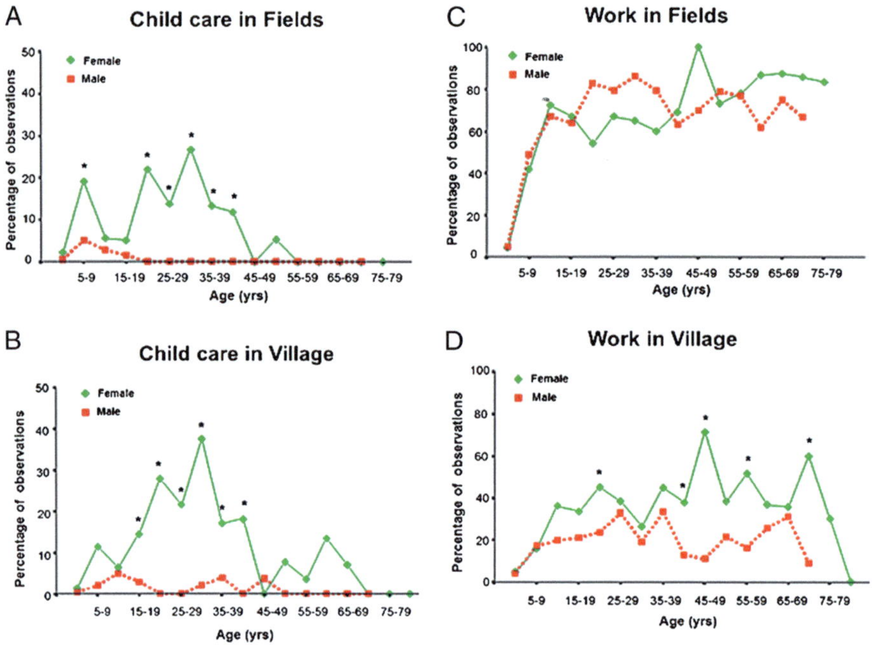

FIGURE 14.2 Childcare and work in the fields and village by sex and age. Solid lines females; dashed lines males. $^*P < 0.05$ for the sex difference. (*A*) Child care in fields. At age 5–9 years, and 20–44 years females did significantly more child care than males. At other ages the difference was not statistically significant. (*B*) Child care in village. At age 15–44 years females did significantly more child care than males. (*C*) Work in fields. At no age was there a sex difference in the percentage of work observations over total observations for males versus females in the fields. Child care is excluded from the definition of "work." (*D*) Work in village. From age 10 years onward, females generally did more nonchild care work in the village than did males with significant differences in five age groups (asterisks).

Work–Eat Groups

The work–eat group (WEG) is the functional family unit or economic group, and it is composed of the people who work together and eat from the same harvest during the millet growing season (Strassmann and Warner, 1998; Strassmann, 2003). The cooperative breeding hypothesis predicts that WEG members participate in alloparental care, especially sisters and grandmothers (Hrdy, 2009). I investigated the popular adage that it takes a village to raise a child (Hrdy, 2009), which implies diffuse and shared responsibility for children across closer and distant kin or even nonkin. Specifically, I asked if children had better survival in extended families than nuclear families. If extended families are preferable, then children should survive better when there are more married adults in the WEG. Kin selection theory does not make a prediction about WEG size per se, but it does predict increased conflict in WEGs that have more asymmetries in genetic relatedness, such as in polygynous WEGs. Kin selection theory (Hamilton, 1964a) also predicts a linear relationship between the child's relatedness to the WEG boss and his or her survival.

I tested these predictions using longitudinal data on the hazard of death in the first 5 years of life (*Methods*) and found that the number of married adults in the WEG was not significantly associated with child mortality (Table 14.1). I infer that, in the Dogon, extended families are not at any advantage or disadvantage compared with nuclear families. Interestingly, the dependency ratio (number of children less than 10 years of age divided by the number of married adults) was also not a significant predictor of the risk of death. The one person in the WEG

TABLE 14.1 Predictors of Death by Age 5 Years (n = 3,000 person-years and n = 165 deaths)

Variable	P	Hazard Ratio	95% Confidence Limits
Mother[a]	0.0113	0.242	0.081–0.726
Father[a]	0.6899	1.511	0.199–11.471
Year 1[b]	0.0002	2.316	1.495–3.589
Year 2[b]	<0.0001	2.500	1.644–3.802
Male[c]	0.0595	0.740	0.541–1.012
WEG boss			
Father's father[d]	0.9647	1.011	0.610–1.677
Mother's father[d]	0.1425	0.535	0.232–1.234
Father's brother[d]	0.5266	1.234	0.644–2.364
Father's father's brother[d]	0.1349	1.719	0.845–3.497
Unrelated[d]	0.4339	0.553	0.126–2.436
Miscellaneous[d]	0.2434	0.646	0.309–1.347
Grandparents			
Father's mother[a]	0.0007	1.857	1.305–2.693
Father's father[a]	0.0485	1.556	1.003–2.415
Mother's mother[a]	0.7668	0.949	0.672–1.341
Mother's father[a]	0.6570	1.075	0.781–1.479
Number of married adults in WEG	0.5651	0.988	0.948–1.030

Notes: This model includes the number of married adults in the WEG. A hazard ratio greater than 1.0 means an increased risk of death, and a hazard ratio less than 1.0 means a decreaseed risk of death. Because child's age was the time axis in the analysis, it was not used as a predictor in the model.

[a]Reference is dead. [b]Reference is year 3. [c]Reference is female. [d]Reference is father.

TABLE 14.2 Predictors of Death by Age 5 Years (n = 2,933 person-years and n = 161 deaths)

Variable	P	Hazard Ratio	95% Confidence Limits
Mother[a]	0.0097	0.230	0.076–0.701
Father[a]	0.7093	1.468	0.195–11.045
Mother's age	0.3458	0.987	0.962–1.014
Year 1[b]	0.0001	2.420	1.533–3.822
Year 2[b]	<0.0001	2.718	1.756–4.206
Male[c]	0.0805	0.750	0.543–1.036
Wealth	0.0532	1.134	0.998–1.287
WEG polygyny[d]	0.0196	—	—
Village 9[e]	0.3659	—	—
Polygyny × village 9[f]	0.0626	—	—
WEG boss			
Father's father[g]	0.8824	1.038	0.631–1.709
Mother's father[g]	0.0899	0.473	0.199–1.124
Father's brother[g]	0.3400	1.378	0.713–2.662
Father's father's brother[g]	0.0424	1.960	1.023–3.755
Unrelated[g]	0.5591	0.633	0.136–2.937
Miscellaneous[g]	0.3540	0.689	0.313–1.516
Grandparents			
Father's mother[a]	0.0009	1.926	1.309–2.832
Father's father[a]	0.0649	1.544	0.974–2.450
Mother's mother[a]	0.6917	0.931	0.653–1.327
Mother's father[a]	0.6366	1.083	0.778–1.508

Notes: A hazard ratio greater than 1.0 means an increased risk of death, and a hazard ratio less than 1.0 means a decreased risk of death. Because child's age was the time axis in the analysis, it was not used as a predictor in the model. There was no significant interaction between offspring sex and grandparental survival.

[a]Reference is dead. [b]Reference is year 3. [c]Reference is female. [d]This variable shows the effect of polygyny on the hazard of death in villages 1–8. [e]Reference is villages 1–8. [f]This variable shows the interaction of polygyny and village. [g]Reference is father.

who emerged as overwhelmingly important for child survival was the child's own mother (Table 14.2). Children whose mothers were alive faced a hazard of death that was 77% lower than that of children whose mothers were dead ($P < 0.01$). In the Dogon, it is mothers alone who are critical for getting children past the early-life bottleneck in survival. The survival of the father was not a significant predictor of the hazard of death for Dogon children in early childhood ($P = 0.71$) (Table 14.2).

The hazard of death was more than twofold greater in year 1 ($P = 0.0001$) and year 2 ($P < 0.0001$) of the study than in year 3 (the reference year). In year 3, a major drought decreased insectborne illnesses such as malaria. The hazard of death for boys was 25% lower than for girls, a difference that was not quite statistically significant ($P = 0.08$). WEG wealth was a rank variable from 1 (rich) to 6 (poor). A change in wealth of one rank (from richer to poorer) increased the hazard of death by 13% ($P = 0.05$) (Table 14.2). I defined WEG polygyny as the ratio of married women to married men in the WEG. There was an interaction between WEG polygyny and the child's village of residence ($P = 0.06$). In villages 1–8, the hazard of death was significantly higher in polygynous WEGs ($P = 0.02$). Village 9 was exceptionally large and wealthy, and only in this village did polygyny have no adverse impact (Table 14.2). In regard to child survival, it was the polygyny index for the WEG as a whole that mattered and not the marital status (sole wife or first, second, or third wife) of the child's own mother.

To explore the mechanisms that underlie the decrease in child survival under polygyny in the more traditional villages (1–8), I compared the growth of Dogon children to the World Health Organization's healthy international reference population (World Health Organization, 2005). During their first 18 months, the children ($n = 474$ observations) fell behind the international reference population at the rate of ~1 SD per year, which is indicative of severe stunting. In the age group 19–40 months, most of the children ($n = 432$ observations) were continuing to fall behind the reference population, but the rate of stunting had slowed down and was now less than 0.1 SD per year. From 41 to 98 months ($n = 572$ observations), the children were experiencing catch-up growth, and the mean change in z score for height for age between successive years was positive. Controlling for other significant predictors, the rate of stunting was significantly lower for the children of sole wives than the children of first wives and intermediate for the children of second, third, and fourth wives (Fig. 14.3). It is unclear why mother's marital status mattered for children's growth, whereas their survival was only impacted by the polygyny of the WEG as a whole. However, it is clear that one needs to examine fam-

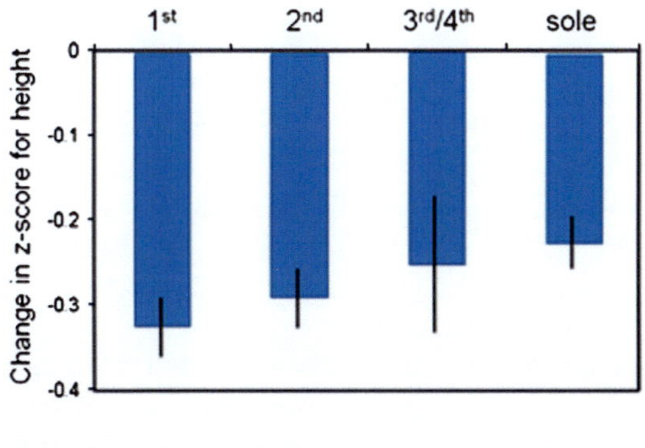

FIGURE 14.3 Child growth by mother's marital rank. Children of first wives were falling behind the healthy reference population at a faster rate than children of sole wives.

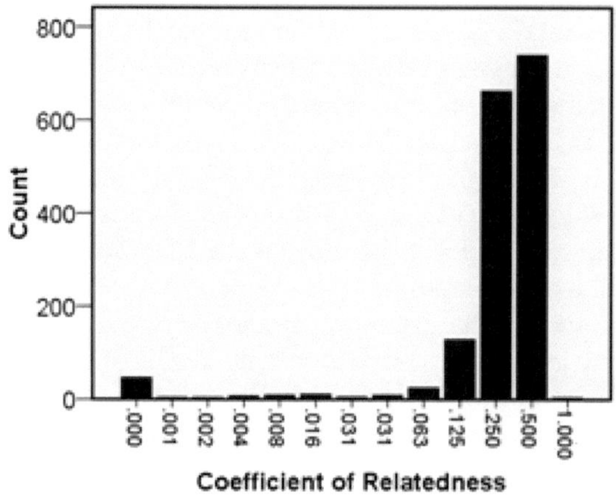

FIGURE 14.4 The number of children by their coefficient of relatedness to their work–eat group (WEG) boss.

ily structure, including such features as polygyny, to understand the social dynamics and patterns of parental and alloparental investment in families.

A major feature of WEG structure is the genetic relatedness among the various family members. The coefficient of relatedness of the children in the prospective cohort study to the boss of their WEGs was quite variable (Fig. 14.4). Most commonly, the WEG boss was the child's father ($r = 0.5$) or grandparent (usually the paternal grandfather; $r = 0.25$), but sometimes, the WEG boss was the paternal great uncle ($r = 0.125$) or rarely, an unrelated man ($r = 0$) who was married to a female relative. Based on kin selection theory (Hamilton, 1964a), I predicted a linear relationship between the child's relatedness to the WEG boss and his or her survival. Instead, there was no significant difference in child survival when the WEG boss was the child's father, paternal grandfather, maternal grandfather, or father's brother (Fig. 14.5 and Table 14.2). Evidently, children whose coefficient of relatedness to the WEG boss was 0.25 (uncles and grandfathers) fared as well as children whose coefficient of relatedness to the WEG boss was 0.50 (fathers). However, if the WEG boss was the child's paternal great uncle (related to the child by 0.125), then the hazard of death by age 5 years was twofold higher than if the WEG boss was the child's father ($P = 0.04$) (Table 14.2).

A possible explanation for these results is that the grandfather is symmetrically related to all of his grandchildren, and he may treat them equally, whereas a paternal great uncle has descendants who are much more closely related to him (0.5 vs. 0.125) and who are better targets for his investment according to Hamilton's rule (Hamilton, 1964a). Grandfathers are only one-half as related to grandchildren as fathers are to children, but the disadvan-

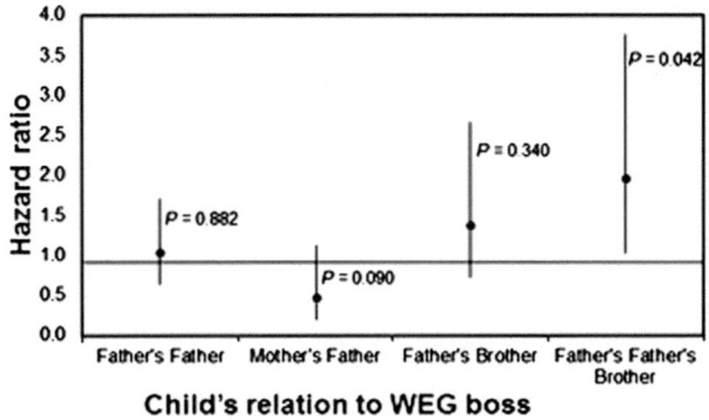

FIGURE 14.5 The hazard of death by age 5 years in relation to the child's relatedness to the WEG boss ($n = 2{,}933$ person-years and $n = 161$ deaths). The child's father as WEG boss is the reference category. A hazard ratio greater than 1.0 means an increased risk of death if the WEG boss is the indicated person instead of the child's father. A hazard ratio less than 1.0 means a decreased risk of death if the WEG boss is the indicated person instead of the child's father. Error bars are 95% confidence limits. Table 14.2 shows the other variables that were controlled.

tage of having a grandfather as WEG boss may be offset by the advantage of having a WEG boss who is older and who commands more authority in this gerontocratic society. Some fields are inherited father to son, but others are owned by the patrilineage under a system of gerontocratic control that gives the eldest WEG boss the choice parcels and the most junior WEG boss the worst parcels (Bouju, 1984). The finding that runs counter to kin selection theory is the lack of difference in survival between children whose father was the WEG boss vs. children whose WEG boss was the paternal uncle (father's brother). This result echoes the finding that the death of the father did not jeopardize child survival.

WEGs are seasonal and do not hold together for the entire year. After the millet has been harvested, WEGs often break up for the onion gardening season. Each person waters their own individual onion garden, and a collaborative effort is needed only for pounding the bulbs and stems. The onion mash is rolled into balls that are spread out in the sun to dry. The desiccated onion balls are easier to store and transport than the fresh bulbs, and they are an export commodity that is shipped to Bamako and Abidjan. People invariably said that they preferred the onion work because there is no problem of free riders: every married adult (male or female) reaps the reward for their own efforts. Children work in the onion gardens as well, but their yield belongs to their fathers.

Over time, larger WEGs tend to fracture into multiple smaller WEGs. Kin selection theory (Hamilton, 1964a) predicts that WEG fission

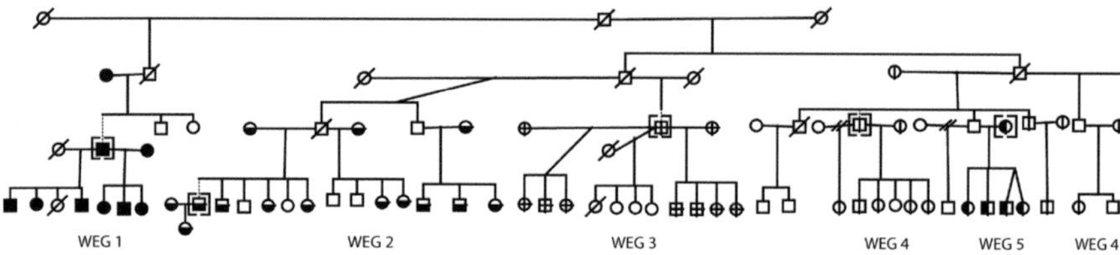

FIGURE 14.6 WEG dissolution over time. In 2000, there were only two WEGs as the members of WEGs 2–5 worked together. By 2010, there were five WEGs. This family segment illustrates the typical pattern wherein half brothers separate after the death of the patriarch, followed ultimately by the separation of full brothers. White icons indicate persons who lived outside the patrilineage in 2010. WEG bosses are indicated in brackets. Slashes through icons indicate that the person was deceased in 2010.

occurs along kinship lines. I tested this hypothesis using data on WEG stability from pedigree data for 29 Dogon patrilineages in nine villages. These patrilineal pedigrees are based on genealogical interviews of 1,285 adult males. Fig. 14.6 shows three generations of descendants of a single Dogon man in one of the patrilineages (the full pedigree has 11 generations and is too large to reproduce here). At the time of the census in the year 2000, the patriarch's son by one wife had split off from his two sons by his other wife, producing a small WEG (Fig. 14.6, WEG1) and a large WEG containing everybody else. By the year 2010, the large WEG had further split into four separate WEGs (WEGs 2–5). The typical pattern is that when the patriarch (grandfather) is alive, he requires his sons and their wives and children to work together. After he dies, there is a brief grace period, and then, the half brothers separate from each other. More time elapses, and then, even the full brothers often split apart (Strassmann, 2003). This pattern is consistent with kin selection theory (Hamilton, 1964a), which predicts greater discord among half than full brothers. It also shows the coercive role of the grandfather in eliciting cooperation from his sons.

Because postmarital residence is patrilocal, women are not genetically related to anyone in the WEG except their own offspring. In interviews (n = 113 women and n = 77 men), women invariably disliked working with their husband's extended family and preferred smaller WEGs. Women were said to instigate WEG fissioning by quarreling with their sisters-in-law. Tensions between cowives extend to their children and often lead to the rupture of the WEG in the next generation (Strassmann, 2003). Although men usually assigned women the blame, it is unlikely that WEG fissioning is entirely caused by the women. A common pattern was for a younger brother who had monetary savings from working in the city to split off from his older brother so as to keep control of his wealth. Brothers are first-degree relatives ($r = 0.5$), but Hamilton's rule takes into consideration not only relatedness but also costs and benefits. Although WEGs are cooperative units, they are unstable and experience internal conflicts of interest. Kin selection theory predicts important aspects of WEG dynamics such as the conflict among cowives and sisters-in-law and the circumstances that trigger WEG fissioning. The person in the WEG who is most important for child survival is the child's own mother, a finding also seen in other demographic datasets (Sear and Mace, 2008) as well as in primate species that are not cooperative breeders.

Grandparents and Grandoffspring

The hypothesis that humans are cooperative breeders has focused, in particular, on grandparental investment (Hawkes et al., 1997, 1998; Beise, 2005; Gibson and Mace, 2005; Hawkes and Jones, 2005; Hrdy, 2005a, 2009; Leonetti et al., 2005; Hill and Hurtado, 2009; Coall and Hertwig, 2010; Kaptijn et al., 2010; Strassmann and Garrard, 2011). The Dogon data provide an opportunity to examine the survival status of children in relation to the survival status of the four kinds of grandparents (father's mother, father's father, mother's mother, and mother's father). In the presence of controls for other predictors of child mortality, the hazard of death for Dogon children was twofold higher if the father's mother was alive rather than dead ($P = 0.0009$) and 54% higher if the father's father was alive ($P = 0.06$) (Fig. 14.7 and Table 14.2). The hazard of death was not influenced by the survival status of the maternal grandparents.

Given that residence is with the husband's and not the wife's parents, I predicted that older paternal grandparents become a drain on family resources and consume more than they produce. In testing this prediction, I used the age of the father as a proxy for the age of his parents. Children whose father's were older (and who presumably had older paternal grandmothers) were at the greatest survival disadvantage if the paternal grandmother was living (Table 14.3). The Dogon seem to take this pattern into account, because they kick the paternal grandmother out of the WEG after the death of her husband (Strassmann, 2000, 2003). In Dogon society, an old woman whose husband is deceased is perceived as a liability rather than as a valuable alloparent. Old women work physically harder than old men (Fig. 14.2*D*), but the old men have a valuable asset: They wield power.

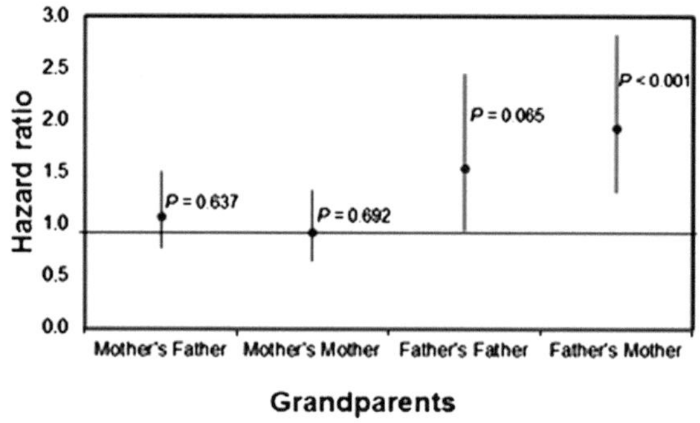

FIGURE 14.7 The hazard of death by age 5 years in relation to the survival status of the child's grandparents (dead is the reference category; n = 2,933 person-years and n = 161 deaths). A hazard ratio greater than 1.0 means an increased risk of death if the grandparent is alive (rather than dead). Error bars are 95% confidence limits. Table 14.2 shows the other variables that were controlled. The hazard of death was twofold greater if the father's mother was alive.

The old ladies must return to their natal patrilineages and work to support themselves with a small parcel of land that they cultivate alone or with the help of a daughter's daughter (Strassmann, 2003). A woman's two firstborn offspring are raised by their maternal grandparents, so boys sometimes also live matrilocally. In the case of boys, this arrangement is usually temporary, and sons eventually go to their fathers. I investigated the growth in height of these matrilocal children (n = 107) in relation to both their sex and the survival status of their four grandparents. The data are adjusted for other significant predictors of growth. In each year of the study, the family structure variables were used to predict the change in a child's height-for-age z score from one year to the next (*Methods*). Girls tended to grow at a faster rate if their maternal grandmother was dead rather than alive (Fig. 14.8), presumably on account of the work that girls perform. Little girls primarily fetch water, but as they grow older, they take on such tasks as weeding gardens, pounding millet, and helping to gather firewood. Boys performed less work for their maternal grandmothers, and their growth rates were unaffected by their maternal grandmothers' survival status (Fig. 14.8).

The relationship between growth and grandparental survival status in the children who lived patrilocally is shown in Fig. 14.9. Because there were more children who lived patrilocally (Fig. 14.10), it was possible to disaggregate the data into three age groups. In the youngest children (ages 0–18 months; n = 474), the survival status of the father's father,

TABLE 14.3 Predictors of Death by Age 5 Years (n = 2,762 person-years and n = 155 deaths)

Variable	P	Hazard Ratio	95% Confidence Limits
Mother[a]	0.0050	0.206	0.068–0.620
Father[a]	0.6597	1.546	0.222–10.755
Father's age	0.0696	—	—
Father's age × father's mother	0.0116	—	—
Year 1[b]	0.0007	2.214	1.395–3.513
Year 2[b]	<0.0001	2.484	1.603–3.851
Male[c]	0.1535	0.786	0.565–1.094
Wealth	0.0616	1.128	0.964–1.281
WEG polygyny[d]	0.0510	—	—
Village 9[e]	0.4676	—	—
Polygyny × village 9[f]	0.0934	—	—
WEG boss			
Father's father[g]	0.6160	1.140	0.683–1.904
Mother's father[g]	0.5642	0.727	0.245–2.152
Father's brother[g]	0.4564	1.294	0.657–2.549
Father's father's brother[g]	0.0274	2.107	1.087–4.086
Unrelated[g]	0.5884	0.554	0.065–4.699
Miscellaneous[g]	0.6710	0.832	0.357–1.941
Grandparents			
Father's mother[a]	0.1026	—	—
Father's father[a]	0.1316	1.443	0.869–2.324
Mother's mother[a]	0.9408	0.987	0.691–1.409
Mother's father[a]	0.5745	1.100	0.788–1.537

Notes: The significant interaction of father's age and survival status of father's mother shows that older paternal grandmothers are more harmful than younger paternal grandmothers. Mother's age was omitted because it is collinear with father's age.

[a]Reference is dead. [b]Reference is year 3. [c]Reference is female. [d]This variable shows the effect of polygyny in villages 1–8 in terms of the ratio of married women to married men in the WEG. [e]Reference is villages 1–8. [f]This variable shows the interaction of polygyny and village. [g]Reference is father.

mother's mother, and mother's father was unrelated to grandchild growth rates (Fig. 14.9*A*). However, the survival status of the father's mother affected girls and boys differently. Girls grew better than boys if the father's mother was alive (P = 0.05). Boys grew better if the father's mother was dead rather than alive (P = 0.04) (Fig. 14.9*A*).

In the next age group (ages 19–40 months; n = 432), the major finding was that boys were catching up in height to the international reference population if the father's mother was dead but continuing to fall further behind if she was alive (P = 0.001) (Fig. 14.9*B*). Additionally, when the father's mother was dead, boys grew at a rate that was significantly better than that of girls (P = 0.0003). In this age group, girls were continuing to fall behind the reference population, regardless of the survival status of the paternal grandmother (Fig. 14.9*B*).

In the oldest age group (ages 41–98 months; $n = 559$), the children were experiencing catch-up growth across the board (Fig. 14.9*C*). The catch-up growth of girls was greater if the father's mother was alive rather than dead ($P = 0.003$). Moreover, if the father's mother was alive, the catch-up growth of girls was greater than that of boys ($P = 0.0002$) (Fig. 14.9*C*). Interestingly, girls grew at a faster rate if their mother's mother was dead rather than alive ($P = 0.003$), echoing the results for the girls who lived matrilocally. When the mother's mother was dead, girls also grew faster than boys ($P = 0.0004$) (Fig. 14.9*C*).

Interactions between sex, grandparental survival, and child survival or growth have also been found in other human populations (Jamison et al., 2002; Gibson and Mace, 2005; Sear, 2008). Rather than being a spurious result, it seems that Dogon paternal grandmothers may prefer girls over boys. A possible evolutionary explanation is local resource competition (Clark, 1978). Girls marry out of the patrilineage, whereas boys remain home to compete against each other for resources, and therefore, it may make sense to produce fewer boys. In no age group did the survival status (alive or dead) of grandfathers make a difference for growth. It could be that grandfathers are not sufficiently involved in the care and feeding of young children for their presence to make a difference.

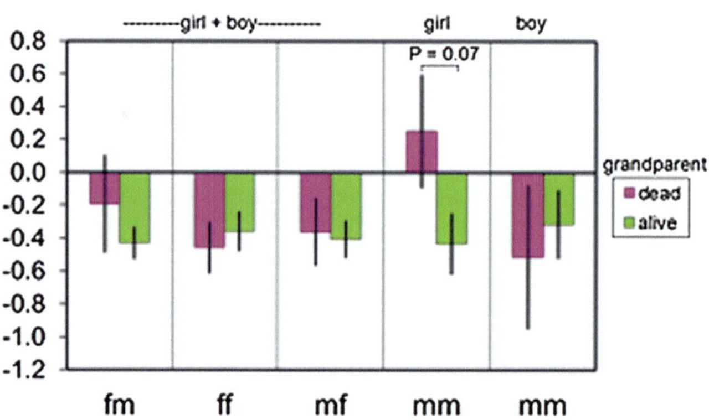

FIGURE 14.8 Growth (change in *z* score for height for age) in children who lived matrilocally ($n = 107$; ages 0–98 months). Other significant predictors of growth were controlled (child's age centered on the mean, year of study, sex, mother's age at child's birth). fm, father's mother; ff; father's father; mf, mother's father; mm, mother's mother.

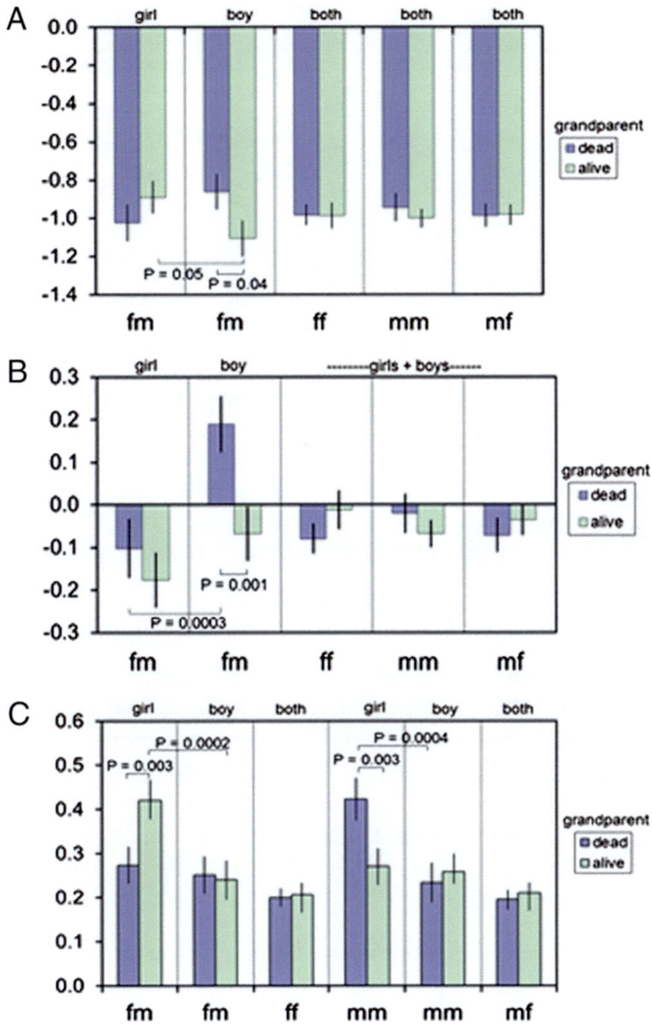

FIGURE 14.9 Growth (change in *z* score for height for age) in children who lived patrilocally in relation to sex and the survival status of the child's grandparents, fm, father's mother; ff, father's father; mm, mother's mother; mf, mother's father. (*A*) Age group 0–18 months (n = 473). (*B*) Age group 19–40 months (n = 432). (C) Age group 41–98 months (n = 559). The analysis controlled for child's age, child's age squared, year, sex, age by sex interaction, nursing vs. weaned, village of residence, birth order (one to seven), number of maternal siblings (one to seven), standardized wealth rank, mother's marital status (fiancée, polygyny, or monogamy), and nursing by mother's marital status interaction.

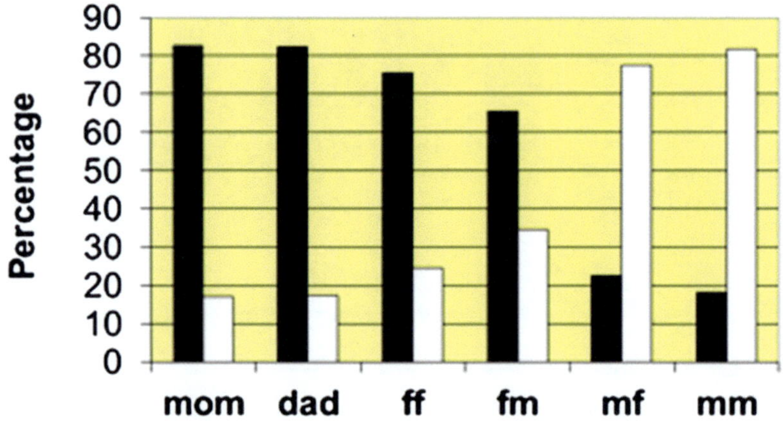

FIGURE 14.10 The percentage of children who lived in the same village as each kind of relative (dead relatives were excluded from the analysis). Black bars represent the children that lived with the relative, and white bars represent the children that did not live with the relative. ff, father's father; fm, father's mother; mf, mother's father; mm, mother's mother.

Maternal grandmothers had an adverse impact on the growth of girls, regardless of whether the girls lived matrilocally or patrilocally. A possible explanation is that growth is slowed by the energetically demanding work that girls perform for their maternal grandmothers. In the literature on grandmothering, it is usually assumed that the most beneficial grandmother is the maternal one (Hawkes et al., 1997, 1998). The data for the growth of Dogon girls point in the opposite direction.

CONCLUSION

If the term cooperative breeding is restricted to human populations in which adult offspring delay or forego reproduction to act as helpers, then the Dogon would not qualify because girls usually initiate reproduction by about age 19 years and boys by age 25 years. It would potentially apply, however, to 19th and early 20th century European farming communities in Ireland and elsewhere that had an unusually high percentage of individuals who postponed reproduction to their thirties or who were permanently celibate. Ireland was akin to a saturated habitat in which opportunities to reproduce were restricted by the availability of farms. It has been suggested that unmarried siblings often emigrated or acted as helpers (Strassmann and Clarke, 1998).

The Dogon pattern is very different. Rather than disqualify the Dogon as cooperative breeders a priori because they do not delay reproduction, I searched for evidence for cooperative breeding behavior in sibling groups, in families, and on the part of grandparents. Siblings competed with each other, producing a tradeoff between the number of maternal siblings and growth and survival. To understand why children work or care for younger siblings, parental coercion is a possibility that needs further investigation. It did not take a village to raise a child; instead, children in nuclear families did as well as children in extended families. The hazard of child mortality to age 5 years was twofold higher if the paternal grandmother was alive rather than dead. In the first 5 years of life, mothers were essential for child survival and could not be replaced by other kinds of kin. These results do not align with the expectations of cooperative breeding theory. Kin selection theory was more helpful for understanding the dialectic between cooperative and competitive interactions in the Dogon, especially when conjoined with parent–offspring conflict and life history theory. The predicted linear relationship between child survival and relatedness of the child to the WEG boss was not supported. However, kin selection theory was helpful for understanding the dynamics of polygyny and WEG instability over time.

METHODS

Subjects

The study population was an entire cohort of 1,700 children in nine Dogon villages who were enrolled between 1998 and 2000. These children have been followed prospectively since their initial enrollment to the present date (2011). The criterion for enrollment was age ≤ 5 years on May 1, 1998, or born during the first 2 years of follow-up study. Subject cooperation for participation in the cohort study was >99.9%. Participation in the anthropometric measurements (height or supine length in centimeters and weight in kilograms) was 91% (1998), 83% (1999), 85% (2000), 81% (2004), 79% (2007), and 85% (2010). The primary reason for nonparticipation was that the child was living outside the study area that year. Approval to conduct this study was obtained from the University of Michigan Internal Review Board, the Malian government, local authorities, and village chiefs and elders. Informed consent (adults) or assent (children) was obtained from the subjects.

WEG Structure

An annual door to door census was conducted of the 652 WEGs in the nine villages. This census asked the identity of the head of the family (WEG boss) and each WEG member's name, parents' names, sex, year of birth, marital status [unmarried, fiancée (tanganu), or married (tanga)], spouses' names, spousal rank order (sole, first, second, third, or fourth wife), wife type [arranged (ya bire), nonarranged (ya kezu), or levirate: married to the deceased husband's brother (ya pani)], and religion (indigenous, Muslim, Catholic, Protestant, or free thinker/agnostic). From these data, other family structure variables were calculated to reflect the social niche of the child (e.g., number of maternal and paternal siblings, ratio of married women to married men in the WEG, and total number of children aged 10 years and younger). Data were also gathered on the survival status of the children's parents and grandparents (dead or alive) at the time the children were born and at the time of the interview.

WEG Wealth

Forty-one informants who lived in the nine villages independently ranked the relative wealth of each WEG in their natal village, producing strong agreement among informants [Cronbach's α: $C = 0.89 \pm 0.06$ (1999) and $C = 0.93 \pm 0.03$ (2000)]. After doing their individual rankings, the informants also met together to produce a consensus ranking.

Children's Survival

Because most child mortality in the Dogon occurs by age 5 years (US Department of State, 2010), I was primarily interested in survival in early childhood. Therefore, in studying the relationship between family structure and child survival, I used the field data from the years 1998–2001. The statistical analysis employs a Cox proportional hazards model to analyze survival as a function of time-varying covariates (Cox, 1972; Allison, 1995). The covariates measured in a given year served as predictors of survival from that year to the next (e.g., covariates measured in 1998 were used to predict survival to 1999). After the initial year (1998), I had 3 years of follow-up (1999, 2000, and 2001). The dependent variable in the statistical model was the duration of survival time for each child across all years that he or she was in the study between 1998 and 2001. If a child survived to the end of a given year, I calculated her survival time as the child's age at the end of the year minus her age at the start of the year. If she died during the year, her survival time was her age at death minus her age at the start of the year. If she was lost to follow-up, her survival time was calculated as her age at midyear minus her age at the start of the year. If she was born and died between annual censuses, her survival time was her age at death. The survival analyses were carried out using Proc Phreg in the statistical software SAS 9.2 (SAS Institute, Inc., 2008), which allowed for control for multiple observations of the same child. The statistical models included a random effect for mother.

Behavioral Data

From 1986 to 1988, an instantaneous behavioral scan was conducted during daylight hours in the fields and the initial study village in all months of the year. The date, time, identification, and activity of each person were recorded (n = 5,097 observations).

Children's Growth

I calculated height-for-age z scores in relation to the international World Health Organization (WHO) standard that is appropriate for breastfed children in developing countries (World Health Organization, 2005). These z scores quantify the distance (measured in SDs) of a given child's height relative to the mean for the reference population. I calculated growth as the change in height-for-age z scores. The covariates in a given year were used to predict the change in the child's height-for-age z score from that year to the next. The statistical analysis used a linear mixed model (Littell et al., 1996) that (*i*) used a repeated measures design to take into account the autocorrelation in the growth of a given child between years and (*ii*) included the mother of the child as a random effect to take into account the correlation among maternal siblings. The analysis was carried out using Proc Mixed in the statistical software SAS 9.2 (SAS Institute, Inc., 2008).

WEG Fissioning

Males (n = 1,218) belonging to 29 patrilineages in 10 villages provided genealogical information for their paternal and maternal ancestry as far back as they could remember. From the oral histories, I made patrilineal pedigrees in the program Progeny (Progeny Software, LLC, 2007) that have a depth of up to 11 generations from the youngest generation to the common ancestor. These pedigrees contain data on WEG composition and changes over time.

ACKNOWLEDGMENTS

I thank the Dogon for their generous participation in this research, including my Dogon field assistants for their valuable help. I am grateful to the Malian government for permission to conduct this study and the following undergraduate and Master degree-level field assistants: Sylvie Moulin, Thomas Stevenson, Betsy Abrams, Sera Young, Brendan Hug, Lindsey Boynton, Joy Rowe, Riley Bove, Anne Lyne, Jean Baptiste, David Plate, Joanna Rogers, Robin Martz, Christine Sauve, and Russell Herrold. I thank Kathy Welch for statistical advice and Claudius Vincenz, Laura Betzig, and two anonymous reviewers for helpful comments. This research was supported by the Louis S. B. Leakey Foundation, National Science Foundation Grants SBR 9727229 and BCS 1029056, the American Philosophical Society, and National Institutes of Health Grant 09-PAF00653.

15

Extent and Limits of Cooperation in Animals

DOROTHY L. CHENEY

Individuals in many animal species are strongly motivated to form close social bonds and to attend to the social interactions of others. Some animals may also recognize other individuals' intentions and simple mental states. Such curiosity appears to be adaptive, because it enables observers to learn about others' status and relationships and to anticipate future events without direct participation. However, many questions remain unresolved. In particular, it remains unclear whether animals keep track of favors given and received when interacting with others, and whether they rely on memory of past cooperative acts when anticipating future ones. Primates appear to possess many of the cognitive abilities required for human-like contingent cooperation. However, most investigations of captive primates have indicated that cooperation is seldom contingency-based, and that interactions are not influenced by inequity aversion or sensitivity to cheaters. In contrast, several experiments with nonprimates have found that animals can take into account recent interactions when supporting others, suggesting that the apparent rarity of contingent cooperation in primates may not stem from cognitive constraints. Instead, individuals may tolerate short-term inequities in favors given and received because most cooperation occurs among long-term reciprocating partners.

Department of Biology, University of Pennsylvania, Philadelphia, PA 19104. E-mail: cheney@sas.upenn.edu.

> An anthropomorphous ape, if he could take a dispassionate view of his own case, ... might insist that they were ready to aid their fellow-apes of the same troop in many ways, to risk their lives for them, and to take charge of their orphans; but they would be forced to acknowledge that disinterested love for all living creatures, the most noble attribute of man, was quite beyond their comprehension.
>
> Charles Darwin, *The Descent of Man*, 1871, p. 105

Humans have for centuries sensed that we share with animals the motivation to form close, enduring social bonds. Recent research has not only confirmed these intuitions but has also begun to uncover the many fitness benefits of such bonds (Silk and House, Chapter 16, this volume). Nevertheless, despite many similarities in patterns of cooperation between humans and other animals, there are also important differences in its quality and scope. The reasons for these differences remain topics of debate, in large part because we still do not understand the full range of animals' cognitive abilities, in what ways these abilities differ from humans', and how these abilities contribute to the formation of cooperative bonds. Many animals share with humans the ability to monitor other individuals' friendships and animosities, to remember the nature of recent interactions, and perhaps also to recognize other individuals' motivations and intentions. Whether they recognize more complex mental attributes like the intent to deceive, however, remains unclear, as does the extent to which animals share humans' sometimes hyperbolic motivation to engage others in cooperative ventures.

RECOGNITION OF OTHER ANIMALS' RELATIONSHIPS

Many social animals live in groups containing both kin and nonkin, in which interactions are simultaneously competitive and cooperative and in which individuals maintain differentiated relationships with a subset of group members. To navigate through this complex network of relationships, it seems essential to be able to monitor not only one's own interactions but also the interactions of others. The ability to acquire and use information about other individuals' social relationships permits individuals to assess the strength of allies and opponents, to reconcile with opponents, and to choose mates, and it appears to be under strong selective pressure. Indeed, there is now an extensive literature indicating that animals are highly motivated to learn about other individuals' relationships and competitive abilities. Knowledge of other individuals' dominance ranks has been demonstrated in a variety of species, including not only primates and other social animals like pinyon jays

[*Gymnorhinus cyanocephalus* (Paz-Y-Miño et al., 2004)] and hyenas [*Crocuta crocuta* (Engh et al., 2005)], but also in less social territorial birds and fish [e.g., Oliveira et al. (1998), Peake et al. (2002), Grosenick et al. (2007)]. When joining a coalition, for example, hyenas and monkeys selectively recruit or support the higher-ranking of two combatants [e.g., Silk (1999), Schino et al. (2006); reviewed in Cheney and Seyfarth (2007)]. Capuchin monkeys (*Cebus capucinus*) selectively recruit allies who both rank higher than their opponents and have a closer bond with themselves than with their opponent, indicating that they are able to compare the bond between the ally and themselves with the bond between the ally and their opponent (Perry et al., 2004). In playback experiments involving wild baboons (*Papio hamadryas ursinus*), a sequence of calls that mimics a higher-ranking opponent threatening a lower-ranking animal elicits little response from listeners, but if the individuals' roles are reversed, the response is significantly stronger—presumably because the rank-reversal sequence violates the listener's expectations (Cheney et al., 1995; Bergman et al., 2003; Kitchen et al., 2005).

The ability to eavesdrop on the social interactions of others enables individuals to acquire knowledge about another's competitive ability and probable allies without directly challenging him. In nonsocial birds and fish, males use information acquired through eavesdropping when deciding whether to challenge an intruder [e.g., Oliveira et al. (1998), Peake et al. (2002); see Paxton et al. (2010) for similar data on rhesus macaques, *Macaca mulatta*]. Similarly, female chickadees (*Poecile atricapillus*) assess their mate's relative dominance status by attending to his singing contests with neighboring males. Females mated to males who are dominated in such contests are subsequently likely to solicit extra-pair copulations from apparently more dominant neighbors (Mennill et al., 2002).

Monkeys also recognize the close bonds that exist among others. In vervets (*Chlorocebus aethiops*) and macaques, an individual who has just been involved in an aggressive interaction will often redirect aggression by attacking a close relative of her opponent (Cheney and Seyfarth, 1990; Judge, 1991). Similarly, if a female baboon hears a call sequence that mimics a fight between one of her own close relatives and the close relative of a more dominant female, she will subsequently avoid that female (Cheney and Seyfarth, 1999). Playback experiments have also demonstrated that low-ranking male baboons monitor the status of other males' sexual consortships to take advantage of opportunities to mate "sneakily" (Crockford et al., 2007).

If a baboon receives aggression from another and then, minutes later, hears a "reconciliatory" grunt from a previously uninvolved animal, the listener's response to the grunt depends on the relationship between the calling animal and the listener's opponent. If the caller is a close matrilineal relative of the opponent, the listener is subsequently more likely to approach her recent opponent and to tolerate her opponent's approach than she is if she hears the grunt of an animal unrelated to her opponent. Subjects act as if they infer that they are the target of the vocalization even though they have not recently interacted with the signaler, but with her relative. They therefore treat the call as a reconciliatory signal that functions as a proxy for reconciliation with the opponent herself (Wittig et al., 2007). They could do so only if they recognize the close bond that exists between the two females. A similar phenomenon occurs among chimpanzees (*Pan troglodytes*), whereby the behavior of bystanders and victims following aggression depends both on their own relationships with the combatants and on their perception of the relationship between the other animals involved (Wittig and Boesch, 2010).

To cite another example, chimpanzees often scream when involved in aggressive disputes. Victims produce acoustically different screams according to the severity of aggression they are receiving. In playback experiments, listeners responded differently to the different scream types (Slocombe et al., 2009). In cases of severe aggression, victims' screams sometimes exaggerated the severity of the attack, but victims gave exaggerated screams only if their audience included at least one listener whose dominance rank was equal to or higher than that of their opponent (Slocombe and Zuberbühler, 2007). Victims seemed to alter their screams depending upon their perception of the relationship between their opponent and their potential allies.

Some progress is beginning to be made in identifying the neural mechanisms underlying knowledge of others' social relationships. In male zebra finch (*Taeniopygia guttata*), for example, hearing another male's song induces activation of a specific group of immediate early genes (Robinson et al., 2008). These genes are activated rapidly and transiently by even brief social experiences, and they influence the transcription of other genes. The genes' expression is linked to the social significance of the song and may function to enable the brain to keep track of the ever-changing social environment (Robinson et al., 2008). Unfamiliar songs elicit a stronger response than familiar songs, and the response is enhanced if the listener is in the presence of another bird. Similarly, when a female cichlid fish (*Astatotilapia burtoni*) observes a preferred mate win a fight against another male, areas in the brain associated with reproduction are activated. If, however, the preferred mate loses a fight, areas in the brain associated with anxiety are activated instead (Desjardins et al., 2010). These changes occur even though the female is only observing the interactions. Such eavesdropping may permit observers to anticipate changes in the social environment without having to experience them first.

Further supporting the hypothesis that social skills have been under strong selective pressure across taxa, there is some indication in mammals that more social species show higher degrees of encephalization than less social species (Schultz and Dunbar, 2010). Sociality may even affect relative brain size within species. In paper wasps (*Polistes dominulus*), for example, there is a significant increase in the size of the antennal lobes and collar in females that nest colonially with other queens, as opposed to solitary breeders (Ehmer et al., 2001). This increase in neural volume may have been favored because sociality places increased demand on the need to discriminate between familiar and unfamiliar individuals and to monitor other females' dominance and breeding status.

In sum, knowledge of other individuals' relationships has been widely documented in many species by using many different techniques [see Shettleworth (2010) for review]. There appears to have been strong selection pressure for passive observational learning in the context of social interactions and for the acquisition of knowledge about other individuals' social relationships. The representations that underlie such recognition undoubtedly differ from one species to the next, and certainly differ from humans' more explicit representations, but there is no doubt that animals acquire and remember information about other animals' relationships and that this knowledge affects their behavior. In principle, this information can be acquired through relatively simple associative processes. The degree to which animals proceed beyond simply recognizing the association between two other animals, however, remains poorly understood. We still do not know, for example, whether a baboon distinguishes among different types of relationships, like "sister" or "daughter," or whether she imbues these relationships with motives and emotions—for example, like "love."

ATTRIBUTION OF INTENTIONS

In the more than 30 years since Premack and Woodruff (1978) posed the question "Does the ape have a theory of mind?" much progress has been made in the study of mental state attribution in animals. Many questions, however, are still unresolved.

Nonhuman primates and other animals are acutely sensitive to other individuals' direction of gaze. When attempting to engage another individual's attention—for example, when recruiting an alliance partner—primates will actively attempt to engage their partner's gaze (Call and Tomasello, 2008). In competitive contexts, rhesus macaques are more likely to attempt to steal food from a human whose eyes are averted than from one whose eyes are not (Flombaum and Santos, 2005), and captive chimpanzees are more likely to approach food that a competitor cannot see than food that the competitor can (Hare et al., 2000). Similarly, when potential competitors are present, ravens (*Corvus corax*) and scrub jays (*Aphelocoma californica*) are more likely to cache food in sites that are out of view or hidden behind barriers than in more open sites [e.g., Emery et al. (2004), Bugnyar and Heinrich (2005), Dally et al. (2006)].

Primates also appear to attribute simple mental states, like intentions and motives, to others. In captivity, apes distinguish between intentional and accidental actions, and they also recognize other individuals' goals (Buttelmann et al., 2007). Under natural conditions, the recognition of others' intentions is most evident in the context of vocalizations, when animals must make inferences about the intended recipient of someone else's calls. Monkey groups are noisy, tumultuous societies, and an individual could not manage her social interactions if she interpreted every vocalization she heard as directed at her. Inferences about the directedness of vocalizations are probably often mediated by gaze direction and relatively simple contingencies. Even in the absence of visual signals, however, monkeys are able to make inferences about the intended recipient of a call based on their knowledge of a signaler's identity and the nature of recent interactions. For example, when female chacma baboons were played the "reconciliatory" grunt of their aggressor within minutes after being threatened, they behaved as if they assumed the call was directed at themselves, as a signal of benign intent. As a result, they were more likely to approach their former opponent and to tolerate their opponent's approaches than after hearing either no grunt or the grunt of another dominant female unrelated to their opponent (Cheney and Seyfarth, 1997). Call type was also important, because subjects avoided their recent opponent if they heard her threat grunt rather than her reconciliatory grunt (Engh et al., 2006). By contrast, if subjects heard a female's threat grunt shortly after grooming with her, they ignored the call and acted as if they assumed that the female was threatening another individual. Thus, baboons use their memory of recent interactions to make inferences about the caller's intention to communicate with them.

In primates, faces and voices are the primary means of transmitting social signals, and monkeys recognize the correspondence between facial and vocal expressions (Ghazanfar and Logothetis, 2003). When rhesus macaques hear one of their own species' vocalizations, they exhibit neural activity not only in areas associated with auditory processing but also in higher-order visual areas, including superior temporal sulcus (Gil-da-Costa et al., 2004). Ghazanfar et al. (2005) explored the neural basis of sensory integration using the coos and grunts of rhesus macaques as stimuli. They found clear evidence that cells in certain areas of the auditory cortex are more responsive to bimodal (visual and auditory) presentation of species-specific calls than to unimodal presentation. Although significant integration of visual and auditory information occurred in trials with both vocalizations, the effect of cross-modal presentation was greater with grunts than with coos. The authors speculate that this may occur because grunts are usually directed toward a specific individual in dyadic interactions, whereas coos tend to be broadcast generally to the group at large. The greater cross-modal integration in the processing of grunts may therefore have arisen because, in contrast to listeners who hear a coo, listeners who hear a grunt must determine whether or not the call is directed at them.

When deciding "Who, me?", then, upon hearing a vocalization or observing an approaching group member, monkeys must take into account the identity of the individual, its direction of gaze (if visible), the type of call given, the nature of their prior interactions with the signaler or her relatives, and the correlation between past interactions and future ones. Learned contingencies doubtless play a role in these assessments. However, because listeners' responses depend on simultaneous consideration of all of these factors, this learning is likely to be both complex and subtle.

ATTRIBUTION OF KNOWLEDGE

Although baboons and other monkeys may be able to recognize other individuals' intentions when inferring, for example, whether or not they are the target of another individual's call, the extent to which animals attribute knowledge, ignorance, and beliefs to others remains controversial. For example, baboons often give "contact" barks when separated from others. When several separated individuals are calling simultaneously, it often appears that they are answering each other's calls to inform each other of their location. Playback experiments suggest, however, that baboons call primarily with respect to their own separation from the group, not their audience's. They "answer" others when they themselves are separated, and they often fail to respond even to the calls of their offspring when they themselves are in close proximity to other group members (Cheney et al., 1996; Rendall et al., 2000). In this respect, the vocalizations of monkeys are very different from human speech, in which we routinely take into account our audience's beliefs and knowledge during conversation.

Some investigators have suggested that animals' attentiveness to gaze direction is an indication that animals recognize what other individuals can and cannot see and hence what they can and cannot know. These arguments are confounded, however, by the possibility that animals use gaze direction to assess not what others know but what they intend to do. As a result, they recognize, for example, that other individuals are motivated to defend food that they are looking at, and less likely to defend food when they are looking away.

Some recent experiments have attempted to avoid this confound by eliminating the possibility that subjects are responding only to their rival's direction of gaze when choosing among food items. Kaminski et al. (2008) presented chimpanzees with the choice of three buckets, two of which contained food. The first bucket was baited in the presence of both the subject and the rival. The second bucket was baited in the presence only of the subject. In the test condition, the subject's view of the apparatus was blocked while the rival was allowed to choose first. In the control condition, the subject chose first. When subjects chose first, they were as likely to choose the bucket that their rival had seen baited as the one he had not. However, when they chose second, they were more likely to choose the bucket that their rival had not seen baited, suggesting they inferred that the rival would have chosen the bucket that he had seen baited. In other words, they acted as if they recognized what their rival knew, based on what he had seen. However, when subjects observed the experimenter mislead the rival by seeming to hide the reward in one bucket but actually putting it in another, they did not distinguish between the rival's true belief and his false belief. The authors conclude that chimpanzees recognize what others know, but not what they believe.

If, as seems likely, chimpanzees and other animals cannot attribute false beliefs to others, this would provide one explanation for their apparent failure to punish cheaters who defect from risky cooperative ventures. If chimpanzees cannot recognize the intent to deceive, they will be unable to distinguish between a partner who fails to cooperate because he is unable to do so, and one who fails to cooperate because he intends to do so, and they will by definition be unable to impose sanctions against such individuals. I return to this question later.

MECHANISMS UNDERLYING COOPERATION

Although it is now clear from a variety of animal species that strong, cooperative bonds enhance longevity and offspring survival (Silk and House, Chapter 16, this volume), the mechanisms that motivate individuals to form such bonds are still far from well understood. Female baboons, for example, do not groom only with close kin and those with whom they share a close social bond; they also groom less regularly with other females. When a close partner dies, they may attempt to establish a close bond with a previously infrequent partner. We hypothesize that knowledge of other individuals' relationships guides the formation of new relationships, but this hypothesis has not yet been tested. Indeed, we still know little about whether or how animals keep track of their social relationships, of cooperative and noncooperative interactions, or of favors given and received.

Furthermore, although it is now clear that both humans and other animals derive reproductive benefits from strong, predictable social relationships, at least some of the mechanisms underlying these relationships are doubtless very different. Human social relationships are imbued with inferences about others' intentions and beliefs, and humans are at times also acutely aware of whether a favor has been returned, or whether a partner has deceived them. The extent to which any animal is capable of similar mental projections remains unknown.

There continues to be debate about the psychological mechanisms that underlie animals' social interactions and relationships. Because we have no direct evidence that animals can plan or anticipate the benefits that might derive from a long-term relationship, a number of investigators have argued that animals' cooperative interactions are motivated only by short-term rewards, such as the opportunity to handle an infant or gain access to food. According to these arguments, social interactions are not founded on long-term patterns of affiliation but are based instead on short-term byproduct mutualism or biological markets motivated by the current value of a potential partner (Noe and Hammerstein, 1994). So, for example, when a female monkey grooms another, she may simply be engaging in a short-term negotiation with a trading partner who controls a desirable commodity, like a young infant (Henzi and Barrett, 2007).

There is no doubt that many social interactions vary with current conditions. Several studies have shown, for instance, that female baboons often groom lactating females to obtain immediate access to their infants (Henzi and Barrett, 2002; Silk et al., 2003b). Similarly, males groom estrous females at higher rates than pregnant or lactating females, and subordinate individuals groom dominant individuals in apparent exchange for tolerance at food sites (de Waal, 1997b; Ventura et al., 2006). In an experiment directly testing the hypothesis that grooming in vervet monkeys is motivated in part by the expectation of immediate reward, Fruteau et al. (2009) manipulated a food container in such a way that it could only be opened by one low-ranking female. Consistent with biological market theory, the rate at which the female subsequently received grooming from others increased significantly. This initial gain, however, decreased after a second subordinate female was allowed to open the container. Thus, grooming appeared to be adjusted according to the relative value of the provider.

Observations indicating that some social interactions are influenced by the current value and supply of alternative trading partners are not inconsistent with evidence that others reflect long-terms patterns of affiliation. Female baboons, for example, form long-term bonds with only a small number of other females; many of their other social interactions may well be initiated or maintained by interactions that depend in part on the current value of commodities. Nevertheless, much grooming occurs in the absence of an immediate reward, and it is seldom evenly balanced between partners within single bouts (Silk et al., 2010a). Despite short-term asymmetries, nonhuman primates form the strongest bonds with those individuals with whom they have the most balanced and reciprocal grooming interactions over extended periods of time (Silk and House, Chapter 16, this volume).

During the past decade, there has also been increasing skepticism about the relevance of contingent-based reciprocity in the social interactions of animals. Because most cooperative interactions like grooming occur between long-term partners (often kin) for whom any single altruistic act may be relatively insignificant, many investigators are now convinced that the sort of reciprocal altruism first proposed by Trivers (1971) may be both rare and fragile in nature (Hammerstein, 2003; Clutton-Brock, 2009). Although there is limited experimental and correlational evidence that animals sometimes rely on memory of recent interactions when behaving altruistically toward others, interpretation has been complicated by a paucity of convincing examples, the absence of important controls in some early tests, and a number of experimental studies seeming to indicate that animals lack the cognitive or empathetic ability to sustain contingent cooperative exchanges.

COGNITIVE CONSTRAINTS

Doubts persist about whether animals possess the cognitive abilities to sustain contingent cooperation. These include the ability to remember specific interactions, to delay reward, to track favors given and returned, to plan and anticipate future outcomes, and to distinguish between cooperators and defectors (Henzi and Barrett, 2007; Stevens et al., 2005; Melis and Semmann, 2010). Some of these objections may be unjustified.

More than 100 years of research on classical conditioning have repeatedly demonstrated that animals are acutely sensitive to contingencies and to the predictive value of different stimuli on outcomes [reviewed in Shettleworth (2010)]. Animals also remember the nature of specific interactions with particular individuals. As already mentioned, for example, if a baboon hears another female's reconciliatory grunt shortly after being threatened by her, she behaves as if the grunt is causally related to the recent fight and directed specifically to her as a signal of benign intent.

Her response appears to be guided by memory of the quality of a specific recent interaction. The extent to which this memory is explicit is as yet unknown.

Other purported cognitive limitations can also be questioned. There is now a large amount of literature on animals' numerical discrimination abilities suggesting that quantity assessments are widespread across many taxa [reviewed in Shettleworth (2010)]. Similarly, although many tests with primates have suggested a general failure to delay rewards beyond short time periods, there appears to be considerable interindividual variation in self-imposed delayed gratification. Moreover, the ability of primates and other animals to delay gratification in contexts that do not involve food rewards remains largely untested. Thus, contingent cooperation in animals is not necessarily constrained by the inability to delay reward or to quantify past cooperative acts.

It has also been assumed that animals are not capable of contingent cooperation because it demands the anticipation of future interactions. Leaving aside for the moment the question of whether mental projections of future outcomes are necessary to sustain contingent cooperation, the assumption that animals are unable to anticipate future events may not be valid. There is a long history in experimental psychology of tests demonstrating that many animals accurately and predictably anticipate future rewards and outcomes [reviewed in Shettleworth (2010)]. Furthermore, a growing number of experiments suggest that primates are able to make prospective decisions based on certainty judgments about their past behavior [e.g., Hampton (2001), Kornell et al. (2007), Paxton and Hampton (2009)]. Indeed, some forms of cooperative behavior in animals—the boundary patrols of chimpanzees in particular (Mitani et al., 2010)—are highly suggestive of shared intentionality, planning, and episodic memory.

It is also doubtful that nonhuman primates are unable to distinguish cooperators from noncooperators. In tests conducted in captivity that require two individuals to work together to obtain a food reward, both capuchin monkeys and chimpanzees are more likely to cooperate with partners with whom rewards are shared more equitably (de Waal and Davis, 2003; Melis et al., 2006b, 2009; Silk and House, in press). Chimpanzees also recognize which partners are most effective (Melis et al., 2006a) and show a limited ability to increase their rate of cooperation with partners who have cooperated with them in the past (Melis et al., 2008). They may also be able to resolve conflicts of interests when working together to achieve a common goal (Melis et al., 2009).

EMOTIONAL CONSTRAINTS

In humans, inequity aversion, tolerance, and the motivation to engage in joint activities are important catalysts for cooperative behavior. Whether primates are motivated by these emotions, however, is a topic of much debate. Some experiments have suggested that primates reject food offered by humans if a rival is receiving a better reward [e.g., Brosnan and de Waal (2003)]. Other studies have failed to replicate these findings, and suggest that the food rejections are caused not by perceived inequality but by frustration at seeing, but not obtaining, a preferred food item [e.g., Bräuer et al. (2006), Dubreuil et al. (2006)]. In some experiments, chimpanzees have appeared to be generally indifferent to inequitable returns to themselves and others. When subjects have the opportunity to deliver food to a partner at no cost to themselves, for example, they show no sensitivity to the consequences for their partner (Silk and House, Chapter 16, this volume). In other experimental paradigms, however, chimpanzees do help partners obtain food rewards, especially when the partner is attempting to reach for the food or soliciting help (Greenberg et al., 2010; Melis et al., 2010). Thus, in some contexts, chimpanzees do seem motivated to help others and to take into consideration others' outcomes, even when they do not derive direct benefits from doing so.

It has also been argued that a lack of social tolerance may contribute to the low levels of cooperation displayed by chimpanzees in many experiments. Bonobos (*Pan paniscus*) achieve higher levels of success in some cooperative tasks than do chimpanzees, seemingly because their willingness to share rewards with their partners prompts continued cooperation (Hare et al., 2007). It remains unclear, however, whether bonobos also show higher degrees of cooperation and tolerance under natural conditions, under which the structure and rewards of the task are not determined by humans. It is not known, for example, whether bonobos show higher levels of cooperation than chimpanzees when hunting, or whether they share their kills more equitably. Similarly, it is not apparent whether bonobos ever engage in any behavior that is as cooperative and potentially costly as chimpanzees' patrolling behavior (Mitani et al., 2010), or if they do, whether they are more likely than chimpanzees to share risks equitably.

Taken together, results suggest that cooperation in animals may be sustained by qualitatively different mechanisms than it is in humans. Indeed, experiments explicitly designed to compare the behavior of children and chimpanzees indicate that humans may be uniquely motivated to engage others' attention, share their intentions, emotions, and knowledge, and impose sanctions on noncooperators (Tomasello et al., 2005; Warneken and Tomasello, 2009; Melis and Semmann, 2010).

MEASURING CONTINGENT COOPERATION

For several reasons, it has proved difficult to investigate contingent cooperation under natural conditions. First, in the absence of experiments, it is almost impossible to determine whether a given altruistic act is causally dependent upon a specific prior interaction. Second, many altruistic acts occur in different currencies—such as grooming and alliance support—whose relative values are difficult to calibrate. Moreover, even altruistic acts that occur in the same currency may not carry equal value for each participant. In species that form dominance hierarchies, a low-ranking individual may value alliance support from a more dominant partner more highly than vice versa. As a result, he may provide substantially more support to the dominant partner than he receives in return, yet still regard the relationship as reciprocal. Given these empirically intractable problems, almost any relationship can be termed reciprocal. Finally, the degree to which interactions are regarded as reciprocal often seems to be a function of the timescale under consideration. As already mentioned, grooming exchanges within single bouts are often unbalanced and asymmetrical. Nonetheless, over longer time periods, partners with close social bonds exhibit a high degree of reciprocity in grooming.

Correlations between grooming and alliance support have been documented in a variety of primates (Silk, 2007b). In a meta-analysis involving 14 primate species, Schino (2007) found a weak but highly significant correlation between grooming and alliances among long-term partners over extended periods, but little evidence that alliance support is motivated by a specific recent grooming bout. Indeed, in one study of captive Japanese macaques (*Macaca fuscata*), kin were never observed to support each other in the 30 minutes after grooming, even when they had the opportunity to do so (Schino et al., 2007). Similarly, although female hyenas form the majority of their alliances with close kin, there is no evidence that this support is reciprocal or based on the memory of a specific recent interaction (Smith et al., 2010b).

Among male chimpanzees, individuals who groom most often are also those who form alliances and share meat at the highest rates. Cooperation thus involves the exchange of services in different currencies, with males reciprocating grooming for support, support for meat, and so on. Although exchanges are often asymmetrical within dyads over short time periods, they become more evenly balanced over longer periods of time and are not simply a byproduct of association frequency or genetic relatedness (Mitani, 2006).

Possibly the most costly cooperative behavior shown by male chimpanzees occurs during boundary patrols, when the males in one community make incursions into the territories of their neighbors (Mitani et al., 2010). These incursions are potentially risky, because a small party is vulnerable to attack if it encounters a larger party; incursions, therefore, are obligately cooperative. Although it remains unclear whether patrols are planned, they appear to involve some degree of shared intentionality. Little is known about the mechanisms that motivate chimpanzees to initiate and participate in these highly cooperative and risky ventures. It is not known, for example, whether chimpanzees take into consideration memory of another individual's behavior during previous patrols when deciding whether or not to join him in a patrol. Whether cooperation in this context is more, or less, contingent upon memory of previous events remains unclear.

Although chimpanzees' interactions with preferred partners become reciprocal over extended periods of time, tests on captive subjects have provided little evidence for contingency-based reciprocity. For example, in one experiment chimpanzees were given a choice of cooperating with either an individual who had previously helped them or one that had not (Melis et al., 2008). Although there was some evidence that subjects increased their cooperation with the more helpful partner, this effect was relatively weak, and subjects did not consistently avoid noncooperators. In another experiment deliberately designed to test whether cooperation was contingency-based, Brosnan et al. (2009) found no evidence that chimpanzees were more likely to provide food to a partner if that partner had previously provided food to them. Melis et al. (2006b) suggest that chimpanzees may be capable of contingent reciprocity, but that long-term partner preferences that develop over repeated interactions may override the decisions that chimpanzees make on the basis of immediate exchanges and rewards.

Curiously, similar experiments conducted with other species have provided more positive evidence for contingent cooperation in the context of food exchange. For example, capuchin monkeys are more likely to share food with a partner who has previously shared food with them (de Waal and Berger, 2000). Similarly, in experiments in which rats were trained to pull a stick to deliver food to another rat in an adjoining cage, subjects were more likely to help a partner who had previously helped them than one who had previously not helped them (Rutte and Taborsky, 2008).

Several investigations conducted under more natural conditions have also provided some indications of contingent cooperation. Unfortunately, however, interpretation has been complicated by the lack of follow-up experiments to correct for potential confounds. For example, in the well-known study of vampire bats (*Desmodus rotundus*) (Wilkinson, 1984), most reciprocal exchanges of blood occurred among close kin. In addition, although some individuals regularly exchanged blood with unrelated partners, it was not clear whether any specific act of regurgitation was contingent upon a specific recent donation.

An investigation of mobbing behavior in pied flycatchers (*Ficedula hypoleuca*) provides more convincing evidence for contingent cooperation (Krams et al., 2008). In this experiment, subjects had the opportunity to help one of two neighbors mob an owl. One of these neighbors had recently helped the subjects to mob an owl at their own nest box, whereas the other had been prevented from doing so by the experimenters. Subjects were significantly more likely to help previous supporters than apparent defectors, suggesting that cooperative behavior was contingent upon memory of the neighbors' behavior.

There is also some evidence that monkeys may sometimes rely on memory of recent interactions when choosing to join another in an alliance, even though most alliances are based on long-term partner preferences. In one experiment conducted with captive long-tailed macaques (*Macaca fascicularis*), females were found to be more willing to support a lower-ranking female in an aggressive dispute after being groomed by that female than in the absence of grooming (Hemelrijk, 1994). In another experiment, female baboons were played the recruitment call of a lower-ranking female at least 10 minutes after either grooming with her or threatening her. Hearing the recruitment call of a recent grooming partner caused subjects to move in the direction of the loudspeaker and approach their former partner, but only when the two were not close relatives (Cheney et al., 2010). Importantly, females' responses were not influenced by any type of recent interaction, because subjects responded to their former partner's recruitment call only after grooming, and not after aggression. Similarly, their responses were not prompted only by the motivation to resume a friendly interaction, because prior grooming alone did not elicit approach. Instead, subjects were most likely to approach their grooming partner when they had also heard her recruitment call. Thus, females' willingness to attend to the recruitment calls of other individuals appeared to be prompted at least in part by memory of a specific friendly interaction.

In sum, several factors may interact to motivate contingent cooperation in animals under natural conditions: the strength of the partners' social relationship, the nature of their recent interactions, and the opportunity to reengage in some form of cooperative behavior. Animals appear to possess many of the cognitive abilities thought to be essential for the emergence of contingent cooperation, if in rudimentary form. Nonetheless, such cooperation appears to be less common than the noncontingent cooperation that develops among kin and long-term partners.

DETECTION OF NONCOOPERATORS

If cooperation depends in part on the memory of previous behavior, why do animals not avoid or punish freeloaders? In captivity, chimpanzees continue to work with noncooperators despite receiving inequitable returns (Melis et al., 2006b, 2009). In addition, although they retaliate against an individual who steals food from them, they do not attempt to punish those who obtain disproportionate rewards, nor are they motivated to damage the welfare of others simply for its own sake (i.e., spite) (Melis et al., 2006b, 2009; Jensen et al., 2007b).

Under natural conditions, too, freeloaders appear to be tolerated. For example, among feral dogs (*Canis lupus familiaris*), participation in intergroup contests declines with increasing group size, and high-ranking individuals are more likely to avoid being at the front of the pack (Bonanni et al., 2010). However, there is no evidence that defectors are punished. Individual lionesses (*Panthera leo*) also vary predictably in their participation in territorial conflicts. In playback experiments that simulated the approach of an aggressive intruder, some females consistently advanced toward the source of the calls, whereas others consistently lagged behind, avoiding the potential cost of a conflict (Heinsohn and Packer, 1995). Advancers appeared to be aware of the laggards' behavior, because they often looked back at them; nonetheless, they did not subsequently avoid or punish them. Similarly, male chimpanzees do not participate equally in boundary patrols (Mitani, 2006). Some individuals are allowed to reap the benefits of territorial integrity without incurring any costs.

There may be several reasons for animals' apparent tolerance of freeloaders. First, in at least some cases, participants may derive inclusive fitness benefits through freeloaders' survival and reproduction. Freeloaders might also cooperate in other currencies, such as hunting. It is also possible that some individuals benefit more than others from the maintenance of the territory's integrity. This explanation may account for the greater participation of dominant individuals in some other primate species, like vervet monkeys (Cheney, 1981) and ring-tailed lemurs (*Lemur catta*) (Pride, 2005). Finally, however, animals may lack the cognitive capacity to foster or infer deceptive intent, or to impose sanctions on perceived cheaters. If true, animals may well not be capable of achieving the sort of extreme cooperation manifested by humans toward nonrelatives and even complete strangers, which is sustained not only by sanctions against inequity, deception, and spite, but also by concerns about reputation and prestige (Jensen et al., 2007b; Melis and Semmann, 2010; Silk and House, Chapter 16, this volume).

This last objection, however, only denies the possibility for human-like contingent cooperation in animals; it does not rule it out entirely. The detection of cheaters does not in principle require the ability to impute complex mental states like deception to others. It could arise through relatively simple associative processes, by which animals learn to avoid individuals whose presence is associated with a negative experience. Such associations may underlie contingent cooperation in flycatchers, for example.

Indeed, mental state attribution may be irrelevant to contingent cooperation in animals. Schino and Aureli (2009) have argued that the focus on cognitive constraints in discussions of contingent cooperation confuses proximate and ultimate explanations for behavior. Altruistic behaviors may be favored by natural selection because of the subsequent benefits they confer, but what motivates animals to behave altruistically are the previous benefits they have received. In this view, the accumulation of multiple, cooperative exchanges over time causes animals to form partner-specific emotional bonds that prompt future altruistic behavior. Thus, reciprocity may be maintained by a kind of partner-specific "emotional bookkeeping" (Schino and Aureli, 2009) that permits long-term tracking of multiple partners and facilitates cooperation in different behavioral currencies. The resulting bonds that develop between preferred partners may motivate future positive interactions without the need for explicit tabulation of favors given and returned, or calculations of anticipated benefits (Aureli and Schaffner, 2002). For unrelated females who interact at low rates, a single grooming bout may temporarily elevate a female's positive emotions toward her partner sufficiently above baseline to influence her immediate interactions with her. In contrast, grooming and support among females with close bonds (who are also usually kin) should be less subject to immediate contingencies and less influenced by single interactions. Many of these proximate mechanisms may also motivate social interactions in humans. It seems unlikely, for example, that the formation of close bonds among humans is driven by expectations that such bonds will enhance health and longevity.

Finally, it is important to emphasize that, although the absence of punishment in animals may derive partly from cognitive constraints, a strict accounting of services given and received is likely maladaptive in groups in which individuals establish close bonds and interact regularly with familiar partners in a variety of contexts. In fact, although the cognitive constraints that supposedly limit contingent cooperation in animals is often contrasted with humans' sensitivity to inequitable exchanges, human friendships are rarely contingency-based. Numerous studies have shown that people seldom keep tabs of costs and benefits in interactions with regular partners (Silk, 2003). Although people become resentful and dissatisfied when exchanges within a friendship are consistently unbalanced, tallying of favors given and received are typically reserved for infrequent associates. There is even some question about the extent to which extreme prosociality reflects an entirely innate human psychological trait. Recent cross-cultural studies have suggested that cultural norms that promote fairness and the punishment of violators may be more common in large-scale industrialized societies—in which people often interact with strangers—than in smaller, less market-based, communities, in which individuals interact primarily with familiar partners, and where mechanisms associated with kin selection and long-term reciprocity may be more relevant to social relations (Henrich et al., 2010a).

These observations emphasize again the importance of distinguishing between proximate and ultimate explanations when considering cooperation in animals. Whether animals have the cognitive capacity to engage in contingent cooperation is one question; whether it is always adaptive for them to do is another. It may well be that the relative rarity of contingent cooperation in animals stems less from the inability to keep track of recent interactions (and even, perhaps, to anticipate future ones) than from the willingness to tolerate short-term inequities with regular partners.

Finally, most studies of mental state attribution in animals to date have been conducted on captive animals, using paradigms and rewards determined by human experimenters. It is to be hoped that future investigations will attempt to address these questions under more natural conditions, on the animals' own terms. Until such experiments are conducted, we can only speculate about the selective forces that might favor the evolution of a theory of mind, and its function in social interactions.

ACKNOWLEDGMENTS

I thank Robert Seyfarth and Joan Silk for comments on this manuscript. I am also grateful to the Office of the President of the Republic of Botswana and the Botswana Department of Wildlife and National Parks for permission to conduct research in the Moremi Reserve. Research was supported by National Institutes of Health Grant MH62249 and was approved by the Animal Care and Use Committee of the University of Pennsylvania (Protocol 19001).

16

Evolutionary Foundations of Human Prosocial Sentiments

JOAN B. SILK*†‡ AND BAILEY R. HOUSE*

A growing body of evidence shows that humans are remarkably altruistic primates. Food sharing and division of labor play an important role in all human societies, and cooperation extends beyond the bounds of close kinship and networks of reciprocating partners. In humans, altruism is motivated at least in part by empathy and concern for the welfare of others. Although altruistic behavior is well documented in other primates, the range of altruistic behaviors in other primate species, including the great apes, is much more limited than it is in humans. Moreover, when altruism does occur among other primates, it is typically limited to familiar group members—close kin, mates, and reciprocating partners. This suggests that there may be fundamental differences in the social preferences that motivate altruism across the primate order, and there is currently considerable interest in how we came to be such unusual apes. A body of experimental studies designed to examine the phylogenetic range of prosocial sentiments and behavior is beginning to shed some light on this issue. In experimental settings, chimpanzees and tamarins do not consistently take advantage of opportunities to deliver food rewards to others, although capuchins and marmosets do deliver food rewards to others in similar kinds of tasks. Although chimpanzees do not satisfy experimental criteria for prosociality in food delivery tasks, they help others complete tasks to obtain a goal. Differences in performance across

*Department of Anthropology and †Center for Society and Genetics, University of California, Los Angeles, CA 90095. ‡To whom correspondence should be addressed. E-mail: jsilk@anthro.ucla.edu.

species and differences in performance across tasks are not yet fully understood and raise new questions for further study.

It is not from the benevolence of the butcher, the brewer, or the baker that we expect our dinner, but from their regard to their own interest.

Adam Smith, *The Wealth of Nations*

How selfish soever man may be supposed, there are evidently some principles in his nature, which interest him in the fortune of others, and render their happiness necessary to him, though he derives nothing from it except the pleasure of seeing it.

Adam Smith, The Theory of Moral Sentiments

As Adam Smith pointed out more than 250 years ago, humans often act out of self-interest but also feel concern for the welfare of others. These sentiments come into conflict when selfish behavior produces negative impacts on others and when concern for others leads to altruistic behavior that reduces one's own welfare. For evolutionary biologists, selfishness is a straightforward consequence of selective forces that favor behaviors that enhance individual fitness. Natural selection is not expected to favor indiscriminate altruism, because altruists bear the costs of the altruistic behaviors that they perform; this reduces their relative fitness. Altruism can only evolve if altruists confer benefits selectively on others who carry the same altruistic alleles. Kin selection (Hamilton, 1964a) and reciprocal altruism (Trivers, 1971; Axelrod and Hamilton, 1981) both rely on this principle. Selection can favor altruism to close relatives, because recent common descent provides a reliable cue of genetic similarity. In the case of reciprocity, past behavior of other group members provides a cue about whether they carry alleles that lead to altruistic behavior. These processes can generate biases in favor of kin and reciprocating partners but not a general predisposition to behave altruistically to others.

Altruism is also paradoxical for many economists. Selfishness is the expected outcome when, as is often assumed, utility functions only include personal consumption. However, as Adam Smith realized, human behavior deviates from the expected behavior for self-interested actors. Experimental studies in behavioral economics that are designed to bring conflicts between self-interest and altruism into sharp relief show that people value their own welfare but also value the welfare of others (Henrich et al., 2004; Fehr and Schmidt, 2006). This body of work provides insight about some of the dimensions of our altruistic social preferences.

DIMENSIONS OF ALTRUISTIC SOCIAL PREFERENCES IN HUMANS

Generosity

In the Dictator Game, subjects are allowed to distribute an endowment between themselves and another player (Camerer and Thaler, 1995). The units of the endowment may take the form of cash or monetary equivalents that will be converted to cash at the end of the experiment. One player, the proposer, is given the opportunity to allocate any amount of his endowment to a second player, the recipient. In the standard form of the game, the proposer's offer is relayed to the recipient anonymously; the two players never meet and never interact again. This eliminates reputational benefits or expectations based on reciprocity. A selfish player would keep the full endowment; an altruistic player would allocate some fraction of the endowment to the recipient. Typically, proposers allocate 20–30% of the endowment to the other player (Camerer and Thaler, 1995), indicating that they value the welfare of others but not as highly as they value their own welfare.

Trust

In the Trust Game, two players are given endowments. Player 1 can allocate any amount of her endowment, e, to Player 2; the experimenter will triple the allocation, and the full amount will be delivered to Player 2. Then, Player 2 is given the opportunity to make an allocation to Player 1. Player 2 can keep all of the money or send some money back to Player 1. If Player 1 sends her whole endowment to Player 2, then Player 2 would receive $3e$. This would be added to Player 2's initial endowment, e, and equal $4e$. If Player 2 sends back one-half, both would get $2e$, which is double their initial endowment. In contrast, if Player 1 sends Player 2 only one-half of the original endowment, then Player 2 will end up with only $2.5e$ ($0.5e \times 3 + e$). If Player 1 trusts Player 2 to repay her, then it is best to send Player 2 the full amount. However, if Player 1 expects Player 2 to defect, then it is best to send nothing. If Player 2 is selfish, she would keep the whole amount; any money sent back to Player 1 is a form of altruism. In fact, the majority of people who take the role of Player 2 do send back money, and the amount that they send is proportional to the amount that they have received (Fehr and Fischbacher, 2003).

Punitive Sentiments

The Ultimatum Game (Camerer and Thaler, 1995) adds a second step to the Dictator Game. The proposer is given an endowment and makes an allocation. Now, the recipient decides whether to accept or reject the proposer's offer. If the recipient accepts the offer, each player gets the designated amount; if the recipient rejects the offer, neither one gets any money. The recipient has little material incentive to reject any nonzero offer, because this will result in a loss of income. Rejections constitute a form of altruistic punishment, because the recipient suffers a cost to punish the proposer; because the players will never interact again, this cannot be a strategy for improving the recipient's payoffs in the future. The Ultimatum Game has now been played by thousands of people in dozens of countries all over the world. Recipients typically reject offers of less than 20%, and the size of the initial endowment has surprisingly little impact on rejections (Hoffman et al., 1996; Camerer, 2003; Henrich et al., 2006, 2010a).

In the Third-Party Punishment Game, a third party is given the opportunity to impose sanctions on the proposer in a Dictator Game (Fehr and Fischbacher, 2004). As in the standard Dictator Game, the proposer receives an endowment and can transfer any fraction of the endowment to a receiver. In this game, however, a third player is given an endowment and informed of the proposer's allocation decision. The third player can spend one unit to reduce the proposer's payoff by three units but cannot have any effect on the recipient's payoff. A majority of subjects imposed sanctions on proposers who made offers of less than one-half of the endowment, and those that offered much less than one-half were punished more severely than those that made offers closer to one-half. It is noteworthy that people are willing to incur costs to punish others for making low offers, although they have not been directly harmed themselves (Fehr and Fischbacher, 2004).

Fairness

Proposers' offers in the Dictator Game, recipients' behavior in the Ultimatum Game, and responses of third parties in the Third-Party Punishment Game all suggest that people have a strong preference for equitable outcomes. Although people are more sensitive to inequities that disadvantage themselves than inequities that benefit themselves, a substantial majority of people are willing to reduce their own payoffs to produce more equitable outcomes for others (Fehr and Schmidt, 1999). Interestingly, people are less bothered by inequitable outcomes that are the product of chance events, such as a coin flip, than inequitable outcomes that are the result of deliberate human action (Blount, 1995; Camerer and Thaler, 1995).

LIMITS ON ALTRUISTIC SOCIAL PREFERENCES IN HUMANS

The evidence for generosity, trust, punitive sentiments, and fairness does not mean that all humans are indiscriminate altruists. Like many other animals, humans show strong nepotistic biases (Flinn et al., 2007; Sear and Mace, 2008; Hrdy, 2009) and also develop long-term relationships with reciprocating partners (Gurven, 2006; Allen-Arave et al., 2008). There is also substantial individual variation in social preferences. For example, in public goods games that continue across multiple rounds, a substantial fraction of subjects are contingent cooperators (Fischbacher et al., 2001; Fehr and Fischbacher, 2003). Conditional cooperators follow cooperative norms as long as other group members cooperate but stop cooperating if others defect. In the presence of contingent cooperators, a small number of selfish, uncooperative individuals can precipitate the collapse of group-level cooperation. Sanctions that make it costly to defect help prevent this from happening. In addition, humans show strong parochial biases, which favor group members over outsiders (Shinada et al., 2004; Bernhard et al., 2006).

CRITIQUES OF INTERPRETATION OF BEHAVIORAL ECONOMICS GAMES

In behavioral economics games, players are paired with strangers in one-shot games to eliminate egoistic motives for altruism, including the opportunity for reciprocity and reputational benefits. Some researchers question the validity of these conditions by claiming that our psychology was designed for a world in which we lived in small groups of close kin and reciprocating partners. In such settings, there may be little opportunity for anonymous, one-shot interactions (Hagen and Hammerstein, 2006; Burnham and Hare, 2007). As a result, participants in experiments may find it hard to believe that their behavior is actually anonymous, and subjects may be influenced by subtle cues that influence their perceptions of being observed and the salience of reputational cues. Such cues matter, because our psychology is "exquisitely sensitive to cues that are (or were, under ancestral conditions) informative with respect to the likely profitability of co-operation in a given situation" (Haley and Fessler, 2005). According to this argument, people behave altruistically in behavioral economics experiments, because they are motivated to enhance others' perception of their value as a cooperative partner. Thus, altruistic behavior is motivated by self-interest not other-regarding preferences.

This claim is partially supported by evidence that levels of contributions increase when subjects are exposed to subtle cues of being watched. For example, a pair of eyes on a sign that instructed users of a university coffee room to pay for their drinks produced more revenue than neutral images (Bateson et al., 2006). A pair of eyes had a similar effect on littering in a university cafeteria (Ernest-Jones et al., 2010). Players in a public goods game that were faced with an image of an anthropomorphic robot on the computer screen contributed more than players faced with a blank screen (Burnham and Hare, 2007). Haley and Fessler (2005) found that contributions in an anonymous Dictator Game were higher when the computer monitor displayed a pair of stylized eyes than when it displayed plain text. The effects of eyes in the Dictator Game have been replicated (Rigdon et al., 2009; Mifune et al., 2010; Oda et al., 2011) and seem to be a function of the expectation of future benefits, not fear of punishment (Oda et al., 2011).

However, cues of being watched do not have the same effects in all games. Fehr and Schneider (2010) found that the stylized eyes that Haley and Fessler (2005) used had no effect on the amount that Player 1 transferred in an anonymous trust game. In contrast, when players were told that their partners would be informed about the amount that they had transferred in previous rounds (but not their identity), transfer amounts doubled. Thus, people were strongly influenced by explicit reputational information but not by the kinds of subtle cues that might have suggested that they were being watched.

If subtle cues of being watched affect cooperative behavior, then the actual presence of others ought to amplify cooperative behavior. To assess this, Lamba and Mace (2010) conducted a series of Ultimatum Games in which they manipulated the degree of anonymity that subjects experienced. In one condition, subjects played an anonymous double-blind game alone in a room (Anonymous/Private). In a second condition, subjects played an anonymous double-blind game in a room with other subjects (Anonymous/Public). In the third condition, subjects' offers were announced to all participants who were together in the same room (Public/Public). Proposers' offers did not differ in the Anonymous/Private and Anonymous/Public conditions, suggesting that players were confident that their offers were anonymous and were not affected by subtle cues of being watched by others. In contrast, knowledge that their offers would be made public significantly increased offers in the Public/Public condition, and this effect was enhanced when proposers were acquainted with others in the room when the experiment was conducted.

At this point, there is no consensus about the importance of subtle cues of being watched on decision making in behavioral economic games or the impact of such cues on prosocial behavior in more naturalistic settings. It is not yet clear whether differences in the effectiveness of cues of being watched depend on the game being played or details of the experimental procedures. In contrast, there is abundant evidence that explicit reputational information has clear and unambiguous effects on prosocial behavior across games and settings.

MOTIVES UNDERLYING HUMAN ALTRUISM

Behavioral economic experiments are designed to elicit preferences that guide choices about payoff outcomes, but they do not provide direct information about the psychological mechanisms that produce these preferences. This is important, because behaviors that have similar outcomes can be the product of very different mechanisms. Human altruism might be motivated by empathy and concern for the welfare of others. Alternatively, altruism might be prompted by more selfish, egoistic concerns, such as improving one's reputation for generosity. If people are motivated by empathy and concern for the welfare of others, their ultimate goal is to provide benefits to others, and any benefits that individuals accrue are incidental by-products. However, if people are motivated by egoistic motives, then the benefits that they deliver to others may be incidental to their primary goals. People might be motivated to help others, because helping brings rewards to themselves (including reputational benefits or future material gains), prevents punishment (including material sanctions), or reduces aversive arousal that comes from observing others in need.

Batson (1991, 2011) has conducted a long series of experiments that were designed to assess the relative importance of egoistic and empathetic motives in altruistic predispositions [reviewed in Batson (1991) and (2011)]. For example, to assess the hypothesis that altruism is prompted by the desire to reduce aversive arousal, Batson (1991, 2011) conducted a series of experiments in which subjects observed a worker receiving electric shocks and were told that they could help by volunteering to take the shocks themselves. (In reality, the workers were confederates, and no electric shocks were administered to anyone.) The experimenters manipulated empathetic responses and how easy it was for subjects to avoid the aversive stimulus: In the easy condition, subjects would not see the worker being shocked after they made their decision, but in the difficult condition, subjects would continue to watch the worker being shocked. If altruism is the product of egoistic motives, then subjects should help more when escape is difficult than when escape is easy under the high-empathy condition. However, if altruism is the product of altruistic motives, then subjects should be equally likely to help under both conditions. Subjects generally conformed to the latter pattern, suggesting that they were motivated to help for altruistic, not egoistic, reasons. The results of the full series of experiments and related experimental work by other investigators (Piliavin and Charng, 1990) are consistent with the hypothesis that altruistic behavior is shaped by empathic concern for the welfare of others, and it is not motivated entirely by self-interest or reputational concerns.

PHYLOGENETIC FOUNDATIONS OF HUMAN ALTRUISM

Social relationships play an important role in the daily lives of both human and nonhuman primates. Social bonds seem to enhance the ability to cope with chronic stressors, such as low social status, or acute stressors, such as the recent loss of preferred partners or immediate risk of infanticide [reviewed in Cheney and Seyfarth (2009). In humans, social support is correlated with better physical and mental health as well as lower mortality risks (Thorsteinsson and James, 1999; Cacioppo et al., 2000; Taylor et al., 2000; Kendler et al., 2005). Similar kinds of findings are accumulating for a range of nonhuman species, including rodents (Weidt et al., 2008; Yee et al., 2008), dolphins (Frère et al., 2010), wild horses (E. Z. Cameron et al., 2009), female baboons (Silk et al., 2003a, 2009, 2010b), and male macaques (Schülke et al., 2010).

There is good evidence that, like humans, monkeys and apes form strong and lasting ties, particularly with close kin (Silk, 2009) and reciprocating partners (Cheney, Chapter 15, this volume), and close social bonds are the foundation for cooperation in nonhuman primate groups (Mitani, 2009; Silk et al., 2010a). Like humans, nonhuman primates also have strong in-group biases. Responses to strangers and members of neighboring groups range from passive avoidance to active hostility (Crofoot and Wrangham, 2010). However, despite these intriguing parallels in the patterns of cooperation and the correlates of social bonds among humans and other primates, there are also important differences in the scope of cooperation. In most primate species, there is no sexual division of labor and little active food sharing. Primates do not cooperate with members of other groups in collective activities, such as warfare, territorial defense, or trade.

DIMENSIONS OF ALTRUISTIC SOCIAL PREFERENCES IN NONHUMAN PRIMATES

Differences in the scope and pattern of cooperation between humans and other primates may be reflected in differences in the nature of their social preferences. Researchers have recently begun to explore the dimensions of altruistic social preferences in nonhuman primates in systematic ways using the same kinds of tools that behavioral economists have used to assess human social preferences. These experimental methods are useful, because the difficulties of identifying the motives underlying altruism are compounded when we extend the analysis to other species

(Silk, 2007a). A chimpanzee who has just caught a colobus monkey might allow another male to take part of his kill, because he feels empathy for the other's hunger, because he can forestall the other's efforts to take the entire carcass by force, or because he received meat from the other male on the previous day. These sorts of ambiguities have prompted a series of experiments that are designed to determine whether other primates have preferences for outcomes that benefit others, a sense of fairness, and punitive sentiments.

GENEROSITY AND PROSOCIAL BEHAVIOR IN CHIMPANZEES

The Prosocial Test is similar to a discrete Dictator Game. Proposers are presented with a choice between two options (Silk et al., 2005). One option delivers a food reward to the proposer and nothing to another individual in an adjacent enclosure. For convenience, this is referred to as the 1/0 option (the reward for the proposer is given on the left side of the slash and the reward for the recipient is given on the right side of the slash). The other option delivers a food reward to the proposer and an identical reward to the other individual (the 1/1 option).

Chimpanzees have prepotent biases for larger number of rewards, regardless of their distribution (Boysen and Berntson, 1995; Uher and Call, 2008). This means that proposers might prefer the prosocial option (1/1) over the selfish option (1/0), because they have biases in favor of larger numbers of rewards. Therefore, a nonsocial control condition was included in which no recipient was present to receive rewards.

Actors' choices in the Prosocial Test provide insights about their social preferences. If individuals are concerned about the welfare of others, they will choose the 1/1 option over the 1/0 option. Moreover, if they are motivated by concern for the welfare of others and not by prepotent biases for a larger number of rewards, their bias in favor of the prosocial option will be stronger when another individual is present (test condition) than when the actor is alone (nonsocial control condition). Alternatively, individuals might view potential recipients as rivals or competitors for a fixed quantity of rewards, and they might be motivated to deprive them of resources. If so, they will choose the 1/0 option over the 1/1 option, and their bias in favor of this selfish option will be stronger in the test condition than in the control condition. Finally, if actors are indifferent to the welfare of others, they will choose at random, and their choices in the control and test conditions will not differ. The major advantage of the Prosocial Test over the Dictator Game is that it costs the proposer nothing to confer a prosocial outcome, meaning that any positive altruistic tendencies should manifest clearly.

The Prosocial Test and several closely related variants have now been conducted with chimpanzees from several different populations using a number of different experimental apparatuses. Jensen et al. (2006) conducted three related experiments. In one, the payoff distribution was 1/1 vs. 1/0; in the second experiment, the actor could provide rewards to the recipient but obtained nothing for herself (0/1 vs. 0/0), and in the third experiment, the actor could prevent the recipient from obtaining rewards. In all three experiments, there was no difference between the test condition and the nonsocial control condition. Yamamoto and Tanaka (2010) trained chimpanzees to associate one colored button with a 1/1 payoff and a differently colored button with 1/0 payoff. The chimpanzees were as likely to choose the 1/1 option in the test condition as in the nonsocial control condition. Vonk et al. (2008) conducted two experiments in which actors could deliver food rewards to themselves and others with independent but identical actions. In both of these studies, the actors were as likely to deliver rewards in the nonsocial control condition as in the social test condition.

Brosnan et al. (2009) and Yamamoto and Tanaka (2010) allowed participants to switch roles within trials, and therefore the proposer in one round became the recipient in the next round. This manipulation had no impact on the level of prosocial responses in either study. The rate of prosocial responses in the iterated Prosocial Test was the same as the rate of prosocial responses in one-shot versions of the Prosocial Test conducted with the same animals.

Thus, in the Prosocial Test, chimpanzees consistently act as if they are indifferent to the welfare of other individuals. This set of findings has been both surprising and controversial, because chimpanzees cooperate in a wide range of contexts, share food in the wild (Muller and Mitani, 2005), collaborate effectively in mutualistic tasks in the laboratory (Melis et al., 2006a,b), and seem to be helpful in other experimental paradigms. Before we turn to experiments in which chimpanzees show helpful behavior, we consider a number of explanations that have been proposed to explain chimpanzees' behavior in the prosocial task.

It is possible that proposers did not differentiate between the test and control conditions in these experiments, because they did not understand how the experimental apparatuses worked. However, subjects' understanding of the experimental apparatuses was explicitly tested in several studies (Jensen et al., 2006; Brosnan et al., 2009; Yamamoto and Tanaka, 2010). Alternatively, the chimpanzees might have found it difficult to track the distribution of food items when rewards were delivered to themselves and their partners simultaneously (Warneken and Tomasello, 2009). However, when Jensen et al. (2006) tested chimpanzees with a payoff distribution that did not provide any rewards for the actor, they found that overall response rates dropped substantially, and proposers still did not distinguish between the test and nonsocial control conditions. In the experiments conducted by Vonk et al. (2008), proposers could deliver food rewards to themselves and others with separate actions. Proposers almost always obtained their own rewards first. After they had obtained rewards for themselves, they sometimes delivered the other reward. However, as the experiment progressed, they were less likely to deliver rewards to the other enclosure in both test and control conditions. Thus, subjects in these experiments became less generous (and no more discriminating), as they gained more familiarity with the test apparatus.

Although chimpanzees seem to have some understanding of others' desires and intentions in competitive situations (Hare et al., 2000, 2001; Kaminski et al., 2008), proposers may have been unaware of their partners' desires for rewards in the prosocial test (Warneken and Tomasello, 2009). However, analyses of recipients' begging gestures in two experiments cast doubt on this possibility (Vonk et al., 2008). Recipients that made begging gestures consistently directed them to the option that contained food for themselves, but begging had no consistent impact on proposers' responses.

It is also possible that proposers did not choose the 1/1 option more often, because chimpanzees often compete over access to food in the wild, and prosocial preferences are muted in the presence of food (Warneken and Tomasello, 2009). However, if chimpanzees view food as a limited, zero-sum resource, they would be expected to show a strong preference for the 1/0 over the 1/1 option. This was not seen in any of the studies.

HELPFUL RESPONSES OF CHIMPANZEES IN OTHER EXPERIMENTAL SETTINGS

The conclusions derived from the prosocial test with chimpanzees conflict with results derived from experimental paradigms in which one individual is given the opportunity to help another individual obtain a goal. The first of these studies was conducted by Warneken and Tomasello (2006) with three young chimpanzees that were paired with their human caretaker in several different task situations. In each task situation, there was one version in which help was needed (test), and a second very similar version in which no help was needed (no-need control). The chimpanzees responded positively to caretakers' requests for help in several tasks that involved retrieving out-of-reach objects and consistently differentiated between the control and test conditions. However, they did not meet this criterion for a number of other kinds of tasks. Warneken and Tomasello (2006) suggest that this may have been because they did not grasp what the recipient needed.

The three young chimpanzees that Warneken and Tomasello (2006) tested had a close relationship with their caretakers and had been rewarded for accommodating behavior in the past (Warneken et al., 2007). To determine whether helpful behavior would extend to unfamiliar humans, Warneken et al. (2007) performed a second set of experiments. In these experiments, two experimenters struggled over a stick, and then, the victor placed the stick out of the loser's reach. In the test condition, the loser stretched out his arm and reached to the stick, and in the control condition, the loser looked at the object but did not reach for it. The chimpanzees were significantly more likely to retrieve the stick in the test condition than in the control condition. In a second experiment, the cost of helping was increased, because the chimpanzees had to climb 2.5 m into an overhead compartment to retrieve the stick. The chimpanzees were actually more likely to retrieve the stick in this experiment than in the first experiment, and they did not distinguish between the test and control conditions. Warneken et al. (2007) speculated that this was "likely due to a carryover effect from experiment 1 in which subjects had possibly learned that the experimenter wanted the object."

Chimpanzees' willingness to help humans complete certain tasks is mirrored in their behavior to conspecifics trying to complete a task. Warneken et al. (2007) conducted a third set of experiments in which one chimpanzee was given an opportunity to help another chimpanzee gain access to a locked room. In this experiment, there were four adjacent rooms. The doors to two rooms were fastened by chains held in place by pegs. The actor was confined to one room and could reach a peg that released one of the two doors, but he could not enter either of the locked rooms. The recipient could not release either of the doors but could enter one of the rooms that the actor could unlock. Recipients were motivated to gain access to the locked room, because it contained food rewards, but these rewards were not visible to the actors. In the test condition, food rewards were placed in the room that the actor could unlock and the recipient could enter; in the control condition, food rewards were placed in the room that the recipient could not enter. Actors were significantly more likely to remove the peg and release the door in the test condition than in the control condition.

Chimpanzees also provide tools that others need to complete food-related tasks (Yamamoto et al., 2009). In this experiment, two chimpanzees were placed in adjacent enclosures, and each was presented with a food task that required a particular tool (stick or straw). In the baseline condition, each chimpanzee was given the appropriate tool. In the test condition, the chimpanzee that needed the stick was given the straw, and the chimpanzee that needed the straw was given the stick. The chimpanzees were significantly more likely to transfer tools when they were given the wrong tools than when they were given the right tools, and tool transfers generally occurred in response to explicit requests.

As noted earlier, there has been speculation that results in the Prosocial Test might reflect chimpanzees' reluctance to provide others with food or their lack of understanding of others' needs and desires. To test these two hypotheses, Melis et al. (2011) constructed an apparatus in which one chimpanzee could release a peg and cause a reward to roll down a ramp to the recipient in a separate enclosure. In one set of trials, the reward was connected to a chain so that the recipient could pull the reward to them after it was released. In another set of trials, the food rolled directly down the ramp and into their enclosure, and there was no chain for recipients to pull. These conditions were meant to encourage active and passive responses by recipients, respectively, but this manipulation was not entirely effective; therefore, analyses were based on the recipients' level of activity across the reach and no-reach conditions. To evaluate whether helpful responses were inhibited when food rewards were present, some trials were conducted with food rewards, and other trials were conducted with tokens, which recipients could trade for food rewards out of the actor's sight. The chimpanzees were significantly more likely to release the peg when food rewards were present than when tokens were present, and they were more likely to release the peg when recipients responded actively than when they responded passively. Melis et al. (2011) concluded that "the main finding of the present study is that recipients' signaling is necessary to elicit helping behaviour."

However, this conclusion is inconsistent with results derived from another study of helping behavior in chimpanzees. In this experiment, conducted by Greenberg et al. (2010), two chimpanzees had to pull on a rope to move a sliding platform along a set of parallel tracks. The apparatus was designed so that it could be baited to dispense rewards at the beginning, middle, or end position of each of the tracks. In the baseline condition, both chimpanzees obtained rewards when the platform was pulled to the middle condition. In one of two altruism conditions, one chimpanzee obtained a reward when the platform was pulled to the middle position, and the other chimpanzee obtained a reward when the platform was pulled to the end position. In the other altruism condition, one chimpanzee obtained a reward when the platform was pulled to the end position, but the reward for the other chimpanzee was simply placed on the apparatus. The participants were significantly more likely to pull the platform to the end position in the two altruism conditions than in the baseline condition, suggesting that the chimpanzees were willing to continue pulling even after they had obtained their own reward. In this setting, direct solicitations or requests for help were rarely observed.

RECONCILING RESULTS FROM EXPERIMENTAL STUDIES OF PROSOCIAL BEHAVIOR IN CHIMPANZEES

In the Prosocial Test, chimpanzees are as likely to choose the prosocial option when another individual is present as when they are alone, and therefore, they do not meet the experimental criterion for prosocial preferences. In several helping tasks, chimpanzees are more likely to provide help when it is needed than when it is not needed and thus, satisfy the criterion for prosocial behavior within these protocols. A number of factors have been invoked in an effort to determine why chimpanzees pass some tests but fail others, including cognitive demands of the task, actors' preoccupation with their own rewards, competitive attitudes to food, limited understanding of what others want, and salience of others' requests. However, none of these explanations seem to be consistent with the full body of evidence.

The focus on whether chimpanzees pass or fail various experimental tests has obscured an important feature of the data. Across the full range of experiments and test treatments, chimpanzees behave prosocially to their partners about one-half of the time (median = 0.51, interquartile range = 0.23). On average, the rates of prosocial responses tend to be higher in the Prosocial Tests than in the helping tasks (Wilcoxon rank sum test, $z = -1.913$, $P = 0.0558$, $n_1 = 10$, $n_2 = 12$) (Fig. 16.1).

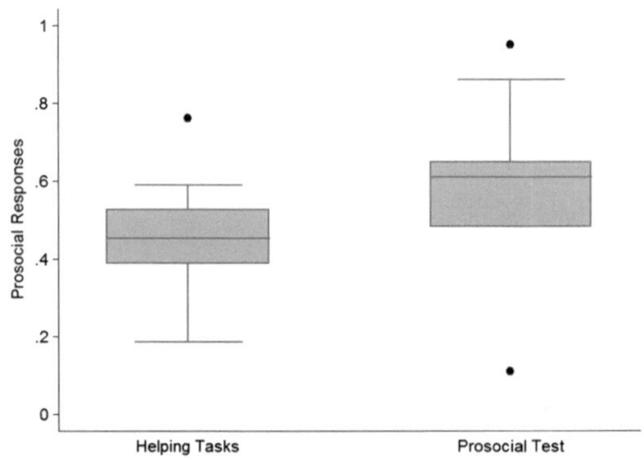

FIGURE 16.1 Rate of prosocial responses. The median and interquartile range of prosocial response rates in the test condition of all studies of prosocial behavior are plotted here. For studies that reported response rates for different categories of actor–recipient pairings or different categories of behavior of recipient, all values are included.

PROSOCIAL BEHAVIOR IN OTHER PRIMATES

If ecological factors shape predispositions about the fitness benefits of cooperation, then prosocial preferences in food distribution tasks might be expected to emerge in species that share food more willingly than chimpanzees do, such as cooperatively breeding marmosets and tamarins as well as capuchins. Marmosets and tamarins are cooperative breeders, and adults and immature helpers often provision younger individuals with food (Brown et al., 2004). In the laboratory, capuchins are remarkably tolerant of others' efforts to take portions of their food (de Waal, 1997a, 2000).

Burkart et al. (2007) conducted a modified version of the prosocial test with marmosets. These animals are significantly more likely to choose 0/1 over 0/0 when others are present than when they are alone. This pattern held for male donors paired with adult female and immature recipients and for adult females paired with immature recipients, but it did not hold for subadult females paired with immature recipients. The scope of prosocial responses roughly parallels food-sharing patterns in naturalistic settings.

Tests for prosocial behavior in tamarins, which also breed cooperatively, have produced mixed results. Cronin et al. (2009) paired tamarins with their long-term mates and offered proposers a choice between 1/0 and 1/1 in one experiment and a choice between 0/0 and 0/1 in a second experiment. Proposers' choices were not influenced by the presence or absence of their mates. In another study of tamarins, the proposer was placed in the middle of three adjacent enclosures (Stevens, 2010). The proposer was able to pull one handle that brought a tray within reach of itself and the occupant of an adjacent enclosure (with a one-to-three payoff structure) or a second handle that brought a tray to within reach of itself and an empty enclosure (also with a one-to-three payoff). Proposers nearly always pulled to obtain food for themselves when the adjacent enclosures were empty, but the rate of pulling declined sharply when they were unable to obtain food for themselves. In trials in which they could deliver food to an empty cage or their mates, they chose at random.

However, in a third experiment, tamarins showed more positive responses to their partners (Cronin et al., 2010). In this experiment, tamarins could deliver food rewards to their mates (but not themselves) after their mates had delivered food to them (reciprocity condition) or had not delivered food to them (no reciprocity condition). Their behavior to their mates was compared with their behavior when the adjacent cage was empty (nonsocial control). The tamarins were significantly more likely to deliver food rewards to their mates in the reciprocity condition than in the no-reciprocity condition. Moreover, they were significantly more likely to deliver rewards to their partners in the reciprocity condition than in the nonsocial control condition. Cronin et al. (2010) argue that the tamarins' behavior is evidence for prosocial preferences and not contingent reciprocity, because the effects only emerged in the last one-third of the 5-minute trials. However, this leaves open the question of why the tamarins did not provide rewards to their partners in the no-reciprocity condition.

There have also been several studies of prosocial preferences in capuchins. Lakshminarayanan and Santos (2008) offered capuchins a choice between two options. In some test trials, the proposer received a low-quality reward (L) and could deliver a low- or high-quality (H) reward to the recipient (L/L vs. L/H). In other test trials, the proposer received a high-quality reward and could deliver a low- or high-quality reward to the recipient (H/L vs. H/H). Behavior in test trials was compared with control trials in which no recipient was present and control trials in which no recipient was present and the proposer had access to the other compartment. Overall, the capuchins were significantly more likely to choose the option that delivered the high-value reward when another monkey was present than when they were alone (and could not obtain the reward themselves). However, there was considerable variability across subjects, and the magnitude of the difference between the test and control conditions was small. Four of seven subjects showed a preference for the prosocial option when they chose between H/L and H/H, whereas three of seven subjects showed a preference for the prosocial option when they chose between L/L and L/H.

de Waal et al. (2008) trained capuchin monkeys to associate tokens with particular payoffs for themselves and another individual in a separate enclosure, and they monitored their choices of tokens across a series of trials. If monkeys chose one token, the experimenter delivered one reward to them and another reward to the other individual (1/1 payoff); if they chose the other token, the experimenter delivered one reward to them and nothing to the other individual (1/0 payoff). This experiment did not include a nonsocial treatment to control for the effects of prepotent biases for larger numbers of rewards. Monkeys were increasingly likely to choose the token associated with the 1/1 payoff as the experiment progressed, and all statistical analyses were limited to the last one-third of the experimental trials. In these trials, capuchins chose the 1/1 option significantly more often than expected by chance when they were paired with kin or nonkin, but their choices dropped to chance levels when they were paired with monkeys from another group. The monkeys' preferences for the prosocial option also declined to chance levels when the 1/1 option provided a more highly valued reward to the recipient than the proposer. This experiment also included a condition in which visual contact between the proposer and recipient was blocked by an opaque partition. The partition had a small window, and therefore, the proposer knew that another individual was present and the identity of that individual. When the partition was in place, monkeys behaved antisocially—they chose the 1/1 option significantly less often than would be expected by chance.

In the third study on capuchins, Takimoto et al. (2010) manipulated the relative rank of the proposer and recipient and the relative value of rewards delivered to both parties. Recipients were either the alpha male or the lowest ranking female in the group. In some trials, proposers received rewards of medium value (M) and could deliver rewards of high (H) or low (L) value to recipients (M/H vs. M/L). In other trials, proposers received rewards of high value and could deliver rewards of high or low value (H/H vs. H/L). Proposers' behaviors were compared when recipients were present or absent. Overall, proposers were more likely to choose generous options when they were paired with the subordinate recipient than when they were alone. However, this effect disappeared when they were paired with the dominant recipient.

Takimoto et al. (2010) also conducted a set of trials in which proposers could not see recipients during trials. When they were paired with a dominant recipient, they were significantly less likely to choose the generous option in the recipient-present condition than in the no-recipient condition. In contrast, when proposers were paired with the subordinate recipient, they did not differentiate between the two conditions.

The results from these three studies suggest that capuchin monkeys have preferences for outcomes that benefit other group members. However, their preferences for generous outcomes disappear or are reversed when proposers and recipients cannot see each other. de Waal et al. (2008) suggest that this is because capuchins derive rewards from "seeing the partner receive or consume food." However, this does not explain why the proposers chose the 1/1 option significantly less often than expected by chance when visual access was blocked (Barnes et al., 2008; de Waal et al., 2008). It is possible that actors have preferences for antisocial outcomes, but antisocial preferences are suppressed when others can see them and can potentially take punitive action.

There has been only one instrumental helping task conducted with monkeys. The experiment was designed to replicate the experiments in which a human experimenter reaches for an inaccessible object (Warneken et al., 2007; Yamamoto et al., 2009). Unlike the chimpanzees, the capuchins did not consistently distinguish between the test and control conditions (Barnes et al., 2008). The capuchins were strongly motivated by the availability of rewards for themselves but not by the opportunity to provide help to others.

PUNISHMENT

Although primates and other animals often use aggression to manipulate the behavior of others to their own advantage (Clutton-Brock and Parker, 1995), evidence for third-party punishment is scarce (Jensen, 2010). Jensen et al. (2007a) adapted the Ultimatum Game for chimpanzees to examine their propensity to impose punishment on conspecifics. In this case, one chimpanzee that played the role of the proposer was able to choose between two set distributions of rewards by pulling one of two rods. To accept the proposer's offer, the responder pulled another rod that brought the food rewards to within reach of both individuals, allowing each to claim their respective rewards. If the responder did not pull the rod, neither one got any food. The chimpanzees were offered a series of different options across trials. One option in every trial provided eight pieces of food to the proposer and two pieces to the responder (eight-to-two payoff). The other option provided a distribution of five to five, eight to two, or ten to zero. Proposers strongly preferred offers that benefited themselves (e.g., eight to two over five to five), but responders rarely rejected any nonzero offers. Moreover, responders showed little evidence of arousal in any of the trials.

FAIRNESS

Research on fairness and inequity aversion in primates was initiated by Brosnan and de Waal (2003), who trained tufted capuchins to exchange tokens for food rewards. The monkeys consistently offered experimenters tokens in exchange for small pieces of cucumbers, but some individuals refused to complete exchanges after they saw other group members receive more highly valued rewards in exchange for tokens or saw other group members obtain more highly valued rewards without exchanging tokens.

Brosnan and de Waal (2003) suggested that their findings were evidence of inequity aversion, although critics pointed out that the monkeys increased the extent of inequity by refusing to complete exchanges (Henrich, 2004a). Brosnan and de Waal (2003) acknowledged this point and argued that monkeys are averse to inequities that disadvantage themselves but not inequities that favor themselves (Brosnan and de Waal, 2004). Others argued that monkeys' responses might have been prompted by frustration at seeing more preferred foods that were inaccessible to themselves (Dubreuil et al., 2006), frustration at being offered less preferred foods after seeing more desirable foods (Roma et al., 2006), or violation of expectation and loss aversion (Chen and Santos, 2006).

At this point, more than a dozen experimental studies of inequity aversion have been conducted in monkeys and apes. S. F. Brosnan et al.

(2010) found that negative responses to inequity were consistently found when subjects were required to perform a task to obtain food, whereas no responses to inequity were found when subjects were able to obtain food without performing a task. However, this generalization does not fit at least one study of great apes that used a token exchange task modeled on the task that S. F. Brosnan et al. (2010) used. Bräuer et al. (2009) found no evidence for inequity aversion among orangutans, bonobos, or chimpanzees.

CONCLUSIONS

Altruism and mutualistic cooperation play important roles in the lives of nonhuman primates, but there are important differences in the scope of altruistic behavior between humans and other primates. In other primates, altruism is strongly biased in favor of kin and reciprocating partners, and it is never extended to strangers. Primates use aggression to deter competitors and rivals, but there is no compelling evidence of third-party punishment. Unlike humans, nonhuman primates show no aversion to inequitable distributions of resources that favor themselves.

It is important to continue efforts to chart the size and dimensions of the gap between humans and other primates if we want to understand the evolutionary forces that have shaped human social preferences. Evidence that closely related primates, particularly great apes, have altruistic social preferences would suggest that our social preferences were built on a set of ancestral motivations that facilitated altruism to kin and reciprocating partners, mutualistic activities with group members, punitive behavior to competitors, antagonistic attitudes to strangers, and concern for reputational status. At the same time, evidence that closely related primates lack the kinds of altruistic social preferences that characterize modern humans would suggest that emergent forces, possibly including cultural group selection (Richerson and Boyd, 2005), demands of raising slow-growing offspring (Hrdy, 2005b, 2007; Jaeggi et al., 2010), knowledge-intensive human foraging strategy (Kaplan et al., 2000, 2003), or risks associated with uncertain hunting returns (Winterhalder, 1986), have transformed us in consequential ways and given rise to important differences in the scope and scale of cooperation, our capacity for empathy and compassion, the development of moral sentiments, and the willingness to enforce culturally specified social norms.

We believe that well-designed experimental studies of social preferences in other primates provide important insights about the nature of social preferences. Such studies should be designed to test hypotheses that are grounded in evolutionary theory and our knowledge of the natural history, social organization, and cognitive capacities of our study subjects.

It would be desirable to establish collaborative efforts to standardize methods and procedures and to replicate experiments in different test populations and species. Such studies would be useful for several reasons. First, they would strengthen comparative analyses by providing a richer database to test functional hypotheses about the factors that contribute to variation in prosocial preferences across species. Second, by using standardized methods, we can generate larger samples of behavior within species. This would generate more robust characterizations of behavior and enable us to begin assessing the extent of intraspecific variation in behavioral responses.

17

The Cultural Niche: Why Social Learning Is Essential for Human Adaptation

ROBERT BOYD,[*§] PETER J. RICHERSON,[†§] AND JOSEPH HENRICH[‡§]

In the last 60,000 years humans have expanded across the globe and now occupy a wider range than any other terrestrial species. Our ability to successfully adapt to such a diverse range of habitats is often explained in terms of our cognitive ability. Humans have relatively bigger brains and more computing power than other animals, and this allows us to figure out how to live in a wide range of environments. Here we argue that humans may be smarter than other creatures, but none of us is nearly smart enough to acquire all of the information necessary to survive in any single habitat. In even the simplest foraging societies, people depend on a vast array of tools, detailed bodies of local knowledge, and complex social arrangements and often do not understand why these tools, beliefs, and behaviors are adaptive. We owe our success to our uniquely developed ability to learn from others. This capacity enables humans to gradually accumulate information across generations and develop well-adapted tools, beliefs, and practices that are too complex for any single individual to invent during their lifetime.

---

*Department of Anthropology, University of California, Los Angeles, CA 90095; †Department of Environmental Science and Policy, University of California, Davis, CA 95616; and ‡Departments of Psychology and Economics, University of British Columbia, Vancouver, BC, Canada V6T 1Z4. §To whom correspondence may be addressed. E-mail: rboyd@anthro.ucla.edu, pjricherson@ucdavis.edu, or henrich@psych.ubc.ca.

In its brief evolutionary history, *Homo sapiens* has come to occupy a larger range than any other terrestrial vertebrate species. Earlier hominins, such as *Homo heidelbergensis* and Neanderthals, were limited to Africa and the temperate regions of southern Eurasia. Behaviorally modern humans were living in Africa by 70,000 years ago (Mourre et al., 2010). Between 50,000 and 60,000 years ago, people left Africa, crossing into southwest Asia (Klein, 2009). From there they spread rapidly through southern Eurasia, reaching Australia by 45,000 years ago, a feat that only one other terrestrial mammal (a murid rodent) was able to accomplish (Rowe et al., 2008). Soon after this, people penetrated far north, reaching the latitude of Moscow by 40,000 years ago and the Arctic Ocean by 30,000 years ago. People had spread almost as far south as the southern tip of South America 13,000 years ago, and by 5,000 years ago humans occupied virtually every terrestrial habitat except Antarctica and some islands in Oceania (Klein, 2009). Even the most cosmopolitan bird and mammal species have substantially smaller ranges (White et al., 1994; Bruce, 1999; Wozencraft, 2005).

This global expansion required the rapid development of a vast range of new knowledge, tools, and social arrangements. The people who moved out of Africa were tropical foragers. Northern Eurasia was an immense treeless steppe, relatively poor in plant resources and teeming with unfamiliar prey species. The people that roamed the steppe confronted a hostile climate—temperatures fell to –20 °C for months at a time, and there were often high winds. Surviving in such environments requires a whole new suite of adaptations—tailored clothing (Gilligan, 2010), well-engineered shelters, local knowledge about game, and techniques for creating light and heat. This is just the northern Eurasian steppe; each of the other environments occupied by modern human foragers presented a different constellation of adaptive problems. Ethnographic and historical accounts of 19th and 20th century foraging peoples make it clear that these problems were solved through a diverse array of habitat-specific adaptations (Kaplan et al., 2000). Although these adaptations were complex and functionally integrated, they were mainly cultural, not genetic, adaptations. Much evidence indicates, in fact, that local genetic changes have played only a relatively small part in our ability to inhabit such a diverse range of environments (Richerson and Boyd, 2005; Richerson et al., 2010).

Why are humans so much better at adapting to novel environments than other mammals? There have been many different answers to this question, but the most influential are rooted in the idea that people are simply smarter than other creatures. We have bigger brains and more computing power, and this allows us to adapt to a wider range of environments than other animals. One of the clearest statements of this hypothesis comes from a series of papers by Tooby, Cosmides, Pinker, and collaborators (Tooby and DeVore, 1987; Cosmides and Tooby, 2001; Barrett et al., 2007; Pinker, 2010). Other animals, they argue, are limited to what they call "dedicated intelligence," domain-specific learning and decision-making mechanisms that are adapted to particular environments. Humans, by contrast, have evolved "improvisational intelligence," a suite of uniquely flexible cognitive capacities that allow our species to acquire locally adaptive behavior in a wide range of environments. In short, we are adapted to the "cognitive niche" (Tooby and DeVore, 1987; Pinker, 2010). These capacities are augmented by our species' ability to learn from each other, especially using grammatical language.

This hypothesis flows from a nativist, modularist view of cognition. Its central premise is that broad general problems are much more difficult to solve than narrow specialized ones, and therefore the minds of all animals, including humans, are built of many special-purpose mechanisms dedicated to solving specific adaptive problems that face particular species. These mechanisms are modular in that they take inputs and generate outputs relevant to problems in particular domains such as mate choice, foraging, and the management of social relationships. These authors are nativists because they believe that evolved mechanisms depend on a considerable amount of innate information about the relationships between cues and outcomes in particular domains for particular species. For example, mechanisms that regulate decisions about mate choice in human males may be based on the assumption that long-term mating is likely, and thus selection favored a psychology that leads men to be attracted to young women. Analogous mechanisms in chimpanzees, which do not form long-term bonds, have produced a psychology that causes males to prefer older females, perhaps because they are better mothers (Muller et al., 2006). Mechanisms regulating social exchange are specialized in other ways. The innate content is built up because learning and decision mechanisms have been shaped by natural selection to solve the important recurrent adaptive problems that confronted the species.

This view of cognitive evolution seems to preclude flexible, widely applicable cognitive abilities; or, as Cosmides and Tooby put it, "on first inspection, there appear to be only two biologically possible choices for evolved minds: either general ineptitude or narrow competences" (Cosmides and Tooby, 2001). However, these authors believe that humans, and only humans, have undergone an evolutionary breakthrough that gives them "the computational ability to improvise solutions in developmental time to evolutionarily novel problems" (Barrett et al., 2007). The key ability is the use of cause-and-effect reasoning to make inferences about local environmental contingencies. As Pinker puts it,

> These inferences are played out internally in mental models of the world. . . . It allows humans to invent tools, traps, and weapons, to extract poisons and drugs from other animals and plants. . . . These cognitive stratagems are devised on the fly in endless combinations suitable to the local ecology. They arise by mental design and are deployed, tested, and fine-tuned by feedback in the lifetime of individuals. . . . (Pinker, 2010, pp 8993–8994)

These inferential capacities are augmented by a second evolutionary innovation, the ability to learn from each other, a capacity that dramatically lowers the cost of acquiring information necessary for local, contingent adaptations.

It seems likely that the average human is smarter than the average chimpanzee, at least in domains like planning, causal reasoning, and theory of mind. However, we do not think this is sufficient to explain our ecological success. The cognitive niche hypothesis overestimates the extent to which individual human cognitive abilities allow people to succeed in diverse environments and misunderstands the role that culture plays in a number of important ways. We suggest, instead, that our uniquely developed ability to learn from others is absolutely crucial for human ecological success. This capacity enables humans to gradually accumulate information across generations and develop well-adapted tools, beliefs, and practices that no individual could invent on their own. We have entered the "cultural niche," and our exploitation of this niche has had a profound impact on the trajectory of human evolution. In the remainder of this chapter, we will develop this argument in more detail.

CULTURE IS ESSENTIAL FOR HUMAN ADAPTATION

It is easy to underestimate the scope, sophistication, and importance of the pool of culturally transmitted information that supports human subsistence, even in what seem to be the "simplest" foraging societies. The archaeological record makes it clear that modern humans adapted to life above the Arctic Circle early in their expansion but tells us little about their way of life. However, ethnographic studies of the Netsilik and Copper Inuit, collectively known as the Central Inuit, give us a sense of the complexity of the adaptations that allow foragers to thrive in the Arctic. These people occupy a habitat that is harsh and unproductive, even by Arctic standards. Their groups were small, and their lifeways were simple compared with foragers living on the coasts of Alaska and Greenland. To focus your mind on the crucial adaptive challenges, imagine that you are marooned on a beach on the coast of King William Island (68.935N, 98.89W). It is November and it is very cold.

Your first problem is to stay warm. Monthly average temperatures in the winter months are between –25 °C and –35 °C. Even well-acclimatized people rapidly succumb to hypothermia below –1 °C, so you need warm clothes. If there were no wind and you could remain motionless, a cloak would do, but this is a windy place and you need to hunt, so you will need well-tailored clothes (Gilligan, 2010). In the winter, the Central Inuit wore elaborately constructed parkas and pants (Issenman, 1997). The best were made from caribou skins harvested in the fall. Caribou skins insulate better than seal or polar bear fur because the individual hairs have an unusual air-filled structure, something like bubble wrap (Otak, 2005). Caribou skins harvested in autumn have fur that is just the right thickness. Hides were repeatedly stretched, scraped, moistened, and then stretched again to yield pliable skins (Meeks and Cartwright, 2005). Parkas were assembled from multiple pieces to create a bell shape that captures heat, while also allowing moisture to dissipate when the hood is thrown back. Hoods were ruffed with a strip of fur taken from a wolverine's shoulders because its variable length makes it easier to clear the hoarfrost. Winter footwear was constructed with many layers: first the *alirsiik,* furlined caribou stockings, then the *ilupirquk,* short lightweight stockings with the fur outside, then a pair of *pinirait,* heavier stockings with the fur to the outside, then *kamiik,* boots with the fur outside, and finally *tuqtuqutiq,* short heavy double-soled boots of caribou skin. Clothing was stitched together with fine thread made from sinew taken from around the vertebrae of caribou. The sinew had to be cleaned, scraped, shredded, and twisted to make thread. Several different kinds of stitches were used for different kinds of seams. A complicated double stitch was used to make footwear waterproof. To make these stitches, Central Inuit women used fine bone needles that made holes that were smaller in diameter than the thread (Issenman, 1997).

Not even the best clothing is enough to protect you from winter storms, so you need shelter. During the winter most Inuit lived in substantial driftwood and sod houses, but the Central Inuit wintered on the sea ice, living in snow houses. These round vaulted structures were ≈3 m high, made of snow blocks cut with a serrated bone knife. The central room was built above a pit, with platforms for sleeping, and a long entrance tunnel below the level of the main room with several low doors to prevent heat loss. The walls were usually lined with skins suspended from toggles on the outside of the snow house. This design allowed the snow walls to stay near freezing, while the inside of the snow house could reach temperatures of 10–20 °C (Damas, 1984).

You need a source of heat and light in your snow house, for cooking and for melting sea ice for water. You cannot use wood fires because there are no trees. Instead, Arctic peoples carved lamps from soapstone and fueled them with rendered seal fat. These lamps were made from oblong stones between 30 cm and 1 m long; a shallow, sharp-sided depression was carved from the surface of the stone, and the lamp was equipped with a long, curtain-like wick made of moss. A well-managed lamp burned without producing any soot (Issenman, 1997).

You also need food. Plants are easy to gather, but for most of the year this is not an option in the Arctic. During the winter, the Central Inuit hunted seals, mainly by ambushing them at their breathing holes. When the sea ice begins to freeze, seals claw a number of breathing holes in the ice within their home ranges. As the ice thickens, they maintain these openings, which form conical chambers under the ice. The Inuit camped in snowy spots near the seals' breathing holes. The ice must be covered with snow to prevent the seals from hearing the hunters' footsteps and evading them. Inuit hunted in teams, monitoring as many holes as possible. The primary tool was a harpoon approximately 1.5 m long. Both the main shaft and foreshaft were carved from antler. On the tip was a detachable toggle harpoon head connected to a heavy braided sinew line. The other end of the harpoon was made from polar bear bone honed to a sharp point. At each hole, the hunter opened the hard icy covering using the end of the harpoon, smelled the interior to make sure it was still in use, and then used a long, thin, curved piece of caribou antler with a rounded nob on one end to investigate the chamber's shape and plan his thrust. The hunter carefully covered most of the hole with snow and tethered a bit of down over the remaining opening. Then, the hunter waited motionless in the frigid darkness, sometimes for hours. When the seal's arrival disturbed the down, the hunter struck downward with all his weight. If he speared the seal, he held fast to the line connected to his harpoon's point; the seal soon tired and could be hauled onto the ice (Balikci, 1989).

During the high summer, the Central Inuit used the leister, a special three-pronged spear with a sharp central spike and two hinged, backward-facing points, to harvest Arctic char in large numbers. Later in summer and the fall, they shifted to caribou hunting. On land, caribou were mainly stalked or driven into ambush, and kills had to be made from a substantial distance. This required a bow with the power to propel a heavy arrow at high velocity. The simplest way to accomplish this is to make a long bow using a dense elastic wood like yew or osage orange, a design common in South America, Eastern North America, Africa, and Europe. This solution was not available to the Inuit, who had only driftwood (mainly spruce), horn, and antler available. Instead, they made short bows and used every bowyer's trick to increase their power. A bow can be made more powerful by adding wood to the limbs. However, making the bow thicker increases the stress within the bow, leading to catastrophic and dangerous failure. This problem is exacerbated in short bows because the curvature is greater. Instead, the Inuit made bows that were thin front to back, wide near the center, and tapering toward the tips. These bows were also recurved, meaning that the unbraced bow formed a backward "C" shape. Bracing the bow leads to a compound curve, a geometry that stores more potential energy. Finally, the Inuit constructed a unique form of composite bow. When a bow is bent, the back (the side away from the archer) is stretched, whereas the belly (the side closer to the archer) is compressed. Wood, horn, and antler are stronger in compression than tension, so the ability of a bow to sustain strong bending forces can be enhanced by adding a material that is strong in tension to the back of the bow. In central Asia and western North America, sinew was glued to the back of the bow to strengthen short bows for use on horseback. The Inuit lashed a woven web of sinew to the backs of their bows, probably because they had no glues that would work in the moist, cold conditions of the Arctic (Mason, 2007).

This sampler of Inuit lifeways represents only a tiny fraction of the immense amount of habitat-specific knowledge that is necessary for humans to survive and prosper in the Central Arctic. To stay warm and get enough to eat, you have to know how to make and use clothes, snow houses, lamps, harpoons, leisters, and bows. We have omitted other crucial tools like kayaks, dog sleds, and sun goggles, and of course, we have had to omit most of the details necessary to make and use the tools we did mention. Moreover, there is still much more you have to know to stay alive. Predicting storms, understanding the habits of game species, making baskets, building sledges, and managing dogs—all require extensive knowledge. Traveling on ice is essential, but also treacherous, and there is much to know about how the current temperature, recent weather, and the color and texture of the ice tell you where and when it is safe to travel. [Nelson (1969) devotes four chapters to ice lore in his book on hunting among the Inupiaq of northern Alaska.]

So, here is the question: Do you think that you could acquire all of the local knowledge necessary to survive in the Arctic on your own? If superior cognitive ability alone is what allows humans to adapt to diverse habitats, then it should be possible. Moreover, to a first approximation, this is the only way that other animals have to learn about their environments—they must rely mainly on innate information and individual experience to figure out how to find food, build shelters, and in some cases to make tools. It is true that some species have simple traditions, probably maintained by learning mechanisms like stimulus enhancement and emulation. However, in every case, the traditions involve behaviors that individuals can and do learn on their own, or combine a handful of elements learned by multiple individuals (Tennie et al., 2009). There are no convincing examples in which social learning allows the gradual cumulative cultural evolution of complex, locally adaptive behaviors that individuals could not learn on their own.

Could you make it? We don't think so.

Two different kinds of natural experiments support the intuition that forager adaptations are beyond the inventive capacities of individuals. The first, which might be called "the lost European explorer experiment," has been repeated many times during the past several centuries. Typically some explorers get stranded in an unfamiliar habitat in which an indigenous population is flourishing. Despite desperate efforts and ample learning time, the explorers die or suffer terribly owing to the lack of crucial information about how to adapt to the habitat. If they survive, it is often due to the hospitality of the indigenous population. The Franklin Expedition of 1845–1846 provides a good example (Lambert, 2011). Sir John Franklin, a Fellow of the Royal Society and an experienced Arctic traveler, set out with two ships to explore the northern coast of North America and find the Northwest Passage. It was the best-equipped expedition in the history of British polar exploration, furnished with an extensive library, manned by a select crew, and stocked with a 3-year supply of food. The expedition spent the winter of 1846 at King William Island, where it became trapped in the ice. When food ran short, the explorers abandoned their ships and attempted to escape on foot. Everyone eventually perished from starvation and scurvy, perhaps exacerbated by lead poisoning from their tinned food.

King William Island is the heart of Netsilik territory, and the Netsilik have lived there for almost a millennium. King William Island is rich in animal resources—the main harbor is named *Uqsuqtuuq* which means "lots of fat." The British sailors starved because they did not have the necessary local knowledge and, despite being endowed with the same improvisational intelligence as the Inuit and having 2 years to use this intelligence, failed to learn the skills necessary to subsist in this habitat. Interestingly, the Norwegian explorer Roald Amundsen spent two winters on King William Island in 1903–1904. Amundsen sought out the Netsilik and learned from them how to make skin clothing, hunt seals, and manage dog sleds. He and his crew survived and completed the first successful traverse of the Northwest Passage. Later he would put these Inuit skills to good use in his race with Scott to the South Pole. Results from this lost European explorer experiment, and many others, suggest that intelligence alone is not enough. For a similar discussion of the ill-fated Burke and Wills expedition into the Australian outback, see Henrich and McElreath (2003).

A second line of evidence comes from the loss of beneficial technologies in small, isolated populations. For instance, the Tasmanian tool kit gradually lost complexity after isolation from mainland Australia at the end of the Holocene (Henrich, 2004b). Other Pacific island groups have apparently lost useful technologies, such as canoes, pottery, and the bow and arrow (Kline and Boyd, 2010). The best documented example comes from the isolated Polar Inuit of northwest Greenland. Explorers Elisha Kane and Isaac Hayes wintered with the Polar Inuit in 1853 and 1861, respectively, and reported that the Polar Inuit lacked kayaks, leisters, and bows and arrows and that their snow houses did not have the long heat-saving entryways that were seen among other Inuit populations. They could not hunt caribou, could only hunt seals during part of the year, and were unable to harvest Arctic char efficiently, although char were plentiful in local streams (Mary-Rousselière, 1996). Apparently the population was struck by an epidemic in the 1820s that carried away the older, knowledgeable members of the group, and according to custom, their possessions had to be buried with them (Rasmussen, 1908). The Polar Inuit lived without these tools until about 1862, when they were visited by a group of Inuit who migrated to Greenland from Baffin Island (Rasmussen, 1908; Mary-Rousselière, 1996). There is every reason to believe that these tools would have been useful between 1820 and 1862. The Polar Inuit population declined during this period, and the tools were immediately adopted once they were reintroduced. After their introduction, population size increased. It is also telling that the kayaks used by the Polar Inuit around the turn of the century closely resemble the large, beamy kayaks used by Baffin Island Inuit and not the small sleek kayaks of the West Greenland Inuit. Over the next half century the Polar Inuit kayak design converged back to the West Greenland design (Golden, 2006). If this inference is correct it means that for 40 years (nearly two generations) the Polar Inuit could have benefitted from the lost knowledge. Moreover, they collectively remembered kayaks, leisters, and bows and arrows, but did not know how to make them and could not recreate that knowledge.

CULTURAL ADAPTATION IS A POPULATION PROCESS

We think that this body of evidence rules out the idea that superior cognitive ability *alone* explains human adaptability; the ability to cumulatively learn from others must play a crucial role. Although advocates of the cognitive niche hypothesis focus on cognition, they do not ignore social learning. They argue that the ability to learn from others reduces the average cost of acquiring locally adaptive information. For example, Barrett et al. (2007) write:

> Cognitive mechanisms underlying cultural transmission coevolved with improvisational intelligence, distributing the costs of the acquisition of nonrivalrous information over a much greater number of individuals, and

> allowing its cost to be amortized over a much greater number of advantageous events and generations. Unlike other species, cultural transmission in humans results in a ratchet-like accumulation of knowledge. (p 244)

On the surface this seems to be a logical argument. It may be costly for individuals using improvisational intelligence to discover locally adaptive information, but once it is acquired, others can get it by teaching or imitation at relatively low cost. As a result, social learning acts to spread the cost of innovations over all who benefit. Innovations accumulate, leading to an accumulation of knowledge.

However, this reasoning is mistaken. It is probably true that learning from others either by teaching or imitation is usually cheaper than learning on your own. It is like cheating on a test: you do as well as the person you copy from but avoid all that tedious studying. However, evolutionary models show that if this is the only benefit of social learning, there will be no increase in the ability of the population to adapt (Rogers, 1988; Boyd and Richerson, 1995; Lehmann et al., 2010; Rendell et al., 2010). This surprising result emerges from the coevolutionary processes that affect the kinds of behaviors that are available to imitate and the psychology that controls learning and imitation. These evolutionary models of social learning rest on two assumptions. First, the propensities to learn and to imitate are part of an evolved psychology shaped by natural selection. This means that the balance between learning and imitating will be governed by the relative fitness of the two modes of behavior—the average fitness of the population is irrelevant. When few individuals imitate, imitators will acquire the locally adaptive behavior with the same probability as individual learners. Because they do not pay the cost of learning, imitators have higher fitness, and the propensity to imitate spreads. As the number of imitators increases, some imitate individuals who imitated other individuals, who imitated other individuals, and so on until the chain is rooted in someone who extracted the information from the environment. As the fraction of imitators in the population increases, these chains extend further.

The second assumption is that the environment varies in time or space. This means that as chains of imitation get longer, there is a greater chance that the learner who roots the chain learned in a different environment than the current environment, either because the environment has changed since then or because someone along the chain migrated from a different environment. The upshot is that on average imitators will be less likely to acquire the locally adaptive behavior than learners. The propensity to imitate will continue to increase until this reduction in fitness exactly balances the benefit of avoiding the costs of learning. At evolutionary equilibrium, the population has the same average fitness as a population without any imitation. There will be no increase in the ability to adapt to varying environments, and cumulative cultural adaptation will not occur.

Although this treatment is very simple, the basic result holds in more realistic models. The primary insight that emerges from these models is that imitation is a form of free riding—imitators scrounge information without producing anything of value. Free riders increase until they destroy the benefits of free riding. Realistic levels of relatedness among models and imitators do not qualitatively change the result (Lehmann et al., 2010). The advocates of the cognitive niche hypothesis err because they take it as unproblematic that once a beneficial innovation arises, it will spread, and as a result, the capacities for imitation will be favored by selection. However, to understand the evolution of social learning psychology you have to know what is available to learn, and this in turn is affected by the nature of the learning psychology. If imitators are simply information scroungers, then they will spread until selection no longer favors imitation.

Thinking about the coevolution of the cultural pool of observable behavior and the genes that control the individual and cultural learning suggests that cultural learning can increase average fitness only if it increases the ability of the population to create adaptive information (Boyd and Richerson, 1995). The propensity to imitate evolves because it is directly beneficial to the individual, but it may, nonetheless, also benefit the population as a side effect. We have thought of three ways in which this could happen. First, cultural learning can allow individuals to learn selectively—using environmental cues when they provide clear guidance and learning from others when they do not. Second, cultural learning allows the gradual accumulation of small improvements, and if small improvements are cheaper than big ones, cultural learning can reduce the population's learning costs. Finally, by comparing "teachers" and learning selectively from those that seem most successful, "pupils" can acquire adaptive information without making any inferences based on environmental cues. If individuals acquire information from multiple teachers and recombine this information, this process can create complex cultural adaptations without any intelligence, save that required to distinguish among more- and less-successful teachers.

The ability to learn or imitate selectively is advantageous because opportunities to learn from experience or by observation of the world vary. For example, a rare chance observation might allow a hunter to associate a particular spoor with a wounded polar bear, or to link the color and texture of ice with its stability on windy days just after a thaw. Such rare cues allow accurate low-cost inferences about the environment. However, most individuals will not observe these cues, and thus making the same inference will be much more difficult for them. Organisms that cannot imitate must rely on individual learning, even when it is difficult and error prone. They are stuck with whatever information that nature offers. In contrast, an organism capable of cultural learning can afford to be choosy, learning individually when it is cheap and accurate, and relying on cultural learning when environmental information is costly or inaccurate. We have shown (Boyd and Richerson, 1988, 1995) that selection can lead to a psychology that causes most individuals to rely on cultural learning most of the time, and also simultaneously increases the average fitness of the population relative to the fitness of a population that does not rely on cultural information. These models assume that our learning psychology has a genetically heritable "information quality threshold" that governs whether an individual relies on inferences from environmental cues or learns from others. Individuals with a low information quality threshold rely on even poor cues, whereas individuals with a high threshold usually imitate. As the mean information quality threshold in the population increases, the fitness of learners increases because they are more likely to make accurate or low-cost inferences. At the same time, the frequency of imitators also increases. As a consequence, the population does not keep up with environmental changes as well as a population of individual learners. Eventually, an equilibrium emerges in which individuals deploy both individual and cultural learning in an optimal mix. At this equilibrium, the average fitness of the population is higher than in an ancestral population lacking cultural learning. When most individuals in the population observe accurate environmental cues, the equilibrium threshold is low, individual learning predominates, and culture plays little role. However, when it is usually difficult for people to learn individually, the equilibrium threshold is high, and most imitate, even when the environmental cues that they do observe indicate a different behavior than the one they acquire by cultural learning. We take the evidence on Inuit adaptations as indicating that many of the problems that faced the Inuit are far too difficult for most individuals to solve. As a result, we interpret this logic as predicting that selection should have favored a psychology that causes individuals to rely heavily on cultural learning.

The ability to learn culturally can also raise the average fitness of a population by allowing acquired improvements to accumulate from one generation to the next. Many kinds of traits admit successive improvements toward some optimum. Bows vary in many dimensions that affect performance—such as length, width, cross section, taper, and degree of recurve. It is typically more difficult to make large improvements by trial and error than small ones for the same reasons that Fisher (1930) identified in his "geometric model" of genetic adaptation. In a small neighborhood in design space, the performance surface is approximately flat, so that even if small changes are made at random, half of them will increase the payoff (unless the design is already at the optimum). Large changes will improve things only if they are in the small cone that includes the distant optimum. Thus, we expect it to be much harder to design a useful bow from scratch than to tinker with the dimensions of a reasonably good bow. Now, imagine that the environment varies, so that different bows are optimal in different environments, perhaps because the kind of wood available varies. Sometimes a long bow with a round cross section is best, other times a short flat wide bow is best. Organisms that cannot imitate must start with whatever initial guess is provided by their genotype. Over their lifetimes, they can learn and improve their bow. However, when they die, these improvements disappear with them, and their offspring must begin again at the genetically inherited initial guess. In contrast, cultural species can learn how to make bows from others after these have been improved by experience. Therefore, cultural learners start their search closer to the best design than pure individual learners and can invest in further improvements. Then, they can transmit *those* improvements to the grandkids, and so on down through the generations until quite sophisticated artifacts evolve. Historians of technology have demonstrated how this step-by-step improvement gradually diversifies and improves tools and other artifacts (Basalla, 1988; Petroski, 1992). Even "great insights" often result from lucky accidents or the recombination of elements from different technological traditions rather than the work of a creative genius who buckles down and racks his brain (Henrich, 2010; S. Johnson, 2010).

The evolution of kayak keels by West Greenland Inuit provides an instructive example of how innovations arise and spread (Scavenius, 1975). When hunting marine mammals from a kayak, Inuit hunters always paddled their kayak hard toward the prey, then picked up their harpoon and hurled it directly over the bow. This increased the momentum transferred to the harpoon and prevented capsizing. When firearms first spread in West Greenland, the Inuit found that they could not pick up and aim their guns before the kayak veered off course, and thus could only use them from land or ice floes. In 1824, a prominent Inuit hunter named Jens Reimer began to experiment with methods to stabilize kayaks for firearm use. He tried trailing a line behind the kayak, but this did not work. He then fastened a partially submerged wooden plate to the kayak's stern, in imitation of the rudders of European ships. This did not work very well either—it was noisy, and the fastenings tended to fail. Nonetheless, a number of younger hunters imitated Reimer, perhaps owing to his local success and prestige. They were not able to produce a quality *ayût* (the Greenlandic word for both a ship's rudder and a kayak keel), and out of "bashfulness" (Scavenius, 1975, p 27) hid their crude rudders under the waterline. They soon discovered that this unintentional innovation allowed them to use guns from their kayaks, and over the next 50 years the *ayût* underwent a series of further small improvements, eventually creating the modern form.

Finally, if learners can compare the success of individuals modeling different behaviors, then a propensity to imitate the successful can lead to the spread of traits that are correlated with success, even though imitators have no causal understanding of the connection. This is obvious when the scope of traits being compared is narrow. You see that your uncle's bow shoots farther than yours and notice that it is thicker, but less tapered, and uses a different plait for attaching the sinew. You copy all three traits, even though in reality it was just the plaiting that made the difference. As long as there is a reliable statistical correlation between plaiting and power, plaiting form trait will change so as to increase power. Causal understanding is helpful because it permits the exclusion of irrelevant traits like the bow's color. However, causal understanding need not be very precise as long as the correlation is reliable. Copying irrelevant traits like thickness or color will only add noise to the process. By recombining different components of technology from different but still successful individuals, copiers can produce both novel and increasingly adaptive tools and techniques over generations, without any improvisational insights. An Inuit might copy the bow design from the best bowyer in his community but adopt the sinew plaiting used by the best hunter in a neighboring community. The result could be a better bow than anyone made in the previous generation without anyone inventing anything new.

Consistent with this, laboratory and field evidence suggests that both children and adults are predisposed to copy a wide range of traits from successful or prestigious people (Henrich and Gil-White, 2001). Advertisers clearly know this. After all, what does Michael Jordan really know about underwear? Recent work in developmental psychology shows that young children readily attend to cues of reliability, success, confidence, and attention when choosing who to learn from (Birch et al., 2008, 2010). Even infants selectively attend to knowledgeable adults rather than their own mothers in novel situations (Stenberg, 2009). This feature of our cultural learning psychology fits a priori evolutionary predictions, emerges spontaneously in experiments, develops early without instruction, and operates largely outside conscious awareness.

These models predict that an adaptive evolved psychology will often cause individuals to acquire the behaviors they observe used by in others even though inferences based on environmental cues suggest that alternative behaviors would be better. In a species capable of acquiring behavior by teaching or imitation, individuals are exposed to two different kinds of cues that they can use to solve local adaptive problems. Like any other organism, they can make inferences based on cues from the environment.

However, they also observe the behaviors of a sample of their population. When most individuals can solve the adaptive problem using environmental cues alone, the models predict that an optimal learning psychology will result in social learning playing a significant but relatively modest role. Many people will rely on their own inferences, but some will copy to avoid learning costs. However, often only a minority will be able to solve the adaptive problem on the basis of environmental cues alone, because the appropriate environmental cues are rare or the adaptive problem is too complex. Then, if the environment is not too variable, an adaptive psychology will evolve in which most people ignore environmental cues and adopt behaviors that are common in the sample of the population they observe. They modify these behaviors rarely, or only at the margin, and as a result local adaptations evolve gradually often over many generations.

EVIDENCE FOR CULTURAL ADAPTATION

The cultural niche hypothesis and the cognitive niche hypothesis make sharply different predictions about how local adaptations are acquired and understood. The cognitive niche hypothesis posits that technologies are adaptive because improvisational intelligence allows some individuals to figure out how they work and why they are better than alternatives. These acquired understandings of the world are then shared, allowing others to acquire the same causal understanding without costly individual investigation. In contrast, we argue that cultural evolution operating over generations has gradually accumulated and recombined adaptive elements, eventually creating adaptive packages beyond the causal understanding of the individuals who use them. In some cases elements of causal understanding may be passed along, but this is not necessary. Often individuals will have no idea why certain elements are included in a design, nor any notion of whether alternative designs would be better. We expect cultural learners to first acquire the local practices and occasionally experiment or modify them. At times this will mean that cultural learning will overrule their direct experience, evolved motivations, or reliably developing intuitions.

Several lines of evidence support the cultural learning hypothesis.

The anthropological literature on child development (Lancy, 1996, 2009, 2010) indicates that children and adolescents acquire most of their cultural information by learning from older individuals who typically discourage questions from young learners and rarely provide causal explanations of their behavior. Kids practice adult behaviors, often using toy versions of adult tools, during mixed-age play, and little experimentation is observed, except that necessary to master the adult repertoire (MacDonald, 2007; Hewlett et al., 2011).

The reliance of young learners on carefully observing and imitating the local repertoires revealed in the anthropological record converges with recent experiments on imitation (Lyons et al., 2007; Nielsen and Tomaselli, 2010). In these experiments, an adult performs a behavior like opening a complex puzzle box to get a reward. The adult's behavior includes both necessary and unnecessary actions. A subject, either a child or a chimpanzee, observes the behavior. Children's performance on such tasks in both western and small-scale societies differs in important ways from that of chimpanzees. Children accurately copy all steps, including steps that direct visual inspection would suggest are unnecessary. Children seem to implicitly assume that if the model performed an action, it was probably important, even if they do not understand why. Chimpanzees do not seem to make this assumption; they mainly skip the unnecessary steps, leading them to develop more efficient repertoires than children (Whiten et al., 2009) in these experimental settings.

Many examples indicate that people often do not understand how adaptive practices work or why they are effective. For example, in the New World, the traditional use of chili peppers in meat recipes likely protected people from foodborne pathogens (Billing and Sherman, 1998). This use of chili peppers is particularly interesting because they are inherently unpalatable. Peppers contain capsaicin, a chemical defense evolved in the genus *Capsicum* to prevent mammals (especially rodents) from eating their fruits. Nonhuman primates and human infants find peppers aversive because capsaicin stimulates pain receptors in the mouth. Efforts to inculcate a taste for chilies in rats using reinforcement procedures have failed (Rozin et al., 1979). However, human food preferences are heavily influenced by the preferences of those around us (Birch, 1987), so we overcome our innate aversion and actually learn to enjoy chilies. Psychological research indicates that people do not get accustomed to the chemical burning sensation. Instead, observational learning leads people to reinterpret their pain as pleasure or excitement (Rozin et al., 1981). So, New World peoples learned to appropriately use and enjoy chili peppers without understanding their antimicrobial properties, and to do this they had to overcome an instinctive aversion that we share with other mammals.

Fijian food taboos provide another example of this process. Many marine species in the Fijian diet contain toxins, which are particularly dangerous for pregnant women and perhaps nursing infants. Food taboos targeting these species during pregnancy and lactation prohibit women from eating these species and reduce the incidence of fish poisoning during this period. Although women in these communities all share the same food taboos, they offer quite different causal explanations for them, and little information is exchanged among women save for the taboos themselves (Henrich and Henrich, 2011). The taboos are learned and are not related to pregnancy sickness aversions. Analyses of the transmission pathways for these taboos indicate the adaptive pattern is sustained by selective learning from prestigious women.

CULTURE AND MALADAPTATION

Cultural adaptation comes with a built-in tradeoff. The cumulative cultural evolution of complex, hard-to-learn adaptations requires individuals to adopt the behavior of those around them even if it conflicts with their own inferences. However, this same propensity will cause individuals to acquire *any* common behavior as long as it is not clearly contradicted by their own inferences. This means that if there are cognitive or social processes that make maladaptive ideas common, and these ideas are not patently false or harmful, people will adopt these ideas as well. Moreover, it is clear that several such processes exist. Here are a couple of examples. For a longer discussion, see Richerson and Boyd (2005).

Weak Cognitive Biases Can Favor the Spread of Maladaptive Beliefs or Practices over Generations

Laboratory diffusion chain studies clearly document that biases that have undetectable effects on individual decisions can have very strong effects when iterated over "generations" in the laboratory (Beppu and Griffiths, 2009). The same effect may lead to the spread of false beliefs in natural populations. For example, Boyer (2002) argues that a number of cognitive biases explain the spread of supernatural beliefs and account for the widespread occurrence of folktales about ghosts and zombies.

Adaptive Social Learning Biases Can Lead to Maladaptive Outcomes

A model's attributes provide indirect evidence about whether it is useful to imitate her. If she is successful, then by imitating her you can increase your chances of acquiring traits that gave rise to her success. If she is more similar to you than alternative models, her behavior may work better in your situation. If her behavior is more common than alternatives, then it is likely to be adaptive because learning increases the frequency of adaptive behaviors. An evolved cultural learning psychology that incorporates such biases increases the chance of acquiring beneficial beliefs and behaviors. However, these same biases can sometimes lead to the spread of maladaptive beliefs and practices. For example, the tendency to imitate the prestigious, or those making credibility-enhancing displays of commitment, can lead to a "runaway" process analogous to sexual selection (Richerson and Boyd, 2005), and this may explain the cultural evolution of maladaptive cultural systems in which people risk life and limb to summit icy peaks or achieve spiritual perfection in celibate seclusion (Henrich, 2009).

Culture Is Part of Human Biology and Has Profoundly Shaped Human Evolution

We have recounted two contrasting accounts of the nature and origins of human uniqueness. On the one hand, there is a widespread view that people are like other mammals, just a lot smarter—in essence, we are brainy, hairless chimpanzees. We have a uniquely flexible cognitive system that lets us make causal inferences in a wide range of environments and use that information to create much better tools, and these differences have allowed us to spread across the world, dominating the world's biota like no other creature. By contrast, we argue that individuals are not nearly smart enough to solve the myriad adaptive problems they face in any of their many habitats. Even experts lack a detailed causal understanding of the tools and techniques that permit them to survive. High-fidelity cultural learning allows human populations to solve these problems because it allows selective learning and the accumulation of small improvements over time. Of course, sophisticated, flexible cognition is important too. However, the degree of cognitive flexibility varies widely in nature—chimpanzees can solve problems that baffle monkeys, and monkeys are geniuses compared with opossums. Nonetheless, no species occupies as wide a range of habitats as *Homo sapiens*. In contrast, there is a sharp break between human cultural learning capacities and those of even our closest relatives. As a result, it is more apt to think of humans occupying a cultural niche than a cognitive niche.

The evolution of the psychological capacities that give rise to cumulative cultural evolution is one of the key events in our evolutionary history. The availability of large amounts of valuable cultural information would have favored the evolution of bigger brains equipped to acquire, store, organize, and retrieve cultural information, a fact that may explain the rapid increase in human encephalization over the last 500,000 years and the evolution of specialized cognitive abilities that emerge early in life, such as theory of mind, selective social referencing (Stenberg, 2009), overimitation (Lyons et al., 2007), a functional understanding of artifacts (Wohlgelernter et al., 2010), and the use of taxonomic inheritance and category-based induction for living kinds (Atran and Medin, 2008). The presence of culturally evolved techniques and products—such as fire, cooking, weapons, and tools—created new selection pressures acting on our bones, muscles, teeth, and guts (Richerson et al., 2010).

Culture has opened up a vast range of evolutionary vistas not available to noncultural species. Nonetheless, culture is as much a part of human biology as our peculiar pelvis. This approach contrasts with the common view that culture and biology are in a tug-of-war for control of human behavior. This common view probably taps into a deep vein of Western thought, which itself may be the result of evolved cognitive biases (Bloom, 2004), but it makes little sense. The ancestral condition in the human lineage is a psychology that does not permit cumulative cultural evolution. Despite earnest efforts, chimpanzees cannot be socialized to become humans and have little or no cumulative cultural evolution. Beginning early in human ontogeny, our psychology allows us to learn from others, powerfully and unconsciously motivates us to do so, and shapes the kind of traits that evolve. So it does not make sense to ask, does culture overcome biology? The right question to ask is, how do genetic and cultural inheritance interact to produce the observed patterns of human psychology and behavior (Henrich et al., 2010b)?

ACKNOWLEDGMENTS

We thank Clark Barrett for very useful comments on a previous draft of this article, and two anonymous referees for their help. This work was supported in part by National Institutes of Health Grant RC1TW008631-02 (to R.B.) and the Canadian Institute for Advanced Research (J.H.).

References

Abbot P, et al. (2011) Inclusive fitness theory and eusociality. *Nature* 471:E1–E4, author reply E9–E10.

Alexander RD (1974) The evolution of social behavior. *Annu Rev Ecol Syst* 5:325–383.

Alexander RD (1979) *Darwinism and Human Affairs* (Univ of Washington, Seattle).

Alexander RD, Borgia G (1978) Group selection, altruism and the levels of organization of life. *Annu Rev Ecol Syst* 9:449–474.

Allen-Arave W, Gurven M, Hill K (2008) Reciprocal altruism, rather than kin selection, maintains nepotistic food transfers on an Ache reservation. *Evol Hum Behav* 29:305–318.

Allesina S, Levine JM (2011) A competitive network theory of species diversity. *Proc Natl Acad Sci USA* 108:5638–5642.

Allison PD (1995) *Survival Analysis Using the SAS System: A Practical Guide* (SAS Institute, Cary, NC).

Alvarez HP (2004) Residence groups among hunter-gatherers. In *Kinship and Behavior in Primates,* eds Chapais B, Berman CM (Oxford Univ Press, New York), pp 420–442.

Alvarez-Curto E, Rozen DE, Ritchie AV, Fouquet C, Baldauf SL, Schaap P (2005) Evolutionary origin of cAMP-based chemoattraction in the social amoebae. *Proc Natl Acad Sci USA* 102:6385–6390.

Alvarez-Ponce D, Aguadé M, Rozas J (2009) Network-level molecular evolutionary analysis of the insulin/TOR signal transduction pathway across 12 *Drosophila* genomes. *Genome Res* 19:234–242.

Amdam GV, Norberg K, Hagen A, Omholt SW (2003) Social exploitation of vitellogenin. *Proc Natl Acad Sci USA* 100:1799–1802.

Ament SA, Corona M, Pollock HS, Robinson GE (2008) Insulin signaling is involved in the regulation of worker division of labor in honey bee colonies. *Proc Natl Acad Sci USA* 105:4226–4231.

American Academy of Microbiology (2009) *Antibiotic Resistance: An Ecological Perspective on an Old Problem* (American Academy of Microbiology, Washington, DC). Available at http://academy.asm.org/index.php/colloquium-program/environmental-microbiology-ecology-a-evolution-/190-antibiotic-resistance-an-ecological-perspective-on-an-old-problem-september-2009-b. Accessed February 12, 2011.

Andersson DI (2006) The biological cost of mutational antibiotic resistance: Any practical conclusions? *Curr Opin Microbiol* 9:461–465.

Andersson DI, Levin BR (1999) The biological cost of antibiotic resistance. *Curr Opin Microbiol* 2:489–493.

Antia R, Yates A, de Roode JC (2008) The dynamics of acute malaria infections. I. Effect of the parasite's red blood cell preference. *Proc Biol Sci* 275:1449–1458.

Apostolou M (2007a) Elements of parental choice: The evolution of parental preferences in relation to in-law selection. *Evol Psychol* 5:70–83.

Apostolou M (2007b) Sexual selection under parental choice: The role of parents in the evolution of human mating. *Evol Hum Behav* 28:403–409.

Apraiz AG (1999) Influence of family size and birth order on menarcheal age of girls from Bilbao city (Biscay, Basque country). *Am J Hum Biol* 11:779–783.

Araki T, Nakao H, Takeuchi I, Maeda Y (1994) Cell-cycle-dependent sorting in the development of *Dictyostelium* cells. *Dev Biol* 162:221–228.

Archetti M, Ubeda F, Fudenberg D, Green J, Pierce NE, Yu DW (2011) Let the right one in: A microeconomic approach to partner choice in mutualisms. *Am Nat* 177:75–85.

Arkhipova I, Meselson M (2000) Transposable elements in sexual and ancient asexual taxa. *Proc Natl Acad Sci USA* 97:14473–14477.

Arnot D (1998) Clone multiplicity of *Plasmodium falciparum* infections in individuals exposed to variable levels of disease transmission. *Trans R Soc Trop Med Hyg* 92:580–585.

Arnqvist G, Rowe L (2005) *Sexual Conflict* (Princeton Univ Press, Princeton, NJ).

Atran S, Medin DL (2008) *The Native Mind and the Cultural Construction of Nature* (MIT Press, Cambridge, MA).

Atzmony D, Zahavi A, Nanjundiah V (1997) Altruistic behaviour in *Dictyostelium discoideum* explained on the basis of individual selection. *Curr Sci* 72:142–145.

Augustine DJ (2003) Long-term, livestock-mediated redistribution of nitrogen and phosphorus in an East African savanna. *J Appl Ecol* 40:137–149.

Aureli F, Schaffner CM (2002) Relationship assessment through emotional mediation. *Behaviour* 139:393–420.

Austin AT, Yahdjian L, Stark JM, Belnap J, Porporato A, Norton U, Ravetta DA, Schaeffer SM (2004) Water pulses and biogeochemical cycles in arid and semiarid ecosystems. *Oecologia* 141:221–235.

Avise JC (2001) Evolving genomic metaphors: A new look at the language of DNA. *Science* 294:86–87.

Avise JC, Ayala FJ, eds (2007) *In the Light of Evolution I: Adaptation and Complex Design. Proc Natl Acad Sci USA* 104(Suppl):8563–8676.

Avise JC, Ayala FJ, eds (2009a) *In the Light of Evolution III: Two Centuries of Darwin. Proc Natl Acad Sci USA* 106(Suppl):9933–10066.

Avise JC, Ayala FJ, eds (2009b) *In the Light of Evolution IV: The Human Condition. Proc Natl Acad Sci USA* 107(Suppl):8897–9022.

Avise JC, Hubbell SP, Ayala FJ, eds (2008) *In the Light of Evolution II: Biodiversity and Extinction. Proc Natl Acad Sci USA* 105(Suppl.):11453–11586.

Axelrod R, Hamilton WD (1981) The evolution of cooperation. *Science* 211:1390–1396.

Axelsson E, Albrechtsen A, van As P, Li L, Megens HJ, Vereijken AL, Crooijmans RP, Groenen MA, Ellegren H, Willerslev E, Nielsen R (2010) Segregation distortion in chicken and the evolutionary consequences of female meiotic drive in birds. *Heredity* 105:290–298.

Azhar M, Kennady PK, Pande G, Espiritu M, Holloman W, Brazill D, Gomer RH, Nanjundiah V (2001) Cell cycle phase, cellular Ca^{2+} and development in *Dictyostelium discoideum*. *Int J Dev Biol* 45:405–414.

Aziz RK, Breitbart M, Edwards RA (2010) Transposases are the most abundant, most ubiquitous genes in nature. *Nucleic Acids Res* 38:4207–4217.

Babiker HA, Ranford-Cartwright LC, Walliker D (1999) The epidemiology of multiple *Plasmodium falciparum* infections. 3. Genetic structure and dynamics of *Plasmodium falciparum* infections in the Kilombero region of Tanzania. *Trans R Soc Trop Med Hyg* 93(Suppl 1):11–14.

Babiker HA, Hastings IM, Swedberg G (2009) Impaired fitness of drug-resistant malaria parasites: Evidence and implication on drug-deployment policies. *Expert Rev Anti Infect Ther* 7:581–593.

Badcock C (2010) *The Imprinted Brain: How Genes Set the Balance of the Mind Between Autism and Psychosis* (Jessica Kingsley, London).

Badcock C, Crespi B (2006) Imbalanced genomic imprinting in brain development: An evolutionary basis for the aetiology of autism. *J Evol Biol* 19:1007–1032.

Baldauf SL, Roger AJ, Wenk-Siefert I, Doolittle WF (2000) A kingdom-level phylogeny of eukaryotes based on combined protein data. *Science* 290:972–977.

Balikci A (1989) *The Netsilik Eskimo* (Waveland Press, Long Grove, IL).

Baliraine FN, Afrane YA, Amenya DA, Bonizzoni M, Vardo-Zalik AM, Menge DM, Githeko AK, Yan G (2010) A cohort study of *Plasmodium falciparum* infection dynamics in Western Kenya Highlands. *BMC Infect Dis* 10:283.

Barke J, Seipke RF, Grüschow S, Heavens D, Drou N, Bibb MJ, Goss RJ, Yu DW, Hutchings MI (2010) A mixed community of actinomycetes produce multiple antibiotics for the fungus farming ant *Acromyrmex octospinosus*. *BMC Biol* 8:109.

Barnes JL, Hill T, Langer M, Martinez M, Santos LR (2008) Helping behaviour and regard for others in capuchin monkeys (*Cebus apella*). *Biol Lett* 4:638–640.

Baron-Cohen S (2009) Autism: The empathizing-systemizing (E-S) theory. *Ann N Y Acad Sci* 1156:68–80.

Barrett C, Cosmides L, Tooby J (2007) The hominid entry into the cognitive niche. In *Evolution of Mind, Fundamental Questions and Controversies,* eds Gangestad S, Simpson J (Guilford Press, New York), pp 241–248.

Bartolomé C, Bello X, Maside X (2009) Widespread evidence for horizontal transfer of transposable elements across *Drosophila* genomes. *Genome Biol* 10:R22.

Basalla G (1988) *The Evolution of Technology* (Cambridge Univ Press, Cambridge, UK).

Baskar R, Chhabra P, Mascarenhas P, Nanjundiah V (2000) A cell type-specific effect of calcium on pattern formation and differentiation in *Dictyostelium discoideum*. *Int J Dev Biol* 44:491–498.

Bateson M, Nettle D, Roberts G (2006) Cues of being watched enhance cooperation in a real-world setting. *Biol Lett* 2:412–414.

Batson CD (1991) *The Altruism Question: Toward a Social-Psychological Answer* (Erlbaum, Hillsdale, NJ).

Batson CD (2011) *Altruism in Humans* (Oxford Univ Press, New York).

Baumgartner LK, Spear JR, Buckley DH, Pace NR, Reid RP, Dupraz C, Visscher PT (2009) Microbial diversity in modern marine stromatolites, Highborne Cay, Bahamas. *Environ Microbiol* 11:2710–2719.

Beaumont HJE, Gallie J, Kost C, Ferguson GC, Rainey PB (2009) Experimental evolution of bet hedging. *Nature* 462:90–93.

Becking JH (1970) Plant-endophyte symbiosis in non-leguminous plants. *Plant Soil* 32:611–654.

Beekman M, Ratnieks FLW (2003) Power over reproduction in social hymenoptera. *Philos Trans R Soc Lond B Biol Sci* 358:1741–1753.

Beise J (2005) The helping and the helpful grandmother: The role of maternal and paternal grandmothers in child mortality in the seventeenth- and eighteenth-century population of French settlers in Québec, Canada. In *Grandmotherhood: The Evolutionary Significance of the Second Half of Female Life,* eds Voland E, Chasiotis A, Schiefenhovel W (Rutgers Univ Press, Brunswick, NJ), pp 215–238.

Belsky J, Steinberg L, Draper P (1991) Childhood experience, interpersonal development, and reproductive strategy: An evolutionary theory of socialization. *Child Dev* 62:647–670.

Ben-Jacob E, Schochet O, Tenenbaum A, Cohen I, Czirók A, Vicsek T (1994) Generic modelling of cooperative growth patterns in bacterial colonies. *Nature* 368:46–49.

Benabentos R, Hirose S, Sucgang R, Curk T, Katoh M, Ostrowski EA, Strassmann JE, Queller DC, Zupan B, Shaulsky G, Kuspa A (2009) Polymorphic members of the *lag* gene family mediate kin discrimination in *Dictyostelium. Curr Biol* 19:567–572.

Bennetzen JL (2005) Transposable elements, gene creation and genome rearrangement in flowering plants. *Curr Opin Genet Dev* 15:621–627.

Beppu A, Griffiths TL (2009) Iterated learning and the cultural ratchet. *Proceedings of the 31st Annual Conference of the Cognitive Science Society* (Cognitive Science Society, Austin, TX).

Bereczkei T, Dunbar RIM (2002) Helping-at-the-nest and sex-biased parental investment in a Hungarian Gypsy population. *Curr Anthropol* 43:804–809.

Bergman TJ, Beehner JC, Cheney DL, Seyfarth RM (2003) Hierarchical classification by rank and kinship in baboons. *Science* 302:1234–1236.

Berleman JE, Kirby JR (2009) Deciphering the hunting strategy of a bacterial wolfpack. *FEMS Microbiol Rev* 33:942–957.

Bernhard H, Fischbacher U, Fehr E (2006) Parochial altruism in humans. *Nature* 442:912–915.

Bernoulli D (1954) Exposition of a new theory on the measurement of risk. *Econometrica* 22:23–36 [Translation of Bernoulli D (1738) Specimen theoriae novae demensura sortis. *Papers Imp Acad Sci St. Petersburg* 5:175–192].

Beye M, Hasselmann M, Fondrk MK, Page RE, Omholt SW (2003) The gene *csd* is the primary signal for sexual development in the honeybee and encodes an SR-type protein. *Cell* 114:419–429.

Biémont C (2010) A brief history of the status of transposable elements: From junk DNA to major players in evolution. *Genetics* 186:1085–1093.

Biémont C, Vieira C (2006) Genetics: Junk DNA as an evolutionary force. *Nature* 443: 521–524.

Bijma P, Wade MJ (2008) The joint effects of kin, multilevel selection and indirect genetic effects on response to genetic selection. *J Evol Biol* 21:1175–1188.

Billing J, Sherman PW (1998) Antimicrobial functions of spices: Why some like it hot. *Q Rev Biol* 73:3–49.

Birch LL (1987) Children's food preferences: Developmental patterns and environmental influence. *Ann Child Dev* 4:171–208.

Birch SAJ, Vauthier SA, Bloom P (2008) Three- and four-year-olds spontaneously use others' past performance to guide their learning. *Cognition* 107:1018–1034.

Birch SAJ, Akmal N, Frampton KL (2010) Two-year-olds are vigilant of others' nonverbal cues to credibility. *Dev Sci* 13:363–369.

Birkhead TR, Chaline N, Biggins JD, Burke T, Pizzari T (2004) Nontransitivity of paternity in a bird. *Evolution* 58:416–420.

Blissett J, Harris G, Kirk J (2001) Feeding problems in Silver-Russell syndrome. *Dev Med Child Neurol* 43:39–44.

Bloom P (2004) *Decartes' Baby: How the Science of Child Development Explains What Makes Us Human* (Basic Books, New York).

Bloomfield G, Skelton J, Ivens A, Tanaka Y, Kay RR (2010) Sex determination in the social amoeba *Dictyostelium discoideum*. *Science* 330:1533–1536.

Blount S (1995) When social outcomes aren't fair: The effect of causal attribution on preferences. *Organ Behav Hum Decis Process* 63:131–144.

Blumenstiel JP (2011) Evolutionary dynamics of transposable elements in a small RNA world. *Trends Genet* 27:23–31.

Boerlijst MC, Hogeweg P (1991) Spiral wave structure in prebiotic evolution: Hypercycles stable against parasites. *Physica D* 48:17–28.

Bolker BM, Pacala SW, Neuhauser C (2003) Spatial dynamics in model plant communities: What do we really know? *Am Nat* 162:135–148.

Bonanni R, Valsecchi P, Natoli E (2010) Pattern of individual participation and cheating in conflicts between groups of free-ranging dogs. *Anim Behav* 79:957–968.

Bonasio R, Zhang G, Ye C, Mutti NS, Fang X, Qin N, Donahue G, Yang P, Li Q, Li C, Zhang P, Huang Z, Berger SL, Reinberg D, Wang J, Liebig J (2010) Genomic comparison of the ants *Camponotus floridanus* and *Harpegnathos saltator. Science* 329:1068–1071.

Bonati MT, Russo S, Finelli P, Valsecchi MR, Cogliati F, Cavalleri F, Roberts W, Elia M, Larizza L (2007) Evaluation of autism traits in Angelman syndrome: A resource to unfold autism genes. *Neurogenetics* 8:169–178.

Bonduriansky R, Chenoweth SF (2009) Intralocus sexual conflict. *Trends Ecol Evol* 24:280–288.

Bonner JT (1967) *The Cellular Slime Molds* (Princeton Univ Press, Princeton, NJ).

Bonner JT (2006) Migration in *Dictyostelium polycephalum*. *Mycologia* 98:260–264.

Bonsall MB, Klug H (2011) The evolution of parental care in stochastic environments. *J Evol Biol* 24:645–655.

Boominathan L (2007) Some facts and thoughts: *p73* as a tumor suppressor gene in the network of tumor suppressors. *Mol Cancer* 6:27.

Boomsma JJ (2007) Kin selection versus sexual selection: Why the ends do not meet. *Curr Biol* 17:R673–R683.

Boomsma JJ (2009) Lifetime monogamy and the evolution of eusociality. *Philos Trans R Soc Lond B Biol Sci* 364:3191–3207.

Boomsma JJ, Grafen A (1990) Intraspecific variation in ant sex-ratios and the Trivers-Hare hypothesis. *Evolution* 44:1026–1034.

Boots M, Mealor M (2007) Local interactions select for lower pathogen infectivity. *Science* 315:1284–1286.

Bordenstein SR (2003) Symbiosis and the origin of species. In *Insect Symbiosis* (CRC Press, Boca Raton, FL), pp 283–304.

Bouju J (1984) *Graine de L'homme, Enfant du Mil* (Société d'Ethnographie, Paris).

Bouma JE, Lenski RE (1988) Evolution of a bacteria/plasmid association. *Nature* 335:351–352.

Bourke AFG (2011) *Principles of Social Evolution* (Oxford Univ Press, Oxford), p 288.

Bourke AFG, Franks NR (1995) *Social Evolution in Ants* (Princeton Univ Press, Princeton, NJ).

Boussau B, Karlberg EO, Frank AC, Legault BA, Andersson SGE (2004) Computational inference of scenarios for alpha-proteobacterial genome evolution. *Proc Natl Acad Sci USA* 101:9722–9727.

Boyd R, Richerson PJ (1988) An evolutionary model of social learning: The effects of spatial and temporal variation. In *Social Learning: Psychological and Biological Approaches,* eds Zentall T, Galef BG (Erlbaum, Hillsdale, NJ), pp 29–48.

Boyd R, Richerson PJ (1995) Why does culture increase human adaptability? *Ethol Sociobiol* 16:125–143.

Boyer P (2002) *Religion Explained* (Basic Books, New York).

Boysen ST, Berntson GG (1995) Responses to quantity: Perceptual versus cognitive mechanisms in chimpanzees (*Pan troglodytes*). *J Exp Psychol Anim Behav Process* 21:82–86.

Bracco E, Pergolizzi B, Peracino B, Ponte E, Balbo A, Mai A, Ceccarelli A, Bozzaro S (2000) Cell-cell signaling and adhesion in phagocytosis and early development of *Dictyostelium. Int J Dev Biol* 44:733–742.

Bradley PL, Myat MM, Comeaux CA, Andrew DJ (2003) Posterior migration of the salivary gland requires an intact visceral mesoderm and integrin function. *Dev Biol* 257:249–262.

Brady SG, Sipes S, Pearson A, Danforth BN (2006) Recent and simultaneous origins of eusociality in halictid bees. *Proc Biol Sci* 273:1643–1649.
Brandt M, Van Wilgenburg E, Tsutsui ND (2009) Global-scale analyses of chemical ecology and population genetics in the invasive Argentine ant. *Mol Ecol* 18:997–1005.
Brandvain Y (2010) Matrisibs, patrisibs, and the evolution of imprinting on autosomes and sex chromosomes. *Am Nat* 176:511–521.
Bräuer J, Call J, Tomasello M (2006) Are apes really inequity averse? *Proc Biol Sci* 273:3123–3128.
Bräuer J, Call J, Tomasello M (2009) Are apes inequity averse? New data on the token-exchange paradigm. *Am J Primatol* 71:175–181.
Bright M, Bulgheresi S (2010) A complex journey: Transmission of microbial symbionts. *Nat Rev Microbiol* 8:218–230.
Bright M, Sorgo A (2003) Ultrastructural reinvestigation of the trophosome in adults of *Riftia pachyptila* (Annelida, Siboglinidae). *Invertebr Biol* 122:347–368.
Bringaud F, Müller M, Cerqueira GC, Smith M, Rochette A, El-Sayed NM, Papadopoulou B, Ghedin E (2007) Members of a large retroposon family are determinants of post-transcriptional gene expression in *Leishmania*. *PLoS Pathog* 3:1291–1307.
Britten RJ (2010) Transposable element insertions have strongly affected human evolution. *Proc Natl Acad Sci USA* 107:19945–19948.
Britten RJ, Davidson EH (1971) Repetitive and non-repetitive DNA sequences and a speculation on the origins of evolutionary novelty. *Q Rev* Biol 46:111–138.
Brock DA, Douglas TE, Queller DC, Strassmann JE (2011) Primitive agriculture in a social amoeba. *Nature* 469:393–396.
Brockhurst MA (2007) Population bottlenecks promote cooperation in bacterial biofilms. *PLoS ONE* 2:e634.
Brockhurst MA, Buckling A, Gardner A (2007) Cooperation peaks at intermediate disturbance. *Curr Biol* 17:761–765.
Brosnan M, Ashwin C, Walker I, Donaghue J (2010) Can an "extreme female brain" be characterized in terms of psychosis? *Pers Individ Dif* 49:738–742.
Brosnan SF, de Waal FBM (2003) Monkeys reject unequal pay. *Nature* 425:297–299.
Brosnan SF, de Waal FBM (2004) Reply to Henrich. *Nature* 428:140.
Brosnan SF, Silk JB, Henrich J, Mareno MC, Lambeth SP, Schapiro SJ (2009) Chimpanzees (*Pan troglodytes*) do not develop contingent reciprocity in an experimental task. *Anim Cogn* 12:587–597.
Brosnan SF, Talbot C, Ahlgren M, Lambeth S, Schapiro SJ (2010) Mechanisms underlying responses to inequitable outcomes in chimpanzees, *Pan troglodytes*. *Anim Behav* 79:1229–1237.
Brown EM, Nathwani D (2005) Antibiotic cycling or rotation: A systematic review of the evidence of efficacy. *J Antimicrob Chemother* 55:6–9.
Brown GR, Almond REA, van Bergen Y (2004) Begging, stealing, and offering food: Food transfer in nonhuman primates. *Adv Stud Behav* 34:265–295.
Brown JL (1983) Cooperation-a biologist's dilemma. *Adv Stud Behav* 13:1–37.
Brown JL (1987) *Helping and Communal Breeding in Birds: Ecology and Evolution* (Princeton Univ Press, Princeton, NJ).
Brown SP (1999) Cooperation and conflict in host-manipulating parasites. *Proc R Soc Lond B Biol Sci* 266:1899–1904.
Brown WM, Consedine NS (2004) Just how happy is the happy puppet? An emotional signaling and kinship theory perspective on the behavioral phenotype of Angelman syndrome children. *Med Hypoth* 63:377–385.
Bruce MC, Donnelly CA, Alpers MP, Galinski MR, Barnwell JW, Walliker D, Day KP (2000) Cross-species interactions between malaria parasites in humans. *Science* 287:845–848.

Bruce MD (1999) Family Tytonidae (barn-owls). In *Handbook of Birds of the World: Vol. 5: Barn-owls to Hummingbirds,* eds del Hoyo J, Elliott A, Sargatal J (Lynx Edicions, Barcelona), pp 34–75.

Buchner P (1921) *Tier Und Pflanze in Intracellular Symbiose* (Borntraeger, Berlin).

Buckling A, Harrison F, Vos M, Brockhurst MA, Gardner A, West SA, Griffin A (2007) Siderophore-mediated cooperation and virulence in *Pseudomonas aeruginosa. FEMS Microbiol Ecol* 62:135–141.

Bugnyar T, Heinrich B (2005) Ravens, *Corvus corax,* differentiate between knowledgeable and ignorant competitors. *Proc Biol Sci* 272:1641–1646.

Buiting K (2010) Prader-Willi syndrome and Angelman syndrome. *Am J Med Genet C Semin Med Genet* 154C:365–376.

Buiting K, Kanber D, Horsthemke B, Lohmann D (2010) Imprinting of *RB1* (the new kid on the block). *Brief Funct Genomics* 9:347–353.

Bull JJ, Rice WR (1991) Distinguishing mechanisms for the evolution of co-operation. *J Theor Biol* 149:63–74.

Bulmer MS, Crozier RH (2004) Duplication and diversifying selection among termite antifungal peptides. *Mol Biol Evol* 21:2256–2264.

Bulmer MS, Crozier RH (2006) Variation in positive selection in termite GNBPs and *Relish. Mol Biol Evol* 23:317–326.

Bulmer MS, Lay F, Hamilton C (2010) Adaptive evolution in subterranean termite antifungal peptides. *Insect Mol Biol* 19:669–674.

Burkart JM, Fehr E, Efferson C, van Schaik CP (2007) Other-regarding preferences in a non-human primate: Common marmosets provision food altruistically. *Proc Natl Acad Sci USA* 104:19762–19766.

Burnham T, Hare B (2007) Engineering human cooperation. *Hum Nat* 18:88–108.

Burt A, Trivers R (2006) *Genes in Conflict: The Biology of Selfish Genetic Elements* (Belknap, Cambridge, MA).

Buss LW (1987) *The Evolution of Individuality* (Princeton Univ Press, Princeton, NJ).

Buss LW, Jackson JBC (1979) Competitive networks: Nontransitive competitive relationships in cryptic coral reef environments. *Am Nat* 113:223–234.

Buttelmann D, Carpenter M, Call J, Tomasello M (2007) Enculturated chimpanzees imitate rationally. *Dev Sci* 10:F31–F38.

Buttery NJ, Rozen DE, Wolf JB, Thompson CR (2009) Quantification of social behavior in *D. discoideum* reveals complex fixed and facultative strategies. *Curr Biol* 19:1373–1377.

Buunk AP, Park JH, Dubbs SL (2008) Parent-offspring conflict in mate preferences. *Rev Gen Psychol* 12:47–62.

Cacioppo JT, Ernst JM, Burleson MH, McClintock MK, Malarkey WB, Hawkley LC, Kowalewski RB, Paulsen A, Hobson JA, Hugdahl K, Spiegel D, Berntson GG (2000) Lonely traits and concomitant physiological processes: The MacArthur social neuroscience studies. *Int J Psychophysiol* 35:143–154.

Caldwell HK, Stephens SL, Young WS, 3rd (2009) Oxytocin as a natural antipsychotic: A study using oxytocin knockout mice. *Mol Psychiatry* 14:190–196.

Call J, Tomasello M (2008) Does the chimpanzee have a theory of mind? 30 years later. *Trends Cogn Sci* 12:187–192.

Callan AC, Milne E (2009) Involvement of the IGF system in fetal growth and childhood cancer: An overview of potential mechanisms. *Cancer Causes Control* 20:1783–1798.

Camacho JP, Sharbel TF, Beukeboom LW (2000) B-chromosome evolution. *Philos Trans R Soc Lond B Biol Sci* 355:163–178.

Camerer C (2003) *Behavioral Game Theory: Experiments on Strategic Interaction* (Princeton Univ Press, Princeton, NJ).

Camerer C, Thaler RH (1995) Anomalies: Ultimatums, dictators and manners. *J Econ Perspect* 9:209–219.

Cameron DD, White A, Antonovics J (2009) Parasite-grass-forb interactions and rock-paper-scissor dynamics: Predicting the effects of the parasitic plant *Rhinanthus minor* on host plant communities. *J Ecol* 97:1311–1319.

Cameron EZ, Setsaas TH, Linklater WL (2009) Social bonds between unrelated females increase reproductive success in feral horses. *Proc Natl Acad Sci USA* 106:13850–13853.

Cameron SA, Mardulyn P (2001) Multiple molecular data sets suggest independent origins of highly eusocial behavior in bees (Hymenoptera: Apinae). *Syst Biol* 50:194–214.

Canario F, Matos S, Soler M (2004) Environmental constraints and cooperative breeding in the azure-winged magpie. *Condor* 106:608–617.

Cant MA, Johnstone RA (2008) Reproductive conflict and the separation of reproductive generations in humans. *Proc Natl Acad Sci USA* 105:5332–5336.

Cantu D, Vanzetti LS, Sumner A, Dubcovsky M, Matvienko M, Distelfeld A, Michelmore RW, Dubcovsky J (2010) Small RNAs, DNA methylation and transposable elements in wheat. *BMC Genomics* 11:408–422.

Cardinal S, Straka J, Danforth BN (2010) Comprehensive phylogeny of apid bees reveals the evolutionary origins and antiquity of cleptoparasitism. *Proc Natl Acad Sci USA* 107:16207–16211.

Carneiro M, Blanco-Aguiar JA, Villafuerte R, Ferrand N, Nachman MW (2010) Speciation in the European rabbit (*Oryctolagus cuniculus*): Islands of differentiation on the X chromosome and autosomes. *Evolution* 64:3443–3460.

Carrel L, Willard HF (2005) X-inactivation profile reveals extensive variability in X-linked gene expression in females. *Nature* 434:400–404.

Carson CA, Christiansen JM, Yampara-Iquise H, Benson VW, Baffaut C, Davis JV, Broz RR, Kurtz WB, Rogers WM, Fales WH (2005) Specificity of a *Bacteroides thetaiotaomicron* marker for human feces. *Appl Environ Microbiol* 71:4945–4949.

Carvalho FM, Souza RC, Barcellos FG, Hungria M, Vasconcelos ATR (2010) Genomic and evolutionary comparisons of diazotrophic and pathogenic bacteria of the order Rhizobiales. *BMC Microbiol* 10:37.

Cascales E, Buchanan SK, Duché D, Kleanthous C, Lloubès R, Postle K, Riley M, Slatin S, Cavard D (2007) Colicin biology. *Microbiol Mol Biol Rev* 71:158–229.

Casey BJ, Duhoux S, Malter Cohen M (2010) Adolescence: What do transmission, transition, and translation have to do with it? *Neuron* 67:749–760.

Castillo DI, Switz GT, Foster KR, Queller DC, Strassmann JE (2005) A cost to chimerism in *Dictyostelium discoideum* on natural substrates. *Evol Ecol Res* 7:263–271.

Castillo DI, Queller DC, Strassmann JE (2011) Cell condition, competition, and chimerism in the social amoeba *Dictyostelium discoideum. Ethol Ecol Evol* 23:262–273.

Chang BY, Dworkin M (1994) Isolated fibrils rescue cohesion and development in the Dsp mutant of *Myxococcus xanthus. J Bacteriol* 176:7190–7196.

Chao L, Levin BR (1981) Structured habitats and the evolution of anticompetitor toxins in bacteria. *Proc Natl Acad Sci USA* 78:6324–6328.

Chapais B (2008) *Primeval Kinship* (Harvard Univ Press, Cambridge, MA).

Chapman T (2006) Evolutionary conflicts of interest between males and females. *Curr Biol* 16:R744–R754.

Chapman T (2008) The soup in my fly: Evolution, form and function of seminal fluid proteins. *PLoS Biol* 6:e179.

Chapuis M, Keller L (1999) Testing kin selection with sex allocation data in eusocial hymenoptera. *Heredity* 82:473–478.

Charlesworth B, Sniegowski P, Stephan W (1994) The evolutionary dynamics of repetitive DNA in eukaryotes. *Nature* 371:220.

Charnov EL (1978) Evolution of eusocial behavior: Offspring choice or parental parasitism? *J Theor Biol* 75:451–465.

Chattwood A, Thompson CRL (2011) Non-genetic heterogeneity and cell fate choice in *Dictyostelium discoideum. Dev Growth Differ* 58:553–566.

Chen CH, Huang H, Ward CM, Su JT, Schaeffer LV, Guo M, Hay BA (2007) A synthetic maternal-effect selfish genetic element drives population replacement in *Drosophila. Science* 316:597–600.

Chen G, Zhuchenko O, Kuspa A (2007) Immune-like phagocyte activity in the social amoeba. *Science* 317:678–681.

Chen MK, Santos LR (2006) Some thoughts on the adaptive function of inequity aversion: An alternative to Brosnan's social hypothesis. *Soc Justice Res* 19:201–207.

Cheney DL (1981) Intergroup encounters among free-ranging vervet monkeys. *Folia Primatol (Basel)* 35:124–146.

Cheney DL, Seyfarth RM (1990) *How Monkeys See the World: Inside the Mind of Another Species* (Univ of Chicago Press, Chicago).

Cheney DL, Seyfarth RM (1997) Reconciliatory grunts by dominant female baboons influence victims' behaviour. *Anim Behav* 54:409–418.

Cheney DL, Seyfarth RM (1999) Recognition of other individuals' social relationships by female baboons. *Anim Behav* 58:67–75.

Cheney DL, Seyfarth RM (2007) *Baboon Metaphysics: The Evolution of a Social Mind* (Univ of Chicago Press, Chicago).

Cheney DL, Seyfarth RM (2009) Stress and coping mechanisms in female primates. *Adv Stud Behav* 39:1–44.

Cheney DL, Seyfarth RM, Silk JB (1995) The responses of female baboons (*Papio cynocephalus ursinus*) to anomalous social interactions: Evidence for causal reasoning? *J Comp Psychol* 109:134–141.

Cheney DL, Seyfarth RM, Palombit R (1996) The function and mechanisms underlying baboon "contact" barks. *Anim Behav* 52:507–518.

Cheney DL, Moscovice LR, Heesen M, Mundry R, Seyfarth RM (2010) Contingent cooperation between wild female baboons. *Proc Natl Acad Sci USA* 107:9562–9566.

Chenoweth SF, Rundle HD, Blows MW (2008) Genetic constraints and the evolution of display trait sexual dimorphism by natural and sexual selection. *Am Nat* 171:22–34.

Clark AB (1978) Sex ratio and local resource competition in a prosimian primate. *Science* 201:163–165.

Clark AG (2006) Genomics of the evolutionary process. *Trends Ecol Evol* 21:316–321.

Clark AG, Dermitzakis ET, Civetta A (2000) Nontransitivity of sperm precedence in *Drosophila. Evolution* 54:1030–1035.

Clark AG, Glanowski S, Nielsen R, Thomas PD, Kejariwal A, Todd MA, Tanenbaum DM, Civello D, Lu F, Murphy B, Ferriera S, Wang G, Zheng X, White TJ, Sninsky JJ, Adams MD, Cargill M (2003) Inferring nonneutral evolution from human-chimp-mouse orthologous gene trios. *Science* 302:1960–1963.

Clutton-Brock TH (2009) Cooperation between non-kin in animal societies. *Nature* 462:51–57.

Clutton-Brock TH, Parker GA (1995) Punishment in animal societies. *Nature* 373:209–216.

Clutton-Brock TH, Brotherton PN, Russell AF, O'Riain MJ, Gaynor D, Kansky R, Griffin A, Manser M, Sharpe L, McIlrath GM, Small T, Moss A, Monfort S (2001) Cooperation, control, and concession in meerkat groups. *Science* 291:478–481.

Coall DA, Hertwig R (2010) Grandparental investment: Past, present, and future. *Behav Brain Sci* 33:1–19.

Cockburn A (1998) Evolution of helping behavior in cooperatively breeding birds. *Annu Rev Ecol Syst* 29:141–177.

Cockburn A (2006) Prevalence of different modes of parental care in birds. *Proc Biol Sci* 273:1375–1383.

Cockburn A, Russell AF (2011) Cooperative breeding: A question of climate? *Curr Biol* 21:R195–R197.

Cohan FM (2001) Bacterial species and speciation. *Syst Biol* 50:513–524.

Cohen MM, Jr. (2005) Beckwith-Wiedemann syndrome: Historical, clinicopathological, and etiopathogenetic perspectives. *Pediatr Dev Pathol* 8:287–304.

Connallon T, Clark AG (2010) Sex linkage, sex-specific selection, and the role of recombination in the evolution of sexually dimorphic gene expression. *Evolution* 64:3417–3442.

Connor RC (1986) Pseudoreciprocity: Investing in mutualism. *Anim Behav* 34:1562–1566.

Connor RC (1995) The benefits of mutualism: A conceptual framework. *Biol Rev Camb Philos Soc* 70:427–457.

Cook JM (1993) Sex determination in the Hymenoptera: A review of models and evidence. *Heredity* 71:421–435.

Cornwallis CK, West SA, Davis KE, Griffin AS (2010) Promiscuity and the evolutionary transition to complex societies. *Nature* 466:969–972.

Corona M, Velarde RA, Remolina S, Moran-Lauter A, Wang Y, Hughes KA, Robinson GE (2007) Vitellogenin, juvenile hormone, insulin signaling, and queen honey bee longevity. *Proc Natl Acad Sci USA* 104:7128–7133.

Corsaro D, Venditti D, Padula M, Valassina M (1999) Intracellular life. *Crit Rev Microbiol* 25:39–79.

Cosmides LM, Tooby J (1981) Cytoplasmic inheritance and intragenomic conflict. *J Theor Biol* 89:83–129.

Cosmides LM, Tooby J (2001) Unraveling the enigma of human intelligence: Evolutionary psychology and the multimodular mind. In *The Evolution of Intelligence,* eds Sternberg RJ, Kaufman JC (Erlbaum, Hillsdale, NJ), pp 145–198.

Costa JT (2006) *The Other Insect Societies* (Belknap Press, Cambridge, MA).

Covas R, Griesser M (2007) Life history and the evolution of family living in birds. *Proc Biol Sci* 274:1349–1357.

Covas R, Doutrelant C, du Plessis MA (2004) Experimental evidence of a link between breeding conditions and the decision to breed or to help in a colonial cooperative bird. *Proc Biol Sci* 271:827–832.

Covas R, du Plessis MA, Doutrelant C (2008) Helpers in colonial cooperatively breeding sociable weavers *Philetairus socius* contribute to buffer the effects of adverse breeding conditions. *Behav Ecol Sociobiol* 63:103–112.

Cox DR (1972) Regression models and life-tables. *J R Stat Soc Series B Stat Methodol* 34:187–220.

Coyne JA (1992) Genetics and speciation. *Nature* 355:511–515.

Coyne JA, Orr HA (1993) Further evidence against meiotic-drive models of hybrid sterility. *Evolution* 47:685–687.

Coyne JA, Orr HA (2004) *Speciation* (Sinauer, Sunderland, MA).

Cravo P, Culleton R, Hunt P, Walliker D, Mackinnon MJ (2001) Antimalarial drugs clear resistant parasites from partially immune hosts. *Antimicrob Agents Chemother* 45:2897–2901.

Crean AJ, Marshall DJ (2009) Coping with environmental uncertainty: Dynamic bet hedging as a maternal effect. *Philos Trans R Soc Lond B Biol Sci* 364:1087–1096.

Crespi BJ (2001) The evolution of social behavior in microorganisms. *Trends Ecol Evol* 16:178–183.

Crespi BJ (2008) Genomic imprinting in the development and evolution of psychotic spectrum conditions. *Biol Rev Camb Philos Soc* 83:441–493.

Crespi BJ (2011) The strategies of the genes: Genomic conflicts, attachment theory and development of the social brain. In *Brain Behavior and Epigenetics,* eds Petronis, A, Mill J (Springer, New York), pp 143–168.

Crespi BJ, Badcock C (2008) Psychosis and autism as diametrical disorders of the social brain. *Behav Brain Sci* 31:241–320.

Crockford C, Wittig RM, Seyfarth RM, Cheney DL (2007) Baboons eavesdrop to deduce mating opportunities. *Anim Behav* 73:885–890.

Crofoot MC, Wrangham RW (2010) Intergroup aggression in primates: The case for a unified theory. In *Mind the Gap*, eds Kappeler PM, Silk JB (Springer, Berlin), pp 171–195.

Cronin KA, Schroeder KKE, Rothwell ES, Silk JB, Snowdon CT (2009) Cooperatively breeding cottontop tamarins (*Saguinus oedipus*) do not donate rewards to their long-term mates. *J Comp Psychol* 123:231–241.

Cronin KA, Schroeder KKE, Snowdon CT (2010) Prosocial behaviour emerges independent of reciprocity in cottontop tamarins. *Proc Biol Sci* 277:3845–3851.

Crow JF, Kimura M (1970) *An Introduction to Population Genetics Theory* (Harper and Row, New York).

Crozier RH (1986) Genetic clonal recognition abilities in marine invertebrates must be maintained by selection for something else. *Evolution* 40:1100–1101.

Cruz-López L, Malo EA, Morgan ED, Rincon M, Guzmán M, Rojas JC (2005) Mandibular gland secretion of *Melipona beecheii:* Chemistry and behavior. *J Chem Ecol* 31:1621–1632.

Cui H, Cruz-Correa M, Giardiello FM, Hutcheon DF, Kafonek DR, Brandenburg S, Wu Y, He X, Powe NR, Feinberg AP (2003) Loss of IGF2 imprinting: A potential marker of colorectal cancer risk. *Science* 299:1753–1755.

Currie CR, Scott JA, Summerbell RC, Malloch D (1999) Fungus-growing ants use antibiotic-producing bacteria to control garden parasites. *Nature* 398:701–704.

Currie CR, Poulsen M, Mendenhall J, Boomsma JJ, Billen J (2006) Coevolved crypts and exocrine glands support mutualistic bacteria in fungus-growing ants. *Science* 311:81–83.

Curry RL (1989) Geographic variation in social organization of Galapágos mockingbirds: Ecological correlates of group territoriality and cooperative breeding. *Behav Ecol Sociobiol* 25:147–160.

Curry RL, Grant PR (1990) Galápagos mockingbirds: Terrotorial cooperative breeding in a climatically variable environment. In *Cooperative Breeding in Birds: Long-Term Studies of Ecology and Behaviour,* eds Stacey PB, Koenig WD (Cambridge Univ Press, Cambridge, UK), pp 289–332.

Dale C, Moran NA (2006) Molecular interactions between bacterial symbionts and their hosts. *Cell* 126:453–465.

Dale C, Young SA, Haydon DT, Welburn SC (2001) The insect endosymbiont *Sodalis glossinidius* utilizes a type III secretion system for cell invasion. *Proc Natl Acad Sci USA* 98:1883–1888.

Dallosso AR, Hancock AL, Brown KW, Williams AC, Jackson S, Malik K (2004) Genomic imprinting at the *WT1* gene involves a novel coding transcript (*AWT1*) that shows deregulation in Wilms' tumours. *Hum Mol Genet* 13:405–415.

Dally JM, Emery NJ, Clayton NS (2006) Food-caching western scrub-jays keep track of who was watching when. *Science* 312:1662–1665.

Damas D (1984) Central Eskimo: Introduction. In *Handbook of North American Indians: Vol* 5. *Arctic,* ed Damas D (Smithsonian Institution, Washington, DC), pp 391–396.

Dan B (2009) Angelman syndrome: Current understanding and research prospects. *Epilepsia* 50:2331–2339.

Dann TC, Roberts DF (1993) Menarcheal age in University of Warwick young women. *J Biosoc Sci* 25:531–538.

Daubersies P, Sallenave-Sales S, Magne S, Trape JF, Contamin H, Fandeur T, Rogier C, Mercereau-Puijalon O, Druilhe P (1996) Rapid turnover of *Plasmodium falciparum* populations in asymptomatic individuals living in a high transmission area. *Am J Trop Med Hyg* 54:18–26.

Davies PSW, Valley R, Preece MA (1988) Adolescent growth and pubertal progression in the Silver-Russell syndrome. *Arch Dis Child* 63:130–135.

Dawkins R (1976a) *The Selfish Gene* (Oxford Univ Press, New York).

Dawkins R (1976b) *The Selfish Gene* (Oxford Univ Press, Oxford).

Dawkins R (1976c) *The Selfish Gene* (Paladin, London).

Day KP, Koella JC, Nee S, Gupta S, Read AF (1992) Population genetics and dynamics of *Plasmodium falciparum:* An ecological view. *Parasitology* 104(Suppl):S35–S52.

de Roode JC, Read AF, Chan BHK, Mackinnon MJ (2003) Rodent malaria parasites suffer from the presence of conspecific clones in three-clone *Plasmodium chabaudi* infections. *Parasitology* 127:411–418.

de Roode JC, Culleton R, Bell AS, Read AF (2004a) Competitive release of drug resistance following drug treatment of mixed *Plasmodium chabaudi* infections. *Malar J* 3:33.

de Roode JC, Culleton R, Cheesman SJ, Carter R, Read AF (2004b) Host heterogeneity is a determinant of competitive exclusion or coexistence in genetically diverse malaria infections. *Proc R Soc B* 271:1073–1080.

de Roode JC, Helinski MEH, Anwar MA, Read AF (2005a) Dynamics of multiple infection and within-host competition in genetically diverse malaria infections. *Am Nat* 166:531–542.

de Roode JC, Pansini R, Cheesman SJ, Helinski ME, Huijben S, Wargo AR, Bell AS, Chan BH, Walliker D, Read AF (2005b) Virulence and competitive ability in genetically diverse malaria infections. *Proc Natl Acad Sci USA* 102:7624–7628.

de Waal FBM (1997a) Food transfers through mesh in brown capuchins. *J Comp Psychol* 111:370–378.

de Waal FBM (1997b) The chimpanzee's service economy: Food for grooming. *Evol Hum Behav* 18:375–386.

de Waal FBM (2000) Attitudinal reciprocity in food sharing among brown capuchin monkeys. *Anim Behav* 60:253–261.

de Waal FBM, Berger ML (2000) Payment for labour in monkeys. *Nature* 404:563.

de Waal FBM, Davis JM (2003) Capuchin cognitive ecology: Cooperation based on projected returns. *Neuropsychologia* 41:221–228.

de Waal FBM, Leimgruber K, Greenberg AR (2008) Giving is self-rewarding for monkeys. *Proc Natl Acad Sci USA* 105:13685–13689.

Dean AM, Thornton JW (2007) Mechanistic approaches to the study of evolution: The functional synthesis. *Nat Rev Genet* 8:675–688.

DeBaun MR, Tucker MA (1998) Risk of cancer during the first four years of life in children from the Beckwith-Wiedemann Syndrome Registry. *J Pediatr* 132:398–400.

Denison RF (2000) Legume sanctions and the evolution of symbiotic cooperation by rhizobia. *Am Nat* 6:567–576.

Desjardins JK, Klausner JQ, Fernald RD (2010) Female genomic response to mate information. *Proc Natl Acad Sci USA* 107:21176–21180.

Dethlefsen L, Eckburg PB, Bik EM, Relman DA (2006) Assembly of the human intestinal microbiota. *Trends Ecol Evol* 21:517–523.

Diggle SP, Gardner A, West SA, Griffin AS (2007a) Evolutionary theory of bacterial quorum sensing: When is a signal not a signal? *Philos Trans R Soc Lond B Biol Sci* 362:1241–1249.

Diggle SP, Griffin AS, Campbell GS, West SA (2007b) Cooperation and conflict in quorum-sensing bacterial populations. *Nature* 450:411–414.

Dobzhansky T (1973) Nothing in biology makes sense except in the light of evolution. *Amer Biol Teacher* 35:125–129.

Doebeli M, Knowlton N (1998) The evolution of interspecific mutualisms. *Proc Natl Acad Sci USA* 95:8676–8680.

Dolgin ES, Charlesworth B (2006) The fate of transposable elements in asexual populations. *Genetics* 174:817–827.

Doolittle WF, Sapienza C (1980) Selfish genes, the phenotype paradigm and genome evolution. *Nature* 284:601–603.

Douglas AE (1989) Mycetocyte symbiosis in insects. *Biol Rev Camb Philos Soc* 64:409–434.

Douglas AE (2008) Conflict, cheats and the persistence of symbioses. *New Phytol* 177:849–858.

Drake JW, Charlesworth B, Charlesworth D, Crow JF (1998) Rates of spontaneous mutation. *Genetics* 148:1667–1686.

Drapeau MD, Albert S, Kucharski R, Prusko C, Maleszka R (2006) Evolution of the Yellow/Major Royal Jelly Protein family and the emergence of social behavior in honey bees. *Genome Res* 16:1385–1394.

Draper P, Howell N (2005) The growth and kinship resources of Ju/'hoansi children. In *Hunter-Gatherer Childhoods: Evolutionary, Developmental, and Cultural Perspectives,* eds Hewlett BS, Lamb ME (de Gruyter, New York), pp 262–281.

Drosophila 12 Genomes Consortium (2007) Evolution of genes and genomes on the *Drosophila* phylogeny. *Nature* 450:203–218.

Drummond AJ, Rambaut S (2007). *BEAST: Bayesian Evolutionary Analysis by Sampling Trees. BMC Evol Biol* 7:214.

Dubreuil D, Gentile MS, Visalberghi E (2006) Are capuchin monkeys (*Cebus apella*) inequity averse? *Proc Biol Sci* 273:1223–1228.

Dunbar RIM, Shultz S (2007) Evolution in the social brain. *Science* 317:1344–1347.

Dunning Hotopp JC, Clark ME, Oliveira DC, Foster JM, Fischer P, Muñoz Torres MC, Giebel JD, Kumar N, Ishmael N, Wang S, Ingram J, Nene RV, Shepard J, Tomkins J, Richards S, Spiro DJ, Ghedin E, Slatko BE, Tettelin H, Werren JH (2007) Widespread lateral gene transfer from intracellular bacteria to multicellular eukaryotes. *Science* 317:1753–1756.

Duret L, Galtier N (2009) Biased gene conversion and the evolution of mammalian genomic landscapes. *Annu Rev Genomics Hum Genet* 10:285–311.

Durkheim E (1963) *Incest. The Nature and Origin of the Taboo* (Lyle Stuart, NewYork).

Durrett R, Levin S (1997) Allelopathy in spatially distributed populations. *J Theor Biol* 185:165–171.

Dye C (2009) Doomsday postponed? Preventing and reversing epidemics of drug-resistant tuberculosis. *Nat Rev Microbiol* 7:81–87.

Eberhard MJW (1975) The evolution of social behavior by kin selection. *Q Rev Biol* 50:1–33.

Eberhard WG (1980) Evolutionary consequences of intracellular organelle competition. *Q Rev Biol* 55:231–249.

Eberhard WG (1996) *Female Control: Sexual Selection by Cryptic Female Choice* (Princeton Univ Press, Princeton, NJ).

Eckburg PB, Bik EM, Bernstein CN, Purdom E, Dethlefsen L, Sargent M, Gill SR, Nelson KE, Relman DA (2005) Diversity of the human intestinal microbial flora. *Science* 308:1635–1638.

Edwards KF, Schreiber SJ (2010) Preemption of space can lead to intransitive coexistence of competitors. *Oikos* 119:1201–1209.

Eggermann T (2010) Russell-Silver syndrome. *Am J Med Genet C Semin Med Genet* 154C:355–364.

Eggermann T, Eggermann K, Schönherr N (2008) Growth retardation versus overgrowth: Silver-Russell syndrome is genetically opposite to Beckwith-Wiedemann syndrome. *Trends Genet* 24:195–204.

Ehmer B, Reeve HK, Hoy RR (2001) Comparison of brain volumes between single and multiple foundresses in the paper wasp *Polistes dominulus*. *Brain Behav Evol* 57:161–168.

Ehrlich P (1913) Address in pathology on chemotherapeutics: Scientific principles, methods, and results. *Lancet* 182:445–451.

Eickbush TH, Eickbush DG (2007) Finely orchestrated movements: Evolution of the ribosomal RNA genes. *Genetics* 175:477–485.

Ellis BJ (2004) Timing of pubertal maturation in girls: An integrated life history approach. *Psychol Bull* 130:920–958.

Ember CR (1978) Myths about hunter-gatherers. *Ethnology* 17:439–448.

Emery NJ, Dally JM, Clayton NS (2004) Western scrub-jays (*Aphelocoma californica*) use cognitive strategies to protect their caches from thieving conspecifics. *Anim Cogn* 7:37–43.

Emlen ST (1982a) The evolution of helping. 1. An ecological constraints model. *Am Nat* 119:29–39.

Emlen ST (1982b) The evolution of helping. 2. The role of behavioral conflict. *Am Nat* 119:40–53.

Emlen ST (1994) Benefits, constraints and the evolution of the family. *Trends Ecol Evol* 9:282–285.

Emlen ST (1995) An evolutionary theory of the family. *Proc Natl Acad Sci USA* 92:8092–8099.

Emlen ST (1997a) Predicting family dynamics in social vertebrates. In *Behavioural Ecology: An Evolutionary Approach,* eds Krebs JR, Davies NB (Blackwell Science, Oxford), 4th Ed., pp 228–253.

Emlen ST (1997b) The evolutionary study of human family systems. *Soc Sci Inf* (*Paris*) 36:563–589.

Emlen ST, Reeve HK, Sherman PW, Wrege PH, Ratnieks FLW, Shellman-Reeve J (1991) Adaptive versus nonadaptive explanations of behavior: The case of alloparental helping. *Am Nat* 138:259–270.

Enard W, Przeworski M, Fisher SE, Lai CS, Wiebe V, Kitano T, Monaco AP, Pääbo S (2002) Molecular evolution of *FOXP2,* a gene involved in speech and language. *Nature* 418:869–872.

Engh AL, Siebert ER, Greenberg DA, Holekamp K (2005) Patterns of alliance formation and postconflict aggression indicate spotted hyenas recognize third party relationships. *Anim Behav* 69:209–217.

Engh AE, Hoffmeier RR, Cheney DL, Seyfarth RM (2006) Who me? Can baboons infer the target of vocalisations? *Anim Behav* 71:381–387.

Engsontia P, Sanderson AP, Cobb M, Walden KK, Robertson HM, Brown S (2008) The red flour beetle's large nose: An expanded odorant receptor gene family in *Tribolium castaneum. Insect Biochem Mol Biol* 38:387–397.

Ennis HL, Dao DN, Pukatzki SU, Kessin RH (2000) *Dictyostelium* amoebae lacking an F-box protein form spores rather than stalk in chimeras with wild type. *Proc Natl Acad Sci USA* 97:3292–3297.

Ernest-Jones M, Nettle D, Bateson M (2010) Effects of eye images on everyday cooperative behavior: A field experiment. *Evol Hum Behav* 32:172–178.

Eshel I, Cavalli-Sforza LL (1982) Assortment of encounters and evolution of cooperativeness. *Proc Natl Acad Sci USA* 79:1331–1335.

Eshelman CM, Vouk R, Stewart JL, Halsne E, Lindsey HA, Schneider S, Gualu M, Dean AM, Kerr B (2010) Unrestricted migration favours virulent pathogens in experimental metapopulations: Evolutionary genetics of a rapacious life history. *Philos Trans R Soc Lond B Biol Sci* 365:2503–2513.

Evans JD, Aronstein K, Chen YP, Hetru C, Imler JL, Jiang H, Kanost M, Thompson GJ, Zou Z, Hultmark D (2006) Immune pathways and defence mechanisms in honey bees *Apis mellifera. Insect Mol Biol* 15:645–656.

Everitt BS, Hothorn T (2009). *A Handbook of Statistical Analyses Using R,* 2nd Ed. (Chapman and Hall/CRC, Boca Raton, FL).

Ewald PW (1987) Transmission modes and evolution of the parasitism-mutualism continuum. *Ann N Y Acad Sci* 503:295–306.

Eyre-Walker A, Keightley PD (2007) The distribution of fitness effects of new mutations. *Nat Rev Genet* 8:610–618.

Fablet M, Bueno M, Potrzebowski L, Kaessmann H (2009) Evolutionary origin and functions of retrogene introns. *Mol Biol Evol* 26:2147–2156.

Färnert A (2008) *Plasmodium falciparum* population dynamics: Only snapshots in time? *Trends Parasitol* 24:340–344.

Feder ME, Mitchell-Olds T (2003) Evolutionary and ecological functional genomics. *Nat Rev Genet* 4:651–657.

Fehr E, Fischbacher U (2003) The nature of human altruism. *Nature* 425:785–791.

Fehr E, Fischbacher U (2004) Third-party punishment and social norms. *Evol Hum Behav* 25:63–87.

Fehr E, Schmidt KM (1999) A theory of fairness, competition, and cooperation. *Q J Econ* 114:817–868.

Fehr E, Schmidt KM (2006) The economics of fairness, reciprocity and altruism—experimental evidence and new theories. In *Handbook on the Economics of Giving, Reciprocity and Altruism*, eds Kolm S-C, Ythier JM (North-Holland, Amsterdam), Vol 1.

Fehr E, Schneider F (2010) Eyes are on us, but nobody cares: Are eye cues relevant for strong reciprocity? *Proc Biol Sci* 277:1315–1323.

Feinberg AP, Ohlsson R, Henikoff S (2006) The epigenetic progenitor origin of human cancer. *Nat Rev Genet* 7:21–33.

Feinberg R, Ottenheimer M, eds (2001) *The Cultural Analysis of Kinship* (Univ of Illinois Press, Urbana).

Feldgarden M, Riley MA (1998) High levels of colicin resistance in *Escherichia coli. Evolution* 52:1270–1276.

Feldgarden M, Riley MA (1999) The phenotypic and fitness effects of colicin resistance in *Escherichia coli* K-12. *Evolution* 53:1019–1027.

Ferguson LC, Green J, Surridge A, Jiggins CD (2011) Evolution of the insect *yellow* gene family. *Mol Biol Evol* 28:257–272.

Ferree PM, Frydman HM, Li JM, Cao J, Wieschaus E, Sullivan W (2005) *Wolbachia* utilizes host microtubules and Dynein for anterior localization in the *Drosophila* oocyte. *PLoS Pathog* 1:e14.

Ferrière R, Gauduchon M, Bronstein JL (2007) Evolution and persistence of obligate mutualists and exploiters: competition for partners and evolutionary immunization. *Ecol Lett* 10:115–126.

Feschotte C (2008) Transposable elements and the evolution of regulatory networks. *Nat Rev Genet* 9:397–405.

Feschotte C, Pritham EJ (2007) DNA transposons and the evolution of eukaryotic genomes. *Annu Rev Genet* 41:331–368.

Fiegna F, Velicer GJ (2003) Competitive fates of bacterial social parasites: Persistence and self-induced extinction of *Myxococcus xanthus* cheaters. *Proc Biol Sci* 270:1527–1534.

Fiegna F, Velicer GJ (2005) Exploitative and hierarchical antagonism in a cooperative bacterium. *PLoS Biol* 3:e370.

Fiegna F, Yu YT, Kadam SV, Velicer GJ (2006) Evolution of an obligate social cheater to a superior cooperator. *Nature* 441:310–314.

Filoche S, Wong L, Sissons CH (2010) Oral biofilms: Emerging concepts in microbial ecology. *J Dent Res* 89:8–18.

Fine PEM (1975) Vectors and vertical transmission: An epidemiologic perspective. *Ann N Y Acad Sci* 266:173–194.

Finkel T, Holbrook NJ (2000) Oxidants, oxidative stress and the biology of ageing. *Nature* 408:239–247.

Fischbacher U, Gächter S, Fehr E (2001) Are people conditionally cooperative? Evidence from a public goods experiment. *Econ Lett* 71:397–404.

Fisher RA (1930) *The Genetical Theory of Natural Selection* (Oxford Univ Press, Oxford).

Fletcher JA, Doebeli M (2009) A simple and general explanation for the evolution of altruism. *Proc Biol Sci* 276:13–19.

Fletcher JA, Zwick M (2006) Unifying the theories of inclusive fitness and reciprocal altruism. *Am Nat* 168:252–262.

Flinn MV, Quinlan RJ, Coe K, Ward CV (2007) Evolution of the human family: Cooperative males, long social childhoods, smart mothers, and extended kin networks. In *Family Relationships: An Evolutionary Perspective,* eds Salmon CA, Shackelford TK (Oxford Univ Press, Oxford), pp 16–38.

Flombaum JI, Santos LR (2005) Rhesus monkeys attribute perceptions to others. *Curr Biol* 15:447–452.

Flowers JM, Li SI, Stathos A, Saxer G, Ostrowski EA, Queller DC, Strassmann JE, Purugganan MD (2010) Variation, sex, and social cooperation: Molecular population genetics of the social amoeba *Dictyostelium discoideum. PLoS Genet* 6:e1001013.

Fortunato A, Queller DC, Strassmann JE (2003a) A linear dominance hierarchy among clones in chimeras of the social amoeba *Dictyostelium discoideum. J Evol Biol* 16:438–445.

Fortunato A, Strassmann JE, Santorelli LA, Queller DC (2003b) Co-occurrence in nature of different clones of the social amoeba, *Dictyostelium discoideum. Mol Ecol* 12:1031–1038.

Foster KR (2005) Hamiltonian medicine: Why the social lives of pathogens matter. *Science* 308:1269–1270.

Foster KR (2010) Social behavior in microorganisms. In *Social Behaviour: Genes, Ecology and Evolution,* eds Székely T, Moore AJ, Komdeur J (Cambridge Univ Press, Cambridge, UK), pp 331–356.

Foster KR (2011) The sociobiology of molecular systems. *Nat Rev Genet* 12:193–203.

Foster KR, Grundmann H (2006) Do we need to put society first? The potential for tragedy in antimicrobial resistance. *PLoS Med* 3:e29.

Foster KR, Wenseleers T (2006) A general model for the evolution of mutualisms. *J Evol* Biol 19:1283–1293.

Foster KR, Xavier JB (2007) Cooperation: Bridging ecology and sociobiology. *Curr Biol* 17:R319–R321.

Foster KR, Fortunato A, Strassmann JE, Queller DC (2002) The costs and benefits of being a chimera. *Proc Biol Sci* 269:2357–2362.

Foster KR, Shaulsky G, Strassmann JE, Queller DC, Thompson CR (2004) Pleiotropy as a mechanism to stabilize cooperation. *Nature* 431:693–696.

Foster KR, Wenseleers T, Ratnieks FLW (2006) Kin selection is the key to altruism. *Trends Ecol Evol* 21:57–60.

Fowden AL, Coan PM, Angiolini E, Burton GJ, Constancia M (2011) Imprinted genes and the epigenetic regulation of placental phenotype. *Prog Biophys Mol Biol* 106:281–288.

Frank AC, Wolfe KH (2009) Evolutionary capture of viral and plasmid DNA by yeast nuclear chromosomes. *Eukaryot Cell* 8:1521–1531.

Frank AC, Alsmark CM, Thollesson M, Andersson SGE (2005) Functional divergence and horizontal transfer of type IV secretion systems. *Mol Biol Evol* 22:1325–1336.

Frank SA (1991) Divergence of meiotic drive-suppression systems as an explanation for sex-biased hybrid sterility and inviability. *Evolution* 45:262–267.

Frank SA (1994) Genetics of mutualism: The evolution of altruism between species. *J Theor Biol* 170:393–400.

Frank SA (1996a) Host control of symbiont transmission: The separation of symbionts into germ and soma. *Am Nat* 148:1113–1124.

Frank SA (1996b) Host-symbiont conflict over the mixing of symbiotic lineages. *Proc Biol Sci* 263:339–344.

Frank SA (1998) *Foundations of Social Evolution* (Princeton Univ Press, Princeton, NJ).

Frank SA, Crespi BJ (2011) Pathology from evolutionary conflict, with a theory of X chromosome versus autosome conflict over sexually antagonistic traits. *Proc Natl Acad Sci USA* 108(Suppl 2):10886–10893.

Frank SA, Slatkin M (1990) Evolution in a variable environment. *Am Nat* 136:244–260.

Frean M, Abraham ER (2001) Rock-scissors-paper and the survival of the weakest. *Proc Biol Sci* 268:1323–1327.

Frère CH, Krützen M, Mann J, Connor RC, Bejder L, Sherwin WB (2010) Social and genetic interactions drive fitness variation in a free-living dolphin population. *Proc Natl Acad Sci USA* 107:19949–19954.

Fruteau C, Voelkl B, van Damme E, Noë R (2009) Supply and demand determine the market value of food providers in wild vervet monkeys. *Proc Natl Acad Sci USA* 106:12007–12012.

Fry JD (2010) The genomic location of sexually antagonistic variation: Some cautionary comments. *Evolution* 64:1510–1516.

Furstenberg FF, Cherlin AJ (1991) *Divided Families* (Harvard Univ Press, Cambridge, MA).

Furuta Y, Abe K, Kobayashi I (2010) Genome comparison and context analysis reveals putative mobile forms of restriction-modification systems and related rearrangements. *Nucleic Acids Res* 38:2428–2443.

Gabory A, Ripoche MA, Le Digarcher A, Watrin F, Ziyyat A, Forné T, Jammes H, Ainscough JF, Surani MA, Journot L, Dandolo L (2009) *H19* acts as a trans regulator of the imprinted gene network controlling growth in mice. *Development* 136:3413–3421.

Gage DJ (2002) Analysis of infection thread development using Gfp- and DsRed-expressing *Sinorhizobium meliloti*. *J Bacteriol* 184:7042–7046.

Gamboa GJ (2004) Kin recognition in eusocial wasps. *Ann Zool Fenn* 41:789–808.

Gandon S, Day T (2009) Evolutionary epidemiology and the dynamics of adaptation. *Evolution* 63:826–838.

Gans J, Wolinsky M, Dunbar J (2005) Computational improvements reveal great bacterial diversity and high metal toxicity in soil. *Science* 309:1387–1390.

Gardner A, West SA (2006) Demography, altruism, and the benefits of budding. *J Evol Biol* 19:1707–1716.

Gardner A, West SA (2010) Greenbeards. *Evolution* 64:25–38.

Gatenby RA (2009) A change of strategy in the war on cancer. *Nature* 459:508–509.

Gatenby RA, Silva AS, Gillies RJ, Frieden BR (2009) Adaptive therapy. *Cancer Res* 69:4894–4903.

Geronimus AT (1997) Teenage childbearing and personal responsibility: An alternative view. *Polit Sci Q* 112:405–430.

Gershenson S (1928) A new sex-ratio abnormality in *Drosophila obscura*. *Genetics* 13:488–507.

Ghazanfar AA, Logothetis NK (2003) Neuroperception: Facial expressions linked to monkey calls. *Nature* 423:937–938.

Ghazanfar AA, Maier JX, Hoffman KL, Logothetis NK (2005) Multisensory integration of dynamic faces and voices in rhesus monkey auditory cortex. *J Neurosci* 25:5004–5012.

Gibson MA, Mace R (2005) Helpful grandmothers in rural Ethiopia: A study of the effect of kin on child survival and growth. *Evol Hum Behav* 26:469–482.

Gilbert OM, Foster KR, Mehdiabadi NJ, Strassmann JE, Queller DC (2007) High relatedness maintains multicellular cooperation in a social amoeba by controlling cheater mutants. *Proc Natl Acad Sci USA* 104:8913–8917.

Gilbert OM, Queller DC, Strassmann JE (2009) Discovery of a large clonal patch of a social amoeba: Implications for social evolution. *Mol Ecol* 18:1273–1281.

Gil-da-Costa R, Braun A, Lopes M, Hauser MD, Carson RE, Herscovitch P, Martin A (2004) Toward an evolutionary perspective on conceptual representation: Species-specific calls activate visual and affective processing systems in the macaque. *Proc Natl Acad Sci USA* 101:17516–17521.

Gillespie JH (1974) Natural selection for within-generation variance in offspring number. *Genetics* 76:601–606.

Gillespie JH (1975) Natural selection for within-generation variance in offspring number II. Discrete haploid models. *Genetics* 81:403–413.

Gillespie JH (1977) Natural selection for variances in offspring numbers: A new evolutionary principle. *Am Nat* 111:1010–1014.

Gilligan I (2010) The prehistoric development of clothing: Archaeological implications of a thermal model. *J Arch Method Theory* 17:15–80.
Gilpin ME (1975) Limit cycles in competition communities. *Am Nat* 109:51–60.
Giraldeau L-A, Caraco T (1993) Genetic relatedness and group size in an aggregation economy. *Evol Ecol* 7:429–438.
Godfrey-Smith P, Kerr B (2009) Selection in ephemeral networks. *Am Nat* 174:906–911.
Goettler W, Kaltenpoth M, Herzner G, Strohm E (2007) Morphology and ultrastructure of a bacteria cultivation organ: The antennal glands of female European beewolves, *Philanthus triangulum* (Hymenoptera, Crabronidae). *Arthropod Struct Dev* 36:1–9.
Golden H (2006) *Kayaks of Greenland, The History and Development of the Greenlandic Hunting Kayak, 1600–2000* (Whitehouse Grocery Press, Portland, OR).
Goldman BS, Nierman WC, Kaiser D, Slater SC, Durkin AS, Eisen JA, Ronning CM, Barbazuk WB, Blanchard M, Field C, Halling C, Hinkle G, Iartchuk O, Kim HS, Mackenzie C, Madupu R, Miller N, Shvartsbeyn A, Sullivan SA, Vaudin M, Wiegand R, Kaplan HB (2006) Evolution of sensory complexity recorded in a myxobacterial genome. *Proc Natl Acad Sci USA* 103:15200–15205.
Gomer RH, Firtel RA (1987) Cell-autonomous determination of cell-type choice in *Dictyostelium* development by cell-cycle phase. *Science* 237:758–762.
Goncalves BPA, Paul REL (2011) Sub-clearance treatment to slow malaria drug resistance? *Trends Parasitol* 27:50–51.
González J, Petrov DA (2009) The adaptive role of transposable elements in the *Drosophila* genome. *Gene* 448:124–133.
Goos LM, Ragsdale G (2008) Genomic imprinting and human psychology: Cognition, behavior and pathology. In *Genomic Imprinting,* ed Wilkins JF (Landes Bioscience, New York), pp 71–88.
Gore J, Youk H, van Oudenaarden A (2009) Snowdrift game dynamics and facultative cheating in yeast. *Nature* 459:253–256.
Gotzek D, Ross KG (2009) Current status of a model system: The gene *Gp-9* and its association with social organization in fire ants. *PLoS ONE* 4:e7713.
Gotzek D, Shoemaker DD, Ross KG (2007) Molecular variation at a candidate gene implicated in the regulation of fire ant social behavior. *PLoS ONE* 2:e1088.
Gourley SA, Liu R, Wu J (2011) Slowing the evolution of insecticide resistance in mosquitoes: A mathematical model. *Proc R Soc A,* 10.1098/rspa.2010.0413.
Grafen A (1984) Natural selection, kin selection and group selection. In *Behavioral Ecology: An Evolutionary Approach,* 2nd Ed, eds Krebs JR, Davies NB (Sinauer, Sunderland, MA), pp 62–84.
Greenberg JR, Hamann K, Warneken F, Tomasello M (2010) Chimpanzee helping in collaborative and noncollaborative contexts. *Anim Behav* 80:873–880.
Gregg C, Zhang J, Weissbourd B, Luo S, Schroth GP, Haig D, Dulac C (2010) High-resolution analysis of parent-of-origin allelic expression in the mouse brain. *Science* 329:643–648.
Greig D, Travisano M (2004) The Prisoner's Dilemma and polymorphism in yeast *SUC* genes. *Proc R Soc Lond B Biol Sci* 271:S25–S26.
Griffin AS, West SA (2002) Kin selection: Fact and fiction. *Trends Ecol Evol* 17:15–21.
Griffin AS, West SA, Buckling A (2004) Cooperation and competition in pathogenic bacteria. *Nature* 430:1024–1027.
Gronenberg W, Riveros AJ (2009) Social brains and behavior, past and present. In *Organization of Insect Societies: From Genome to Sociocomplexity,* eds Gadau J, Fewell J (Harvard Univ Press, Cambridge, MA), pp 377–401.
Grönke S, Clarke DF, Broughton S, Andrews TD, Partridge L (2010) Molecular evolution and functional characterization of *Drosophila* insulin-like peptides. *PLoS Genet* 6:e1000857.
Grosberg RK, Strathmann RR (1998) One cell, two cell, red cell, blue cell: The persistence of a unicellular stage in multicellular life histories. *Trends Ecol Evol* 13:112–116.

Grosberg RK, Rice WR, Palumbi SR (1985) Graft compatibility and clonal identity in invertebrates. *Science* 229:487–488.
Grosenick L, Clement TS, Fernald RD (2007) Fish can infer social rank by observation alone. *Nature* 445:429–432.
Guarner F, Malagelada JR (2003) Gut flora in health and disease. *Lancet* 361:512–519.
Gudelj I, Weitz JS, Ferenci T, Claire Horner-Devine M, Marx CJ, Meyer JR, Forde SE (2010) An integrative approach to understanding microbial diversity: From intracellular mechanisms to community structure. *Ecol Lett* 13:1073–1084.
Guilford T (1985) Is kin selection involved in the evolution of warning coloration. *Oikos* 45:31–36.
Guindon S, Gascuel O (2003) A simple, fast, and accurate algorithm to estimate large phylogenies by maximum likelihood. *Syst Biol* 52:696–704.
Gurven M (2006) The evolution of contingent cooperation. *Curr Anthropol* 47:185–192.
Gurven M, Kaplan H, Gutierrez M (2006) How long does it take to become a proficient hunter? Implications for the evolution of extended development and long life span. *J Hum Evol* 51:454–470.
Hagen DC, Bretscher AP, Kaiser D (1978). Synergism between morphogenetic mutants of *Myxococcus xanthus. Dev Biol* 64:284–296.
Hagen EH, Hammerstein P (2006) Game theory and human evolution: A critique of some recent interpretations of experimental games. *Theor Popul Biol* 69:339–348.
Haig D (1993) Genetic conflicts in human pregnancy. *Q Rev Biol* 68:495–532.
Haig D (1996) Placental hormones, genomic imprinting, and maternal-fetal communication. *J Evol Biol* 9:357–380.
Haig D (1997) Parental antagonism, relatedness asymmetries, and genomic imprinting. *Proc Biol Sci* 264:1657–1662.
Haig D (1999) Asymmetric relations: Internal conflicts and the horror of incest. *Evol Hum Behav* 20:83–98.
Haig D (2000a) Genomic imprinting, sex-biased dispersal, and social behavior. *Ann N Y Acad Sci* 907:149–163.
Haig D (2000b) The kinship theory of genomic imprinting. *Annu Rev Ecol Syst* 31:9–32.
Haig D (2004) Genomic imprinting and kinship: How good is the evidence? *Annu Rev Genet* 38:553–585.
Haig D (2006a) Intragenomic politics. *Cytogenet Genome Res* 113:68–74.
Haig D (2006b) Self-imposed silence: Parental antagonism and the evolution of X-chromosome inactivation. *Evolution* 60:440–447.
Haig D (2008) Conflicting messages: Genomic imprinting and internal communication. In *Sociobiology of Communication,* eds D'Ettorre P, Hughes DP (Oxford Univ Press, Oxford), pp 209–223.
Haig D (2010) Transfers and transitions: Parent-offspring conflict, genomic imprinting, and the evolution of human life history. *Proc Natl Acad Sci USA* 107(Suppl 1):1731–1735.
Haig D, Graham C (1991) Genomic imprinting and the strange case of the insulin-like growth factor II receptor. *Cell* 64:1045–1046.
Haig D, Wharton R (2003) Prader-Willi syndrome and the evolution of human childhood. *Am J Hum Biol* 15:320–329.
Haley KJ, Fessler DMT (2005) Nobody's watching? Subtle cues affect generosity in an anonymous economic game. *Evol Hum Behav* 26:245–256.
Hallatschek O, Hersen P, Ramanathan S, Nelson DR (2007) Genetic drift at expanding frontiers promotes gene segregation. *Proc Natl Acad Sci USA* 104:19926–19930.
Hall-Stoodley L, Costerton JW, Stoodley P (2004) Bacterial biofilms: From the natural environment to infectious diseases. *Nat Rev Microbiol* 2:95–108.
Hamilton WD (1964a) The genetical evolution of social behaviour. I. *J Theor Biol* 7:1–16.
Hamilton WD (1964b) The genetical evolution of social behaviour. II. *J Theor Biol* 7:17–52.

Hamilton WD (1971) Geometry for the selfish herd. *J Theor Biol* 31:295–311.
Hamilton WD (1975) Innate social aptitudes of man: An approach from evolutionary genetics. In *Biosocial Anthropology,* ed Fox R (Wiley, New York), pp 133–155.
Hammerstein P (2003) Why is reciprocity so rare in social animals? A protestant appeal. In *Genetic and Cultural Evolution of Cooperation,* ed Hammerstein P (MIT Press, Cambridge, MA), pp 83–93.
Hammond RL, Keller L (2004) Conflict over male parentage in social insects. *PLoS Biol* 2:E248.
Hampton RR (2001) Rhesus monkeys know when they remember. *Proc Natl Acad Sci USA* 98:5359–5362.
Hansen SK, Rainey PB, Haagensen JAJ, Molin S (2007) Evolution of species interactions in a biofilm community. *Nature* 445:533–536.
Harcombe W (2010) Novel cooperation experimentally evolved between species. *Evolution* 64:2166–2172.
Hardin G (1968) The tragedy of the commons. The population problem has no technical solution; it requires a fundamental extension in morality. *Science* 162:1243–1248.
Hare B, Call J, Agnetta B, Tomasello M (2000) Chimpanzees know what conspecifics do and do not see. *Anim Behav* 59:771–785.
Hare B, Call J, Tomasello M (2001) Do chimpanzees know what conspecifics know? *Anim Behav* 61:139–151.
Hare B, Melis AP, Woods V, Hastings S, Wrangham RW (2007) Tolerance allows bonobos to outperform chimpanzees on a cooperative task. *Curr Biol* 17:619–623.
Harrington WE, Mutabingwa TK, Muehlenbachs A, Sorensen B, Bolla MC, Fried M, Duffy PE (2009) Competitive facilitation of drug-resistant *Plasmodium falciparum* malaria parasites in pregnant women who receive preventive treatment. *Proc Natl Acad Sci USA* 106:9027–9032.
Harris KE, Schnittke N, Beckendorf SK (2007) Two ligands signal through the *Drosophila* PDGF/VEGF receptor to ensure proper salivary gland positioning. *Mech Dev* 124:441–448.
Harrison F, Buckling A (2009) Siderophore production and biofilm formation as linked social traits. *ISME J* 3:632–634.
Harrison F, Paul J, Massey RC, Buckling A (2008) Interspecific competition and siderophore-mediated cooperation in *Pseudomonas aeruginosa. ISME J* 2:49–55.
Hasselmann M, Gempe T, Schiøtt M, Nunes-Silva CG, Otte M, Beye M (2008a) Evidence for the evolutionary nascence of a novel sex determination pathway in honeybees. *Nature* 454:519–522.
Hasselmann M, Vekemans X, Pflugfelder J, Koeniger N, Koeniger G, Tingek S, Beye M (2008b) Evidence for convergent nucleotide evolution and high allelic turnover rates at the complementary sex determiner gene of Western and Asian honeybees. *Mol Biol Evol* 25:696–708.
Hastings IM (1997) A model for the origins and spread of drug resistance. *Parasitology* 115:133–141.
Hastings IM (2003) Malaria control and the evolution of drug resistance: An intriguing link. *Trends Parasitol* 19:70–73.
Hastings IM (2006) Complex dynamics and stability of resistance to antimalarial drugs. *Parasitology* 132:615–624.
Hastings IM (2011a) How artemisinin-containing combination therapies slow the spread of antimalarial drug resistance. *Trends Parasitol* 27:67–72.
Hastings IM (2011b) Why we should effectively treat malaria. *Trends Parasitol* 27:51–52.
Hastings IM, D'Alessandro U (2000) Modelling a predictable disaster: The rise and spread of drug-resistant malaria. *Parasitol Today* 16:340–347.

Hastings IM, Donnelly MJ (2005) The impact of antimalarial drug resistance mutations on parasite fitness, and its implications for the evolution of resistance. *Drug Resist Updat* 8:43–50.

Hastings IM, Watkins WM (2006) Tolerance is the key to understanding antimalarial drug resistance. *Trends Parasitol* 22:71–77.

Hatchwell BJ (2009) The evolution of cooperative breeding in birds: Kinship, dispersal and life history. *Philos Trans R Soc Lond B Biol Sci* 364:3217–3227.

Hatchwell BJ, Komdeur J (2000) Ecological constraints, life history traits and the evolution of cooperative breeding. *Anim Behav* 59:1079–1086.

Hatchwell BJ, Sharp SJ (2006) Kin selection, constraints, and the evolution of cooperative breeding in long-tailed tits. *Adv Stud Behav* 36:355–395.

Hauffa BP, Schlippe G, Roos M, Gillessen-Kaesbach G, Gasser T (2000) Spontaneous growth in German children and adolescents with genetically confirmed Prader-Willi syndrome. *Acta Paediatr* 89:1302–1311.

Hawkes K, Jones N (2005) Human age structures, paleodemography, and the grandmother hypothesis. In *Grandmotherhood: The Evolutionary Significance of the Second Half of Female Life,* eds Voland E, Chasiotis A, Schiefenhovel W (Rutgers Univ Press, Brunswick, NJ), pp 118–140.

Hawkes K, O'Connell J, Jones N (1997) Hadza women's time allocation, offspring provisioning, and the evolution of long postmenopausal life spans. *Curr Anthropol* 38:551–577.

Hawkes K, O'Connell JF, Jones NG, Alvarez H, Charnov EL (1998) Grandmothering, menopause, and the evolution of human life histories. *Proc Natl Acad Sci USA* 95:1336–1339.

Haydon DT, Matthews L, Timms R, Colegrave N (2003) Top-down or bottom-up regulation of intra-host blood-stage malaria: Do malaria parasites most resemble the dynamics of prey or predator? *Proc R Soc B* 270:289–298.

Heath KD, Tiffin P (2007) Context dependence in the coevolution of plant and rhizobial mutualists. *Proc Biol Sci* 274:1905–1912.

Heath KD, Stock AJ, Stinchcombe JR (2010) Mutualism variation in the nodulation response to nitrate. *J Evol Biol* 23:2494–2500.

Heinsohn R, Packer C (1995) Complex cooperative strategies in group-territorial African lions. *Science* 269:1260–1262.

Helanterä H, Bargum K (2007) Pedigree relatedness, not greenbeard genes, explains eusociality. *Oikos* 116:217–220.

Helanterä H, Strassmann JE, Carrillo J, Queller DC (2009) Unicolonial ants: Where do they come from, what are they and where are they going? *Trends Ecol Evol* 24:341–349.

Hellriegel B (1992) Modelling the immune response to malaria with ecological concepts: Short-term behaviour against long-term equilibrium. *Proc R Soc B* 250:249–256.

Hemelrijk CK (1994) Support for being groomed in long-tailed macaques, *Macaca fascicularis. Anim Behav* 48:479–481.

Henrich J (2004a) Animal behaviour: Inequity aversion in capuchins? *Nature* 428:139.

Henrich J (2004b) Demography and cultural evolution: Why adaptive cultural processes produced maladaptive losses in Tasmania. *Am Antiq* 69:197–218.

Henrich J (2009) The evolution of costly displays, cooperation, and religion: Credibility enhancing displays and their implications for cultural evolution. *Evol Hum Behav* 30:244–260.

Henrich J (2010) The evolution of innovation-enhancing institutions. In *Innovation in Cultural Systems: Contributions from Evolutionary Anthropology,* eds Shennan S, O'Brien M (MIT Press, Cambridge, MA), pp 99–210.

Henrich J, Gil-White FJ (2001) The evolution of prestige: Freely conferred deference as a mechanism for enhancing the benefits of cultural transmission. *Evol Hum Behav* 22:165–196.

Henrich J, Henrich N (2011) The evolution of cultural adaptations: Fijian taboos during pregnancy and lactation protect against marine toxins. *Proc Biol Sci* 366:1139–1148.

Henrich J, McElreath R (2003) The evolution of cultural evolution. *Evol Anthropol* 12:123–135.

Henrich J, Boyd R, Bowles S, Camerer C, Fehr E, Gintis H, eds (2004) *Foundations of Human Sociality: Economic Experiments and Ethnographic Evidence from Fifteen Small-Scale Societies* (Oxford Univ Press, New York).

Henrich J, McElreath R, Barr A, Ensminger J, Barrett C, Bolyanatz A, Cardenas JC, Gurven M, Gwako E, Henrich N, Lesorogol C, Marlowe F, Tracer D, Ziker J (2006) Costly punishment across human societies. *Science* 312:1767–1770.

Henrich J, Ensminger J, McElreath R, Barr A, Barrett C, Bolyanatz A, Camilo Cardenas J, Gurven M, Gwako E, Henrich N, Lesorogol C, Marlowe F, Tracer D, Ziker J (2010a) Markets, religion, community size, and the evolution of fairness and punishment. *Science* 327:1480–1484.

Henrich J, Heine SJ, Norenzayan A (2010b) The weirdest people in the world? *Behav Brain Sci* 33:61–135.

Henzi SP, Barrett L (2002) Infants as a commodity in a baboon market. *Anim Behav* 63:915–921.

Henzi SP, Barrett L (2007) Coexistence in female-bonded primate groups. *Adv Stud Behav* 37:107–132.

Herre EA, Wcislo WT (2011) In defence of inclusive fitness theory. *Nature* 471:E8–E9, author reply E9–E10.

Herron MD, Michod RE (2008) Evolution of complexity in the volvocine algae: Transitions in individuality through Darwin's eye. *Evolution* 62:436–451.

Hetzel C, Anderson RM (1996) The within-host cellular dynamics of bloodstage malaria: Theoretical and experimental studies. *Parasitology* 113:25–38.

Hewlett BS (1991) Demography and childcare in preindustrial societies. *J Anthropol Res* 47:1–37.

Hewlett BS, Fouts HN, Boyette AH, Hewlett BL (2011) Social learning among Congo Basin hunter-gatherers. *Philos Trans R Soc Lond B Biol Sci* 366:1168–1178.

Higashi M, Yamamura N (1993) What determines animal group size? Insider-outsider conflict and its resolution. *Am Nat* 142:553–563.

Hilgenboecker K, Hammerstein P, Schlattmann P, Telschow A, Werren JH (2008) How many species are infected with *Wolbachia*?—A statistical analysis of current data. *FEMS Microbiol Lett* 281:215–220.

Hill K, Hurtado AM (2009) Cooperative breeding in South American hunter-gatherers. *Proc Biol Sci* 276:3863–3870.

Hillesland KL, Velicer GJ (2005) Resource level affects relative performance of the two motility systems of *Myxococcus xanthus*. *Microb Ecol* 49:558–566.

Hillesland KL, Velicer GJ, Lenski RE (2009) Experimental evolution of a microbial predator's ability to find prey. *Proc Biol Sci* 276:459–467.

Hines HM, Hunt JH, O'Connor TK, Gillespie JJ, Cameron SA (2007) Multigene phylogeny reveals eusociality evolved twice in vespid wasps. *Proc Natl Acad Sci USA* 104:3295–3299.

Hirose E (2000) Plant rake and algal pouch of the larvae in the tropical ascidian *Diplosoma similis*: An adaptation for vertical transmission of photosynthetic symbionts *Prochloron* sp. *Zoolog Sci* 17:233–240.

Hirose E, Neilan BA, Schmidt EW, Murakami A (2009) Enigmatic life and evolution of *Prochloron* and related cyanobacteria inhabiting colonial ascidians. In *Handbook on Cyanobacteria: Biochemistry, Biotechnology and Applications*, eds Gault PM, Marler HJ (Nova Science, Hauppauge, NY), pp 161–189.

Hodgkin JDK (1979) Genetics of gliding motility in *Myxococcus xanthus*. *Mol Gen Genet* 171:177–191.

Hodgkin J, Kaiser D (1977) Cell-to-cell stimulation of movement in nonmotile mutants of *Myxococcus*. *Proc Natl Acad Sci USA* 74:2938–2942.

Hoffman E, McCabe K, Smith VL (1996) Social distance and other-regarding behavior in dictator games. *Am Econ Rev* 86:653–660.

Hoffmann JA (2003) The immune response of *Drosophila. Nature* 426:33–38.

Holland B, Rice WR (1998) Chase-away sexual selection: Antagonistic seduction versus resistance. *Evolution* 52:1–7.

Holland JN, DeAngelis DL, Schultz ST (2004) Evolutionary stability of mutualism: Interspecific population regulation as an evolutionarily stable strategy. *Proc Biol Sci* 271:1807–1814.

Holmes CW (1974) The Massey grass meter. In *Dairy Farming Annual* (Mass Univ, Palmerston North, New Zealand), pp 26–30.

Honeybee Genome Sequencing Consortium (2006) Insights into social insects from the genome of the honeybee *Apis mellifera. Nature* 443:931–949.

Hooper LV, Midtvedt T, Gordon JI (2002) How host-microbial interactions shape the nutrient environment of the mammalian intestine. *Annu Rev Nutr* 22:283–307.

Hopper KR, Rosenheim JA, Prout T, Oppenheim SJ (2003) Within-generation bet hedging: A seductive explanation? *Oikos* 101:219–222.

Horn M, Collingro A, Schmitz-Esser S, Beier CL, Purkhold U, Fartmann B, Brandt P, Nyakatura GJ, Droege M, Frishman D, Rattei T, Mewes HW, Wagner M (2004) Illuminating the evolutionary history of chlamydiae. *Science* 304:728–730.

Hrdy S (2005a) Comes the child before man: How cooperative breeding and prolonged post-weaning dependence shaped human potential. In *Hunter-Gatherer Childhoods: Evolutionary, Developmental, and Cultural Perspectives,* eds Hewlett BS, Lamb ME (de Gruyter, New York), pp 65–91.

Hrdy SB (2005b) The evolutionary context of human development. In *Attachment and Bonding: A New Synthesis,* eds Carter CS, et al. (MIT Press, Cambridge, MA), pp 9–32.

Hrdy SB (2007) Evolutionary context of human development: The cooperative breeding model. In *Family Relationships: An Evolutionary Perspective,* eds Salmon CA, Shackelford TK (Oxford Univ Press, Oxford), pp 16–38.

Hrdy SB (2009) *Mothers and Others: The Evolutionary Origins of Mutual Understanding* (Belknap Press, Cambridge, MA).

Huang HJ, Takagawa D, Weeks G, Pears C (1997) Cells at the center of *Dictyostelium* aggregates become spores. *Dev Biol* 192:564–571.

Hugenholtz P, Goebel BM, Pace NR (1998) Impact of culture-independent studies on the emerging phylogenetic view of bacterial diversity. *J Bacteriol* 180:4765–4774.

Hughes WOH, Oldroyd BP, Beekman M, Ratnieks FLW (2008) Ancestral monogamy shows kin selection is key to the evolution of eusociality. *Science* 320:1213–1216.

Huijben S, Nelson WA, Wargo AR, Sim DG, Drew DR, Read AF (2010) Chemotherapy, within-host ecology and the fitness of drug-resistant malaria parasites. *Evolution* 64:2952–2968.

Humphrey LT (2010) Weaning behaviour in human evolution. *Semin Cell Dev Biol* 21:453–461.

Hunt BG, Wyder S, Elango N, Werren JH, Zdobnov EM, Yi SV, Goodisman MA (2010) Sociality is linked to rates of protein evolution in a highly social insect. *Mol Biol Evol* 27:497–500.

Hunt JH, Wolschin F, Henshaw MT, Newman TC, Toth AL, Amdam GV (2010) Differential gene expression and protein abundance evince ontogenetic bias toward castes in a primitively eusocial wasp. *PLoS ONE* 5:e10674.

Hurst GDD, Werren JH (2001) The role of selfish genetic elements in eukaryotic evolution. *Nat Rev Genet* 2:597–606.

Hurst GDD, Hurst LD, Johnstone RA (1992) Intranuclear conflict and its role in evolution. *Trends Ecol Evol* 7:373–378.

Hurst LD, Pomiankowski A (1991) Causes of sex ratio bias may account for unisexual sterility in hybrids: A new explanation of Haldane's rule and related phenomena. *Genetics* 128:841–858.

Huss-Danell K, Frej AK (1986) Distribution of *Frankia* in soils from forest and afforestation sites in northern Sweden. *Plant Soil* 90:407–417.

Hyde JE (2005) Drug-resistant malaria. *Trends Parasitol* 21:494–498.

International Aphid Genomics Consortium (2010) Genome sequence of the pea aphid *Acyrthosiphon pisum*. *PLoS Biol* 8:e1000313.

Ishida K, Hata T, Urushihara H (2005) Gamete fusion and cytokinesis preceding zygote establishment in the sexual process of *Dictyostelium discoideum*. *Dev Growth Differ* 47:25–35.

Isles AR, Davies W, Wilkinson LS (2006) Genomic imprinting and the social brain. *Philos Trans R Soc Lond B Biol Sci* 361:2229–2237.

Issenman BK (1997) *The Sinews of Survival* (UBC Press, Vancouver).

Ivey PK (2000) Cooperative reproduction in Ituri forest hunter-gatherers: Who cares for Efe infants? *Curr Anthropol* 41:856–866.

Ivey PK, Morelli G, Tronick E (2005) Child caretakers among Efe foragers of the Ituri forest. In *Hunter-Gatherer Childhoods: Evolutionary, Developmental, and Cultural Perspectives*, eds Hewlett BS, Lamb ME (de Gruyter, New York), pp 191–213.

Jack CN, Ridgeway JG, Mehdiabadi NJ, Jones EI, Edwards TA, Queller DC, Strassmann JE (2008) Segregate or cooperate—A study of the interaction between two species of *Dictyostelium*. *BMC Evol Biol* 8:8.

Jackson JBC, Buss L (1975) Alleopathy and spatial competition among coral reef invertebrates. *Proc Natl Acad Sci USA* 72:5160–5163.

Jaeggi AV, Burkart JM, Van Schaik CP (2010) On the psychology of cooperation in humans and other primates: Combining the natural history and experimental evidence of prosociality. *Philos Trans R Soc Lond B Biol Sci* 365:2723–2735.

Jaenike J (2001) Sex chromosome meiotic drive. *Annu Rev Ecol Syst* 32:25–49.

Jaenike J, Dyer KA, Cornish C, Minhas MS (2006) Asymmetrical reinforcement and *Wolbachia* infection in *Drosophila*. *PLoS Biol* 4:e325.

Jafari S, Le Bras J, Bouchaud O, Durand R (2004) *Plasmodium falciparum* clonal population dynamics during malaria treatment. *J Infect Dis* 189:195–203.

James R, Kleanthous C, Moore GR (1996) The biology of E colicins: Paradigms and paradoxes. *Microbiology* 142:1569–1580.

Jamieson IG (1989) Behavioral heterochrony and the evolution of birds helping at the nest: An unselected consequence of communal breeding. *Am Nat* 133:394–406.

Jamison CS, Cornell LL, Jamison PL, Nakazato H (2002) Are all grandmothers equal? A review and a preliminary test of the "grandmother hypothesis" in Tokugawa Japan. *Am J Phys Anthropol* 119:67–76.

Jang W, Gomer RH (2011) Initial cell type choice in *Dictyostelium*. *Eukaryot Cell* 10:150–155.

Jankowiak W, Diderich M (2000) Sibling solidarity in a polygamous community in the USA. Unpacking inclusive fitness. *Evol Hum Behav* 21:125–139.

Jarman PJ (1974) Social-organization on antelope in relation to their ecology. *Behaviour* 48:215–267.

Jarra W, Brown KN (1985) Protective immunity to malaria: Studies with cloned lines of *Plasmodium chabaudi* and *P. berghei* in CBA/Ca mice. I. The effectiveness and inter- and intra-species specificity of immunity induced by infection. *Parasite Immunol* 7:595–606.

Jeffreys AJ, Neumann R (2002) Reciprocal crossover asymmetry and meiotic drive in a human recombination hot spot. *Nat Genet* 31:267–271.

Jensen K (2010) Punishment and spite, the dark side of cooperation. *Philos Trans R Soc Lond B Biol Sci* 365:2635–2650.

Jensen K, Hare B, Call J, Tomasello M (2006) What's in it for me? Self-regard precludes altruism and spite in chimpanzees. *Proc Biol Sci* 273:1013–1021.

Jensen K, Call J, Tomasello M (2007a) Chimpanzees are rational maximizers in an ultimatum game. *Science* 318:107–109.

Jensen K, Call J, Tomasello M (2007b) Chimpanzees are vengeful but not spiteful. *Proc Natl Acad Sci USA* 104:13046–13050.

Jetz W, Rubenstein DR (2011) Environmental uncertainty and the global biogeography of cooperative breeding in birds. *Curr Biol* 21:72–78.

Johnson BR, Linksvayer TA (2010) Deconstructing the superorganism: Social physiology, groundplans, and sociogenomics. *Q Rev Biol* 85:57–79.

Johnson CR, Seinen I (2002) Selection for restraint in competitive ability in spatial competition systems. *Proc Biol Sci* 269:655–663.

Johnson GR (1987) In the name of the fatherland: An analysis of kin term usage in patriotic speech and literature. *Int Polit Sci Rev* 8:165–174.

Johnson LJ (2007) The genome strikes back: The evolutionary importance of defence against mobile elements. *Evol Biol* 34:121–129.

Johnson LJ (2008) Selfish genetic elements favor the evolution of a distinction between soma and germline. *Evolution* 62:2122–2124.

Johnson NA (2010) Hybrid incompatibility genes: Remnants of a genomic battlefield? *Trends Genet* 26:317–325.

Johnson S (2010) *Where Good Ideas Come From: The Natural History of Innovation* (Riverhead Hardcover, New York).

Johnstone KA, DuBose AJ, Futtner CR, Elmore MD, Brannan CI, Resnick JL (2006) A human imprinting centre demonstrates conserved acquisition but diverged maintenance of imprinting in a mouse model for Angelman syndrome imprinting defects. *Hum Mol Genet* 15:393–404.

Judge PG (1991) Dyadic and triadic reconciliation in pigtailed macaques (*Macaca nemestrina*). *Am J Primatol* 23:225–237.

Juliano JJ, Kwiek JJ, Cappell K, Mwapasa V, Meshnick SR (2007) Minority-variant *pfcrt* K76T mutations and chloroquine resistance, Malawi. *Emerg Infect Dis* 13:872–877.

Juliano JJ, Porter K, Mwapasa V, Sem R, Rogers WO, Ariey F, Wongsrichanalai C, Read A, Meshnick SR (2010) Exposing malaria in-host diversity and estimating population diversity by capture-recapture using massively parallel pyrosequencing. *Proc Natl Acad Sci USA* 107:20138–20143.

Kadam SV, Velicer GJ (2006) Variable patterns of density-dependent survival in social bacteria. *Behav Ecol* 17:833–838.

Kaiser D (1979) Social gliding is correlated with the presence of pili in *Myxococcus xanthus*. *Proc Natl Acad Sci USA* 76:5952–5956.

Kaltenpoth M, Göttler W, Herzner G, Strohm E (2005) Symbiotic bacteria protect wasp larvae from fungal infestation. *Curr Biol* 15:475–479.

Kaltenpoth M, Goettler W, Dale C, Stubblefield JW, Herzner G, Roeser-Mueller K, Strohm E (2006) "*Candidatus* Streptomyces philanthi," an endosymbiotic streptomycete in the antennae of *Philanthus* digger wasps. *Int J Syst Evol Microbiol* 56:1403–1411.

Kaltenpoth M, Winter SA, Kleinhammer A (2009) Localization and transmission route of *Coriobacterium glomerans*, the endosymbiont of pyrrhocorid bugs. *FEMS Microbiol Ecol* 69:373–383.

Kaltenpoth M, Goettler W, Koehler S, Strohm E (2010) Life cycle and population dynamics of a protective insect symbiont reveal severe bottlenecks during vertical transmission. *Evol Ecol* 24:463–477.

Kamakura M (2011) Royalactin induces queen differentiation in honeybees. *Nature* 473:478–483.

Kaminski J, Call J, Tomasello M (2008) Chimpanzees know what others know, but not what they believe. *Cognition* 109:224–234.

Kaneda A, Feinberg AP (2005) Loss of imprinting of *IGF2*: A common epigenetic modifier of intestinal tumor risk. *Cancer Res* 65:11236–11240.

Kaneda A, Wang CJ, Cheong R, Timp W, Onyango P, Wen B, Iacobuzio-Donahue CA, Ohlsson R, Andraos R, Pearson MA, Sharov AA, Longo DL, Ko MS, Levchenko A, Feinberg AP (2007) Enhanced sensitivity to IGF-II signaling links loss of imprinting of *IGF2* to increased cell proliferation and tumor risk. *Proc Natl Acad Sci USA* 104:20926–20931.

Kaplan HS, Hill KR, Lancaster JB, Hurtado AM (2000) A theory of human life history evolution: Diet, intelligence, and longevity. *Evol Anthropol* 9:156–185.

Kaplan HS, Lancaster JB, Robson A (2003) Embodied capital and the evolutionary economics of the human lifespan. *Popul Dev Rev* 29(Suppl):152–182.

Kaptijn R, Thomese F, van Tilburg TG, Liefbroer AC (2010) How grandparents matter: Support for the cooperative breeding hypothesis in a contemporary Dutch population. *Hum Nat* 21:393–405.

Kay RR (1998) The biosynthesis of differentiation-inducing factor, a chlorinated signal molecule regulating *Dictyostelium* development. *J Biol Chem* 273:2669–2675.

Kay RR, Thompson CRL (2009) Forming patterns in development without morphogen gradients: Scattered differentiation and sorting out. *Cold Spring Harb Perspect Biol* 1: a001503.

Keeler KH (1985) Cost:benefit models of mutualism. In *The Biology of Mutualism, Ecology and Evolution,* ed Boucher DH (Oxford Univ Press, London), pp 100–127.

Keeling PJ (2010) The endosymbiotic origin, diversification and fate of plastids. *Philos Trans R Soc Lond B Biol Sci* 365:729–748.

Keller L (1997) Indiscriminate altruism: Unduly nice parents and siblings. *Trends Ecol Evol* 12:99–103.

Keller L (2007) Uncovering the biodiversity of genetic and reproductive systems: Time for a more open approach. American Society of Naturalists E. O. Wilson Award winner address. *Am Nat* 169:1–8.

Keller L, Ross KG (1998) Selfish genes: A green beard in the red fire ant. *Nature* 394:573–575.

Kendler KS, Myers J, Prescott CA (2005) Sex differences in the relationship between social support and risk for major depression: A longitudinal study of opposite-sex twin pairs. *Am J Psychiatry* 162:250–256.

Kennedy GE (2005) From the ape's dilemma to the weanling's dilemma: Early weaning and its evolutionary context. *J Hum Evol* 48:123–145.

Kent L, Bowdin S, Kirby GA, Cooper WN, Maher ER (2008) Beckwith-Weidemann syndrome: A behavioral phenotype-genotype study. *Am J Med Genet B Neuropsychiatr Genet* 147B:1295–1297.

Kerr B (2009) Theoretical and experimental approaches to the evolution of altruism and the levels of selection. In *Experimental Evolution: Concepts, Methods, and Applications of Selection Experiments,* eds Garland T, Rose MR (Univ of California Press, Berkeley, CA), 1st Ed, pp 585–630.

Kerr B, Riley MA, Feldman MW, Bohannan BJ (2002) Local dispersal promotes biodiversity in a real-life game of rock-paper-scissors. *Nature* 418:171–174.

Kerr B, Neuhauser C, Bohannan BJM, Dean AM (2006) Local migration promotes competitive restraint in a host-pathogen "tragedy of the commons." *Nature* 442:75–78.

Kessin RH (2001) *Dictyostelium: Evolution, Cell Biology, and the Development of Multicellularity* (Cambridge Univ Press, Cambridge, UK).

Kessler DA, Levine H (1998) Fluctuation-induced diffusive instabilities. *Nature* 394:556–558.

Khare A, Santorelli LA, Strassmann JE, Queller DC, Kuspa A, Shaulsky G (2009) Cheater-resistance is not futile. *Nature* 461:980–982.

Kidwell MG, Lisch DR (2001) Perspective: Transposable elements, parasitic DNA, and genome evolution. *Evolution* 55:1–24.

Kiers ET, Rousseau RA, West SA, Denison RF (2003) Host sanctions and the legume-rhizobium mutualism. *Nature* 425:78–81.

Kikuchi Y, Hosokawa T, Fukatsu T (2007) Insect-microbe mutualism without vertical transmission: A stinkbug acquires a beneficial gut symbiont from the environment every generation. *Appl Environ Microbiol* 73:4308–4316.

Kikuchi Y, Hosokawa T, Nikoh N, Meng XY, Kamagata Y, Fukatsu T (2009) Host-symbiont co-speciation and reductive genome evolution in gut symbiotic bacteria of acanthosomatid stinkbugs. *BMC Biol* 7:2.

Kikuchi Y, Hosokawa T, Fukatsu T (2011) An ancient but promiscuous host-symbiont association between *Burkholderia* gut symbionts and their heteropteran hosts. *ISME J* 5:446–460.

Killingback T, Bieri J, Flatt T (2006) Evolution in group-structured populations can resolve the tragedy of the commons. *Proc Biol Sci* 273:1477–1481.

Kirkup BC, Riley MA (2004) Antibiotic-mediated antagonism leads to a bacterial game of rock-paper-scissors in vivo. *Nature* 428:412–414.

Kitchen DM, Cheney DL, Seyfarth RM (2005) Male chacma baboons (*Papio hamadryas ursinus*) discriminate loud call contests between rivals of different relative ranks. *Anim Cogn* 8:1–6.

Klein R (2009) *The Human Career,* 3rd Ed. (Univ of Chicago Press, Chicago), .

Kline MA, Boyd R (2010) Population size predicts technological complexity in Oceania. *Proc Biol Sci* 277:2559–2564.

Kneitel JM, Chase JM (2004) Trade-offs in community ecology: Linking spatial scales and species coexistence. *Ecol Lett* 7:69–80.

Kobayashi I (2001) Behavior of restriction-modification systems as selfish mobile elements and their impact on genome evolution. *Nucleic Acids Res* 29:3742–3756.

Koch R (1893). Über den augenblicklichen stand der bakteriologischen choleradiagnose. *Z Hyg Infectionskr* 14:319–333. doi:10.1007/BF02284324.

Kochin BF, Yates AJ, de Roode JC, Antia R (2010) On the control of acute rodent malaria infections by innate immunity. *PLoS ONE* 5:e10444.

Koella JC, Lynch PA, Thomas MB, Read AF (2009) Towards evolution-proof malaria control with insecticides. *Evol Appl* 2:469–480.

Koenig WD, Pitelka FA (1981) *Natural Selection and Social Behavior,* eds Alexander RD, Tinkle DW (Chiron Press, New York), pp 261–280.

Koenig WD, Pitelka FA, Carmen WJ, Mumme RL, Stanback MT (1992) The evolution of delayed dispersal in cooperative breeders. *Q Rev Biol* 67:111–150.

Köhler T, Buckling A, van Delden C (2009) Cooperation and virulence of clinical *Pseudomonas aeruginosa* populations. *Proc Natl Acad Sci USA* 106:6339–6344.

Kojima A, Hirose E (2010) Transfer of prokaryotic algal symbionts from a tropical ascidian (*Lissoclinum punctatum*) colony to its larvae. *Zoolog Sci* 27:124–127.

Komdeur J (1992) Importance of habitat saturation and territory quality for evolution of cooperative breeding in the Seychelles warbler. *Nature* 358:493–495.

Komdeur J (1996) Variation in individual investment strategies among social animals. *Ethology* 112:729–747.

Koons DN, Metcalf CJE, Tuljapurkar S (2008) Evolution of delayed reproduction in uncertain environments: A life-history perspective. *Am Nat* 172:797–805.

Korb J, Foster KR (2010) Ecological competition favours cooperation in termite societies. *Ecol Lett* 13:754–760.

Korb J, Weil T, Hoffmann K, Foster KR, Rehli M (2009) A gene necessary for reproductive suppression in termites. *Science* 324:758.

Kornell N, Son LK, Terrace HS (2007) Transfer of metacognitive skills and hint seeking in monkeys. *Psychol Sci* 18:64–71.

Korobeinikov A, Wake GC (1999) Global properties of the three-dimensional predator-prey Lotka-Volterra systems. *J Appl Math Decis Sci* 3:155–162.
Kost C, Lakatos T, Böttcher I, Arendholz WR, Redenbach M, Wirth R (2007) Non-specific association between filamentous bacteria and fungus-growing ants. *Naturwissenschaften* 94:821–828.
Kotzot D (2004) Maternal uniparental disomy 14 dissection of the phenotype with respect to rare autosomal recessively inherited traits, trisomy mosaicism, and genomic imprinting. *Ann Genet* 47:251–260.
Kraemer SA, Toups MA, Velicer GJ (2010) Natural variation in developmental life-history traits of the bacterium *Myxococcus xanthus. FEMS Microbiol Ecol* 73:226–233.
Kramer KL (2002) Variation in juvenile dependence. Helping behavior among Mayan children. *Hum Nat* 13:299–325.
Kramer KL (2005a) Children's help and the pace of reproduction: Cooperative breeding in humans. *Evol Anthropol* 14:224–237.
Kramer KL (2005b) *Maya Children: Helpers at the Farm* (Harvard Univ Press, Cambridge, MA).
Kramer KL, Ellison PT (2010) Pooled energy budgets: Resituating human energy allocation trade-offs. *Evol Anthropol* 19:136–147.
Krams I, Krama T, Igaune K, Mand R (2008) Experimental evidence of reciprocal altruism in the pied flycatcher. *Behav Ecol Sociobiol* 62:599–605.
Krebs J, Davies N (1997) *Ecology: An Evolutionary Approach* (Blackwell Science, Oxford).
Kreft JU (2004) Biofilms promote altruism. *Microbiology* 150:2751–2760.
Kremer N, Voronin D, Charif D, Mavingui P, Mollereau B, Vavre F (2009) *Wolbachia* interferes with ferritin expression and iron metabolism in insects. *PLoS Pathog* 5:e1000630.
Krieger MJB, Ross KG (2005) Molecular evolutionary analyses of the odorant-binding protein gene *Gp-9* in fire ants and other *Solenopsis* species. *Mol Biol Evol* 22:2090–2103.
Kroos L, Kuspa A, Kaiser D (1986). A global analysis of developmentally regulated genes in *Myxococcus xanthus. Dev Biol* 117:252–266.
Krug D, Zurek G, Revermann O, Vos M, Velicer GJ, Müller R (2008) Discovering the hidden secondary metabolome of *Myxococcus xanthus:* A study of intraspecific diversity. *Appl Environ Microbiol* 74:3058–3068.
Kunieda T, Fujiyuki T, Kucharski R, Foret S, Ament SA, Toth AL, Ohashi K, Takeuchi H, Kamikouchi A, Kage E, Morioka M, Beye M, Kubo T, Robinson GE, Maleszka R (2006) Carbohydrate metabolism genes and pathways in insects: Insights from the honey bee genome. *Insect Mol Biol* 15:563–576.
Kuspa A, Loomis WF (1992) Tagging developmental genes in *Dictyostelium* by restriction enzyme-mediated integration of plasmid DNA. *Proc Natl Acad Sci USA* 89:8803–8807.
Kuzdzal-Fick J, Foster K, Queller D, Strassmann J (2007) Exploiting new terrain: An advantage to sociality in the slime mold *Dictyostelium discoideum. Behav Ecol* 18: 433–437.
Kuzdzal-Fick JJ, Queller DC, Strassmann JE (2010) An invitation to die: Initiators of sociality in a social amoeba become selfish spores. *Biol Lett* 6:800–802.
Lacy RC, Sherman PW (1983) Kin recognition by phenotype matching. *Am Nat* 121:489–512.
Lahdenperä M, Lummaa V, Helle S, Tremblay M, Russell AF (2004) Fitness benefits of prolonged post-reproductive lifespan in women. *Nature* 428:178–181.
Lakshminarayanan VR, Santos LR (2008) Capuchin monkeys are sensitive to others' welfare. *Curr Biol* 18:R999–R1000.
Lamba S, Mace R (2010) People recognize when they are really anonymous in an economic game. *Evol Hum Behav* 31:271–278.
Lambert A (2011) *The Gates of Hell: Sir John Franklin's Tragic Quest for the North West Passage* (Yale Univ Press, New Haven, CT).
Lambowitz AM, Zimmerly S (2004) Mobile group II introns. *Annu Rev Genet* 38:1–35.
Lancy D (1996) *Playing on Mother Ground: Cultural Routines for Children's Development* (Guilford Press, New York).

Lancy D (2009) *The Anthropology of Childhood: Cherubs, Chattel and Changlings* (Cambridge Univ Press, Cambridge, UK).
Lancy D (2010) Learning from nobody: The limited role of teaching in folk models of children's development. *Childhood in the Past* 3:79–106.
Lande R (1980) Sexual dimorphism, sexual selection, and adaptation in polygenic characters. *Evolution* 34:292–305.
Lande R, Arnold SJ (1983) The measurement of selection on correlated characters. *Evolution* 37:1210–1226.
Lankau RA, Strauss SY (2007) Mutual feedbacks maintain both genetic and species diversity in a plant community. *Science* 317:1561–1563.
Laue BE, Gill RE (1994) Use of a phase variation-specific promoter of *Myxococcus xanthus* in a strategy for isolating a phase-locked mutant. *J Bacteriol* 176:5341–5349.
Laue BE, Gill RE (1995) Using a phase-locked mutant of *Myxococcus xanthus* to study the role of phase variation in development. *J Bacteriol* 177:4089–4096.
Laurvick CL, Milne E, Blair E, de Klerk N, Charles AK, Bower C (2008) Fetal growth and the risk of childhood non-CNS solid tumours in Western Australia. *Br J Cancer* 99:179–181.
Lazzaro BP (2008) Natural selection on the *Drosophila* antimicrobial immune system. *Curr Opin Microbiol* 11:284–289.
Le Conte Y, Hefetz A (2008) Primer pheromones in social hymenoptera. *Annu Rev Entomol* 53:523–542.
Le Rouzic A, Boutin TS, Capy P (2007) Long-term evolution of transposable elements. *Proc Natl Acad Sci USA* 104:19375–19380.
Leach CK, Ashworth JM, Garrod DR (1973) Cell sorting out during the differentiation of mixtures of metabolically distinct populations of *Dictyostelium discoideum. J Embryol Exp Morphol* 29:647–661.
Leadbeater E, Carruthers JM, Green JP, van Heusden J, Field J (2010) Unrelated helpers in a primitively eusocial wasp: Is helping tailored towards direct fitness? *PLoS ONE* 5:e11997.
Leal WS, Ishida Y (2008) GP-9s are ubiquitous proteins unlikely involved in olfactory mediation of social organization in the red imported fire ant, *Solenopsis invicta. PLoS ONE* 3:e3762.
Lehmann L, Balloux F (2007) Natural selection on fecundity variance in subdivided populations: Kin selection meets bet hedging. *Genetics* 176:361–377.
Lehmann L, Keller L (2006) The evolution of cooperation and altruism—a general framework and a classification of models. *J Evol Biol* 19:1365–1376.
Lehmann L, Keller L, West SA, Roze D (2007) Group selection and kin selection: Two concepts but one process. *Proc Natl Acad Sci USA* 104:6736–6739.
Lehmann L, Feldman MW, Kaeuffer R (2010) Cumulative cultural dynamics and the coevolution of cultural innovation and transmission: An ESS model for panmictic and structured populations. *J Evol Biol* 23:2356–2369.
Leigh EG, Jr. (2010) The group selection controversy. *J Evol Biol* 23:6–19.
Leonetti D, Nath D, Hemam N, Neill D (2005) Kinship organization and the impact of grandmothers on reproductive success among the matrilineal Khasi and patrilineal Bengali of northeast India. In *Grandmotherhood: The Evolutionary Significance of the Second Half of Female Life,* eds Voland E, Chasiotis A, Schiefenhovel W (Rutgers Univ Press, Brunswick, NJ), pp 194–214.
Levin BR, Rozen DE (2006) Non-inherited antibiotic resistance. *Nat Rev Microbiol* 4:556–562.
Levin BR, Perrot V, Walker N (2000) Compensatory mutations, antibiotic resistance and the population genetics of adaptive evolution in bacteria. *Genetics* 154:985–997.
Levins R (1966) The strategy of model building in population biology. *Am Sci* 54:421–431.
Lévi-Strauss C (1980) *The Elementary Structures of Kinship,* eds trans Bell JH, von Sturmer JR, Needham R (Beacon Press, Boston). French.

Ley RE, Peterson DA, Gordon JI (2006a) Ecological and evolutionary forces shaping microbial diversity in the human intestine. *Cell* 124:837–848.

Ley RE, Turnbaugh PJ, Klein S, Gordon JI (2006b) Microbial ecology: Human gut microbes associated with obesity. *Nature* 444:1022–1023.

Ley RE, Hamady M, Lozupone C, Turnbaugh PJ, Ramey RR, Bircher JS, Schlegel ML, Tucker TA, Schrenzel MD, Knight R, Gordon JI (2008) Evolution of mammals and their gut microbes. *Science* 320:1647–1651.

Lew TA, Morrow EH, Rice WR (2006) Standing genetic variance for female resistance to harm from males and its relationship to intralocus sexual conflict. *Evolution* 60:97–105.

Li WH (1997) *Molecular Evolution* (Sinauer, Sunderland, MA).

Liao Z, Jia Q, Li F, Han Z (2010) Identification of two piwi genes and their expression profile in honeybee, *Apis mellifera. Arch Insect Biochem Physiol* 74:91–102.

Ligon RA, Hill GE (2010) Feeding decisions of eastern bluebirds are situationally influenced by fledgling plumage color. *Behav Ecol* 21:456–464.

Lim AK, Kai T (2007) Unique germ-line organelle, nuage, functions to repress selfish genetic elements in *Drosophila melanogaster*. *Proc Natl Acad Sci USA* 104:6714–6719.

Lim DH, Maher ER (2010) Genomic imprinting syndromes and cancer. *Adv Genet* 70:145–175.

Linksvayer TA, Wade MJ (2009) Genes with social effects are expected to harbor more sequence variation within and between species. *Evolution* 63:1685–1696.

Lipsitch M, Samore MH (2002) Antimicrobial use and antimicrobial resistance: A population perspective. *Emerg Infect Dis* 8:347–354.

Littell RC, Milliken GA, Stroup WW, Wolinger RD (1996) *SAS System for Mixed Models* (SAS Institute, Cary, NC).

Little AEF, Robinson CJ, Peterson SB, Raffa KF, Handelsman J (2008) Rules of engagement: Interspecies interactions that regulate microbial communities. *Annu Rev Microbiol* 62:375–401.

Lorenzen MD, Gnirke A, Margolis J, Garnes J, Campbell M, Stuart JJ, Aggarwal R, Richards S, Park Y, Beeman RW (2008) The maternal-effect, selfish genetic element *Medea* is associated with a composite *Tc1* transposon. *Proc Natl Acad Sci USA* 105:10085–10089.

Lu X, Shapiro JA, Ting CT, Li Y, Li C, Xu J, Huang H, Cheng YJ, Greenberg AJ, Li SH, Wu ML, Shen Y, Wu CI (2010) Genome-wide misexpression of X-linked versus autosomal genes associated with hybrid male sterility. *Genome Res* 20:1097–1102.

Luciani F, Sisson SA, Jiang HL, Francis AR, Tanaka MM (2009) The epidemiological fitness cost of drug resistance in *Mycobacterium tuberculosis. Proc Natl Acad Sci USA* 106:14711–14715.

Lyko F, Foret S, Kucharski R, Wolf S, Falckenhayn C, Maleszka R (2010) The honey bee epigenomes: Differential methylation of brain DNA in queens and workers. *PLoS Biol* 8:e1000506.

Lyon BE, Eadie JM, Hamilton LD (1994) Parental choice selects for ornamental plumage in American coot chicks. *Nature* 371:240–243.

Lyons DE, Young AG, Keil FC (2007) The hidden structure of overimitation. *Proc Natl Acad Sci USA* 104:19751–19756.

Lyttle TW (1991) Segregation distorters. *Annu Rev Genet* 25:511–557.

Ma WB, Dong FFT, Stavrinides J, Guttman DS (2006) Type III effector diversification via both pathoadaptation and horizontal transfer in response to a coevolutionary arms race. *PLoS Genet* 2:e209.

MacColl ADC, Hatchwell BJ (2002) Temporal variation in fitness payoffs promotes cooperative breeding in long-tailed tits *Aegithalos caudatus. Am Nat* 160:186–194.

MacDonald K (2007) Cross-cultural comparison of learning in human hunting: Implications for life history evolution. *Hum Nat* 18:386–402.

Macfarlane S, Dillon JF (2007) Microbial biofilms in the human gastrointestinal tract. *J Appl Microbiol* 102:1187–1196.

Mackinnon MJ (2005) Drug resistance models for malaria. *Acta Trop* 94:207–217.

Mackinnon MJ, Hastings IM (1998) The evolution of multiple drug resistance in malaria parasites. *Trans R Soc Trop Med Hyg* 92:188–195.

MacWilliams H, Doquang K, Pedrola R, Dollman G, Grassi D, Peis T, Tsang A, Ceccarelli A (2006) A retinoblastoma ortholog controls stalk/spore preference in *Dictyostelium*. *Development* 133:1287–1297.

Maddison WP, Maddison DR (2010) Mesquite: A modular system for evolutionary analysis. Version 2.74. Available at http://mesquiteproject.org. Accessed on November 2010.

Madigan MT, Martinko JM, Dunlap PV, Clark DP (2009) *Brock Biology of Microorganisms* (Pearson Benjamin-Cummings, San Francisco).

Maestripieri D, Roney JR, DeBias N, Durante KM, Spaepen GM (2004) Father absence, menarche and interest in infants among adolescent girls. *Dev Sci* 7:560–566.

Malik HS, Henikoff S (2009) Major evolutionary transitions in centromere complexity. *Cell* 138:1067–1082.

Manhes P, Velicer GJ (2011) Experimental evolution of selfish policing in social bacteria. *Proc Natl Acad Sci USA* 108:8357–8362.

Manichanh C, Rigottier-Gois L, Bonnaud E, Gloux K, Pelletier E, Frangeul L, Nalin R, Jarrin C, Chardon P, Marteau P, Roca J, Dore J (2006) Reduced diversity of faecal microbiota in Crohn's disease revealed by a metagenomic approach. *Gut* 55:205–211.

Marco EJ, Skuse DH (2006) Autism-lessons from the X chromosome. *Soc Cogn Affect Neurosci* 1:183–193.

Margulis L (1970) *Origin of Eukaryotic Cells* (Yale Univ Press, New Haven, CT).

Marlowe FW (2004) Marital residence among foragers. *Curr Anthropol* 45:277–284.

Marlowe FW (2005) Who tends Hadza children? In *Hunter-Gatherer Childhoods. Evolutionary, Developmental and Cultural Perspectives,* eds Hewlett BS, Lamb ME (Aldine Transaction, New Brunswick, NJ), pp 177–190.

Marsland T, Frank I, eds (2001) *Computers and Games* (Springer, Berlin).

Martin S, Sodergren E, Masuda T, Kaiser D (1978) Systematic isolation of transducing phages for *Myxococcus xanthus*. *Virology* 88:44–53.

Martin W, Herrmann RG (1998) Gene transfer from organelles to the nucleus: How much, what happens, and why? *Plant Physiol* 118:9–17.

Martínez JA, Nicolás JM, Marco F, Horcajada JP, Garcia-Segarra G, Trilla A, Codina C, Torres A, Mensa J (2006) Comparison of antimicrobial cycling and mixing strategies in two medical intensive care units. *Crit Care Med* 34:329–336.

Martins JR, Nunes FMF, Cristino AS, Simões ZLP, Bitondi MMG (2010) The four hexamerin genes in the honey bee: Structure, molecular evolution and function deduced from expression patterns in queens, workers and drones. *BMC Mol Biol* 11:23.

Mary-Rousselière G (1996) *Qitdlarssuaq, The Story of a Polar Migration* (Wuerz, Winnipeg, MB, Canada).

Masento MS, Morris HR, Taylor GW, Johnson SJ, Skapski AC, and Kay RR (1988) Differentiation-inducing factor from the slime mould *Dictyostelium discoideum* and its analogues. Synthesis, structure and biological activity. *Biochem J* 256:23–28.

Mason OT (2007) *North American Bows and Quivers* (Skyhorse, New York).

Matessi C, Jayakar SD (1976) Conditions for the evolution of altruism under Darwinian selection. *Theor Popul Biol* 9:360–387.

Matsushita M, Fujikawa HH (1990) Diffusion-limited growth in bacterial colony formation. *Physica A* 168:498–506.

Maynard Smith J (1964) Group selection and kin selection. *Nature* 201:1145–1147.

Maynard Smith J (1982) *Evolution and the Theory of Games* (Cambridge Univ Press, Cambridge, UK).

Maynard Smith J (1989a) *Evolutionary Genetics* (Oxford Univ Press, Oxford).

Maynard Smith J (1989b) Evolutionary progress and the levels of selection. In *Evolutionary Progress*, ed Nitecki MH (Univ of Chicago Press, Chicago), pp 219–230.

Maynard Smith J (1991) The population genetics of bacteria. *Proc R Soc Lond B Biol Sci* 245:37–41.

Maynard Smith J, Price GR (1973) The logic of animal conflict. *Nature* 246:15–18.

Maynard Smith J, Szathmáry E (1995) *The Major Transitions in Evolution* (Oxford Univ Press, Oxford).

Mayr E (2004) Happy birthday: 80 years of watching the evolutionary scenery. *Science* 305:46–47.

McCollum AM, Basco LK, Tahar R, Udhayakumar V, Escalante AA (2008) Hitchhiking and selective sweeps of *Plasmodium falciparum* sulfadoxine and pyrimethamine resistance alleles in a population from central Africa. *Antimicrob Agents Chemother* 52:4089–4097.

McDermott SR, Noor MAF (2010) The role of meiotic drive in hybrid male sterility. *Philos Trans R Soc Lond B Biol Sci* 365:1265–1272.

McGarr SE, Ridlon JM, Hylemon PB (2005) Diet, anaerobic bacterial metabolism, and colon cancer: A review of the literature. *J Clin Gastroenterol* 39:98–109.

McGlothlin JW, Moore AJ, Wolf JB, Brodie ED, III (2010) Interacting phenotypes and the evolutionary process. III. Social evolution. *Evolution* 64:2558–2574.

Medina M, Sachs JL (2010) Symbiont genomics, our new tangled bank. *Genomics* 95:129–137.

Meeks N, Cartwright CR (2005) Caribou and seal hair: Examination by scanning electron microscope. *Arctic Clothing,* eds King JC, Pauksztat B, Storrie R (McGill-Queens Univ Press, Montreal).

Mehdiabadi NJ, Jack CN, Farnham TT, Platt TG, Kalla SE, Shaulsky G, Queller DC, Strassmann JE (2006) Social evolution: Kin preference in a social microorganism. *Nature* 442:881–882.

Melis AP, Semmann D (2010) How is human cooperation different? *Philos Trans R Soc Lond B Biol Sci* 365:2663–2674.

Melis AP, Hare B, Tomasello M (2006a) Chimpanzees recruit the best collaborators. *Science* 311:1297–1300.

Melis AP, Hare B, Tomasello M (2006a) Engineering chimpanzee cooperation: Social tolerance constrains cooperation. *Anim Behav* 72:275–286.

Melis AP, Hare B, Tomasello M (2008) Do chimpanzees reciprocate favours? *Anim Behav* 76:951–962.

Melis AP, Hare B, Tomasello M (2009) Chimpanzees coordinate in a negotiation game. *Evol Hum Behav* 30:381–392.

Melis AP, Warneken F, Jensen K, Schneider AC, Call J, Tomasello M (2010) Chimpanzees help conspecifics obtain food and non-food items. *Proc R Soc Lond B,* 10.1098/rspb.2010.1735.

Melis AP, Warneken F, Jensen K, Schneider AC, Call J, Tomasello M (2011) Chimpanzees help conspecifics obtain food and non-food items. *Proc Biol Sci* 278:1405–1413.

Mendle J, Turkheimer E, D'Onofrio BM, Lynch SK, Emery RE, Slutske WS, Martin NG (2006) Family structure and age at menarche: A children-of-twins approach. *Dev Psychol* 42:533–542.

Mendle J, Harden KP, Turkheimer E, Van Hulle CA, D'Onofrio BM, Brooks-Gunn J, Rodgers JL, Emery RE, Lahey BB (2009) Associations between father absence and age of first sexual intercourse. *Child Dev* 80:1463–1480.

Mennill DJ, Ratcliffe LM, Boag PT (2002) Female eavesdropping on male song contests in songbirds. *Science* 296:873.

Mercereau-Puijalon O (1996) Revisiting host/parasite interactions: Molecular analysis of parasites collected during longitudinal and cross-sectional surveys in humans. *Parasite Immunol* 18:173–180.

Merhej V, Royer-Carenzi M, Pontarotti P, Raoult D (2009) Massive comparative genomic analysis reveals convergent evolution of specialized bacteria. *Biol Direct* 4:13.

Meunier J, West SA, Chapuisat M (2008) Split sex ratios in the social Hymenoptera: A meta-analysis. *Behav Ecol* 19:382–390.

Meunier J, Delaplace L, Chapuisat M (2010) Reproductive conflicts and egg discrimination in a socially polymorphic ant. *Behav Ecol Sociobiol* 64:1655–1663.

Michener CD (1974) *The Social Behavior of Bees* (Belknap Press, Cambridge, MA).

Michod RE, Hamilton WD (1980) Coefficients of relatedness in sociobiology. *Nature* 288:694–697.

Mideo N, Barclay VC, Chan BH, Savill NJ, Read AF, Day T (2008) Understanding and predicting strain-specific patterns of pathogenesis in the rodent malaria *Plasmodium chabaudi. Am Nat* 172:214–238.

Mifune N, Hashimoto H, Yamagishi T (2010) Altruism toward in-group members as a reputation mechanism. *Evol Hum Behav* 31:109–117.

Migliano AB, Vinicius L, Lahr MM (2007) Life history trade-offs explain the evolution of human pygmies. *Proc Natl Acad Sci USA* 104:20216–20219.

Miller MR, Råberg L, Read AF, Savill NJ (2010) Quantitative analysis of immune response and erythropoiesis during rodent malaria infection. *PLoS Comput Biol* 6:e1000946.

Milne E, Laurvick CL, Blair E, Bower C, de Klerk N (2007) Fetal growth and acute childhood leukemia: Looking beyond birth weight. *Am J Epidemiol* 166:151–159.

Mitani J (2006) Reciprocal exchange in chimpanzees and other primates. In *Cooperation in Primates and Humans: Mechanisms and Evolution,* eds Kappeler PM, van Schaik CP (Springer-Verlag, Berlin), pp 107–119.

Mitani JC (2009) Male chimpanzees form enduring and equitable social bonds. *Anim Behav* 7:633–640.

Mitani JC, Watts DP, Amsler SJ (2010) Lethal intergroup aggression leads to territorial expansion in wild chimpanzees. *Curr Biol* 20:R507–R508.

Mitteldorf J, Croll DH, Ch S, Ravela U (2002) Multilevel selection and the evolution of predatory restraint. *Artif Life* 8:146–152.

Molina Y, Harris RM, O'Donnell S (2009) Brain organization mirrors caste differences, colony founding and nest architecture in paper wasps (Hymenoptera: Vespidae). *Proc Biol Sci* 276:3345–3351.

Monds RD, O'Toole GA (2009) The developmental model of microbial biofilms: Ten years of a paradigm up for review. *Trends Microbiol* 17:73–87.

Monier JM, Lindow SE (2005) Spatial organization of dual-species bacterial aggregates on leaf surfaces. *Appl Environ Microbiol* 71:5484–5493.

Monk D (2010) Deciphering the cancer imprintome. *Brief Funct Genomics* 9:329–339.

Moore AJ, Brodie ED, Wolf JB (1997) Interacting phenotypes and the evolutionary process. 1. Direct and indirect genetic effects of social interactions. *Evolution* 51:1352–1362.

Moran NA (2003) Tracing the evolution of gene loss in obligate bacterial symbionts. *Curr Opin Microbiol* 6:512–518.

Moran NA, Wernegreen JJ (2000) Lifestyle evolution in symbiotic bacteria: Insights from genomics. *Trends Ecol Evol* 15:321–326.

Moran NA, McLaughlin HJ, Sorek R (2009) The dynamics and time scale of ongoing genomic erosion in symbiotic bacteria. *Science* 323:379–382.

Morgan AD, MacLean RC, Hillesland KL, Velicer GJ (2010) Comparative analysis of myxococcus predation on soil bacteria. *Appl Environ Microbiol* 76:6920–6927.

Morrison G, Barbosa P (1987) Spatial heterogeneity, population "regulation" and local extinction in simulated host-parasitoid interactions. *Oecologia* 73:609–614.

Mota MM, Brown KN, Do Rosário VE, Holder AA, Jarra W (2001) Antibody recognition of rodent malaria parasite antigens exposed at the infected erythrocyte surface: Specificity of immunity generated in hyperimmune mice. *Infect Immun* 69:2535–2541.

Mourre V, Villa P, Henshilwood CS (2010) Early use of pressure flaking on lithic artifacts at Blombos Cave, South Africa. *Science* 330:659–662.

Mueller UG, Dash D, Rabeling C, Rodrigues A (2008) Coevolution between attine ants and actinomycete bacteria: A reevaluation. *Evolution* 62:2894–2912.

Mueller UG, Ishak H, Lee JC, Sen R, Gutell RR (2010) Placement of attine ant-associated *Pseudonocardia* in a global *Pseudonocardia* phylogeny (Pseudonocardiaceae, Actinomycetales): A test of two symbiont-association models. *Antonie van Leeuwenhoek* 98:195–212.

Muller MN, Mitani JC (2005) Conflict and cooperation in wild chimpanzees. *Adv Stud Behav* 35:275–331.

Muller MN, Thompson ME, Wrangham RW (2006) Male chimpanzees prefer mating with old females. *Curr Biol* 16:2234–2238.

Münchhoff J, Hirose E, Maruyama T, Sunairi M, Burns BP, Neilan BA (2007) Host specificity and phylogeography of the prochlorophyte *Prochloron* sp., an obligate symbiont in didemnid ascidians. *Environ Microbiol* 9:890–899.

Muscatelli F, Abrous DN, Massacrier A, Boccaccio I, Le Moal M, Cau P, Cremer H (2000) Disruption of the mouse *Necdin* gene results in hypothalamic and behavioral alterations reminiscent of the human Prader-Willi syndrome. *Hum Mol Genet* 9:3101–3110.

Nadell CD, Xavier JB, Levin SA, Foster KR (2008) The evolution of quorum sensing in bacterial biofilms. *PLoS Biol* 6:e14.

Nadell CD, Xavier JB, Foster KR (2009) The sociobiology of biofilms. *FEMS Microbiol Rev* 33:206–224.

Nadell CD, Foster KR, Xavier JB (2010) Emergence of spatial structure in cell groups and the evolution of cooperation. *PLOS Comput Biol* 6:e1000716.

Nagaev I, Björkman J, Andersson DI, Hughes D (2001) Biological cost and compensatory evolution in fusidic acid-resistant *Staphylococcus aureus*. *Mol Microbiol* 40:433–439.

Nam KB, Simeoni M, Sharp SP, Hatchwell BJ (2010) Kinship affects investment by helpers in a cooperatively breeding bird. *Proc Biol Sci* 277:3299–3306.

Nariya H, Inouye M (2008) MazF, an mRNA interferase, mediates programmed cell death during multicellular *Myxococcus* development. *Cell* 132:55–66.

Nei M, Niimura Y, Nozawa M (2008) The evolution of animal chemosensory receptor gene repertoires: Roles of chance and necessity. *Nat Rev Genet* 9:951–963.

Nelson R (1969) *Hunters of the Northern Ice* (Univ of Chicago Press, Chicago).

Newton ILG, Bordenstein SR (2011) Correlations between bacterial ecology and mobile DNA. *Curr Microbiol* 62:198–208.

Nielsen M, Tomaselli K (2010) Overimitation in Kalahari Bushman children and the origins of human cultural cognition. *Psychol Sci* 21:729–736.

Nishiguchi MK, Nair VS (2003) Evolution of symbiosis in the Vibrionaceae: A combined approach using molecules and physiology. *Int J Syst Evol Microbiol* 53:2019–2026.

Noe R, Hammerstein P (1994) Biological markets: Supply and demand determine the effect of partner choice in cooperation, mutualism, and mating. *Behav Ecol Sociobiol* 35:1–11.

Nonacs P (1986) Ant reproductive strategies and sex allocation theory. *Q Rev Biol* 61:1–21.

Nonacs P (2006a) Nepotism and brood reliability in the suppression of worker reproduction in the eusocial Hymenoptera. *Biol Lett* 2:577–579.

Nonacs P (2006b) The rise and fall of transactional skew theory in the model genus *Polistes*. *Ann Zool Fenn* 43:443–455.

Nonacs P (2010) Ground truth is the test that counts. *Nature* 467:661.

Nonacs P (2011) Monogamy and high relatedness do not preferentially favor the evolution of cooperation. *BMC Evol Biol* 11:58.

Nonacs P, Carlin NF (1990) When can ants discriminate the sex of brood? A new aspect of queen-worker conflict. *Proc Natl Acad Sci USA* 87:9670–9673.

Nonacs P, Hager R (2011) The past, present and future of reproductive skew theory and experiments. *Biol Rev Camb Philos Soc* 86:271–298.

Nonacs P, Kapheim KM (2007) Social heterosis and the maintenance of genetic diversity. *J Evol Biol* 20:2253–2265.

Nonacs P, Kapheim KM (2008) Social heterosis and the maintenance of genetic diversity at the genome level. *J Evol Biol* 21:631–635.

Nonacs P, Liebert AE, Starks PT (2006) Transactional skew and assured fitness return models fail to predict patterns of cooperation in wasps. *Am Nat* 167:467–480.

Normand P, Orso S, Cournoyer B, Jeannin P, Chapelon C, Dawson J, Evtushenko L, Misra AK (1996) Molecular phylogeny of the genus *Frankia* and related genera and emendation of the family Frankiaceae. *Int J Syst Bacteriol* 46:1–9.

Nowak MA, May RM (1992) Evolutionary games and spatial chaos. *Nature* 359:826–829.

Nowak MA, Tarnita CE, Wilson EO (2010) The evolution of eusociality. *Nature* 466:1057–1062.

Nur U, Werren JH, Eickbush DG, Burke WD, Eickbush TH (1988) A "selfish" B chromosome that enhances its transmission by eliminating the paternal genome. *Science* 240:512–514.

Nussbaumer AD, Fisher CR, Bright M (2006) Horizontal endosymbiont transmission in hydrothermal vent tubeworms. *Nature* 441:345–348.

Nuzhdin SV, Petrov DA (2003) Transposable elements in clonal lineages: Lethal hangover from sex. *Biol J Linn Soc Lond* 79:33–41.

Nyholm SV, McFall-Ngai MJ (2004) The winnowing: Establishing the squid-vibrio symbiosis. *Nat Rev Microbiol* 2:632–642.

O'Donnell KA, Burns KH (2010) Mobilizing diversity: Transposable element insertions in genetic variation and disease. *Mob DNA* 1:21.

O'Donnell S, Donlan N, Jones T (2007) Developmental and dominance-associated differences in mushroom body structure in the paper wasp *Mischocyttarus mastigophorus*. *Dev Neurobiol* 67:39–46.

O'Toole GA, Kolter R (1998) Flagellar and twitching motility are necessary for *Pseudomonas aeruginosa* biofilm development. *Mol Microbiol* 30:295–304.

Oda R, Niwa Y, Honma A, Hiraishi K (2011) An eye-like painting enhances the expectation of a good reputation. *Evol Hum Behav* 32:166–171.

Okasha S (2008) *Biological Altruism*. Available at http://plato.stanford.edu/entries/altruism-biological/. Accessed December 10, 2010.

Oliveira RF, McGregor PK, Latruffe C (1998) Know thine enemy: Fighting fish gather information from observing conspecific interactions. *Proc Biol Sci* 265:1045–1049.

Oliver KM, Campos J, Moran NA, Hunter MS (2008) Population dynamics of defensive symbionts in aphids. *Proc Biol Sci* 275:293–299.

Oliver KR, Greene WK (2009) Transposable elements: Powerful facilitators of evolution. *Bioessays* 31:703–714.

Orgel LE, Crick FH (1980) Selfish DNA: The ultimate parasite. *Nature* 284:604–607.

Orjuela-Sánchez P, Da Silva-Nunes M, Da Silva NS, Scopel KK, Gonçalves RM, Malafronte RS, Ferreira MU (2009) Population dynamics of genetically diverse *Plasmodium falciparum* lineages: Community-based prospective study in rural Amazonia. *Parasitology* 136:1097–1105.

Oster GF, Wilson EO (1979) *Caste and Ecology in the Social Insects* (Princeton Univ Press, Princeton, NJ).

Östergren G (1945) Parasitic nature of extra fragment chromosomes. *Botan Notiser* 2:157–163.

Ostrowski EA, Katoh M, Shaulsky G, Queller DC, Strassmann JE (2008) Kin discrimination increases with genetic distance in a social amoeba. *PLoS Biol* 6:e287.

Otak LA (2005) Iniqsimajug: Caribou-skin preparation in Igloolik, Nunavut. In *Arctic Clothing*, eds King JC, Pauksztat B, Storrie R (McGill-Queens Univ Press, Montreal).

Owens RC, Jr. (2008) Antimicrobial stewardship: Concepts and strategies in the 21st century. *Diagn Microbiol Infect Dis* 61:110–128.

Owusu-Agyei S, Asante KP, Adjuik M, Adjei G, Awini E, Adams M, Newton S, Dosoo D, Dery D, Agyeman-Budu A, Gyapong J, Greenwood B, Chandramohan D (2009) Epidemiology of malaria in the forest-savanna transitional zone of Ghana. *Malar J* 8:220.

Oxley PR, Spivak M, Oldroyd BP (2010) Six quantitative trait loci influence task thresholds for hygienic behaviour in honeybees (*Apis mellifera*). *Mol Ecol* 19:1452–1461.

Paquin CE, Adams J (1983) Relative fitness can decrease in evolving asexual populations of *S. cerevisiae*. *Nature* 306:368–370.

Parkinson K, Buttery NJ, Wolf JB, Thompson CRL (2011) A simple mechanism for complex behavior. *PLoSBiol*, 9:e1001039.

Patten MM, Haig D (2009) Maintenance or loss of genetic variation under sexual and parental antagonism at a sex-linked locus. *Evolution* 63:2888–2895.

Paxton R, Hampton RR (2009) Tests of planning and the Bischof-Köhler hypothesis in rhesus monkeys (*Macaca mulatta*). *Behav Processes* 80:238–246.

Paxton R, Basile BM, Adachi I, Suzuki WA, Wilson ME, Hampton RR (2010) Rhesus monkeys (*Macaca mulatta*) rapidly learn to select dominant individuals in videos of artificial social interactions between unfamiliar conspecifics. *J Comp Psychol* 124:395–401.

Paz-Y-Miño CG, Bond AB, Kamil AC, Balda RP (2004) Pinyon jays use transitive inference to predict social dominance. *Nature* 430:778–781.

Peake TM, Terry AMR, McGregor PK, Dabelsteen T (2002) Do great tits assess rivals by combining direct experience with information gathered by eavesdropping? *Proc Biol Sci* 269:1925–1929.

Pedersen JS, Krieger MJB, Vogel V, Giraud T, Keller L (2006) Native supercolonies of unrelated individuals in the invasive Argentine ant. *Evolution* 60:782–791.

Penilla RP, Rodríguez AD, Hemingway J, Torres JL, Arredondo-Jiménez JI, Rodríguez MH (2007) Cytochrome P^{450}-based resistance mechanism and pyrethroid resistance in the field *Anopheles albimanus* resistance management trial. *Pestic Biochem Physiol* 89:111–117.

Pepper JW (2000) Relatedness in trait group models of social evolution. *J Theor Biol* 206:355–368.

Pepper JW (2007) Simple models of assortment through environmental feedback. *Artif Life* 13:1–9.

Pepper JW, Smuts BB (2002) A mechanism for the evolution of altruism among nonkin: Positive assortment through environmental feedback. *Am Nat* 160:205–213.

Perry S, Barrett HC, Manson J (2004) White-faced capuchin monkeys show triadic awareness in their choice of allies. *Anim Behav* 67:165–170.

Petroski H (1992) *The Evolution of Useful Things: How Everyday Artifacts—From Forks and Pins to Paper Clips and Zippers—Came to Be as They Are* (Vintage Books, NewYork).

Pflugfelder B, Cary SC, Bright M (2009) Dynamics of cell proliferation and apoptosis reflect different life strategies in hydrothermal vent and cold seep vestimentiferan tubeworms. *Cell Tissue Res* 337:149–165.

Philippi T, Seger J (1989) Hedging one's evolutionary bets, revisited. *Trends Ecol Evol* 4:41–44.

Philippot L, Andersson SG, Battin TJ, Prosser JI, Schimel JP, Whitman WB, Hallin S (2010) The ecological coherence of high bacterial taxonomic ranks. *Nat Rev Microbiol* 8:523–529.

Piliavin JA, Charng H-W (1990) Altruism: A review of recent theory and research. *Am Sociol Rev* 16:27–65.

Pinker S (2010) The cognitive niche: Coevolution of intelligence, sociality, and language. *Proc Natl Acad Sci USA* 107(Suppl 2):8993–8999.

Platt TG, Bever JD (2009) Kin competition and the evolution of cooperation. *Trends Ecol Evol* 24:370–377.

Plettner E, Otis GW, Wimalaratne PDC, Winston ML, Slessor KN, Pankiw T, Punchihewa PWK (1997) Species- and caste-determined mandibular gland signals in honeybees (*Apis*). *J Chem Ecol* 23:363–377.

Poissant J, Coltman DW (2009) The ontogeny of cross-sex genetic correlations: An analysis of patterns. *J Evol Biol* 22:2558–2562.

Poissant J, Wilson AJ, Coltman DW (2010) Sex-specific genetic variance and the evolution of sexual dimorphism: A systematic review of cross-sex genetic correlations. *Evolution* 64:97–107.

Pollard KS, Salama SR, Lambert N, Lambot MA, Coppens S, Pedersen JS, Katzman S, King B, Onodera C, Siepel A, Kern AD, Dehay C, Igel H, Ares M Jr, Vanderhaeghen P, Haussler D (2006) An RNA gene expressed during cortical development evolved rapidly in humans. *Nature* 443:167–172.

Pollitt LC, Mideo N, Drew DR, Schneider P, Colegrave N, Reece SE (2011) Competition and the evolution of reproductive restraint in malaria parasites. *Am Nat* 177:358–367.

Pollock F, Maitland FW (1895) *The History of English Law Before the Time of Edward I* (Cambridge Univ Press, Cambridge, UK), p 8.

Pongtavornpinyo W, Hastings IM, Dondorp A, White LJ, Maude RJ, Saralamba S, Day NP, White NJ, Boni MF (2009) Probability of emergence of antimalarial resistance in different stages of the parasite life cycle. *Evol Appl* 2:52–61.

Ponte E, Bracco E, Faix J, Bozzaro S (1998) Detection of subtle phenotypes: The case of the cell adhesion molecule csA in *Dictyostelium*. *Proc Natl Acad Sci USA* 95:9360–9365.

Port M, Kappeler PM (2010) The utility of reproductive skew models in the study of male primates, a critical evaluation. *Evol Anthropol* 19:46–56.

Prado F, Kerr B (2008) The evolution of restraint in bacterial biofilms under nontransitive competition. *Evolution* 62:538–548.

Premack D, Woodruff G (1978) Does the chimpanzee have a theory of mind? *Behav Brain Sci* 4:515–526.

Presgraves D (2009) Drive and sperm: Evolution and genetics of male meiotic drive. In *Sperm Biology: An Evolutionary Perspective,* eds Birkhead TR, Hosken DJ, Pitnick S (Elsevier, Burlington, MA), pp 471–506.

Presgraves DC (2010) The molecular evolutionary basis of species formation. *Nat Rev Genet* 11:175–180.

Press WH, Flannery BP, Teukolsky SA Vetterling WT (1997) *Numerical Recipes in C: The Art of Scientific Computing* (Cambridge Univ Press, Cambridge, UK), pp 882–888.

Price GR (1970) Selection and covariance. *Nature* 227:520–521.

Price GR (1972) Extension of covariance selection mathematics. *Ann Hum Genet* 35:485–490.

Price TA, Hodgson DJ, Lewis Z, Hurst GD, Wedell N (2008) Selfish genetic elements promote polyandry in a fly. *Science* 322:1241–1243.

Prichard R, Tait A (2001) The role of molecular biology in veterinary parasitology. *Vet Parasitol* 98:169–194.

Pride RE (2005) Foraging success, agonism, and predator alarms: Behavioral predictors of cortisol in *Lemur catta. Int J Primatol* 26:295–319.

Progeny Software, LLC (2007) Progeny 7.6.03 (Delray Beach, FL).

Queller DC (1984) Kin selection and frequency dependence: A game theoretic approach. *Biol J Linn Soc Lond* 23:133–143.

Queller DC (1985) Kinship, reciprocity and synergism in the evolution of social behaviour. *Nature* 318:366–367.

Queller DC (1992a) A general model for kin selection. *Evolution* 46:376–380.

Queller DC (1992b) Does population viscosity promote kin selection? *Trends Ecol Evol* 7:322–324.

Queller DC (1992c) Quantitative genetics, inclusive fitness, and group selection. *Am Nat* 139:540–558.

Queller DC (1994) Genetic relatedness in viscous populations. *Evol Ecol* 8:70–73.

Queller DC (1997) Cooperators since life began. *Q Rev Biol* 72:184–188 (book review).

Queller DC (2000) Relatedness and the fraternal major transitions. *Philos Trans R Soc Lond B Biol Sci* 355:1647–1655.

Queller DC, Strassmann JE (2009) Beyond society: The evolution of organismality. *Philos Trans R Soc Lond B Biol Sci* 364:3143–3155.

Queller DC, Zacchi F, Cervo R, Turillazzi S, Henshaw MT, Santorelli LA, Strassmann JE (2000) Unrelated helpers in a social insect. *Nature* 405:784–787.

Queller DC, Ponte E, Bozzaro S, Strassmann JE (2003) Single-gene greenbeard effects in the social amoeba *Dictyostelium discoideum*. *Science* 299:105–106.

R Development Core Team (2009). *R: A Language and Environment for Statistical Computing*, Vienna, Austria. Available at http://www.r-project.org/.

Raberg L, de Roode JC, Bell AS, Stamou P, Gray D, Read AF (2006) The role of immune-mediated apparent competition in genetically diverse malaria infections. *Am Nat* 168:41–53.

Ran LA, Larsson J, Vigil-Stenman T, Nylander JA, Ininbergs K, Zheng WW, Lapidus A, Lowry S, Haselkorn R, Bergman B (2010) Genome erosion in a nitrogen-fixing vertically transmitted endosymbiotic multicellular cyanobacterium. *PLoS ONE* 5:e11486.

Rankin DJ, Taborsky M (2009) Assortment and the evolution of generalized reciprocity. *Evolution* 63:1913–1922.

Rankin DJ, López-Sepulcre A, Foster KR, Kokko H (2007) Species-level selection reduces selfishness through competitive exclusion. *J Evol Biol* 20:1459–1468.

Raper KB (1984) *The Dictyostelids* (Princeton Univ Press, Princeton, NJ).

Rasmussen K (1908) *People of the Polar North* (Kegan Paul, Trench, Trübner & Co., London).

Ratnieks FLW (1990) Worker policing in social insects. In *Social Insects and the Environment: Proceedings of the Eleventh International Congress of the International Union for the Study of Social Insects*, eds Veeresh GK, Mallik B, Viraktamath CA (Oxford & IBH, New Delhi), pp 365–366.

Ratnieks FL, Foster KR, Wenseleers T (2006) Conflict resolution in insect societies. *Annu Rev Entomol* 51:581–608.

Rawls J (1971) *A Theory of Justice* (Harvard Univ Press, Cambridge, MA).

Raymond J, Siefert JL, Staples CR, Blankenship RE (2004) The natural history of nitrogen fixation. *Mol Biol Evol* 21:541–554.

Read AF, Huijben S (2009) Evolutionary biology and the avoidance of antimicrobial resistance. *Evol Appl* 2:40–51.

Read AF, Taylor LH (2001) The ecology of genetically diverse infections. *Science* 292:1099–1102.

Read AF, Lynch PA, Thomas MB (2009) How to make evolution-proof insecticides for malaria control. *PLoS Biol* 7:e1000058.

Reeve HK, Keller L (1995) Partitioning of reproduction in mother-daughter versus sibling associations—A test of optimal skew theory. *Am Nat* 145:119–132.

Reiches MW, Ellison PT, Lipson SF, Sharrock KC, Gardiner E, Duncan LG (2009) Pooled energy budget and human life history. *Am J Hum Biol* 21:421–429.

Rendall D, Cheney DL, Seyfarth RM (2000) Proximate factors mediating "contact" calls in adult female baboons (*Papio cynocephalus ursinus*) and their infants. *J Comp Psychol* 114:36–46.

Rendell L, Fogarty L, Laland KN (2010) Rogers' paradox recast and resolved: Population structure and the evolution of social learning strategies. *Evolution* 64:534–548.

Reynolds MG (2000) Compensatory evolution in rifampin-resistant *Escherichia coli*. *Genetics* 156:1471–1481.

Rice LB (2008a) Bacteria by the book: Response. *Science* 322:853–854.

Rice LB (2008b) The Maxwell Finland Lecture: For the duration-rational antibiotic administration in an era of antimicrobial resistance and *Clostridium difficile*. *Clin Infect Dis* 46:491–496.

Rice WR (1984) Sex-chromosomes and the evolution of sexual dimorphism. *Evolution* 38:735–742.

Rice WR, Chippindale AK (2001) Intersexual ontogenetic conflict. *J Evol Biol* 14:685–693.

Rice WR, Holland B (1997) The enemies within: Intergenomic conflict, interlocus contest evolution (ICE), and the intraspecific Red Queen. *Behav Ecol Sociobiol* 41:1–10.
Richerson P, Boyd R (2005) *Not by Genes Alone* (Univ of Chicago Press, Chicago).
Richerson PJ, Boyd R, Henrich J (2010) Gene-culture coevolution in the age of genomics. *Proc Natl Acad Sci USA* 107(Suppl 2):8985–8992.
Riddiford LM (2008) Juvenile hormone action: A 2007 perspective. *J Insect Physiol* 54:895–901.
Rigdon M, Ishii K, Watabe M, Kitayama S (2009) Minimal social cues in the dictator game. *J Econ Psychol* 30:358–367.
Riley MA, Gordon DM (1999) The ecological role of bacteriocins in bacterial competition. *Trends Microbiol* 7:129–133.
Riley MA, Wertz JE (2002) Bacteriocins: Evolution, ecology, and application. *Annu Rev Microbiol* 56:117–137.
Robertson HM, Wanner KW (2006) The chemoreceptor superfamily in the honey bee, *Apis mellifera:* Expansion of the odorant, but not gustatory, receptor family. *Genome Res* 16:1395–1403.
Robertson HM, Gadau J, Wanner KW (2010) The insect chemoreceptor superfamily of the parasitoid jewel wasp *Nasonia vitripennis. Insect Mol Biol* 19(Suppl 1):121–136.
Robinson CJ, Bohannan BJM, Young VB (2010) From structure to function: The ecology of host-associated microbial communities. *Microbiol Mol Biol Rev* 74:453–476.
Robinson GE (2002) Genomics and integrative analyses of division of labor in honeybee colonies. *Am Nat* 160(Suppl 6):S160–S172.
Robinson GE, Vargo EL (1997) Juvenile hormone in adult eusocial Hymenoptera: Gonadotropin and behavioral pacemaker. *Arch Insect Biochem Physiol* 35:559–583.
Robinson GE, Fernald RD, Clayton DF (2008) Genes and social behavior. *Science* 322:896–900.
Rodseth L, Wrangham RW, Harrigan AM, Smuts BB (1991) The human community as a primate society. *Curr Anthropol* 32:221–254.
Roesch LFW, Fulthorpe RR, Riva A, Casella G, Hadwin AK, Kent AD, Daroub SH, Camargo FA, Farmerie WG, Triplett EW (2007) Pyrosequencing enumerates and contrasts soil microbial diversity. *ISME J* 1:283–290.
Rogers AR (1988) Does biology constrain culture? *Am Anthropol* 90:819–831.
Roll Back Malaria (2008) *The Global Malaria Action Plan* (Roll Back Malaria Partnership). Available at http://www.rbm.who.int/gmap/gmap.pdf. Accessed November 28, 2008.
Roma PG, Silberberg A, Ruggiero AM, Suomi SJ (2006) Capuchin monkeys, inequity aversion, and the frustration effect. *J Comp Psychol* 120:67–73.
Roper C, Pearce R, Nair S, Sharp B, Nosten F, Anderson T (2004) Intercontinental spread of pyrimethamine-resistant malaria. *Science* 305:1124.
Rosenberg E, Varon M (1984) Antibiotics and lytic enzymes. In *Myxobacteria: Development and Cell Interactions,* ed Rosenberg E (Springer, New York), pp 109–125.
Rosenfeld AJ, Lieberman JA, Jarskog LF (2010) Oxytocin, dopamine, and the amygdala: A neurofunctional model of social cognitive deficits in schizophrenia. *Schizophr Bull* 37:1077–1087.
Ross-Gillespie A, Gardner A, West SA, Griffin AS (2007) Frequency dependence and cooperation: Theory and a test with bacteria. *Am Nat* 170:331–342.
Roubik DW (1992) *Ecology and Natural History of Tropical Bees* (Cambridge Univ Press, Cambridge, United Kingdom).
Rousset F (2004) *Genetic Structure and Selection in Subdivided Populations* (Princeton Univ Press, Princeton, NJ).
Rousset F, Roze D (2007) Constraints on the origin and maintenance of genetic kin recognition. *Evolution* 61:2320–2330.
Rowe KC, Reno ML, Richmond DM, Adkins RM, Steppan SJ (2008) Pliocene colonization and adaptive radiations in Australia and New Guinea (Sahul): Multilocus systematics of the old endemic rodents (Muroidea: Murinae). *Mol Phylogenet Evol* 47:84–101.

Rozin P, Gruss L, Berk G (1979) Reversal of innate aversions: Attempts to induce a preference for chili peppers in rats. *J Comp Physiol Psychol* 93:1001–1014.
Rozin P, Mark M, Schiller D (1981) The role of desensitization to capsaicin in chili pepper ingestion and preference. *Chem Senses* 6:23–31.
Rubenstein DR (2007a) Stress hormones and sociality: Integrating social and environmental stressors. *Proc Biol Sci* 274:967–975.
Rubenstein DR (2007b) Temporal but not spatial environmental variation drives adaptive offspring sex allocation in a plural cooperative breeder. *Am Nat* 170:155–165.
Rubenstein DR (2007c) Territory quality drives intraspecific patterns of extrapair paternity. *Behav Ecol* 18:1058–1064.
Rubenstein DR (2009) The secret lives of starlings. *Nat Hist* 118:28–33.
Rubenstein DR, Lovette IJ (2007) Temporal environmental variability drives the evolution of cooperative breeding in birds. *Curr Biol* 17:1414–1419.
Rubenstein DR, Shen S-F (2009) Reproductive conflict and the costs of social status in cooperatively breeding vertebrates. *Am Nat* 173:650–661.
Rubenstein DR, Parlow AF, Hutch CR, Martin LB, 2nd (2008) Environmental and hormonal correlates of immune activity in a cooperatively breeding tropical bird. *Gen Comp Endocrinol* 159:10–15.
Ruby EG, Urbanowski M, Campbell J, Dunn A, Faini M, Gunsalus R, Lostroh P, Lupp C, McCann J, Millikan D, Schaefer A, Stabb E, Stevens A, Visick K, Whistler C, Greenberg EP (2005) Complete genome sequence of *Vibrio fischeri*: A symbiotic bacterium with pathogenic congeners. *Proc Natl Acad Sci USA* 102:3004–3009.
Rumbaugh KP, Diggle SP, Watters CM, Ross-Gillespie A, Griffin AS, West SA (2009) Quorum sensing and the social evolution of bacterial virulence. *CurrBiol* 19:341–345.
Rump P, Zeegers MP, van Essen AJ (2005) Tumor risk in Beckwith-Wiedemann syndrome: A review and meta-analysis. *Am J Med Genet A* 136:95–104.
Russell JA, Latorre A, Sabater-Muñoz B, Moya A, Moran NA (2003) Side-stepping secondary symbionts: Widespread horizontal transfer across and beyond the Aphidoidea. *Mol Ecol* 12:1061–1075.
Rutte C, Taborsky M (2008) The influence of social experience on cooperative behaviour of rats (*Rattus norvegicus*): Direct vs generalised reciprocity. *Behav Ecol Sociobiol* 62:499–505.
Sachs JL, Bull JJ (2005) Experimental evolution of conflict mediation between genomes. *Proc Natl Acad Sci USA* 102:390–395.
Sachs JL, Simms EL (2006) Pathways to mutualism breakdown. *Trends Ecol Evol* 21:585–592.
Sachs JL, Simms EL (2008) The origins of uncooperative rhizobia. *Oikos* 117:961–966.
Sachs JL, WilcoxTP (2006) A shift to parasitism in the jellyfish symbiont *Symbiodinium microadriaticum*. *Proc Biol Sci* 273:425–429.
Sachs JL, Mueller UG, Wilcox TP, Bull JJ (2004) The evolution of cooperation. *Q Rev Biol* 79:135–160.
Sachs JL, Kembel SW, Lau AH, Simms EL (2009) In situ phylogenetic structure and diversity of wild *Bradyrhizobium* communities. *Appl Environ Microbiol* 75:4727–4735.
Sachs JL, Ehinger MO, Simms EL (2010a) Origins of cheating and loss of symbiosis in wild *Bradyrhizobium*. *J Evol Biol* 23:1075–1089.
Sachs JL, Russell JE, Lii YE, Black KC, Lopez G, Patil AS (2010b) Host control over infection and proliferation of a cheater symbiont. *J Evol Biol* 23:1919–1927.
Sachs JL, Essenberg CJ, Turcotte MM (2011) New paradigms for the evolution of beneficial infections. *Trends Ecol Evol* 26:202–209.
Sagan L (1967) On the origin of mitosing cells. *J Theor Biol* 14:255–274.
Salmon CA (1998) The evocative nature of kin terminology in political rhetoric. *Politics Life Sci* 17:51–57.
Salvador A, Costa R (2009) Coping with competition: Neuroendocrine responses and cognitive variables. *Neurosci Biobehav Rev* 33:160–170.

Samuelsen SO, Bakketeig LS, Tretli S, Johannesen TB, Magnus P (2009) Birth weight and childhood cancer. *Epidemiology* 20:484–487.

Sandoz KM, Mitzimberg SM, Schuster M (2007) Social cheating in *Pseudomonas aeruginosa* quorum sensing. *Proc Natl Acad Sci USA* 104:15876–15881.

Santorelli L, Thompson CRL, Villegas E, Svetz J, Dinh C, Parikh A, Sucgang R, Kuspa A, Strassmann JE, Queller DC, Shaulsky G (2008) Facultative cheater mutants reveal the genetic complexity of cooperation in social amoebae. *Nature* 451:1107–1110.

Sappington TW, Raikhel AS (1998) Molecular characteristics of insect vitellogenins and vitellogenin receptors. *Insect Biochem Mol Biol* 28:277–300.

Sarhan A, Kokko H (2007) Multiple mating in the Glanville fritillary butterfly: A case of within-generation bet hedging? *Evolution* 61:606–616.

SAS Institute, Inc. (2008) SAS 9.2. (Cary, NC).

SAS Institute, Inc. (2010) *Using JMP 9* (SAS Institute, Cary, NC).

Savage DC (1977) Microbial ecology of the gastrointestinal tract. *Annu Rev Microbiol* 31:107–133.

Sawada H, Kuykendall LD, Young JM (2003) Changing concepts in the systematics of bacterial nitrogen-fixing legume symbionts. *J Gen Appl Microbiol* 49:155–179.

Saxer G, Doebeli M, Travisano M (2009) Spatial structure leads to ecological breakdown and loss of diversity. *Proc Biol Sci* 276:2065–2070.

Saxer G, Brock DA, Queller DC, Strassmann JE (2010) Cheating does not explain selective differences at high and low relatedness in a social amoeba. *BMC Evol Biol* 10:76.

Scavenius JP (1975) *The Greenland Kayak and Its Accessories* (Nyt Nordisk Forlag Library, Copenhagen).

Schaack S, Gilbert C, Feschotte C (2010a) Promiscuous DNA: Horizontal transfer of transposable elements and why it matters for eukaryotic evolution. *Trends in Ecology & Evolution* 25:537–546.

Schaack S, Pritham EJ, Wolf A, Lynch M (2010b) DNA transposon dynamics in populations of *Daphnia pulex* with and without sex. *Proc Biol Sci* 277:2381–2387.

Schaap P, Nebl T, Fisher PR (1996) A slow sustained increase in cytosolic Ca^{2+} levels mediates stalk gene induction by differentiation inducing factor in *Dictyostelium*. *EMBO J* 15:5177–5183.

Schaap P, Winckler T, Nelson M, Alvarez-Curto E, Elgie B, Hagiwara H, Cavender J, Milano-Curto A, Rozen DE, Dingermann T, Mutzel R, Baldauf SL (2006) Molecular phylogeny and evolution of morphology in the social amoebas. *Science* 314:661–663.

Scheffer M, van Nes EH (2006) Self-organized similarity, the evolutionary emergence of groups of similar species. *Proc Natl Acad Sci USA* 103:6230–6235.

Schino G (2007) Grooming and agonistic support: A meta-analysis of primate reciprocal altruism. *Behav Ecol* 18:115–120.

Schino G, Aureli F (2009) Reciprocal altruism in primates: Partner choice, cognition, and emotions. *Adv Stud Behav* 39:45–69.

Schino G, Tiddi B, Polizzi di Sorrentino E (2006) Simultaneous classification by rank and kinship in Japanese macaques. *Anim Behav* 71:1069–1074.

Schino G, di Sorrentino EP, Tiddi B (2007) Grooming and coalitions in Japanese macaques (*Macaca fuscata*): Partner choice and the time frame reciprocation. *J Comp Psychol* 121:181–188.

Schrag SJ, Perrot V, Levin BR (1997) Adaptation to the fitness costs of antibiotic resistance in *Escherichia coli*. *Proc Biol Sci* 264:1287–1291.

Schülke O, Bhagavatula J, Vigilant L, Ostner J (2010) Social bonds enhance reproductive success in male macaques. *Curr Biol* 20:2207–2210.

Schultz S, Dunbar R (2010) Encephalization is not a universal macroevolutionary phenomenon in mammals but is associated with sociality. *Proc Natl Acad Sci USA* 107:21582–21586.

Sear R (2008) Kin and child survival in rural Malawi: Are matrilineal kin always beneficial in a matrilineal society? *Hum Nat* 19:277–293.

Sear R, Mace R (2008) Who keeps children alive? A review of the effects of kin on child survival. *Evol Hum Behav* 29:1–18.

Sebens KP (1986) Spatial relationships among encrusting marine organisms in the New England subtidal zone. *Ecol Monogr* 56:73–96.

Seger J (1981) Kinship and covariance. *J Theor Biol* 91:191–213.

Sekita Y, Wagatsuma H, Nakamura K, Ono R, Kagami M, Wakisaka N, Hino T, Suzuki-Migishima R, Kohda T, Ogura A, Ogata T, Yokoyama M, Kaneko-Ishino T, Ishino F (2008) Role of retrotransposon-derived imprinted gene, Rtl1, in the feto-maternal interface of mouse placenta. *Nat Genet* 40:243–248.

Sellen DW (2007) Evolution of infant and young child feeding: Implications for contemporary public health. *Annu Rev Nutr* 27:123–148.

Sharp SP, Simeoni M, McGowan A, Nam K-B, Hatchwell BJ (2011) Patterns of recruitment, relatedness and cooperative breeding in two populations of long-tailed tits. *Anim Behav* 81:843–849.

Shaulsky G, Loomis WF (1993) Cell type regulation in response to expression of ricin A in *Dictyostelium. Dev Biol* 160:85–98.

Shen S-F, Emlen ST (2010) Group benefits, insider-outsider conflict and the evolution of cooperative social groups. Manuscript.

Shettleworth S (2010) *Cognition, Evolution, and Behavior* (Oxford Univ Press, NewYork), 2nd Ed.

Shi WY, Zusman DR (1993) The two motility systems of *Myxococcus xanthus* show different selective advantages on various surfaces. *Proc Natl Acad Sci USA* 90:3378–3382.

Shigenobu S, Watanabe H, Hattori M, Sakaki Y, Ishikawa H (2000) Genome sequence of the endocellular bacterial symbiont of aphids *Buchnera* sp. APS. *Nature* 407:81–86.

Shimkets LJ, Dworkin M, Reichenbach H (2006) The myxobacteria. In *The Prokaryotes,* eds Dworkin M, Rosenberg E, Schleifer K-H, Stackebrandt E, Dworkin M (Springer, New York), pp 31–115.

Shimoyama T, Kato S, Ishii S, Watanabe K (2009) Flagellum mediates symbiosis. *Science* 323:1574.

Shinada M, Yamagishi T, Ohmura Y (2004) False friends are worse than bitter enemies: "Altruistic" punishment of in-group members. *Evol Hum Behav* 25:379–393.

Shizuka D, Lyon BE (2010) Coots use hatch order to learn to recognize and reject conspecific brood parasitic chicks. *Nature* 463:223–226.

Shpak M (2005) Evolution of variance in offspring number: The effects of population size and migration. *Theory Biosci* 124:65–85.

Silk JB (1999) Male bonnet macaques use information about third-party rank relationships to recruit allies. *Anim Behav* 58:45–51.

Silk JB (2003) Cooperation Without Counting: The Puzzle of friendship. In *Genetic and Cultural Evolution of Cooperation,* ed Hammerstein P (MIT Press, Cambridge, MA), pp 37–54.

Silk JB (2007a) Empathy, sympathy, and prosocial preferences in primates. In *Oxford Handbook of Evolutionary Psychology,* eds Dunbar RIM, Barrett L (Oxford Univ Press, Oxford), pp 115–126.

Silk JB (2007b) The strategic dynamics of cooperation in primate groups. *Adv Stud Behav* 37:1–41.

Silk JB (2009) Nepotistic cooperation in nonhuman primate groups. *Philos Trans R Soc Lond B Biol Sci* 364:3243–3254.

Silk JB, House BR (in press) The phylogeny and ontogeny of prosocial behavior. In *The Oxford Handbook of Comparative Evolutionary Psychology,* eds Vonk J, Shackelford T (Oxford Univ Press, Oxford).

Silk JB, Alberts SC, Altmann J (2003a) Social bonds of female baboons enhance infant survival. *Science* 302:1331–1334.

Silk JB, Rendall D, Cheney DL, Seyfarth RM (2003b) Natal attraction in adult female baboons (*Papio cynocephalus ursinus*) in the Moremi Reserve Botswana. *Ethology* 109:627–644.

Silk JB, Brosnan SF, Vonk J, Henrich J, Povinelli DJ, Richardson AS, Lambeth SP, Mascaro J, Schapiro SJ (2005) Chimpanzees are indifferent to the welfare of unrelated group members. *Nature* 437:1357–1359.

Silk JB, Beehner JC, Bergman TJ, Crockford C, Engh AL, Moscovice LR, Wittig RM, Seyfarth RM, Cheney DL (2009) The benefits of social capital: Close social bonds among female baboons enhance offspring survival. *Proc Biol Sci* 276:3099–3104.

Silk JB, Beehner JC, Bergman TJ, Crockford C, Engh AL, Moscovice LR, Wittig RM, Seyfarth RM, Cheney DL (2010a) Female chacma baboons form strong, equitable, and enduring social bonds. *Behav Ecol Sociobiol* 64:1733–1747.

Silk JB, Beehner JC, Bergman TJ, Crockford C, Engh AL, Moscovice LR, Wittig RM, Seyfarth RM, Cheney DL (2010b) Strong and consistent social bonds enhance the longevity of female baboons. *Curr Biol* 20:1359–1361.

Silva AJ, Kogan JH, Frankland PW, Kida S (1998) CREB and memory. *Annu Rev Neurosci* 21:127–148.

Simms EL, Taylor DL (2002) Partner choice in nitrogen-fixing mutualisms of legumes and rhizobia. *Integr Comp Biol* 42:369–380.

Simms EL, Taylor DL, Povich J, Shefferson RP, Sachs JL, Urbina M, Tausczik Y (2006) An empirical test of partner choice mechanisms in a wild legume-rhizobium interaction. *Proc Biol Sci* 273:77–81.

Sinervo B, Calsbeek R (2006) The developmental, physiological, neural, and genetical causes and consequences of frequency-dependent selection in the wild. *Annu Rev Ecol Evol Syst* 37:581–610.

Sinervo B, Lively CM (1996) The rock-paper-scissors game and the evolution of alternative male strategies. *Nature* 380:240–243.

Sinervo B, Chaine A, Clobert J, Calsbeek R, Hazard L, Lancaster L, McAdam AG, Alonzo S, Corrigan G, Hochberg ME (2006) Self-recognition, color signals, and cycles of greenbeard mutualism and altruism. *Proc Natl Acad Sci USA* 103:7372–7377.

Sinervo B, Heulin B, Surget-Groba Y, Clobert J, Miles DB, Corl A, Chaine A, Davis A (2007) Models of density-dependent genic selection and a new rock–paper–scissors social system. *Am Nat* 170:663–680.

Sinzelle L, Izsvák Z, Ivics Z (2009) Molecular domestication of transposable elements: From detrimental parasites to useful host genes. *Cell Mol Life Sci* 66:1073–1093.

Siomi MC, Mannen T, Siomi H (2010) How does the royal family of *Tudor* rule the PIWI-interacting RNA pathway? *Genes Dev* 24:636–646.

Slocombe KE, Zuberbühler K (2007) Chimpanzees modify recruitment screams as a function of audience composition. *Proc Natl Acad Sci USA* 104:17228–17233.

Slocombe KE, Townsend SW, Zuberbühler K (2009) Wild chimpanzees (*Pan troglodytes schweinfurthii*) distinguish between different scream types: Evidence from a playback study. *Anim Cogn* 12:441–449.

Smillie C, Garcillán-Barcia MP, Francia MV, Rocha EPC, de la Cruz F (2010) Mobility of plasmids. *Microbiol Mol Biol Rev* 74:434–452.

Smith CD, Zimin A, Holt C, Abouheif E, Benton R, Cash E, Croset V, Currie CR, Elhaik E, Elsik CG, Fave MJ, Fernandes V, Gadau J, Gibson JD, Graur D, Grubbs KJ, Hagen DE, Helmkampf M, Holley JA, Hu H, Viniegra AS, Johnson BR, Johnson RM, Khila A, Kim JW, Laird J, Mathis KA, Moeller JA, Muñoz-Torres MC, Murphy MC, Nakamura R, Nigam S, Overson RP, Placek JE, Rajakumar R, Reese JT, Robertson HM, Smith CR, Suarez AV, Suen G, Suhr EL, Tao S, Torres CW, van Wilgenburg E, Viljakainen L, Walden KK, Wild AL, Yandell M, Yorke JA, Tsutsui ND (2011) Draft genome of the globally widespread and invasive Argentine ant (*Linepithema humile*). *Proc Natl Acad Sci USA* 108:5673–5678.

Smith CR, Toth AL, Suarez AV, Robinson GE (2008) Genetic and genomic analyses of the division of labour in insect societies. *Nat Rev Genet* 9:735–748.

Smith CR, Smith CD, Robertson HM, Helmkampf M, Zimin A, Yandell M, Holt C, Hu H, Abouheif E, Benton R, Cash E, Croset V, Currie CR, Elhaik E, Elsik CG, Favé MJ, Fernandes V, Gibson JD, Graur D, Gronenberg W, Grubbs KJ, Hagen DE, Viniegra AS, Johnson BR, Johnson RM, Khila A, Kim JW, Mathis KA, Munoz-Torres MC, Murphy MC, Mustard JA, Nakamura R, Niehuis O, Nigam S, Overson RP, Placek JE, Rajakumar R, Reese JT, Suen G, Tao S, Torres CW, Tsutsui ND, Viljakainen L, Wolschin F, Gadau J (2011) Draft genome of the red harvester ant *Pogonomyrmex barbatus. Proc Natl Acad Sci USA* 108:5667–5672.

Smith J, Van Dyken JD, Zee PC (2010a) A generalization of Hamilton's rule for the evolution of microbial cooperation. *Science* 328:1700–1703.

Smith JE, Van Horn RC, Powning KS, Cole AR, Graham KE, Memenis SK, Holekamp KE (2010b) Evolutionary forces favoring intragroup coalitions among spotted hyenas and other animals. *Behav Ecol* 21:284–303.

Smith RL, Evans HL, Chong TW, McElearney ST, Hedrick TL, Swenson BR, Scheld WM, Pruett TL, Sawyer RG (2008) Reduction in rates of methicillin-resistant *Staphylococcus aureus* infection after introduction of quarterly linezolid-vancomycin cycling in a surgical intensive care unit. *Surg Infect (Larchmt)* 9:423–431.

Smith T, Felger I, Tanner M, Beck H-P (1999) The epidemiology of multiple *Plasmodium falciparum* infections. 11. Premunition in *Plasmodium falciparum* infection: Insight from the epidemiology of multiple infections. *Trans R Soc Trop Med Hyg* 93(Suppl 1):59–64.

Smukalla S, Caldara M, Pochet N, Beauvais A, Guadagnini S, Yan C, Vinces MD, Jansen A, Prevost MC, Latgé JP, Fink GR, Foster KR, Verstrepen KJ (2008) *FLO1* is a variable green beard gene that drives biofilm-like cooperation in budding yeast. *Cell* 135:726–737.

Snel B, Bork P, Huynen MA (2002) Genomes in flux: The evolution of archaeal and proteobacterial gene content. *Genome Res* 12:17–25.

Sober E, Wilson DS (1999) *Unto Others: The Evolution and Psychology of Unselfish Behavior* (Harvard Univ Press, Cambridge, MA).

Sogin ML, Morrison HG, Huber JA, Mark Welch D, Huse SM, Neal PR, Arrieta JM, Herndl GJ (2006) Microbial diversity in the deep sea and the underexplored "rare biosphere." *Proc Natl Acad Sci USA* 103:12115–12120.

Solomon NG, French JA (1997) *Cooperative Breeding in Mammals* (Cambridge Univ Press, Cambridge, UK).

Sonnenburg JL, Xu J, Leip DD, Chen CH, Westover BP, Weatherford J, Buhler JD, Gordon JI (2005) Glycan foraging in vivo by an intestine-adapted bacterial symbiont. *Science* 307:1955–1959.

Southwick EE, Heldmaier G (1987) Temperature control in honey bee colonies. *Bioscience* 37:395–399.

Spellberg B, Guidos R, Gilbert D, Bradley J, Boucher HW, Scheld WM, Bartlett JG, Edwards J Jr; Infectious Diseases Society of America (2008) The epidemic of antibiotic-resistant infections: A call to action for the medical community from the Infectious Diseases Society of America. *Clin Infect Dis* 46:155–164.

Sprent JI, Sutherland JM, Faria SM (1987) Some aspects of the biology of nitrogen-fixing organisms. *Philos Trans R Soc Lond B Biol Sci* 317:111–129.

Stacey PB, Ligon JD (1987) Territory quality and dispersal options in the acorn woodpecker, and a challenge to the habitat-saturation model of cooperative breeding. *Am Nat* 130:654–676.

Stacey PB, Ligon JD (1991) The benefits-of-philopatry hypothesis for the evolution of cooperative breeding: Variation in territory quality and group size effects. *Am Nat* 137:831–846.

Stearns SC (1992) *The Evolution of Life Histories* (Oxford Univ Press, Oxford), p 123.

Stearns SC (2000) Daniel Bernoulli (1738): Evolution and economics under risk. *J Biosci* 25:221–228.

Stenberg G (2009) Selectivity in infant social referencing. *Infancy* 14:457–473.

Stepniewska K, White NJ (2008) Pharmacokinetic determinants of the window of selection for antimalarial drug resistance. *Antimicrob Agents Chemother* 52:1589–1596.

Stevens JR (2010) Donor payoffs and other-regarding preferences in cotton-top tamarins (*Saguinus oedipus*). *Anim Cogn* 13:663–670.

Stevens JR, Cushman FA, Hauser MD (2005) Evolving the psychological mechanisms for cooperation. *Annu Rev Ecol Evol Syst* 36:499–518.

Stewart AD, Logsdon JM, Jr., Kelley SE (2005) An empirical study of the evolution of virulence under both horizontal and vertical transmission. *Evolution* 59:730–739.

Stewart AD, Pischedda A, Rice WR (2010) Resolving intralocus sexual conflict: Genetic mechanisms and time frame. *J Hered* 101(Suppl 1):S94–S99.

Stewart KEJ, Bourn NAD, Thomas JA (2001) An evaluation of three quick methods commonly used to assess sward height in ecology. *J Appl Ecol* 38:1148–1154.

Stoddard BL (2011) Homing endonucleases: From microbial genetic invaders to reagents for targeted DNA modification. *Structure* 19:7–15.

Stouthamer R, Breeuwer JAJ, Hurst GDD (1999) *Wolbachia pipientis*: Microbial manipulator of arthropod reproduction. *Annu Rev Microbiol* 53:71–102.

Stouthamer R, Russell JE, Vavre F, Nunney L (2010) Intragenomic conflict in populations infected by parthenogenesis inducing *Wolbachia* ends with irreversible loss of sexual reproduction. *BMC Evol Biol* 10:229.

Stoven S, Silverman N, Junell A, Hedengren-Olcott M, Erturk D, Engstrom Y, Maniatis T, Hultmark D (2003) Caspase-mediated processing of the *Drosophila* NF-κB factor Relish. *Proc Natl Acad Sci USA* 100:5991–5996.

Strassmann BI (1992) The function of menstrual taboos among the Dogon. *Hum Nat* 3:89–131.

Strassmann BI (2000) Polygyny, family structure, and child mortality: A prospective study among the Dogon of Mali. *Adaptation and Human Behavior: An Anthropological Perspective*, eds Cronk L, Chagnon N, Irons W (de Gruyter, New York), pp 49–67.

Strassmann BI (2003) Social monogamy in a human society: Marriage and reproductive success among the Dogon. *Monogamy: Mating Strategies and Partnerships in Birds, Humans and Other Mammals*, eds Reichard UH, Boesch C (Cambridge Univ Press, Cambridge, UK), pp 177–189.

Strassmann BI, Clarke AL (1998) Ecological constraints on marriage in rural Ireland. *Evol Hum Behav* 19:33–55.

Strassmann BI, Garrard WM (2011) Alternatives to the grandmother hypothesis: A meta-analysis of the association between grandparental and grandchild survival in patrilineal populations. *Hum Nat* 22:201–222.

Strassmann BI, Gillespie B (2002) Life-history theory, fertility and reproductive success in humans. *Proc Biol Sci* 269:553–562.

Strassmann BI, Warner JH (1998) Predictors of fecundability and conception waits among the Dogon of Mali. *Am J Phys Anthropol* 105:167–184.

Strassmann JE, Queller DC (2007) Insect societies as divided organisms: The complexities of purpose and cross-purpose. *Proc Natl Acad Sci USA* 104(Suppl 1):8619–8626.

Strassmann JE, Queller DC (2010) The social organism: Congresses, parties, committees. *Evolution* 64:605–616.

Strassmann JE, Zhu Y, Queller DC (2000) Altruism and social cheating in the social amoeba *Dictyostelium discoideum. Nature* 408:965–967.

Strassmann JE, Gilbert OM, Queller DC (2011a) Kin discrimination and cooperation in microbes. *Annu Rev Microbiol* 65:doi:10.1146/annurev.micro.112408.134109.

Strassmann JE, Page RE, Jr, Robinson GE, Seeley TD (2011b) Kin selection and eusociality. *Nature* 471:E5–E6, author reply E9–E10.

Suarez AV, Holway DA, Tsutsui ND (2008) Genetics and behavior of a colonizing species: The invasive Argentine ant. *Am Nat* 172(Suppl 1):S72-S84.

Suarez RK, Darveau CA, Welch KC Jr, O'Brien DM, Roubik DW, Hochachka PW (2005) Energy metabolism in orchid bee flight muscles: Carbohydrate fuels all. *J Exp Biol* 208:3573–3579.

Sucgang R, Kuo A, Tian X, Salerno W, Parikh A, Feasley CL, Dalin E, Tu H, Huang E, Barry K, Lindquist E, Shapiro H, Bruce D, Schmutz J, Salamov A, Fey P, Gaudet P, Anjard C, Babu MM, Basu S, Bushmanova Y, van der Wel H, Katoh-Kurasawa M, Dinh C, Coutinho PM, Saito T, Elias M, Schaap P, Kay RR, Henrissat B, Eichinger L, Rivero F, Putnam NH, West CM, Loomis WF, Chisholm RL, Shaulsky G, Strassmann JE, Queller DC, Kuspa A, Grigoriev IV (2011) Comparative genomics of the social amoebae *Dictyostelium discoideum* and *Dictyostelium purpureum. Genome Biol* 12:R20.

Sullivan JT, Patrick HN, Lowther WL, Scott DB, Ronson CW (1995) Nodulating strains of *Rhizobium loti* arise through chromosomal symbiotic gene transfer in the environment. *Proc Natl Acad Sci USA* 92:8985–8989.

Sun M, Wartel M, Cascales E, Shaevitz JW, Mignot T (2011) Motor-driven intracellular transport powers bacterial gliding motility. *Proc Natl Acad Sci USA* 108:7559–7564.

Surbey MK (1990) Family composition, stress, and the timing of human menarche. In *Socioendocrinology of Primate Reproduction,* eds Ziegler T, Bercovitch FB (Wiley-Liss, New York), pp 11–32.

Surbey MK (1998) Parent and offspring strategies in the transition at adolescence. *Hum Nat* 9:67–94.

Suzuki S, Ono R, Narita T, Pask AJ, Shaw G, Wang C, Kohda T, Alsop AE, Marshall Graves JA, Kohara Y, Ishino F, Renfree MB, Kaneko-Ishino T (2007) Retrotransposon silencing by DNA methylation can drive mammalian genomic imprinting. *PLoS Genet* 3:e55.

Svenning MM, Eriksson T, Rasmussen U (2005) Phylogeny of symbiotic cyanobacteria within the genus *Nostoc* based on 16S rDNA sequence analyses. *Arch Microbiol* 183:19–26.

Swaab DF, Purba JS, Hofman MA (1995) Alterations in the hypothalamic paraventricular nucleus and its oxytocin neurons (putative satiety cells) in Prader-Willi syndrome: A study of five cases. *J Clin Endocrinol Metab* 80:573–579.

Swenson W, Arendt J, Wilson DS (2000) Artificial selection of microbial ecosystems for 3-chloroaniline biodegradation. *Environ Microbiol* 2:564–571.

Swidsinski A, Weber J, Loening-Baucke V, Hale LP, Lochs H (2005) Spatial organization and composition of the mucosal flora in patients with inflammatory bowel disease. *J Clin Microbiol* 43:3380–3389.

Szathmáry E, Smith JM (1995) The major evolutionary transitions. *Nature* 374:227–232.

Tainaka K (1993) Paradoxical effect in a three-candidate voter model. *Phys Lett A* 176:303–306.

Tainaka K (1995) Indirect effect in cyclic voter models. *Phys Lett A* 207:53–57.

Takeuchi I, Noce T, Tasaka M (1986) Prestalk and prespore differentiation during development of *Dictyostelium discoideum. Curr Top Dev Biol* 20:243–256.

Takimoto A, Kuroshima H, Fujita K (2010) Capuchin monkeys (*Cebus apella*) are sensitive to others' reward: An experimental analysis of food-choice for conspecifics. *Anim Cogn* 13:249–261.

Talisuna AO, Langi P, Mutabingwa TK, Van Marck E, Speybroeck N, Egwang TG, Watkins WW, Hastings IM, D'Alessandro U (2003) Intensity of transmission and spread of gene mutations linked to chloroquine and sulphadoxine-pyrimethamine resistance in falciparum malaria. *Int J Parasitol* 33:1051–1058.

Talisuna AO, Bloland P, D'Alessandro U (2004) History, dynamics, and public health importance of malaria parasite resistance. *Clin Microbiol Rev* 17:235–254.

Talisuna AO, Erhart A, Samarasinghe S, Van Overmeir C, Speybroeck N, D'Alessandro U (2006) Malaria transmission intensity and the rate of spread of chloroquine resistant *Plasmodium falciparum:* Why have theoretical models generated conflicting results? *Infect Genet Evol* 6:241–248.

Tamura K, Dudley J, Nei M, Kumar S (2007). MEGA4: Molecular Evolutionary Genetics Analysis (MEGA) software version 4.0. *Mol Biol Evol* 24:1596–1599.

Tarpy DR, Seeley TD (2006) Lower disease infections in honeybee (*Apis mellifera*) colonies headed by polyandrous vs monandrous queens. *Naturwissenschaften* 93:195–199.

Tarpy DR, Gilley DC, Seeley TD (2004) Levels of selection in a social insect: A review of conflict and cooperation during honey bee (*Apis mellifera*) queen replacement. *Behav Ecol Sociobiol* 55:513–523.

Taubes G (2008) The bacteria fight back. *Science* 321:356–361.

Taylor DR, Aarssen LW(1990) Complex competitive relationships among genotypes of three perennial grasses: Implications for species coexistence. *Am Nat* 136:305–327.

Taylor LH, Read AF (1998) Determinants of transmission success of individual clones from mixed-clone infections of the rodent malaria parasite, *Plasmodium chabaudi. Int J Parasitol* 28:719–725.

Taylor LH, Walliker D, Read AF (1997a) Mixed-genotype infections of malaria parasites: Within-host dynamics and transmission success of competing clones. *Proc R Soc B* 264:927–935.

Taylor LH, Walliker D, Read AF (1997b) Mixed-genotype infections of the rodent malaria *Plasmodium chabaudi* are more infectious to mosquitoes than single-genotype infections. *Parasitology* 115:121–132.

Taylor PD, Frank SA (1996) How to make a kin selection model. *J Theor Biol* 180:27–37.

Taylor SE, Klein LC, Lewis BP, Gruenewald TL, Gurung RA, Updegraff JA (2000) Biobehavioral responses to stress in females: Tend-and-befriend, not fight-or-flight. *Psychol Rev* 107:411–429.

Tennie C, Call J, Tomasello M (2009) Ratcheting up the ratchet: On the evolution of cumulative culture. *Philos Trans R Soc Lond B Biol Sci* 364:2405–2415.

Thompson CRL, Kay RR (2000a) Cell-fate choice in *Dictyostelium*: Intrinsic biases modulate sensitivity to DIF signaling. *Dev Biol* 227:56–64.

Thompson CRL, Kay RR (2000b) The role of DIF-1 signaling in *Dictyostelium* development. *Mol Cell* 6:1509–1514.

Thompson CRL, Fu Q, Buhay C, Kay RR, Shaulsky G (2004) A bZIP/bRLZ transcription factor required for DIF signaling in *Dictyostelium*. *Development* 131:513–523.

Thorsteinsson EB, James JE (1999) A meta-analysis of the effects of experimental manipulations of social support during laboratory stress. *Psychol Health* 14:869–886.

Thrall PH, Burdon JJ (2002) Evolution of gene-for-gene systems in metapopulations: The effect of spatial scale of host and pathogen dispersal. *Plant Pathol* 51:169–184.

Toft C, Andersson SGE (2010) Evolutionary microbial genomics: Insights into bacterial host adaptation. *Nat Rev Genet* 11:465–475.

Toh H, Weiss BL, Perkin SA, Yamashita A, Oshima K, Hattori M, Aksoy S (2006) Massive genome erosion and functional adaptations provide insights into the symbiotic lifestyle of *Sodalis glossinidius* in the tsetse host. *Genome Res* 16:149–156.

Tokuda G, Saito H, Watanabe H (2002) A digestive β-glucosidase from the salivary glands of the termite, *Neotermes koshunensis* (Shiraki): Distribution, characterization and isolation of its precursor cDNA by 5′- and 3′-RACE amplifications with degenerate primers. *Insect Biochem Mol Biol* 32:1681–1689.

Tomasello M, Carpenter M, Call J, Behne T, Moll H (2005) Understanding and sharing intentions: The origins of cultural cognition. *Behav Brain Sci* 28:675–691.

Tooby J, DeVore I (1987) The reconstruction of hominid behavioral evolution through strategic modeling. In *Primate Models of Hominid Behavior,* ed Kinzey W (SUNY Press, New York), pp 183–237.

Toro M, Silio L (1986) Assortment of encounters in the two-strategy game. *J Theor Biol* 123:193–204.

Toth AL, Robinson GE (2009) Evo-devo and the evolution of social behavior: Brain gene expression analyses in social insects. *Cold Spring Harb Symp Quant Biol* 74:419–426.

Toth AL, Varala K, Henshaw MT, Rodriguez-Zas SL, Hudson ME, Robinson GE (2010) Brain transcriptomic analysis in paper wasps identifies genes associated with behaviour across social insect lineages. *Proc Biol Sci* 277:2139–2148.

Travisano M, Velicer GJ (2004) Strategies of microbial cheater control. *Trends Microbiol* 12:72–78.

Trivers RL (1971) The evolution of reciprocal altruism. *Q Rev Biol* 46:35–57.

Trivers RL (1974) Parent-offspring conflict. *Am Zool* 14:249–264.

Trivers RL, Hare H (1976) Haploidploidy and the evolution of the social insect. *Science* 191:249–263.

Troisi R, Masters MN, Joshipura K, Douglass C, Cole BF, Hoover RN (2006) Perinatal factors, growth and development, and osteosarcoma risk. *Br J Cancer* 95:1603–1607.

Turke P (1988) Helpers at the nest: Childcare networks on Ifaluk Micronesia. *Human Reproductive Behavior: A Darwinian Perspective,* eds Betzig L, Borgerhoff Mulder M, Turke P (Cambridge Univ Press, Cambridge, UK), pp 173–188.

Turnbaugh PJ, Hamady M, Yatsunenko T, Cantarel BL, Duncan A, Ley RE, Sogin ML, Jones WJ, Roe BA, Affourtit JP, Egholm M, Henrissat B, Heath AC, Knight R, Gordon JI (2009) A core gut microbiome in obese and lean twins. *Nature* 457:480–484.

Turner S, Pryer KM, Miao VPW, Palmer JD (1999) Investigating deep phylogenetic relationships among cyanobacteria and plastids by small subunit rRNA sequence analysis. *J Eukaryot Microbiol* 46:327–338.

Tylor EB (1889) On a method of investigating the development of institutions; Applied to laws of marriage and descent. *J Anthropol Inst Great Brit Ireland* 18:245–272.

Úbeda F (2008) Evolution of genomic imprinting with biparental care: Implications for Prader-Willi and Angelman syndromes. *PLoS Biol* 6:e208.

Úbeda F, Gardner A (2010) A model for genomic imprinting in the social brain: Juveniles. *Evolution* 64:2587–2600.

Úbeda F, Gardner A (2011) A model for genomic imprinting in the social brain: Adults. *Evolution* 65:462–475.

Úbeda F, Wilkins JF (2008) Imprinted genes and human disease: An evolutionary perspective. In *Genomic Imprinting,* ed Wilkins JF (Landes Bioscience, Austin, TX), pp 101–115.

Uher J, Call J (2008) How the great apes (*Pan troglodytes, Pongo pygmaeus, Pan paniscus, Gorilla gorilla*) perform on the reversed reward contingency task II: Transfer to new quantities, long-term retention, and the impact of quantity ratios. *J Comp Psychol* 122:204–212.

Urbanová K, Cahlíková L, Hovorka O, Ptácek V, Valterová I (2008) Age-dependent changes in the chemistry of exocrine glands of *Bombus terrestris* queens. *J Chem Ecol* 34:458–466.

Urushihara H (1992) Sexual development of cellular slime molds. *Dev Growth Differ* 34:1–17.

US Department of State (2010) *Bureau of African Affairs: Mali.* Available at http://www.state.gov/r/pa/ei/bgn/2828.htm. Accessed May 25, 2011.

Vaishnava S, Behrendt CL, Ismail AS, Eckmann L, Hooper LV (2008) Paneth cells directly sense gut commensals and maintain homeostasis at the intestinal host-microbial interface. *Proc Natl Acad Sci USA* 105:20858–20863.

van Baalen M, Rand DA (1998) The unit of selection in viscous populations and the evolution of altruism. *J Theor Biol* 193:631–648.

Van Cleve J, Feldman MW, Lehmann L (2010) How demography, life history, and kinship shape the evolution of genomic imprinting. *Am Nat* 176:440–455.

van Doorn GS (2009) Intralocus sexual conflict. *Ann N Y Acad Sci* 1168:52–71.

Van Dyken JD, Linksvayer TA, Wade MJ (2011) Kin selection-mutation balance: A model for the origin, maintenance, and consequences of social cheating. *Am Nat* 177:288–300.

van Zweden JS, Heinze J, Boomsma JJ, d'Ettorre P (2009) Ant queen egg-marking signals: Matching deceptive laboratory simplicity with natural complexity. *PLoS ONE* 4:e4718.

van Zweden JS, Brask JB, Christensen JH, Boomsma JJ, Linksvayer TA, d'Ettorre P (2010) Blending of heritable recognition cues among ant nestmates creates distinct colony gestalt odours but prevents within-colony nepotism. *J Evol Biol* 23:1498–1508.

Vandermeer JH, Boucher DH (1978) Varieties of mutualistic interaction in population models. *J Theor Biol* 74:549–558.

Velicer GJ (2003) Social strife in the microbial world. *Trends Microbiol* 11:330–337.

Velicer GJ, Lenski RE (1999) Evolutionary trade-offs under conditions of resource abundance and scarcity: Experiments with bacteria. *Ecology* 80:1168–1179.

Velicer GJ, Stredwick KL (2002) Experimental social evolution with *Myxococcus xanthus*. *Antonie van Leeuwenhoek* 81:155–164.

Velicer GJ, Vos M (2009) Sociobiology of the myxobacteria. *Annu Rev Microbiol* 63:599–623.

Velicer GJ, Kroos L, Lenski RE (1998) Loss of social behaviors by *Myxococcus xanthus* during evolution in an unstructured habitat. *Proc Natl Acad Sci USA* 95:12376–12380.

Velicer GJ, Kroos L, Lenski RE (2000) Developmental cheating in the social bacterium *Myxococcus xanthus*. *Nature* 404:598–601.

Ventura R, Majolo B, Koyama NF, Hardie S, Schino G (2006) Reciprocation and interchange in wild Japanese macaques: Grooming, cofeeding, and agonistic support. *Am J Primatol* 68:1138–1149.

Verhoef HA, Morin PJ (2010) *Community Ecology: Processes, Models, and Applications* (Oxford Univ Press, Oxford).

Viljakainen L, Pamilo P (2008) Selection on an antimicrobial peptide *defensin* in ants. *J Mol Evol* 67:643–652.

Viljakainen L, Evans JD, Hasselmann M, Rueppell O, Tingek S, Pamilo P (2009) Rapid evolution of immune proteins in social insects. *Mol Biol Evol* 26:1791–1801.

Vinogradov AE (2004) Genome size and extinction risk in vertebrates. *Proc Biol Sci* 271:1701–1705.

Visick KL, Ruby EG (2006) *Vibrio fischeri* and its host: It takes two to tango. *Curr Opin Microbiol* 9:632–638.

Visick KL, Foster J, Doino J, McFall-Ngai M, Ruby EG (2000) *Vibrio fischeri lux* genes play an important role in colonization and development of the host light organ. *J Bacteriol* 182:4578–4586.

Vonberg RP, Kuijper EJ, Wilcox MH, Barbut F, Tüll P, Gastmeier P; European *C. difficile*-Infection Control Group; European Centre for Disease Prevention and Control (ECDC), van den Broek PJ, Colville A, Coignard B, Daha T, Debast S, Duerden BI, van den Hof S, van der Kooi T, Maarleveld HJ, Nagy E, Notermans DW, O'Driscoll J, Patel B, Stone S, Wiuff C (2008) Infection control measures to limit the spread of *Clostridium difficile*. *Clin Microbiol Infect* 14(Suppl 5):2–20.

Vonk J, Brosnan SF, Silk JB, Henrich J, Richardson AS, Lambeth SP, Schapiro SJ, Povinelli DJ (2008) Chimpanzees do not take advantage of very low cost opportunities to deliver food to unrelated group members. *Anim Behav* 75:1757–1770.

Vos M, Didelot X (2009) A comparison of homologous recombination rates in bacteria and archaea. *ISME J* 3:199–208.

Vos M, Velicer GJ (2006) Genetic population structure of the soil bacterium *Myxococcus xanthus* at the centimeter scale. *Appl Environ Microbiol* 72:3615–3625.

Vos M, Velicer GJ (2008a) Isolation by distance in the spore-forming soil bacterium *Myxococcus xanthus. Curr Biol* 18:386–391.

Vos M, Velicer GJ (2008b) Natural variation of gliding motility in a centimetre-scale population of *Myxococcus xanthus. FEMS Microbiol Ecol* 64:343–350.

Vos M, Velicer GJ (2009) Social conflict in centimeter- and global-scale populations of the bacterium *Myxococcus xanthus. Curr Biol* 19:1763–1767.

Waddington SJ, Santorellia LA, Ryana FR, Hughes WOH (2010) Genetic polyethism in leaf-cutting ants. *Behav Ecol* 21:1165–1169.

Wade MJ (1980) Kin selection: Its components. *Science* 210:665–667.

Wade MJ (1985) Soft selection, hard selection, kin selection, and group selection. *Am Nat* 125:61–73.

Wade MJ (2007) The co-evolutionary genetics of ecological communities. *Nat Rev Genet* 8:185–195.

Wagner GP, Pavlicev M, Cheverud JM (2007) The road to modularity. *Nat Rev Genet* 8:921–931.

Walker R, Gurven M, Hill K, Migliano A, Chagnon N, De Souza R, Djurovic G, Hames R, Hurtado AM, Kaplan H, Kramer K, Oliver WJ, Valeggia C, Yamauchi T (2006) Growth rates and life histories in twenty-two small-scale societies. *Am J Hum Biol* 18:295–311.

Walliker D, Hunt P, Babiker H (2005) Fitness of drug-resistant malaria parasites. *Acta Trop* 94:251–259.

Wang TB, Patel A, Vu F, Nonacs P (2010) Natural history observations on the velvety tree ant (*Liometopum occidentale*): Unicoloniality and mating flights. *Sociobiol* 55:787–794.

Wanner KW, Nichols AS, Walden KK, Brockmann A, Luetje CW, Robertson HM (2007) A honey bee odorant receptor for the queen substance 9-oxo-2-decenoic acid. *Proc Natl Acad Sci USA* 104:14383–14388.

Wargo AR, Huijben S, de Roode JC, Shepherd J, Read AF (2007) Competitive release and facilitation of drug-resistant parasites after therapeutic chemotherapy in a rodent malaria model. *Proc Natl Acad Sci USA* 104:19914–19919.

Warneken F, Tomasello M (2006) Altruistic helping in human infants and young chimpanzees. *Science* 311:1301–1303.

Warneken F, Tomasello M (2009) Varieties of altruism in children and chimpanzees. *Trends Cogn Sci* 13:397–402.

Warneken F, Hare B, Melis AP, Hanus D, Tomasello M (2007) Spontaneous altruism by chimpanzees and young children. *PLoS Biol* 5:e184.

Weatherhead PJ, Boag PT (1997) Genetic estimates of annual and lifetime reproductive success in male red-winged blackbirds. *Ecology* 78:884–896.

Webb T, Maina EN, Soni S, Whittington J, Boer H, Clarke D, Holland A (2008) In search of the psychosis gene in people with Prader-Willi syndrome. *Am J Med Genet A* 146:843–853.

Webster T, Peng Y (1988) The evolution of food-producing glands in eusocial bees (Apoidea, Hymenoptera). *J Evol Biol* 2:165–176.

Weeks AR, Turelli M, Harcombe WR, Reynolds KT, Hoffmann AA (2007) From parasite to mutualist: Rapid evolution of *Wolbachia* in natural populations of *Drosophila. PLoS Biol* 5:e114.

Weidt A, Hofmann SE, König B (2008) Not only mate choice matters: Fitness consequences of social partner choice in female house mice. *Anim Behav* 85:801–808.

Weijer CJ, Duschl G, David CN (1984) A revision of the *Dictyostelium discoideum* cell cycle. *J Cell Sci* 70:111–131.

Weil T, Korb J, Rehli M (2009) Comparison of queen-specific gene expression in related lower termite species. *Mol Biol Evol* 26:1841–1850.

Weinert LA, Werren JH, Aebi A, Stone GN, Jiggins FM (2009) Evolution and diversity of *Rickettsia* basteria. *BMC Biol* 7:6.
Wenseleers T, Ratnieks FLW (2006a) Comparative analysis of worker reproduction and policing in eusocial hymenoptera supports relatedness theory. *Am Nat* 168:E163–E179.
Wenseleers T, Ratnieks FLW (2006b) Enforced altruism in insect societies. *Nature* 444:50.
Werren JH (1998) *Wolbachia* and speciation. In *Endless Forms: Species and Speciation* (Oxford Univ Press, New York), pp 245–260.
Werren JH, Beukeboom LW (1998) Sex determination, sex ratios, and genetic conflict. *Annu Rev Ecol Syst* 29:233–261.
Werren JH, Stouthamer R (2003) PSR (paternal sex ratio) chromosomes: The ultimate selfish genetic elements. *Genetica* 117:85–101.
Werren JH, Nur U, Wu CI (1988) Selfish genetic elements. *Trends Ecol Evol* 3:297–302.
Werren JH, Baldo L, Clark ME (2008) *Wolbachia*: Master manipulators of invertebrate biology. *Nat Rev Microbiol* 6:741–751.
Werren JH, Richards S, Desjardins CA, Niehuis O, Gadau J, Colbourne JK; Nasonia Genome Working Group (2010) Functional and evolutionary insights from the genomes of three parasitoid *Nasonia* species. *Science* 327:343–348.
West SA, Kiers ET, Pen I, Denison RF (2002a) Sanctions and mutualism stability: When should less beneficial mutualists be tolerated? *J Evol Biol* 15:830–837.
West SA, Kiers ET, Simms EL, Denison RF (2002b) Sanctions and mutualism stability: Why do rhizobia fix nitrogen? *Proc Biol Sci* 269:685–694.
West SA, Pen I, Griffin AS (2002c) Cooperation and competition between relatives. *Science* 296:72–75.
West SA, Griffin AS, Gardner A, Diggle SP (2006) Social evolution theory for microorganisms. *Nat Rev Microbiol* 4:597–607.
West SA, Diggle SP, Buckling A, Gardner A, Griffin AS (2007a) The social lives of microbes. *Annu Rev Ecol Evol Syst* 38:53–77.
West SA, Griffin AS, Gardner A (2007b) Evolutionary explanations for cooperation. *Curr Biol* 17:R661–R672.
West SA, Griffin AS, Gardner A (2007c) Social semantics: Altruism, cooperation, mutualism, strong reciprocity and group selection. *J Evol Biol* 20:415–432.
West SA, Griffin AS, Gardner A (2008) Social semantics: How useful has group selection been? *J Evol Biol* 21:374–385.
West-Eberhard MJ (1975) The evolution of social behavior by kin selection. *Q Rev Biol* 50:1–33.
West-Eberhard MJ (1983) Sexual selection, social competition, and speciation. *Q Rev Biol* 158:155–183.
Weyl EG, Frederickson ME, Yu DW, Pierce NE (2010) Economic contract theory tests models of mutualism. *Proc Natl Acad Sci USA* 107:15712–15716.
Wheeler DE, Buck NA (1995) Storage proteins in ants during development and colony founding. *J Insect Physiol* 41:885–894.
White CE, Winans SC (2007) Cell-cell communication in the plant pathogen *Agrobacterium tumefaciens. Philos Trans R Soc Lond B Biol Sci* 362:1135–1148.
White CM, Olsen PD, Kiff LF (1994) New World vultures to guineafowl, family Falconidae (falcons and caracaras). In *Handbook of Birds of the World*, eds del Hoyo J, Ellio A, Sargatal J (Lynx Edicions, Barcelona), Vol 2, pp 216–275.
White NJ (2004) Antimalarial drug resistance. *J Clin Invest* 113:1084–1092.
Whiten A, McGuigan N, Marshall-Pescini S, Hopper LM (2009) Emulation, imitation, over-imitation and the scope of culture for child and chimpanzee. *Philos Trans R Soc Lond B Biol Sci* 364:2417–2428.

Wilder CN, Allada G, Schuster M (2009) Instantaneous within-patient diversity of *Pseudomonas aeruginosa* quorum-sensing populations from cystic fibrosis lung infections. *Infect Immun* 77:5631–5639.

Wilkins JF (2011) Genomic imprinting and conflict-induced decanalization. *Evolution* 65:537–553.

Wilkins JF, Haig D (2003a) Inbreeding, maternal care and genomic imprinting. *J Theor Biol* 221:559–564.

Wilkins JF, Haig D (2003b) What good is genomic imprinting: The function of parent-specific gene expression. *Nat Rev Genet* 4:359–368.

Wilkinson DM (1999) Bacterial ecology, antibiotics and selection for virulence. *Ecol Lett* 2:207–209.

Wilkinson GS (1984) Reciprocal food sharing in vampire bats. *Nature* 308:181–184.

Williams GC (1971) *Group Selection* (Aldine-Atherton, Chicago).

Williams GC (1996) *Adaptation and Natural Selection: A Critique of Some Current Evolutionary Thought* (Princeton Univ Press, Princeton, NJ).

Williams GC, Nesse RM (1991) The dawn of Darwinian medicine. *Q Rev Biol* 66:1–22.

Williams HTP, Lenton TM (2008) Environmental regulation in a network of simulated microbial ecosystems. *Proc Natl Acad Sci USA* 105:10432–10437.

Williams KP, Sobral BW, Dickerman AW (2007) A robust species tree for the alphaproteobacteria. *J Bacteriol* 189:4578–4586.

Williams KP, Gillespie JJ, Sobral BW, Nordberg EK, Snyder EE, Shallom JM, Dickerman AW (2010) Phylogeny of gammaproteobacteria. *J Bacteriol* 192:2305–2314.

Wilson DS, Wilson EO (2007) Rethinking the theoretical foundation of sociobiology. *Q Rev Biol* 82:327–348.

Wilson D, Pollock G, Dugatkin L (1992) Can altruism evolve in purely viscous populations? *Evol Ecol* 6:331–341.

Wilson EO (1971) *The Insect Societies* (Belknap Press, Cambridge, MA).

Wilson EO (1975) *Sociobiology: The New Synthesis* (Belknap Press, Cambridge, MA).

Wilson EO, Hölldobler B (2005) Eusociality: Origin and consequences. *Proc Natl Acad Sci USA* 102:13367–13371.

Wilson-Rich N, Spivak M, Fefferman NH, Starks PT (2009) Genetic, individual, and group facilitation of disease resistance in insect societies. *Annu Rev Entomol* 54:405–423.

Winterhalder B (1986) Diet choice, risk, and food sharing in a stochastic environment. *J Anthropol Archaeol* 5:369–392.

Wittig RM, Boesch C (2010) Receiving post-conflict affiliation from the enemy's friend reconciles former opponents. *PLoS ONE* 5:e13995.

Wittig RM, Crockford C, Wikberg E, Seyfarth RM, Cheney DL (2007) Kin-mediated reconciliation substitutes for direct reconciliation in female baboons. *Proc Biol Sci* 274:1109–1115.

Wohlgelernter S, Diesendruck G, Markson L (2010) What is a conventional object function? The effects of intentionality and consistency of use. *J Cogn Dev* 11:269–292.

Wollenberg MS, Ruby EG (2009) Population structure of *Vibrio fischeri* within the light organs of *Euprymna scolopes* squid from two Oahu (Hawaii) populations. *Appl Environ Microbiol* 75:193–202.

Wollmann HA, Kirchner T, Enders H, Preece MA, Ranke MB (1995) Growth and symptoms in Silver-Russell syndrome: Review on the basis of 386 patients. *Eur J Pediatr* 154:958–968.

Wood SA, Ammann RR, Brock DA, Li L, Spann T, Gomer RH (1996) RtoA links initial cell type choice to the cell cycle in *Dictyostelium*. *Development* 122:3677–3685.

Woodard SH, Fischman BJ, Venkat A, Hudson ME, Varala K, Cameron SA, Clark AG, Robinson GE (2011) Genes involved in convergent evolution of eusociality in bees. *Proc Natl Acad Sci USA* 108:7472–7477.

Wootton JC, Feng X, Ferdig MT, Cooper RA, Mu J, Baruch DI, Magill AJ, Su XZ (2002) Genetic diversity and chloroquine selective sweeps in *Plasmodium falciparum. Nature* 418:320–323.

World Health Organization (2005) *Department of Nutrition* (World Health Organization, Geneva).

World Health Organization (2006) *Guidelines for the Treatment of Malaria* (World Health Organization, Geneva).

World Health Organization (2010a) *Guidelines for the Treatment of Malaria* (World Health Organization, Geneva), 2nd Ed. Available at http://www.who.int/malaria/publications/atoz/9789241547925/en/index.html. Accessed February 5, 2011.

World Health Organization (2010b) *World Malaria Report* (World Health Organization, Geneva). Available at http://www.who.int/malaria/world_malaria_report_2010/en/index.html. Accessed February 12, 2011.

World Health Organization (2011) *Global Plan for Artemisinin Resistance Containment* (World Health Organization, Geneva). Available at http://www.who.int/malaria/publications/atoz/9789241500838/en/index.html. Accessed February 12, 2011.

Wozencraft WC (2005) Order Carnivora. In *Mammal Species of the World: A Taxonomic and Geographic Reference,* 3rd Ed, eds Wilson DE, Reeder DM (Johns Hopkins Univ Press, Baltimore), pp 532–628.

Wright S (1931) Evolution in Mendelian populations. *Genetics* 16:97–159.

Wright S (1932) The roles of mutation, inbreeding, crossbreeding and selection in evolution, *Proceedings of the Sixth International Congress in Genetics,* ed Jones DF (Brooklyn Botanic Garden, New York), pp 356–366.

Wu DY, Hugenholtz P, Mavromatis K, Pukall R, Dalin E, Ivanova NN, Kunin V, Goodwin L, Wu M, Tindall BJ, Hooper SD, Pati A, Lykidis A, Spring S, Anderson IJ, D'haeseleer P, Zemla A, Singer M, Lapidus A, Nolan M, Copeland A, Han C, Chen F, Cheng JF, Lucas S, Kerfeld C, Lang E, Gronow S, Chain P, Bruce D, Rubin EM, Kyrpides NC, Klenk HP, Eisen JA (2009) A phylogeny-driven genomic encyclopaedia of Bacteria and Archaea. *Nature* 462:1056–1060.

Wu M, Eisen JA (2008) A simple, fast, and accurate method of phylogenomic inference. *Genome Biol* 9:R151.

Wu SS, Kaiser D (1995) Genetic and functional evidence that Type IV pili are required for social gliding motility in *Myxococcus xanthus. Mol Microbiol* 18:547–558.

Wurm Y, Wang J, Riba-Grognuz O, Corona M, Nygaard S, Hunt BG, Ingram KK, Falquet L, Nipitwattanaphon M, Gotzek D, Dijkstra MB, Oettler J, Comtesse F, Shih CJ, Wu WJ, Yang CC, Thomas J, Beaudoing E, Pradervand S, Flegel V, Cook ED, Fabbretti R, Stockinger H, Long L, Farmerie WG, Oakey J, Boomsma JJ, Pamilo P, Yi SV, Heinze J, Goodisman MA, Farinelli L, Harshman K, Hulo N, Cerutti L, Xenarios I, Shoemaker D, Keller L (2011) The genome of the fire ant *Solenopsis invicta. Proc Natl Acad Sci USA* 108:5679–5684.

Xavier JB, Foster KR (2007) Cooperation and conflict in microbial biofilms. *Proc Natl Acad Sci USA* 104:876–881.

Xavier JB, Picioreanu C, van Loosdrecht MCM (2004) Assessment of three-dimensional biofilm models through direct comparison with confocal microscopy imaging. *Water Sci Technol* 49:177–185.

Xavier JB, Picioreanu C, van Loosdrecht MCM (2005) A framework for multidimensional modelling of activity and structure of multispecies biofilms. *Environ Microbiol* 7:1085–1103.

Xavier JB, De Kreuk MK, Picioreanu C, Van Loosdrecht MCM (2007) Multi-scale individual-based model of microbial and bioconversion dynamics in aerobic granular sludge. *Environ Sci Technol* 41:6410–6417.

Xavier JB, Kim W, Foster KR (2011) A molecular mechanism that stabilizes cooperative secretions in *Pseudomonas aeruginosa. Mol Microbiol* 79:166–179.

Yamamoto S, Tanaka M (2010) The influence of kin relationship and reciprocal context on chimpanzees' other regarding preferences. *Anim Behav* 79:595–602.

Yamamoto S, Humle T, Tanaka M (2009) Chimpanzees help each other upon request. *PLoS One* 4:e7416.

Yap GS, Stevenson MM (1994) Blood transfusion alters the course and outcome of *Plasmodium chabaudi* AS infection in mice. *Infect Immun* 62:3761–3765.

Yee JR, Cavigelli SA, Delgado B, McClintock MK (2008) Reciprocal affiliation among adolescent rats during a mild group stressor predicts mammary tumors and lifespan. *Psychosom Med* 70:1050–1059.

Yooseph S, Sutton G, Rusch DB, Halpern AL, Williamson SJ, Remington K, Eisen JA, Heidelberg KB, Manning G, Li W, Jaroszewski L, Cieplak P, Miller CS, Li H, Mashiyama ST, Joachimiak MP, van Belle C, Chandonia JM, Soergel DA, Zhai Y, Natarajan K, Lee S, Raphael BJ, Bafna V, Friedman R, Brenner SE, Godzik A, Eisenberg D, Dixon JE, Taylor SS, Strausberg RL, Frazier M, Venter JC (2007) The Sorcerer II Global Ocean Sampling expedition: Expanding the universe of protein families. *PLoS Biol* 5:e16.

Young JPW, Haukka KE (1996) Diversity and phylogeny of rhizobia. *New Phytol* 133:87–94.

Yu F, Mao F, Jianke L (2010) Royal jelly proteome comparison between *A. mellifera* ligustica and *A. cerana* cerana. *J Proteome Res* 9:2207–2215.

Zeh DW, Zeh JA, Ishida Y (2009) Transposable elements and an epigenetic basis for punctuated equilibria. *Bioessays* 31:715–726.

Zera AJ, Harshman LG, Williams TD (2007) Evolutionary endocrinology: The developing synthesis between endocrinology and evolutionary genetics. *Annu Rev Ecol Evol Syst* 38:793–817.

Zhang QG, Buckling A, Ellis RJ, Godfray HC (2009) Coevolution between cooperators and cheats in a microbial system. *Evolution* 63:2248–2256.

Zhong DB, Afrane Y, Githeko A, Cui L, Menge DM, Yan G (2008) Molecular epidemiology of drug-resistant malaria in western Kenya highlands. *BMC Infect Dis* 8:105.

Zhou X, Tarver MR, Bennett GW, Oi FM, Scharf ME (2006) Two hexamerin genes from the termite *Reticulitermes flavipes:* Sequence, expression, and proposed functions in caste regulation. *Gene* 376:47–58.

Zhou X, Tarver MR, Scharf ME (2007) Hexamerin-based regulation of juvenile hormone-dependent gene expression underlies phenotypic plasticity in a social insect. *Development* 134:601–610.

Zhu Z, Zhang Y, Long M (2009) Extensive structural renovation of retrogenes in the evolution of the *Populus* genome. *Plant Physiol* 151:1943–1951.

zur Wiesch PA, Kouyos R, Engelstädter J, Regoes RR, Bonhoeffer S (2011) Population biological principles of drug-resistance evolution in infectious diseases. *Lancet Infect Dis* 11:236–247.

Index

A

E

F

I

J

K

N